ENVIRONMENTAL PHILOSOPHY

VIBS

Volume 273

a volume in
Studies in Applied Ethics
SAE
Edited by Gerhold K. Becker

ENVIRONMENTAL PHILOSOPHY

A Revaluation of Cosmopolitan Ethics from an Ecocentric Standpoint

Hugh P. McDonald

Rodopi

Amsterdam - New York, NY 2014

Cover photo: Hugh P. McDonald

Cover design: Studio Pollmann

The paper on which this book is printed meets the requirements of "ISO 9706:1994, Information and documentation - Paper for documents - Requirements for permanence".

ISBN: 978-90-420-3831-8
E-Book ISBN: 978-94-012-1076-8

Printed in the Netherlands

I affectionately dedicate this book to Prof. Frank Derringh, whose excellent criticisms and gentle encouragement contributed in numerous and subtle ways to its composition

We are still a long way from the point where we could consider ourselves morally mature....and this encourages the hope that, after many revolutions, with all their transforming effects, the highest purpose of nature, a universal *cosmopolitan existence*, will at last be realized as the matrix within which all the original capacities of the human race may develop.

—Kant

The ultimate value in these marshes is wildness, and the crane is wildness incarnate. But all conservation of wildness is self-defeating, for to cherish we must see and fondle, and when enough have seen and fondled there is no wilderness left to cherish.

In short, a land ethic changes the role of *Homo sapiens* from conqueror of the land-community to plain member and citizen of it. It implies respect for his fellow-members, and also respect for the community as such.

—Aldo Leopold

To be viable, the human community must move from its present anthropocentric norm to a geocentric norm of reality and value

—Thomas Berry

When a quarter million birds are stuffed into a single shed, unable even to flap their wings, when more than a million pigs inhabit a single farm, never once stepping into the light of day, when every year tens of millions of creatures go to their death without knowing the least measure of human kindness, it is time to question old assumptions, to ask what we are doing and what spirit drives us on.

—Matthew Scully

CONTENTS

Acknowledgements xiii

Introduction 1

ONE Toward a Deontological Environmental Ethic 17
1. Creative Actualization 17
2. Limitations on Creative Actualization 28
3. Creative Actualization of Moral Values 34
4. Toward a Deontological Environmental Ethic 37

TWO Rethinking Cosmopolitan Ethics 51
1. Cosmopolitanism 51
2. Moral Progress 53
3. The Challenge from Humanism, Humanism as a World-view: The Hierarchy of Nature as Human Chauvinism 55
4. A Critical Evaluation of Anthropocentrism 57
5. A Defense of Ecocentrism 63
6. Humanist Critics of Ecocentrism 67
7. Environmental Ethics as Cosmopolitan: Ecocentric Cosmopolitanism 69
8. Principles of Ethics 73
9. Environmental Justice and Social Justice 80

THREE Cosmopolitan Virtue: a Critical Evaluation of a Generation 85
1. Free Will and the Moral Self: Development 85
2. The Self and Free Will 96
3. The Actuality of Character 98
4. Ways of Life 105
5. Cosmopolitan Virtue 109
6. The Revalulation of Virtues from an Ecocentric Standpoint 113

FOUR The Problem of "Rights" in the Wild 127
1. Wild Rights 127
2. Moral Pluralism 137
3. Situation Ethics 140
4. Wild Value 148

FIVE Value and Obligation 157
1. Consequentialism 158
2. Pragmatic Consequentialism 159
3. Kant 164

	4. The Relation of Value and Obligation	173
	5. Creative Actualization of Moral Value	177
SIX	Rational Self-Interest, an Oxymoron.	185
	1. Rationality	186
	2. A Critique of Environmental Ethics	197
	3. The Lack of Ethics in the Self-Interest Theory	205
	4. The Immorality of Utilitarianism	211
	5. The Theory of Obligation in Utilitarianism: The Greatest Good of the Greatest Number	217
	6. Satisfaction of Interests	220
	7. Economic Utility	225
SEVEN	Population: A Critical Evaluation of the "Right" to Unlimited Procreation	231
	1. Population Ethics	232
	2. The Basis for Encouraging Population: Military, and Strategic	244
	3. Relativism	245
	4. Historical Relativism	255
	5. Individual Relativism	256
	6. Tribalism	259
	7. The Basis for Encouraging Population: Religion	264
	8. The Problem of Coercion	266
EIGHT	In Defense of Species	273
	1. Social Constructivism	273
	2. Cultural Relativism	285
	3. Would Social Construction of Species Make a Practical Difference?	287
	4. The Challenge from Nominalism	288
	5. The Moral Issue	298
	6. Evil	302
NINE	The Virtues of Thrift	307
	1. Economics and Ethics	307
	2. Thrift	316
	3. Consumer Society	322
	4. Social Thrift	326
TEN	The Culture of Death	329
	1. Human Expansion	329
	2. Threats to All Life	338
	3. Technology	344

ELEVEN Practical Program for Implementing the Environmental Era 351
1. Relations to Other Species 352
2. Conservation for Future Generations 359
3. Ecology for Society 362
4. Reform of Institutions 376
5. End "Management" of Wildlands 377
6. Limit or Outlaw Some Types of Technology 382

EPILOGUE: Environmental Philosophy as a System: From Environmental Ethics to Environmental Philosophy 385
1. Environmental Ethics and First Philosophy 385
2. System 388

APPENDIX: Rare, Threatened and Endangered: Reflections upon the Categories of Botanical Scarcity 391
1. Rarity 391
2. Threatened and Endangered 393

NOTES 397

BIBLIOGRAPHY 451

ABOUT THE AUTHOR 479

INDEX 481

Acknowledgements

Chapter one was published originally in *Environmental Ethics*. The chapter has been revised for this book. I presented an earlier version of this chapter at a Philosophy Conference at New Jersey City University in November, 2000. I presented a later version of it at the Department of Philosophy, Addis Ababa University, Ethiopia, April 2009.

A version of chapter two originally appeared in *Free Inquiry* in 2009. I presented another version, "The Challenge of a Global Environmental Philosophy to Humanism: Revaluation of Cosmopolitan Ethics" to a Panel Sponsored by The Humanist Institute at the World Congress of Philosophy, Seoul, Republic of Korea, July, 2008. I also presented a version of it at the Department of Philosophy, Addis Ababa University, Ethiopia, April 2009.

I presented an earlier version of chapter seven, "Population," at the Department of Philosophy, Addis Ababa University, Ethiopia, April 2009.

I presented an earlier version of chapter eight, "In Defense of Species," to a conference on Philosophy and Biology, at Southern Illinois University at Carbondale, organized by Dr. Mark Tschape and Dr. Justin Bell in February 2007.

I presented an earlier version of the epilogue "Can Environmental Ethics Constitute a First Philosophy?" to a panel on "Approaches to Philosophy" at the World Congress of Philosophy, Seoul, Republic of Korea, July, 2008. I also presented a version of it at the Department of Philosophy, Debrecen University, Hungary in April 2011.

Appendix A: Reflections on Rare, Threatened and Endangered appeared in part in *Mariposa*, the Newsletter of the Calochortus Society, in 1997-8

My travels around North America, beginning in California, spreading out to the other Pacific Coast States, the Mountain West, and Mexico, and later in The Eastern United States, hunting for wild stands of plants, shocked me into awareness of the grim condition many plant species face. My thanks to my beloved flora.

I am indebted to Prof. J. Baird Callicott and Prof. Holmes Rolston III, who decisively changed my thinking about the environment.

My thanks to Prof. G. Becker, the editor of this series, Mr. Eric van Broekhuizen, the general editor at Rodopi, as well as an anonymous reviewer, for the daunting and thankless task of going over my somewhat lengthy text and finding typing errors, inconsistencies poor arguments and other potential defects, hopefully now corrected. I also thank Prof. Thomas Avery for critically evaluating the first section of chapter six.

I cannot thank my colleague, Prof. Frank Derringh, enough. He read many of the chapters in the book and made trenchant and valuable criticisms. As always, our conversations have been stimulating, interesting and sometimes contentious, in a productive way. I am lucky to have encountered him in my life, in more ways than one, and thank fortune that we met.

INTRODUCTION

Philosophy as Cosmopolitan

1. Environmental ethics

Environmental ethics is really very simple. As an old saying has it, a person can live a month without food, a week without water but only minutes without air. The principle underlying this saying is the necessity for air, water, and food as necessities of life and with them all other values: of the person, the subject, and of desire. Destruction of the environment is destruction of all life and with it all value. No matter which human-centered values individuals favor and pursue, they depend upon the environment. Economic values of wealth and prosperity require clean air and water, fertile, non-polluted soil, and the conditions that make these possible. Thus pursuing economic self-interest is impossible without a suitable environment. Wall Street ignores this at its peril: if the ice caps melt, the Street may well be underwater, as coastal cities are deluged. Forests that are now productive may die due to acid rain caused by the burning of fossil fuels to run gadgets whose sale is profitable, but whose contribution to human well-being is at best marginal. Soils may not be able to support the eight to twelve billion humans that the world is projected to contain in the near future. Starvation, not prosperity will be our destiny; the overexploitation of the earth will end with human disaster.

Like gamblers, those that lead the economic life hope to get away with it, based on either the resiliency of nature or the vastness of the resources of the planet. But of course the resources are not infinite: oil is already reaching peak production and will soon decline (Roberts, 2005). Experts have estimated that ninety percent of the large fish have already been taken from the seas, leaving little for the future and imperiling many species. Moreover, "markets" and the transactions that take place in them are within an environment. Economic agents and consumers require air, water, and food as much as any other animal: a suitable habitat. As for politics, conservatives who vote for candidates who are in the pockets of polluting corporations require air, water, soil, and the intricate relations of parts to whole in environments if they are to pursue their dreams of wealth. Liberals require air, water, soil, and the other species as much as conservatives to pursue their dreams of social justice. In sum, anthropocentric values, including those of people who pursue "identity politics"—women, gays and lesbians, minorities—require an environment.

Art, religion, and intellectual life also require a healthy environment. Artists cannot paint if the air they breathe is polluted, nor can scientists proceed with experiments. Priests cannot turn polluted water into healthy wine. "Be fruitful and multiply" will be a recipe for disaster when the population exceeds the carrying capacity of the planet. But long before that, the air may be damaged beyond repair by the burning of forests to create more cropland to feed ever more human mouths. The sacredness of life that is recognized by all religions, especially in the East, must be expanded to include the lives of non-humans. The model of Noah's Ark must replace that of conquest, tribalism, and destruction.

Similarly, the values of happiness, of the self (self-interest), and of personal feelings are chimeras without an environment. As John Dewey realized, we are part of an environment at all times, regardless of our technical prowess at altering it to suit us—a habitat, whether natural or artificial. Indeed, the natural-artificial distinction does not apply, since both are environments.[1] Technology cannot do away with the needs of survival of a biological species. We are inseparable from our environment. No human values can be achieved without the most universal condition of all, the environmental conditions of human and all other life. To destroy our environment is to destroy ourselves: it is suicidal.

In short, the environment is not a luxury whose value is debatable, but a requirement of all life, and particularly of human life, and thereby of all human values. Since the environment is required by all life it is a universal requirement, and generates duties. Its value is not instrumental, but inherent. More, it is the condition of all values, and thus prior to all other values as their basis. Environmental ethics is the other that constitutes first philosophy. For the biosphere is the framework for all other values. Thus the distinction of human and environmental ethics ought to be superceded. Universality of obligation must include other self-actualizing agents qua universal: other species.

Ecocentrism is more than just an ethic. Ecocentrism also includes the idea that philosophy should center on environmental concerns and issues, not anthropocentric ones. The mainstream in philosophy, following in René Descartes footsteps, is still debating essentially sixteenth century ideas regarding knowledge: representationalism versus non-representationalism, subjectivism versus objectivism, and the like. Ecocentrism sees such epistemological philosophy as anthropocentric, since human knowledge is at the center of concern. Epistemological philosophy as the central focus is disguised anthropocentrism. Ecocentric philosophy would place knowledge of the relation of humans within their world at the forefront, and knowledge of the interaction of the diverse parts of a habitat functioning in a whole. But more, it would replace epistemological speculations with moral wisdom, in working out the place of a destructive species in its own environment.

Human knowledge is within and a response to environments. Only by interacting with our environment can knowledge arise; all knowledge is knowledge of elements of the environment in some respect. In a sense, empiricism recognizes this point, with its emphasis on what is sensed. But empiricism always borders on solipsism, even in its more communitarian versions.[2] Once knowledge is revalued as within an environment, and as both requiring an environment for the life of the knower, and as responding to the environment, the threat of solipsism disappears. The theoretical stance of scepticism, that we could somehow be locked up inside ourselves, is impossible in a world where practical action in an environment replaces perception alone as the organ of knowledge. The abstract subject who is detached from their environment, like Ibn Sena's "flying man," is at best an abstraction: with every breath we take we are attached to the world around us in which we live. The problematic of subjective and objective is superceded by environmental philosophy, which grounds knowledge in the necessity for an environment and in the response of organisms to their habitat.

I will also engage the debate over rights in political philosophy and ethics in this book from an ecocentric standpoint. Our argument is that rights talk is anthropocentric as well. Rights, as J. Baird Callicott has noted, belong within civilization and cannot be applied to the wild. However, this raises the difficult issue of separating the wild from civilization. Since human settlements are continually and increasingly encroaching upon wild places, the boundaries between them are not fixed or settled. Humans are attracted to woodlands, mountains, and other "wild" places where they are ill at ease with the presence of dangerous predator species, wolves, bears, and lions. In other words, they have a love-hate relationship with the wild, and ultimately want to tame it—or worse, domesticate it and thereby destroy it.

Environmentalism in philosophy was justified by some as the extension of rights from humans to animals, and then to the biosphere. Even the pioneer in ecocentrism, Aldo Leopold touched on this model (Leopold, 1949). However, ecocentrism ultimately rejects any notion of value as proceeding from human masters. On the contrary, humans are part of a larger whole, the planetary biosphere. Humans are dependent upon air, water, soil, and other species. People are "just plain citizens" of the biosphere along with other species, citizens of a larger world. The model provided by the environment is that of parts to a whole. Human civilization is one part of a larger whole, namely, the atmosphere, the oceans and sources of fresh water, the soils upon which our food supply depends, and other species upon which we are codependent (symbiosis), whether fruits, cereal grains or livestock. This larger world is now the horizon of human understanding. Once the notion of moral progress has led us out of a narrow, anthropocentric view of ethics, it can be "discarded," like Ludwig Wittgenstein's ladder. Ecocentrism supercedes the moral progress view and replaces it with a view of humans as within and part of a larger world.

2. Philosophical ethics

Morality is autonomous but not distinct from all other categories; rather, it marks the limit of social actions. Morality is to society what logic is to knowledge—the form of propriety. It follows that there is no such thing as a personal morality, only personal values.[3] Morality is doing justice to others, to whom we have obligations as social agents and moral equals. The goal of doing right acts is social cohesion and harmony, not strife—or the possibility of society as a distinct value. Just as harmony is a relation of notes, morals are a distinct relation of people. A rearrangement of notes will not change the number of parts, only the total effect. Musical harmony cannot be discovered by looking for it as an attribute of just one note, but requires a relation. Moral relations are similar. Morality is a limit on actions done in a social context and thus a limit on all actions in social situations. Subjectivity as the inward turn cannot generate duty. Duty is to others, an external, social relation, not an internal relation.

But is there a distinct moral realm? Morality includes all categories in the sense of a limit on what is moral in all areas of life, and thus a check on them. But refusing to engage in immoral acts or acting creatively with moral ends in view is a way of life. These ways of life first arise or emerge culturally. Cultural ethics is closely connected to a way of life, and philosophy is also a distinct way of life, whether the contemplative model of Aristotle and the hedonists or the active way of the idealists. While philosophy as a way of life is rooted in a cultural way of life, it can also transcend it, as the example of the Sophists and the Cosmopolitan Cynics indicates.

What is the relation of philosophical ethics to culture? The origin of ethical theory lies in philosophical evaluation of cultural rules.[4] The Sophists in particular critically evaluated contemporaneous Greek culture using "nature" as a standard, a standard independent of the culture. Some might argue that their thought was in fact culture bound, but the persecution of some Sophists argues against this view: their critical evaluations were viewed as a threat to the culture. The gist of the critique of cultural norms consists in the abstraction of the purely ethical elements and their use as a standard for critical evaluation of the culture. The form this could take includes using one ethical standard of the culture, for example truth, to criticize other elements. Cultural values are not discarded, but are subject to critical evaluation. The ethical elements form a standard independent of the culture at that point, for its ethical heart is used to critique the cultural inconsistencies. Another form is the importation of standards from other cultures, a process aided by cultural mixing and upheaval. Finally, there is reflection upon ethical values and their character, critical evaluation and refinement of ethical norms, and their use for cultural criticism of the Sophistic sort. Critical ethical reflection often arises due to conflicts within a culture in which what is right is viewed differently by different individuals and thereby becomes problematic. Again, if two cultures have contrasting views on a practice—burial of the dead was one such issue in Hellenic times—which is right?

Ethical behavior is subject to choice. Values differ from some facts because humans have a choice: they are agents. (As agents they can also change some facts.) An individual or even a whole society may reject that culture, in part or wholly. Since norm implies only normal in activity, there is nothing in such norms that is a basis for acceptance or rejection of them by each individual and in each situation.[5] The choice may or may not be affected by cultural values; personal and family values may conflict with those of the culture. In choosing, moral agents escape determinism. Though we learn to choose in accordance with cultural values, we gradually pass from cultural determination to evaluation.

Ethics is not grounded in reason, either, although reasoning is a propaedeutic to ethical theory and articulating ethical issues, a convenient tool in ethical critique. Philosophers and others may reason about ethics to help clarify its issues, but this endeavor presumes ethics, that is, truth telling, and a topic or content to which reason can be applied. Reason, in the sense of logic, is a purely formal study of norms of inference applicable to almost any content. Ethics is only one of these. I will examine this relation in more detail in chapter six. Moreover, ethics arises in culture as a response to habitats. This origin predated formal logic and was required for survival of cultures. The survival of the culture required an ethic; moral values are inextricably linked to human society. Morality cannot be ruled by logic, it can only be served by it—or it becomes corrupted.

3. Critique of subjectivity

The triumph of Stoic ethics is a hidden stream in Western thought: the internalization of ethics as subjective. (Stoic ethics is also cosmopolitan in a sense.) In theory of value this has meant that value is tied to internal states, whether pleasure, sentiment, feeling, emotion, reason, desire, attitudes, interest, or will as the locus of value. Modern ethics derives duty, "ought," from "is," that is, psychological states. The logic follows the Cartesian metaphysics, for which knowledge requires a knower. Thus evaluation requires an evaluator and choice requires a chooser. Modern thought also tends to reduce ethics, in the form of the chooser, to subjective internal states. What modern ethics missed was the necessity of an environment for the evaluator, upending the subjective hierarchy.

Does evaluation require an evaluator? But a "subject" who evaluates is itself an evaluation over alternative, competing psychological hypotheses and models: behaviorism, reductionism, the "flux of forces," in Friedrich Nietzsche, and so on. The model itself represents an evaluation, and thus is meta-subjective. The "subject" is one candidate psychology and has been superceded in contemporary philosophy and psychology. Yet philosophers who would not dream of using Cartesian metaphysics in other areas of philosophy nevertheless retain Cartesian categories, subjective and objective, in ethics and axiology. The model of a subjective evaluator presumes the normative identity of the individual. The "subject" is a normative requirement who is never experienced as such: a product of evaluation, which therefore cannot be the source of evaluation. The model ignores the changes that individuals go through during development and other stages of life.

The "subject" is made by culture, not logically prior to a culture. Ethics and other norms are inculcated into individuals by their parents and other social agents. The model of individual moral subjects counterfactually ignores the social formation of individual character, a cultural universal that sociologists and anthropologists from Emile Durkheim and George Herbert Mead to the present have exhaustively documented. As Virginia Held has argued, autonomy emerges with moral development: character.[6] The moral self is a composite of individual impulses and social norms. This point was obvious to ancient ethicists, who would not have dreamed of ignoring moral education and development. Ancient ethicists also emphasized the role of action, not the subject, since actions affect others, whereas subjective feelings and the like as such do not. Ancient ethics concentrated on the social consequences of action. Action, not mental events, is central. Virtue is manifested and exhibited in action, not feelings, desires, and other internal states.

Ethics is not the rules of one or another culture, either: its origin is cultural in one respect but it is no more bound by culture than other universal principles. How can values be independent of culture if they have their origin in culture? One might raise the same issue with regard to science, which claims universality but had its origin primarily in Western culture.[7] Values are not simply cultural, since they arise also in the cultural response to a habitat. The conditions of survival of the culture are a response to the geographic conditions with which the culture must deal. These are gradually formulated as rules, and are passed on through the generations. With new historical events, the culture may change the rules; the culture may meld with others, combining or modifying the rules. But the rules reflect the world in which the culture arose as much as the culture itself. The subject lives in a culture within a habitat, and is as dependent upon it as any other living creature.

4. Why Ethics? Moral History as Moral Progress: Extension of Rights, Reciprocity, and the Principle of Participation

How can philosophic ethics, universal ethics, claim priority given the numerous cultural taboos and their frequent clash?[8] The answer lies in the notion of moral progress, which, as James Rachels has convincingly argued, is inconsistent with the sanctity of tradition (Rachels, 2002). To argue that slavery, rape, and murder of innocents are wrong is to argue that there be no exceptions for any culture. Moral progress includes the spread of universal moral norms as a "decent minimum" to all cultures and the expansion of rights beyond the "rights of (white) man" to include non-whites, women, children/future generations, sexual minorities, living things, and the biosphere. Moral progress is progress toward creative actualization of a universal ethics. The origin of society as a culture has a moral goal in the expansion of moral norms beyond the in group to include the other. The "other" is not any one member of the society but collective others; moral progress includes breaking out of cultural taboos toward a more cosmopolitan, universal ethics. I am not arguing for cultural genocide: the preservation of cultures is consistent with moral progress—indeed, demanded by it. Expansion of the scope of moral considerability is morally superior since it expands moral good, a corollary of meliorism.

Moral progress is implied in the notion of philosophical ethics, since the latter aims at a universal ethics, and a tribal or cultural ethic can never be universal. Similarly, the ethical notion of rights is universal and cannot achieve universality without progress away from limited privileges, confined to the propertied classes of men, to include all. Limiting moral consideration to humans is similarly inconsistent with the notion of universality in philosophical ethics. True progress is a moral advance. Moral advances move humans from a less moral state to a more moral condition. The alternative to ethical progress, which we are free to choose, is disaster in the form of "omnicide." This could take the form of nuclear Armageddon or the slower death by the poisoning of the biosphere through technology. Nuclear bombs without ethics hardly constitutes progress: the Atomic bombing of Hiroshima, bombing civilians, was technology without ethical restraints. Technology requires ethical limits if it is to improve human life. This principle applies *a fortiori* in the relation of humans to their environment.

A fundamentally moral perspective, far from being unrealistic, is alone truly realistic. For it recognizes how delicate are the moral threads that make the social fabric possible: the condition of a non-authoritarian society. Would any association be possible if everyone murdered at whim, or prudentially, when they got the chance? (Consider the expression, "honor among thieves.") Morality is a commitment to non-violence and non-coercive methods. Thus morality lies between anarchy, which argues against principles for action; and tyranny, the rule of fear and violence.[9] Increasing violence is a reversion to the jungle, to fixed ends without choice: an escape from freedom. Both an atomized individualistic society and a more interdependent one need freedom from coercion, public or private: the right to security. Non-violence is not an ideal but a practice and a *sine qua non* of any value. However, true non-violence must be revalued ecocentrically: non-violence against the environment. This is a far more radical ethic, since it extends the principle of non-violence beyond human society to embrace the biosphere.

Moral history means increasing the extension of the "in-group." Reversions are crime, or reduction of the in-group, cultural limitations on the in-group, or those tied to specific stages or styles of life.[10] Moral association has allowed the ascent of humans—without a moral outlook, and the attendant cooperation it makes possible, humans would be back in the jungle. Morals are the hidden "spirit of the times" that have brought on the human epoch in world history. The human epoch as the voice of "nature" has a moral condition. Technological advances have taken place as an aspect of association; and the last is a reflection of morals. No work can accomplish anything without moral recognition—for example, John Locke's point that the fruits of labor belonging to the laborer by right requires social recognition and rights.[11] Again, no greatness is possible without a moral agreement, and a social condition: *morals*.

5. Cosmopolitan ethics

Friedrich Nietzsche referred to his critical appraisal and reevaluation of the traditional hierarchy of values as the "revaluation of all values hitherto." Nietzsche's standard was ostensibly naturalistic, but still retained anthropocentrism in a disguised form, the "overman," the next development in evolution. Such a revaluation does not actually break with anthropocentrism, it only exaggerates it. The basic problem of human destruction of the environment is not overcome, since the "lower" is treated as an instrument of the "higher" in this philosophy. Given the logic of his approach, the "overman" would probably continue to treat the environment as an instrument. Nietzsche's "revaluation" of morality is actually tribal: its genealogy is the perspective of an elite.

A genuine revaluation would move away from anthropocentric values toward an ecocentric perspective. Such a revaluation would at the same time complete Immanuel Kant's proclaimed Copernican Revolution, without Immanuel Kant's Cartesian baggage. Nicklas Kopernik, as is well known, argued against an earth centered view of the cosmos—which is really a human centered view, making humans the center of the universe, since it is our view of the heavens. A Copernican revolution in ethics would similarly dethrone humans as monarchs of the ethical world and recognize their dependence upon and symbiosis with other parts of the environment. Such an ecocentric cosmopolitanism completes Kant's Copernican Revolution in ethics by supplanting anthropocentrism with environmentalism. Decentering of geocentric views and religious-based anthropocentrism are comparable. Humans are not at the center of the universe, of creation, any more than the earth is. A specific environment gave rise to life, including human life.

Why should we adopt the cosmopolitan ethic, whose roots lie in ancient Greece, despite its Kantian imprimatur? The ethics of a world citizen was articulated long ago: cosmopolitan means a citizen of the world. A citizen of the world is a member of a particular culture, but also recognizes the value of other cultures in that same world, and even their legitimacy.[12] Thus it goes beyond the particularism and parochialism of tribalism, while recognizing different cultural values. Moreover, such cultures are within the world, and thus the world (cosmos) is the background necessary to the actuality of such cultures. The ecological characteristics of the world have not previously been emphasized, despite the standard of "nature" that the ancient Cosmopolitans appealed to. It belongs to our time to have realized the facts of human dependence and thereby the place of humans in a larger world. The world for humans is the environment in which we dwell.

Cosmopolitanism was primarily anthropocentric in its original formulation. A better, more accurate perspective is that humans are indeed citizens of the world, but a much larger world than the narrow world of human cultures surrounding the Mediterranean Sea. Cosmopolitan ethics must expand to include all the elements and parts that make up the world. Only then will it have achieved its true potentiality as an ethic of a citizen of the world. Ecocentrism is thus a new beginning for philosophy, in which cosmopolitanism comes into its own. Ecocentrism provides a new perspective for cosmopolitanism, in which humans can see their place in a much larger, greater whole, the non-human world. Cosmopolitanism always recognized culture, but may have been blind to culture in the originary sense of cultivation of the earth. Culture was built in an environment, which is an essential condition of it. Culture was built in a distinctive habitat that allowed the cultivation of specific plants, and thereby a settled way of life. The world is larger than culture in the narrow sense of distinct human cultures. Genuine cosmopolitanism will not exclude the larger world for the narrow part formed by human culture, but will embrace the entire world.

A revaluation of cosmopolitan ethics incorporates the insights of the environmental ethicists into ethics thereby revaluing ethics. Anthropocentric ethics is revalued as a part of ethics, and not the main part. Humans form one part of a larger whole and the whole, the world, is the world of the environment as a whole, the biosphere. Thus environmental ethics is not "applied ethics" but ethics. Starting from the necessities of life is more universal: the necessity for all life of air, water, food, and other essentials. Thus cosmopolitanism should be revalued in accord with environmental ethics. But more, if the environment is the practically required other, and universal ethics is ecocentric, then ethics is revalued from conditioned to condition: first philosophy.

While our approach will be critical of anthropocentric ethics, it is obviously in the self-interest of humans to preserve their environment. Even a utilitarian perspective, the "greatest good of the greatest number" requires preservation of species, since non-humans are greater in number. Geocentric views are often religiously based, or connected with a religion. Thus our revaluation is likely to encounter opposition from some religious quarters. But religion requires an environment in which to practice rituals, worship and other religious forms. Believers must be able to breathe, to drink and to eat.

However, some religious thinkers realize the obligation to future generations and human stewardship over the creation.[13]

Since environmental ethics concerns consequences for this world, transcendent ethics ("if God wills it") is *irrelevant* to ethics. God *might* will either choice or option so deciding which one is the divine choice is both impossible and of no aid in choosing.[14] Because we would not, unlike Abraham, kill our loved ones based on divine command, divine will is not the basis of right, nor is religion required for law or morals. Indeed, some religions have used supposed divine sanction as an excuse for violence against others. Thus environmental ethics must stand on its own feet. This is another sense in which environmental ethics completes Kant's Copernican Revolution in ethics.

An ecocentric cosmopolitan ethic in a sense is a post-modern ethic, since it does not start from the human subject, Cartesianism, or from anthropocentrism.[15] Environmental philosophy also breaks with logocentrism.[16] The relations in ecocentrism are holistic, instead of linear, as in logocentrism. This is not to say that ecocentrists cannot give reasons, but only that logic as the sole standard is of less importance. On the other hand, philosophy always meant giving reasons, in the broad sense, and any turn to irrationality would be a big mistake.

The implicit stance of philosophy is universal and opposed to partisan ideology. An ideology disguising itself as a philosophy is the part masquerading as the whole, a contradiction in terms. Humanitarian Ethics, the enlightened ethics of the past, is inherently anti-partisan. It attempted to discover universal not partisan views, by looking for the good of humans as humans. It includes the premise of universal moral competence, the equal competence of responsible individual actors. However, from the more universal outlook of environmental ethics, humane ethics has turned out to be partisan not universal, since a part, the human species, is treated as if it were the whole. Environmental philosophy has unmasked this approach as anthropocentrism, an ideology of human domination.

Cosmopolitanism is universal in its perspective, since it embraces the whole world; it is consistent with the universality of philosophical ethics. As John Shook noted, universal ethics has the most general principles, and can thereby allow for cultural variety and individual choice up to a point. Different cultural arrangements may be compatible with general principles. By contrast with some intolerant tribal cultures, cosmopolitanism includes tolerance of cultural differences. But cosmopolitanism is not uncritical: universal without tolerating immorality, it looks to the best in each culture.

Moral and value theories are philosophic constructs at a high level of generality, but they are not arbitrary constructs: they are evaluated superior by deliberation and aim at some goal, which may include proscribing some evil not previously, recognized. Also, the principles aim at universality: universality is the goal, whether we achieve it or not. They capture the universal proper to morals/values as such, their normative character. The evaluation is that universality is superior to tribal particularity. Ethical theory consists in theories or hypotheses about moral universals. Since these theories are speculative, they may be wrong, and then, like any theory, they may need modification or replacement by a better theory: meliorism. The argument of this book is that cosmopolitanism requires modification in the light of the insights of the environmental ethicists. A revalued cosmopolitanism would expand to include an appreciation of different species, similar to its valuation of different cultures.

What are the "elements" of a cosmopolitan ethic that have been revalued by environmental concerns? An ethic must be universal, as I argued above. Anthropocentric ethics has never been universal, as speciesist. (a) The norm of duty requires reciprocity; a cosmopolitan revaluation of reciprocity requires inclusion of other species and their habitats in moral consideration and evaluation. Ethics is the recognition of the intrinsic value of the other, and if other species have intrinsic value, they are morally considerable. Thus (b) moral progress consists in the extension of moral consideration to universality (*universal* other). Environmental ethics is universal as more inclusive and cosmopolitan: humans are included as part of the biosphere. For cosmopolitan ethics, (c) duty is universal and involves moral limits on non-moral values.

Another norm is what I will label the principle of participation. The principle of participation is derived from a common starting point for ethics: what if murder, rape, and so on happened to you? Would you want to be murdered or raped? We are thus participants in moral life, willy nilly. This principle constitutes another reason to be moral. We cannot evade moral responsibility in the modern world, as the underlings of the Nazis tried to do in exculpation. For we are in an environment. (d) Cosmopolitan revaluation of this principle is the recognition that humans are in the environment, that we participate as agents in the world.

Are we free to make moral choices? One answer to this is that we have something called a "free will." I will argue that much more than will is involved in free choice, namely character and a way of life. The model of virtue, character development, is the standard of ancient philosophy for moral agency. (e) Character building is another element that should be revalued in the light of human destruction of the environment and the attempt to liberate human appetite from all bounds. What are cosmopolitan virtues?

Environmental philosophy must include consequences for the environment. Yet one reading of deontological ethics is that it does not make reference to consequences in considering the morality of an action. Thus (f) a reform of deontological ethics is called for in which consequences for the environment are considered. Ecocentric ethics must take consequences into account but also, as we will see, duty and character. Since duty, character, and consideration of consequences are all elements of an ecocentric cosmopolitan ethic, the relation of value and obligation is not foundational. Intrinsic value, properly understood, argues against *summum bonum* theories, and for a kind of holism that is in harmony with environmental relations in the wild.

Since all these elements are involved, ecocentric ethics is holistic in one sense of that word. However, there is a previous ethic that has argued for consideration of all these factors in moral deliberation, namely pragmatism. Pragmatism also marked the beginning of the contemporary critique of Cartesian dualism. John Dewey argued for this holistic treatment of ethics, but de-emphasized duties unduly. Thus cosmopolitan ethics will involve a pragmatic, holistic reading of deontological ethics, as in Clarence. I. Lewis. Moreover, I will argue in chapter five that moral holism is a form of pluralism in our sense, involving the reciprocal relation of different environmental factors. Such an ecocentric cosmopolitanism includes the norm of (g) meliorism, the attempt to improve the future through action as a consequence of action. Another pragmatic element is the acceptance of (h) generality, not absolutism in ethical evaluation.

In the background of the modes of actualization, imagination, creativity in the world, and norms, is the *world as a whole*. Holistic value is the condition of all possible creative actualization, principles, and imaginings. Thus the world has *holistic* moral value, as well as plural sources of value: cosmopolitan ethics. Our problematic, then, is to revalue traditional humane ethics and its issues from an ecocentric perspective. This revaluation of cosmopolitan ethics requires revising it in the light of the inclusion of ecocentric holism. These issues should be revalued by the standards of ecocentric cosmopolitan ethics.

6. Precis of chapters

Given a genuine cosmopolitan view, a radically ecocentric ethic, what do some traditional issues look like? What of virtue, human rights, the treatment of non-human species, reproductive decisions, and the like? We will examine these issues in the chapters of this book from the perspective of environmental values. The standard will be the good of the environment as a whole, not only human good.

In chapter one I will address the theoretical issue of the extension of the deontological ethics of Kant to include the environment. I will first explicate a new theory of value, creative actualization, which includes other species. I will then address limitations on creative actualization, in view of the requirements of universality, especially reciprocity, participation, and meliorism. I will argue that ethics proper concerns a relation to an other; that moral progress is the continuous expansion of this other to universal inclusion. Since I am addressing the issue of value, I will include an examination of the topic of the intrinsic value of other species, as well as the issue of biodiversity, the inherent value of diversity as such

In chapter two, I will examine traditional cosmopolitan ethics and also moral progress. Are humans one species among others or special? I will consider humanist challenges to the extension of ethics to other species, and critically evaluate traditional humanist assumptions, for example, the hierarchy of nature. I will mount a defense of Ecocentrism, and argue for a revaluation of cosmopolitanism in an ecocentric direction. I will contrast environmental ethics and humanist ethics, and argue that the distinction collapses, once the need for the environment is understood. I will argue against one sect of anthropocentrists that although environmental pollution has disproportionately affected the poor, that the demands of species equity, that is universality, require environmental justice before social justice, but do not rule out the latter.

To what extent should humans be willing to make sacrifices to preserve the environment? I will cover the notions of duties and virtues in chapters three, four and five (with a reprise in eleven and twelve).

Perfectionism, development, realizationism, and "virtue ethics" are ultimately a tribal ethic, even where they are not the realization of Greek virtue or English, or Italian (*virtu*) or Chinese. The refinement is the perfection of human characteristics, anthropocentric. It is still the virtue of a part, and not universal, not duty to all (Kant), love of all (Christianity, Moism), non-violence to all (Mahatma Gandhi, Martin L. King Jr.). However, developing moral agency within a more ecocentric outlook should be the goal of social policy and family life. Chapter three will cover the ethics of character development, of the notion of a moral self, and propose an ecocentric cosmopolitan revaluation. I will argue that "free will" requires the development of moral agency, that is, of character. Thus "free will" will be revalued as the choice of a moral agent. Moreover, an environmental ethic includes a way of life. I will then revalue traditional virtues in relation to an ecocentric cosmopolitan ethics.

Do animals have rights? To what extent should humans be willing to make sacrifices to respect such rights, for example in preserving rare species? What are the rights of other species: what of hunting and predation by other species? What of cosmetic and medical testing? What of obligations to future generations? I will address these issues and the issue of rights in the wild in chapter four. I will argue against the extension of the notion of "rights" to the wild. Rights should be respected within civilization, but a more comprehensive ethic must govern our relations to other species and the larger whole of the earth, moral considerability. I will discuss pluralism, the notion of different ethics for the wild and for human society in this context. In one sense, a cosmopolitan ethic is pluralistic, since different cultures have distinctive values. Moreover, different species have a value of their own. Thus while there are plural values, I will argue against a plural ethic. I will also address the issue of the value of and obligations to landscapes.

One issue in environmental ethics has been that of the intrinsic or inherent value of other species, whether as a whole or as individuals. Another issue is whether such value implies ethical obligations. I will cover the relation of value and obligation in chapter five. Thus a more holistic environmental ethic should be developed. An environmental ethic must consider consequences for the environment. In a revaluation of Kantian deontology, I will include pragmatic consequentialism as an essential element in moral evaluation. Pragmatic cosmopolitan ethics has the holistic relation of one part to other parts making up the whole. Thus it challenges the traditional and widespread model of the relation of value and obligation, shared by Aristotle's form of the ethics of character development, and consequentialism. Pragmatism rejects the notion of a "foundation" for ethics. However, Kant's ethics is a corrective to Dewey's radical stress on the situation. Ecocentric cosmopolitan ethics is close to that of a later pragmatist, C. I. Lewis, who combined a pragmatic stress on consequences with a Kantian inspired emphasis on imperatives. I will examine these issues along with the role of principles in ethics and creative actualization of moral values in chapter five.

Hedonism and affectivism are egoistic theories of value. Moreover, self-interest as a foundation for ethics is anthropocentric. I will critically evaluate the "ethics" based on self-interest in chapter six as well as specific forms of utilitarianism that appeal to self-interest. I will defend environmental ethics from a critique by a proponent of self-interest. I will argue that the universality of reason argues against the oxymoronic notion of "rational self-interest" some ethicists have borrowed from economics. Utilitarianism in its many forms can be, as majoritarian, tribal but as democratic can be a pragmatic form of universal cosmopolitan ethics. However, both the "ethics" of self-interest and Utilitarianism could be used to justify gross immorality. I will also critically evaluate economic forms of Utilitarianism.

I will examine the threat to all life of human expansion in the final chapters. They address more specific issues from the standpoint of an ecocentric cosmopolitan ethics. Should human population be limited? I will critically evaluate the so-called "right" to reproduce and thus to overpopulate in chapter seven, including arguments for individual and cultural relativism and strategic arguments for increasing population and the problem of coercion.

I will defend the notion of "species" in chapter eight from attacks by social constructivists and nominalists.

I will contrast the economics of thrift with consumer society in chapter nine, and critically evaluate consumerism. Would everyone driving a large car mean environmental catastrophe? Rising expectations and affluence are evaluated as contributing factors in destruction. I will argue for the virtue of thrift as model and duty to conserve wealth, as well as the priority of morals over economics. I will critically evaluate the notions of demand and needs in the economic sense. I will also argue for the thrift economy over consumer economics and for the social responsibility of economic institutions.

I will examine other threats to all life of human expansion in chapter ten, including the effects of global warming; and its result, rising ocean levels. The costs of technology will be examined: the destruction of forests from logging and acid rain, the effects of pollution, including reduced air quality, the effects of synthetics, including ozone depletion, and soil destruction. I will also consider the hidden ecological costs of cheap electricity, including acid rain, global warming, and nuclear waste. Finally I will consider threats to other species from habitat destruction and the resulting species extinction.

I will then provide practical suggestions to accompany such a revaluation utilizing traditional moral principles, in chapter eleven. I will also offer examples of how to alleviate the causes and effects of human's destruction of their environment.

In the epilogue I will examine the relation of environmental philosophy and first philosophy.

In the appendix I will examine the notions of rare, threatened, and endangered, as a supplement to the chapter on species.

Revaluation of cosmopolitan ethics is the simultaneous revaluation of many issues in ethics. The book as a whole will revalue moral progress through the expansion of moral considerability, and thereby the relation of humans to the non-human world in light of such a revaluation of moral progress: a more moral world that is more virtuous.

One

TOWARD A DEONTOLOGICAL ENVIRONMENTAL ETHICS[1]

It is not the sense of duty to a neighbor, but the practical answer to the question 'Who is my neighbor' that has varied. Theodore H. Green

To put ourselves in the place of others, to see things from the standpoint of their purposes and values, to humble...our own pretensions and claims till they reach the level they would assume in the eye of an impartial sympathetic observer, is the surest way to attain objectivity of moral knowledge...the tool, par excellence for resolving complex situations. John Dewey[2]

1. Creative Actualization

Value theory in Western Philosophy generally has the structure of the *summum bonum* model, in which a supreme value, the source of value, gives value to all other values. The *summum bonum* alone has value and all other values have value only in relation to and by virtue of the supreme value. Only the relation to a supreme value gives value to the many values.[3] A hierarchy of values is thereby created by the *summum bonum* in which the rank of subordinate values is fixed in relation to the *summum bonum*.[4] Since their value is conditional upon the one supreme value, the *summum bonum*, this kind of theory contains less value in principle, as it devalues all but the source of value. If each good partakes strictly of "one principle," a good cannot have a unique inherent principle or be used for its own principle, as ruled by another principle. The value of the different is subordinated to the value of the one. I would argue, however, the value of food is different than the value of clothing or of a beautiful sunset. If they were the same, we could neither eat nor be warm.

Such hierarchical models are based on Plato's "form of the Good" in the *Republic* VI. This is the metaphysical basis for later hierarchies of value, including *summum bonum* theories. Lower values derive their value from the *summum bonum*, just as copies derive their reality from the form of the good. The model refers back to the "One," the origin as unity in metaphysics. As I noted in two previous works (McDonald 2004, 2011), the *summum bonum* model has persisted throughout almost all of the history of Western Ethics.

Different theories of value may substitute other values, such as pleasure as the ultimate value, the *summum bonum* (the Cyreneacs, Epicureans, Jeremy Bentham, and the other Utilitarians). But they retain hierarchy from Platonic Idealism, the origin of *summum bonum* theories, the idea that there is one supreme value which encompasses or grounds the rest. Theocratic value theories are also hierarchical, in that value derives from the divine will.

By extension, instrumental value refers back to intrinsic value for its value.[5] Echoes of this relation can be seen in theories of value that retain this relation to the human subject.[6] The good refers back to a highest good, either human good or the human subject, as its instrument. Generally these theories have been combined in the modern period, for which the human subject is foundational for value. All of reality is instrumental to the intrinsic value of the subject who in some sense is identified with the *summum bonum*. Instrumental value derives its value from a subject on which its claim to have value is based: a hierarchical relation. The valuing relation back to a subject as an intrinsically valuable first creates a hierarchy of value. The value of objects is for a subject to which it is subjected. Thus good is subordinated to human good, the anthropocentric view. This is the ultimate basis of human domination of the environment, for the good of the world apart from its value to humans is not considered in the subordination of good to the human as an attribute.

A theory of value that avoids subordination of the environment to human exploitation must avoid the value hierarchy relation. Hierarchy places anthropocentric values at the top and other values, for example the value of other species, are subordinate. Such hierarchical relations are intrinsically anti-pluralist, as only values compatible with the supreme value have value. They must somehow reflect the supreme value, either in the relation of instrumentality or of participation. The relation of instrumentality is subtly inserted into hierarchy, as the good is conditional on a first, a *summum bonum*. Thus the good of the environment can be subordinated in a similar instrumental relation: treated as deriving its value from a human subject and subordinate to human desire, feeling, or will. A theory of value that is non-hierarchical and thereby pluralist should avoid subordination relations.

The value theory I am proposing would meet this standard. Value may be characterized as creative actualization. Creativity encompasses the novel and pragmatic elements of value. Creativity is the differentiating principle in the sense that what is novel or creative is in some sense different from what was previously created or what has come before. It is inherently pluralist as value involves differentiation: creation of new goods. The creation of new goods is a pluralistic process. Creativity is also pragmatic, as creation of the novel may involve trial and error as much as planning.[7] The creation of flying machines was a novelty at the turn of the century, which involved much experimentation and many initial failures before the first successes at Kitty Hawk. This trial and error mirrors the adaptation of pioneering plants to new soils or to environmental changes. The pace of technological innovation may have numbed us to the wonder of actualization: that we can actually bring new and marvelous inventions to fruition. Creation of the biosphere is equally wondrous and innovative.

Actualization is also pragmatic but has less to do with differentiation than with the world. A plane that won't actually fly is *no good.* It is no good as a flying machine in the world. The phrase "it won't fly" is shorthand for this aspect of actualization. Flying has not been made actual in the world by such a machine. Because the world is not defined in terms of a subject or the human, this aspect of value avoids anthropocentrism and subjectivism. Subjectivity of values is avoided with an acid pragmatic test: the notion must be actualized in the world or it is no good. But actualization is not hierarchical in value, for there are different kinds of actualization, including techniques, prediction, survival of life, normative, and relational to mention a few. And different categories of value involve creative actualizing for distinct ends, including beauty, wealth, health, intimacy, and knowledge. Even the world is not a *summum bonum*, for "world" is transformed and differentiated as new creations, not previously actualized, are manifested within it. Plants and animals are part of this transformative process.

"Actualization" and "actual" have the same root. To actualize is to act. As good is achieved by acting and must involve actualization in the world, value is removed from human subjectivity. Values are not only perceived by a subject either as attributes of a substance[8] or as subjective reactions: feelings, desires, and volitions. We can observe that a screw is rusty, defective or inappropriate for the circumstances at hand. But this is grounded in action in some sense, such as the previous action of having used a defective screw. Action is required to demonstrate value and bring about value. Action is publicly observable and knowable, not private and subjective.[9] Action may be a means of actualization or an end in itself.

Action itself is grounded in value in the sense that creative actualization, the good, completes or ends action and thus gives it meaning. Action is performed in view of a valued end and *completed when grounded in the world,* that is, when the goal has been achieved or actualized. Otherwise the action falls short and is no good. The failures of action may contribute to the eventual success but the final achievement is, critically, an achievement *in the world.* The accomplishment is not an achievement in the subject and does not refer back to subjective states for its warrant. It is not enough to wish for a flying machine, feel it would make us happy, or even to build one. It must successfully function as one in the world. Only then has the good been attained. The mode of value is achievement through such action and this has nothing to do with the subject, since it is accomplished through action and actually achieves some result in the world or falls short. Actualization as achievement in this mode is thus doubly removed from subjectivity, since the action that brought it about is public and observable and the achievement of action, the actual value achieved, is an achievement in the world. (There are modes of the achievement of subjective values such as happiness as well, but these are distinct modes of value.) As the circumstances under which this can be brought about change, sometimes even as a result of previous actualizations, actualization involves creative adaptation to such changing circumstances. Thus "creative" is normatively generic as a characterization of actualization in a changing world.

Combined, creative actualization as a theory of value involves the corollary that value is the most general term of differentiation, for each novelty in the world is a value or disvalue. This theory of value must be qualified in several respects—namely, moral conditions for creative actualization must be given, which I will examine in the next two sections. However, the hierarchy of subordinating *summum bonum* theories is avoided. With respect to the dialogue about intrinsic value, a debate that can be traced back to Plato, creative actualization can involve intrinsic value. However, intrinsic value is not defined in the relation of good as an attribute of a substance, the Aristotelian relation, or in relation to the subject. Rather, intrinsic is defined by contrast with extrinsic. Value is located with respect to a distinctive form of actualization in the world, intrinsic to this distinct (creative) actualization, not extrinsic to it. The value does not refer beyond in an extrinsic relation to a hierarchical *summum bonum*. It is intrinsic or inherent to the creative actualization as such, as a distinctive good. Thus each good which is distinctive, however this is defined or determined, has some inherent value; self-actualizing creatures are intrinsically valuable. There is not one good but many goods. The value hierarchy of *summum bonum* theories, including subjective *summum bonum* theories is thus avoided, but not at the expense of intrinsic value. This view of intrinsic value follows the logic of "intrinsic" to the full, since defining intrinsic value as the end and then arguing that "lesser ends" can be instruments for "higher ends," as Aristotle does, involves, at the least, a paradox.

Why is creative actualization good? Creative actualization is good because of the addition to the good in the world of novel goods. Human values are already plural: different categories or kinds of value require that they be insubordinate to any hierarchical first, any *summum bonum*. What is good in itself, or inherently valuable is not reducible to other species of goods, by definition. It is good by itself, apart from any hierarchical value relation. I add the qualification "value" to relation in response to John Dewey's critique of inherent value. Dewey noted that inherent values are not out of all relations, and I would agree. Indeed, intrinsic value requires a world in which its value contained and thereby a relation to that world and its other inventory.

A species of good is unique, different from all other kinds. Creative actualization involves creating the unique, each of the different goods. Value pluralism includes the totality of different and unique goods. The preservation of unique differences is good as a unique good, a differentiation. Creative actualization is also good because of pragmatic feasibility. An unreliable tool is "no good." It does not function as a tool properly. Using the wrong tool for the job is also no good. Rocks are not screwdrivers; screws require the creation of screwdrivers or vice versa. These are new goods in the world's inventory. Creative actualization is compatible to a degree with older theories of value such as hedonism and self-realization. The creative actualization of pleasures in new recipes or other novel pleasures is good, as is the creative actualization of virtues in self-development. However, these are not the only goods.

Plants and animals also creatively actualize. This point was touched on briefly above, in the example of pioneer plants creatively adapting to a new soil, created by geologic action or an environmental change. Each seed must adapt to the world to survive: it must actualize itself in the world each moment or from season to season. Herbivores must be creative in avoiding predators and not over-grazing. Carnivores must create strategies to catch enough herbivores to survive. Each species considered as a whole, to survive, must creatively deal with both routine and novel circumstances.

Animals and plants are creative. This is not a metaphor from human art: speciation, according to Charles Darwin, often involves adaptation to new conditions not previously encountered by the species. Plants gradually create footholds even in mostly barren conditions (sagebrush, creosote). Animals must also creatively actualize: they must adapt to changing circumstances in the world each day to survive as a species. This adaptation is creative as individuals must constantly deal with new situations over the course of their lifetime, and bring their experiences, habits, and species resources to bear on such circumstances. My own cat once jumped up to a ledge to catch some rays and warm herself. This could not have been taught to her, as it was context laden: her action was creative in dealing with particular circumstances to enhance her warmth. Thus the social contribution to inner life can be over-estimated. No one explained to the dog to paw the door to go out or taught the cat to jump to *this* ledge to get sun at *this* time of day.

Indeed, I would argue that, contrary to the mechanistic view of René Descartes[10] or the modern stimulus-response view, animals exhibit actual deliberation and practical reasoning. They actually consider different means to ends.[11] In a well-known experiment, a monkey piled up crates to get some bananas that had been hung from the ceiling: acting for a valued goal by using a creative means. This experiment provides evidence against the notion that humans alone use reason. Creatively reasoning in this way may require experience and habit to some degree, but also involves consideration of circumstance, of means, of diverse goals, and creative responses to the various alternatives. According to William James, animals can also conceive (James, 1892, p. 107), abstract common elements or "classes" (*ibid.*, also p. 234 of

Jimmy) and even reason, albeit at a more simplified level than humans (ibid., p. 258 and 262). As animals demonstrably have memories, there is no basis to believe they cannot compare experiences. Dogs are not so much irrational as less talented in abstract rationality. But hounds using their sense of smell to track in ways that humans cannot argues for different types of intelligence, even apart from differences of degree.

The examples also constitute evidence against the degree to which language, with its universals, is required for practical reasoning, as animals reason and are creative without it. Yet the growl of a dog is sufficient to warn people away from private property, as well as warn other animals. To avoid confrontation, other animals will retreat, a kind of practical reasoning. I concede that without speaking their "language" we cannot establish their motives in any complete sense. Animals, as outside the community of human language users, can only be interpreted not described.

According to Richard Rorty, language is confined to humans, and, therefore, so is awareness (2007, ch. 1). But animals communicate in their own way with their rudimentary language of barks, meows, and other sounds, which are quite effective in communicating warnings and feelings. Although their language differs, and is simpler, language is not confined to humans. Moreover, they understand concepts, as that a door is always the way out to relieve themselves. Rorty also argues that there are no understandings, meanings, or awareness without concepts, and so only humans can have these. But does this mean that there would be no orgasm without the "concept" of orgasm? I doubt anyone who believes in such linguistic determinism ever had one.

Whether monkeys can be described as "deliberating" in regard to ideas of potential stimuli and not just actual stimuli is less clear. The problem of attributing states to animal "consciousness" is the general problem of behaviorism: behavior usually does not tell us what an animal is "thinking," feeling, desiring, or believing. We identify Rover's pawing at a bowl and its interpretation as wanting food, as "a desire," and that this is occurring now in his "mind." This is rational or logical insofar as he does not paw bowl unless he wants food (rational behavior and rationality as interpreted by us.) It cannot be "instinct", as bowls were not around when genes evolved. "Stimulus-response" cannot tell us what is on Rover's "mind." The evidence? Rover does not paw at the bowl continually, but when he is "hungry."

Unfortunately, psychology does not give a clear answer to questions of animal reasoning and consciousness. However, psychologists assume some sort of continuity between animal psychology and humans in their experiments; psychological experiments on animals apply in some way to humans. Darwin also posits this continuity.[12] However, this psychological assumption is modified by "evolutionist" view that human intelligence is more advanced (Daniel Gleitman et al.). But given evolutionary links this can only be a difference in degree, not in kind. The main difference is that we can "describe" our inner experiences, but animals cannot ("What is it like to be a bat?"—Thomas Nagel)—due to lack of ability in our language. But there is no reason to assume their inner life is lacking, given the facts of neuro-anatomy, obvious ability to feel (pain is used to train), perceive, and especially to learn. Ironically, introspection is of little use in contemporary psychology.

Critics may object that we do not know an animal's internal processes. Animals act but whether their acts are habitual, thus lawlike, is interpretive and speculative. Nevertheless, they exhibit what is described in humans as goal-directed or teleological actions. Since action is only intelligible teleologically—for otherwise it is merely motion—it requires the notion of animal mind: anticipating goals and choosing the seemingly superior one.[13] Goal directed behavior is widespread and includes motion, especially movements to gather food, drink, and mate. Such action is also a form of practical reasoning of the means to ends type. Signaling is also teleological and argues for understanding of a sort on the part of others, whether of the same species (marking territory) or other species (threats and alarm calls).[14] Indeed, some observers even see different cultures in one animal species.[15]

Animals facing capture by predators exhibit self-consciousness, in the attempt to preserve themselves as a living unit by flight; reasoning, by fleeing to escape danger; and sentience, since pain is used to train pets and to experiment on rats. Indeed, psychologists depend on these responses in order to conduct psychological experiments. If animals could not perceive, sense, learn, remember, and know, psychological tests would be futile. Even fish can learn.[16] Clams dig deeper to avoid capture; flies avoid swats. Why should the prey respond? Animals do not exhibit "fleeing behavior" when there is no predator present but instead when a predator is present, that is, appropriately. How is this "conditioned" if it involves new situations, which could not possibly be pre-learned?[17] It could not be a "stimulus-response generalization" if it is situational or contextual and thus individuated. Each situation has some new or different elements which animals must cope with. But even if it were, it would reveal the ability to generalize: a form of reasoning. Similarly, animals select appropriate foods. Animals make choices and must do so if they are to survive. As Robert Hinde puts it, "In some cases a single stimulus situation (e.g. a model of a rival) can elicit a number of responses (e.g., attack, threat, flight) which do not depend on qualitatively different aspects of the stimulus" (1970, p. 21).

Similar arguments apply to "instinct," or genetic "programming." We do not know whether animals value food or only "instinctively" appropriate it. Yet as Ed Messemer noted, "instinct is a word for our ignorance," that is, the word we use to explain how a species without language can be motivated (James, 1892, ch. 16). Even if instinctual, behavior still involves judgment of appropriate situations for behavior. David Hume argues it is not blind instinct but "custom alone which engages animals from every object that strikes their senses, to infer its usual attendant, and carries their imagination, from the appearance of the one, to conceive of the other, in that particular manner which we denominate belief." (Hume; quoted Matthew Scully, 2002, p. 196).

No genetic program could possibly take all novel situations into account. Nor could the response to every situation be genetically programmed: genes have been passed down for thousands of years but situations and circumstances change and new ones could not possibly have been genetically pre-programmed, socio-biology style, thousands or millions of years ago. Neither stimulus-response nor genetics explain such goal-directed behavior. Survival demands some prudential awareness. Animals must "imagine" their future escape through present means. Flexibility argues for intelligence not programming. Moreover, the environmental requirements of all life are similar, which refutes the anthropocentric view. We need water and dogs, cats, and other species require water. Physiological similarity argues for similarity in mentality: differences in degree, not kind.[18]

Matthew Scully notes that "We are prohibited by the terms of modern behavioral science, the reigning school of animal research, from saying that animals can *believe, think, desire, want, intend, try, hope, feel, or suffer* anything…or even experience conscious pain." But he also argues that,

> Like us, animals are subject to the operations of instinct and appetite. Like us too, they must learn individually to apply these drives and appetites and fears to their world. In their cruder way, they must also form basic concepts and beliefs with which to make sense of their surroundings, grasping categories and patterns and sequential laws.[19]

He adds,

> Studies by Jane Goodall, Dian Fossey, and many other researchers have shown very clearly that primates such as chimpanzees and gorillas can think, though few scientists would also concede that farm animals have thinking abilities…. My experience is that these animals think in discrete visual images. They are able to make an association between a visual image stored in their memory and what they are seeing in the present.[20]

We can examine creative actualization by other species as well, for evidence against human exclusivity. Animals and plants creatively actualize in the ways distinctive to their species. When a plant "grows toward" the sunlight (phototropism) this is without, as far as we know, conscious or rational consideration on the part of the plant. The light may well act as a stimulus but this is not a cause of "response towards" in the strict sense; the sunlight does not push, move or force the plant to grow "towards." The plant must respond creatively, through growth. At a minimum, plants exhibit sensitivity: in response to light, both daylight and day length, to temperature, and in the ability of roots to grow around soil barriers, such as rocks. The product of this creative actualization is its own life, survival as an individual and member of a distinct species: actualization over time or self-actualization. By creatively interacting with soil, plants, which evolved from one-celled organisms not rooted in soil, have created food for themselves and thus their own actualization, their survival. Different species have created niches in different habitats (xerophytes, geophytes, and hydrophytes).[21] Thus they have intrinsic value, as their creative actualization is distinctive to themselves, not dependent upon any human subject (Ian Smith, 2010). Animals also adapt creatively to their world, to plants, and to each other (carnivore and prey) or perish. If they perish their survival strategy is of inferior value, for they have been unable to actualize themselves or creatively adapt to circumstances.

Ecological norms provide a value structure, independent of utility. Wild flowers, insects, "weeds," and predators all have a place regardless of their utility for humans. Thus biodiversity as such has inherent value, since greater biodiversity means more intrinsic goods of more kinds with greater diversity. More species in different habitats results in—or reflects—the achievement of greater survival skills and greater adaptation, consistent with meliorism. Such diversity reflects an increase in strategies of creative actualization and thereby new and different modes of intrinsic value. Since relations in such environments are symbiotic, they involve association, and thereby the value of different forms of association, including reciprocity and compossibility. Lest association between plant species seem too much like a metaphor, I should note that botanists always note the "association" of plants on herbarium sheets that form a community. This information is sometimes crucial in determining species identity and relations, as "associations" will differ—for instance among xerophytes and hydrophytes.

Donella Meadows has pointed out that, apart from the varied use to humans of biodiversity, this diversity performs "environmental services beyond price," including pollination, decomposition, photosynthesis, cleansing of air and water, flood control, pest control, temperature regulation, and "maintenance of the world's most valuable library—the genes of all living organisms."[22] However, biodiversity also has value in a more holistic sense, since a unique holistic good has established itself as a changing whole that is larger than any part, species, or individual. Biodiversity is a mark of the biosphere and in that sense holistic. Such diversity is the basis for adaptation to niches and thereby the expansion of the biosphere. Since biodiversity is a mark of the biosphere, and the biosphere is a condition of all life, biodiversity is a condition of life. The oxygen in the air we breathe was liberated by organisms and is refreshed and renewed in the symbiosis of plants and animals in oxygen-carbon dioxide exchange. The conditions of all life were creatively actualized by the symbiosis that characterizes biodiversity. Similarly, plants creatively actualized soils, both in their life and by their death, since the organic matter in soils—essential to rich soils—is primarily a plant product. Moreover, communities of plants hold the soil in place. Air and soil are creatively actualized in and through biodiversity.

In the face of new conditions, a species can either retreat to the tried and true, which may often result in decline (and "relict" species) or expand its range by learning to adapt. The adaptation involves novel strategies of behavior, which the species must develop pragmatically. Thus they create novel actualizations in the world, and, in some cases, novel species. These add to the world's inventory new values, distinctive to the new species. Every triumph of a species, that is, every creative actualization in the specific sense of the emergence and survival of a new species, including humans, constitutes a qualitative change. Though not necessarily an epochal change, a new species does mark a change in the way the world looks and functions: the world is dynamic. The slow accretion of such qualitative changes transforms the world over time from epoch to epoch over time through creative actualization. The world has been transformed from barren rock to biosphere and there have been periods within the time span of the biosphere divided by disasters and creative responses to them, namely, the age of reptiles, of mammals, and so on. The transformation from barren to lush, the creation of biosphere, must be ranked as intrinsically good by *any* theory of values as its condition: superior to rocky sterility and the condition of all life.

The biosphere includes the whole sphere of life. I would argue that this has meant the creative actualization of an entirely new mode of value, derived from the holistic relation of parts to whole. Obviously, earlier authors, notably Aldo Leopold, and perhaps even Darwin, realized that the environment functioned holistically in some sense. I am arguing here that the *value* relation in the environment is a new mode, which goes beyond means-ends (although Holmes Rolston may have anticipated this point). This holistic interrelation has parts that create and maintain the whole and the whole, parts. Parts function in the whole. Functional roles can be of instrumental value for others. Bees function and thereby have value as pollinators. Or is this to confuse "functioning as" with value, function in a habitat the same as functional value? Normatively each function is the condition of a healthy whole; the whole, of each part. They are co-conditional, since each part is required for a healthy whole (a normative notion or standard). But this benefits bees as well. The predatory function does not include looking to the welfare of others, for example, lions to prey, but functions that way inadvertently. Indeed, functional value is not necessarily conscious, as for example water functions to keep whole healthy. But the function is not only mechanical, since its benefits to each organism are manifest. Plants will die without water, but can thrive with it. Thus organic functions involve more than mere mechanical functioning. In mechanisms, each part is essential for the whole to work. In organic, it is necessary for the good of the whole. I am not arguing that "functions for" is equivalent to "good for" in all cases, but that functions in an organic whole are more than conditions. Organisms can often compensate without specific health producing factors, but do better with them, for example, trace amounts of elements in the soil for plants, as opposed to readily available amounts. The plants will thrive with greater amounts, whether of water, soil elements or other factors.

Martin Gorke raises the issue of what is a "healthy" state of a dynamic community/habitat? (2003, p. 298) "Healthy" is of course a metaphor to medicine, but the word health is applied to animals by vets and, perhaps less securely, to thriving plants. I would respond that a healthy state of a dynamic community is one where potentially all animals and plants that could adapt to such a habitat have a chance of doing so. This rules out water lilies in the desert, but also alteration exclusively for human purposes. The changes are distinct from human activity and involve only interaction between the species themselves. It also argues against a complete takeover by alien plants and animals (Hettinger and Throop in Pojman, 2001; Daniel Simberloff, 2012).

Intrinsically valuable agents are within this whole even as they alter and help maintain it. Thus their autonomy is limited. They are also mortal, thus their intrinsic value is finite. But they still have some freedom of action. As self-actualizing they still have intrinsic value, but also interact. They also function within the whole as a token. Thus they are instrumental as a functional token—unless functional value is distinct from a mere means. The parts organisms play are the condition of all life, as in the relation of organs in the body: normative.

The environment is a complex relation of intrinsically value agents, inherent value elements and parts, for example, water, predators, and other functions. Thus the environment is a very distinct type of whole, involving intrinsically valuable agents. But is the value of parts within a whole reducible to functional value? Function is not simply value but required for creation of a healthy whole, the normative aspect of the creation of the whole from functional parts. The relation involves the interaction of parts and whole: the parts creatively actualize the whole and whole allows creative actualization—which takes place within whole as its condition. The relation can be seen as special sort of reciprocity. Perhaps we can call this "mutual" relation "systemic value?"

Gorke argues that "all existence" has intrinsic value in a holistic system (2003, p. 261, 267 et al.). I would argue for the systemic and instrumental value of non-living elements/functions, but not for their intrinsic value. Water can be, following Clarence Irving Lewis, inherently valuable as water, with systemic value. But one difference between holism and cosmopolitanism is that while water has systemic value, instrumental value and even inherent value, it does not have intrinsic value. Water does not creatively actualize itself, nor is it an agent in any sense, alive, or any other mark of intrinsic value. If Gorke only means that the whole has a value separate from its parts, then he is on firmer ground.

The functional/systemic value of intrinsically value agents is as tokens of their function that serves the whole. Individual organisms performing a function are "tokens" of the function. A lion is concrete token of the predator function. Thus intrinsically value agents are both ends in themselves and function as a systemic token for others, conscious or not. Intrinsically value agents have plural value: they actualize themselves while sometimes functioning for others. Bees function as pollinators for plants while actualizing their own lives in the process. Humans and other species are in a larger whole, which is their condition; but species also help actualize the whole. Species also have systemic value. Systemic value, involving the relation of token functions, is independent of changes in the local community of plants and animals.[23]

The environmental community is dynamic. Change is continual, but the historical element does not imply anti-realism in the sense of constructivism. Instead, it dissolves the distinction between realism and anti- or irrealism, and with it anthropocentrism as a metaphysical-epistemological stance, subject vs. object, and so on. "Subjects" are in the world and reflect development by other factors outside of any self. Subjects are within and totally dependent upon an environment. We are organisms in an environment that allows creative actualization.

2. Limitations on Creative Actualization

Are all creative actualizations equally good? Evidently they are not, for if some types of values are actualized, others may perish. Atomic war would result in destruction of the entire biosphere. Increased pollution would contribute to a slower form of the same denouement. Thus the creative actualization of nuclear weapons in war or pollution are bad, indeed, evil. They do not add to the world's inventory of good as they diminish that inventory and would be incompossible with other goods. Quality of good is affected as much as quantity. As there is no one good that can justify such destruction of many distinct goods, atomic weapons can be adjudged bad, for their use would result in what John Somervell has called "omnicide," in which all life could perish. Omnicide, the deliberate destruction of the biosphere as a whole is the ultimate evil. Destruction of rare species is similar, for it reduces the kinds and varieties of creative actualization. Both of these examples involve the subordination of other goods that is justified in terms of a hierarchy based on a *summum bonum*, a single good. Pollution subordinates all other goods to economic goods in a hierarchy based on wealth as the highest good. In this view, all goods are good by virtue of the profit that can be wrung out of them. The end of economic gain justifies the means of pollution, since economic profits are the sole source of value. Similarly, all other values are subordinated to the highest good of national interests in atomic war. The destruction of the earth would not be morally considerable in this view, as value derives from the nation and the value of people, animals and other life are only valuable insofar as they reflect the nation: fascism.

From these examples two limitations on creative actualization may be derived. One is meliorism: creative actualization must improve the world somehow, not worsen it (I have kidnapped the term "meliorism" from William James.) A melioristic actualization must be either novel and worthwhile or an improvement on previously actualized alternatives. Adding to the good in the world through creative actualization is a melioristic process—the world is even better after Wolfgang A. Mozart than before—but destructive actualization is not melioristic. Novelties whose actualization destroys other values are not melioristic, as the amount of value in the world is not increased. Judging if or how the world is better, evaluation, may be difficult; but destructive actualization is clearly not for the better. Meliorism argues for ultimate rather than immediate value and against creative actualizations that have beneficial short-term effects (temporary economic benefits) but disastrous longer term ones, as the effects are not melioristic. It is supportive of ecocentric values as ultimate and better in the long term than immediate gratification.

The problem that value and morals don't coincide is as old as ethics. Any theory of value must cope with moral evil, for example, bad pleasures (sadism) for hedonism, innocent victims of bad events for theological ethics, and so on. The examples are meant to show that the problem is that the structure of justification in terms of subordination to a first, supreme value is the primary source of the problem. Reduction to a first, whether hierarchical or reflecting some other *summum bonum*, reduces value and is thus anti-meliorist. Reduction also undermines reciprocity with other values. Reciprocity of values rules out their subordination as instruments in the relation of values.

A critic has argued that meliorism must be grounded in a subject who judges what is better, involving a relation to a subject. However, a theory of creative actualization avoids such a relation on several grounds. The first is that creative actualization is not subjective. Either the plane flies or it doesn't. Subjective wishes or desires that it flies are irrelevant to whether it flies. As I noted in the last section, achievement or creative actualization is generally in the world, not the subject, even in the distinct case of actualization of positive states of mind; and involves action as its agency, not subjective states. But how is this better, that is, melioristic? Creative actualization is better as it adds a new kind or good to the inventory of values in the world, again without reference to the subject. A new species is actually new whether or not we wish it, desire it, are interested in it, take pleasure in it, or have emotions and attitudes about it. One measure of meliorism is quantitative and differential: the actualization of new goods which are good by virtue of functioning and an improvement in terms of a new kind, thus a different, quantitatively distinct good. This kind of good or differentiation of good is independent of subjective valuation, for it is marked by an increase in the quantity of kinds of goods in the world. In this respect meliorism is a pluralistic condition. Another measure is qualitative: an improvement on an older good. This criterion or test is not subjective either: a newer model plane is faster, more efficient, more reliable, and so on. These criteria are distinct, which alleviates the sometimes-difficult problem of evaluation. New works of art may be incomparable to older works as singular kinds. The works are, strictly, incomparable as singular intrinsic goods, like comparing eagles and beans. Some technical comparisons can be made, both of art works (technique) and animals (adaptation). Such improvements *within* a kind are also measurable apart from subjective factors, especially in terms of impersonal standards. Some adaptations by a species may increase its chances of survival by comparison with older adaptations whether human subjects wish it or not, for example the increasing resistance of cockroaches to pesticides.[24] In this respect meliorism is qualitative. Meliorism as qualitative also rules out creative actualization of the trivial, since this is not an improvement in any sense, just as actualization of the defective is no good.

Another reason is that intrinsic, inherent, and instrumental and other kinds of value belong to the bearer of such value. Our value judgments and evaluations do not affect the bearer. The subjective view confuses value and evaluation.

As I noted in the Introduction, the triumph of Stoic ethics is a hidden stream in Western thought: the internalization of ethics as subjective. Ethics began as the study of practice, of action, but in the modern period has been subjectivized. In theory of value this has meant that value is tied to internal states, whether pleasure, sentiment, feeling, emotion, reason, desire, attitudes, interest, or will as the locus of value. Modern thought also tends to reduce ethics to subjective internal states, deriving duty, the “ought,” from “is” in the sense of psychological states. The logic follows the Cartesian metaphysics, for which knowledge requires a knower. Thus evaluation requires an evaluator and choice requires a chooser.

The view that value requires a valuer confuses value with evaluation, a major fallacy of modern value theory. Almost all categories of value involve establishing value apart from personal evaluations. Several vitamins, and the foods that contain them, are healthy regardless of personal or subjective evaluations. Many facts are established—and thereby have been creatively actualized—regardless of subjective wishes, desires, etc, for example evolution, despite its invalidation of the account of species creation in Genesis that many wished was true. Evaluation creates a hierarchy of values, whether subjective or not, based on a supreme value, a *summum bonum*. But as I have argued above, inherent value, instrumental value, and intrinsic value are independent of such a *summum bonum*, for they are either good in themselves, apart from any hierarchy, or have properties that are inherently good for instrumental use. A plane flies or it does not fly, whether it accords with voluntarism, hedonism, agapism or any other *summum bonum* value theory.[25]

What animals consider good for themselves may not accord with human wishes and desires at all. Indeed, animals must be fenced in to keep them from pursuing their own perceived good. Animals have their own relations to alien species and their own evaluations of safety and danger regardless of how humans evaluate them. Similarly, what is good for plants is independent of our desires, feelings and attitudes; no amount of willing it otherwise will change such norms. Raising water lilies on a regimen fit for desert cacti will kill the plant.

As I also mentioned, the "subject" is made by culture, not logically prior to a culture. Ethics and other norms are inculcated into individuals by their parents and other social agents. Agents creatively actualize values by means of action, not subjective states. Subjective states are derivative in several senses, not originary (McDonald, 2011).

Ironically, Stoic ethics itself did not have a subjective theory of value and even appealed to a natural law, universally valid for all. Modern philosophy exaggerated the inward turn begun in ancient times by Neo-Platonism as the source of wisdom by subjectivizing ethics and values. Ironically, again, Platonism rejected the subject as the source of value (Gorgias, Republic).

Another limitation on creative actualization is reciprocity of value. Reciprocity is a condition of pluralism and differentiation of value, the mutual distinctive modes of creative actualization. Destruction is not compatible with other goods, as they are destroyed. The number and kinds of goods requires their reciprocal differentiation, for if they are not distinctive, no increase or plurality of goods accrues. The relation must be reciprocal to avoid subordination to a *summum bonum* (value monism), which would constitute devaluation of the many: subordination would work against meliorism. Reciprocity of value is a rejection of the doctrine that value is inherently hierarchical. Moreover, reciprocity is a condition of novel forms of actualization emerging in non-destructive ways. If new actualizations are not possible without destroying or subordinating previous actualizations, they are not melioristic. The world's goods are not increased, the world is not bettered. Reciprocity in actualization means that creative actualization must not interfere with other values (as with dams in the West and salmon). Different goods must be compossible, must stand together in reciprocal relations in which their intrinsic value is differentiated. The condition of such differentiation of intrinsic values is a reciprocal, insubordinate relation. Intrinsic worth cannot derive its value from a subordinate relation by definition. Intrinsic good cannot be extrinsic, and does not derive value from a subordinating, legitimating *summum bonum* such as the human subject. Intrinsic good is necessarily autonomous from other values: thereby reciprocally different. It is distinctive to mutually different kinds of creative actualization. Together, reciprocity and meliorism justify creative actualization. Grounding in the world alone is not sufficient for *justification*, since bad actualizations are feasible. From these justifying conditions duties can be derived.

The question of whether we ought to actualize an end is a moral question, an issue for ethics, not value theory. Not everything that has value is right, or moral. Stolen money still retains its value as money, but is not morally right or, in another formulation, has no moral value. This paradox of value is an argument against grounding morals in value, foundational arguments in which obligation is defined in terms of a *summum bonum*. An example is Utilitarianism, in which the highest good, pleasure, is the basis for obligation. Since bad values or bad use of values cannot ground moral values, or non-moral value cannot ground moral value, ethics ought to be independent in some respect from value theory, or at least constitute a distinct kind of value. Obligation should not be grounded in non-moral values, since obligation must limit value in some respect. Creative actualization of morals is required, in which duty as moral value is practiced as the creative actualization of norms. However, this does not rule out consideration of consequences, since the consequences of bad acts can be a reason, or even the main reason in forbidding such acts. On a more positive reading, creative actualization of moral values is an element of the better, more complete good, that is, meliorism.

Ethics in the narrow sense, apart from issues of value, involves questions of obligation to the other person, social obligations.[26] The other in human association is morally compelling. For firstly, we are already a member of society or, if you will, a community or particular culture, which is one form of association.[27] The "opposition" of self and society is a false opposition, since self and others are members of one society. Second, all humans require society, that is a community, even if only by virtue of being born and nurtured by a society, learning its culture, and thus attaining membership in the culture. All the benefits of being alive, and of individual achievement require association. Society, then, is a condition of the individual, and required for the well-being of individuals. Third, association is also to the good, since it benefits individuals with the values unique to human association, whether economic, intellectual, spiritual, or other social goods. The development of humans requires what Dewey called a social "environment," which is only superceded by the physical environment in its importance for human flourishing.

Why do our duty at all? One reason is that we do our duty because it is right and just because it is right. It is inherently right and this outweighs any benefit of doing wrong, whether personal gain or pleasure or other self-regarding motives. An act is right in itself, self-justifying. But one could ask why is it right? In one sense, such a question misses the point, for doing what is right is self-justifying. In another sense, however, doing what is right for its own sake can be further justified. Doing what is right is right because it is morally best in the sense of fulfilling the golden rule and other norms of reciprocity, that bring self and other into a relation under principles. I fulfill my obligations to a generalized other because I expect such treatment in return, or at least would prefer to be treated rightly.[28]

The golden rule and similar norms are paradigms of following rules with respect to the other. The Golden Rule is an application of the principle of reciprocity. It has its echo in the Confucian principle, "Treat others as you would like to be treated." The other is recognized as a moral equal to the self in such formulas and treated as well as the self. Imagination is involved in projection of empathy or sympathy to the other.[29] Thus doing one's duty best achieves strictly moral values such as justice, equality, and so on. Moreover, by following rules, I am morally consistent.[30] Further, if everyone does what is right, the result will be the best overall consequences

Another respect of doing what is right to the other means not wronging the other, that is, not treating them unjustly. Our actions may wrong the other and this would be wrong in itself, apart from consequences. It may also be wrong in its consequences. The moral contrary of doing what is wrong is doing what is right. We should do what is right because it avoids doing what is wrong, but also prevents consequences that would be wrong to the other, injustice.

Moreover, harming the other treats the other inconsistently with oneself, an attempt to subordinate others in a hierarchy, violating norms of equality, and justice. Others are morally equal as members of society. As a member of society a self has a reciprocal relation and thereby obligations to other agents.[31] It is wrong to violate this relation, for it is anti-social. Society, the condition of oneself in all respects, is subordinated to personal whims, disrupting social relations for freeloading self-aggrandizement. Among humans, harming another self violates human bond. But for a more progressive, more universal ethic, wherein the field of moral considerability is extended to all self-actualizing agents, association with other species, symbiosis, is recognized as an obligation. For a cosmopolitan ethic, needless harm to other species is morally wrong. Finally, causing needless pain to others cannot be a good reason to act.

If we are to make accusations of, for example, injustice, and the wrong of injustice, then we must ourselves be moral, if we are consistent, not hypocritical. If we expect justice or duty from others, we should ourselves be moral (reciprocity). We cannot exempt ourselves from a universal obligation, or we are immoral: we are participants in moral society, and this participation is itself a duty. The principle of participation follows from reciprocity in the abstract and social membership in the concrete.[32] The principle involves the recognition that anyone could be the victim of injustice. Thus when the question arises, "why be moral?" we can respond: due to the principle of participation, which formulates the general ethical quandary, "What if it happened to you?" What if I myself was the victim of injustice? What if I were in their circumstances?[33] What if I were to be murdered or raped? How would I want to be treated? The principle of participation means that we cannot exempt ourselves from ethics, since we are in society. If we are to make accusations of injustice, for example, then we must ourselves be moral. Our appeal to morals places us within the moral life. We also are participants in society and require justice for ourselves. In other words, the principle extends justice to all: social justice. By tying duties to intrinsically valuable agents in a reciprocal relation, the obligation is not contractual but unconditional. That is, if x lies to y, y may not in turn lie. The reciprocity is abstract not concrete. Obligations to self-actualizing agents, ends in themselves, are unconditional. It might be argued that some duties are not reciprocal, for example, duties of doctors to patients are not reciprocally obligatory for patients. But that the duties of doctors are unconditional follows from the professional status of doctors and their duties as professionals. Some duties are limited to agents: they are not reciprocal. However, patients are obligated in specific ways, that are not professional, for example, payment, whether through taxes or fee for service. Means are relative to an end, desire, or context: means.

Compossibility is the condition of environmental relations, the relations of different functions in the larger whole, whether local niches or the larger biosphere. Different functional tokens must work together for the whole to be creatively actualized and thereby to allow further functioning by individuals and species. As I noted, this functioning need not be conscious, but is the condition of all life, including human life.

Equal applicability of moral rules implies moral equality: justice and universal rights in political life are an extention of this equality. The basis of law as a universal obligation of citizens lies in ethics. Moreover, reciprocity at the national level requires reciprocity between institutions of different kinds. (I am not endorsing the nation-state. A cosmopolitan ethic is the hope for the end to nationalism, while recognizing the value of cultures.) There can be no subordination of institutions in a hierarchy, one to another, whether religion to politics or politics to religion, or either to economics.[34] With regard to politics, the structure of reciprocity rules out domination by hierarchies, whether political or economic. Reciprocity at international level means live and let live, each nation entitled to live and prosper by itself, and not at the expense of another.

3. Creative Actualization of Moral Values

The topic of normative "oughts" raises the issue of ethical theory in relation to practice. Since ethical theory is a theory about a practical endeavor, ethical actions or behavior, it can be interpreted as a theory of practice. This was

Aristotle's view, duly qualified. However, Kant had another view of the relation of theory and practice in ethics (which I will examine in chapter five). John Ladd presented a widespread contemporary view in his *Ethical Relativism*,[35] where he argues that the study of ethics is a practical study, distinct from theoretical studies of a body of knowledge. The pragmatists argued against a sharp separation of theory and practice. However, my position is distinct, since I argue, following the British pragmatist F. C. S. Schiller, for an ethical basis for knowledge. In practice, gathering knowledge requires an environment and a moral order to support scientists and other investigators. In theory, knowledge requires the normative standard of truth, as well as norms for knowledge. This work consists in part in a defense of values and ethics against philosophies from mysticism to flux that deny or downplay values and ethics.[36]

While knowledge may be useful for making ethical choices, it is not essential, nor the ground of morals. Knowledge does not cover what ought to be, the area of potentiality. Technology can use actualities to create new goods from the potentialities inherent in the actual, but science describes and explains actualities. Some might argue that ethics is actualized in actions in society and culture, and in part this is correct. But critical evaluation of potentialities and which of these ought to be creatively actualized is distinct from what is actual, except in the cases where all is as it should be. What "is," like the "given" is only one factor in evaluation and not the most important consideration, since it may be immoral. Potentiality is the realm of evaluation: critical appraisal of different potential choices for an improved future. What ought to be is creatively actualized in the future and thereby becomes part of the world, of actuality, not mere potentiality; the distinction of potential and actual is not absolute. The world can be changed through creative actualization in accord with how it should be, in accord with value. What ought to be and actuality are distinct in the abstract and in time, but combined in creative actualization of possibilities, moral or not. Norms are actualized as goods. Thus "meta-ethics" as the epistemology of ethics is misguided, for it confuses ground and consequent—or a part, knowledge, with the whole: the moral cosmos. Ethics is basic in theory and the environment is basic in practice. Since we cannot separate theory entirely from a suitable environment, environmental ethics is the new philosophical ethics.

Charles Peirce argued that ethics is grounded in aesthetics, perhaps following ancient Greek theories that appeal to the beauty of character. Peirce argued that esthetics is the normative science that examines what is good in itself. I would respond that beauty is only one form of value, and not the foundation or ground of the others. Beauty is not a condition of other values, and is thereby a distinct category or kind of value. Immanuel Kant argued that a good will is good in itself and thus for a moral model of intrinsic value. However, intrinsic value is a wider notion than either morals or esthetics. Morals are a condition of other values, but not the content of these others. A diamond has inherent value of many sorts: economic, esthetic, and social, and thus the value of a diamond is contextual. This value does not derive from its morality. Nevertheless, morals are the condition of the others.

Wisdom, not knowledge is the practical virtue, but knowledge of good moral judgments is still possible—that is, epistemological considerations are not impossible, including knowledge of first principles, of logic and truth values. There is some knowledge of "first principles" but not in the same respect as in a science. With regard to the environment, we "know" some aspects of the environment but hardly all the complex relationships in an ecosystem. But that does not stop us from living. The relation of knowledge and ethics is complex, but knowledge is not "beyond" ethics. Knowledge, like reason, is a propaedeutic in ethics. Knowledge of past moral decisions and their results can be of use in making new decisions.

However, ethics does not only concern the ought, it also critically describes the "is," in the form of practice. Because value is creative actualization, what humans do describes their values, as well as their idea what we ought to creatively actualize. A scientist values truth or physics or the explanation of sub-atomic particles. The scientist believes that she or he ought to be engaged in such research. "Oughts" are not a priori but telic—they involve time and the future, if they are practical and truly valued. Genuine oughts predict what *will be* but *are not*, unless an "is" coincides with an "ought" already, for example in acts of bravery (Iris Murdoch, 1970), or previous creative actualizations, which took place in the past. If we ought to we will and this making actual or true of the ought confirms it. Reflection on and commitment to values generates oughts, and thus not a "metaphysics of morals," since potentialities are actualized, attenuating the distinction of metaphysics and values. Values imply duties or they are not valued. Thus the distinction of "is" and "ought," which some authors treat as absolute, has been overdrawn. The separation of "is" and "ought" is attenuated by creative actualization, since creative actualization of moral values creates new actualities that "were not," but now "are." What "is" reflects past actualizations, and thus what "is" is derivative from creative actualization.

Ethics is not simply concerned with how we "feel" since it involves our relations with others, whether intimates or more casual associates. It is not a matter of "opinion," since it is learned as part of socialization and acculturation. The actuality of ethics consists of ethical actions, which take place in space and time, according to ends. Thus ethics connects mind to the world and is neither completely idealistic nor realistic. Reality is changed or at least affected by actions that creative actualize moral norms.[37] Moral deliberation is manifested in actions in the world. Were it not it would be utopian.

4. Toward a Deontological Environmental Ethic

Reciprocity involves equality of a rough sort, as reciprocity is non-subordinate. By "roughly equal," I mean that species stand in an equal relation of reciprocity, but that there are fewer predators than prey. They are not quantitatively equal, but equal as distinct kinds of creative actualization, actualizations of value-types. Predators are symbiotes and thus in a special sense reciprocal with prey. Moreover, they are governed by the same principles of natural selection.

Mutual differentiation is not a relation to a highest good but the equal relation of intrinsic values. The basis for an axiological ethic can be determined from such a relation. This ethic is a rejection of non-equalitarian ethical theories, which include any egoistic or majoritarian theories. Ethical egoism, in the form of interest or self-interest and even self-realization theories, subordinate all good to a self or ego, thus denying reciprocity to the other. Subjectivist theories include this subordination, namely, subordination of the world to a human subject. Affective, eudaemonistic, intellectualist, emotive, conative, interest, attitudinal, and voluntarist theories subordinate good to, respectively, the feelings, happiness, knowledge, feelings, desires, interests, attitudes, or will of a (human) subject. In effect, subjective theories of value arbitrarily reduce all value outside the whimsical occurrence of subjective states to instrumentality: value for a subject. The world is devalued by such theories as it is without intrinsic value.

Utilitarian theories, whose theory of duty involves "the greatest good for the greatest number," are majoritarian.[38] Value is subordinated to a (human) majority in such theories—which in the world taken as a whole is actually a minority—in a relation in which all value reduces to majority good. (Anthropocentrism is oligarchy. Utilitarianism is anthropocentric.) These value theories are, therefore, non-egalitarian. For they do not recognize the intrinsic value of the other, but only its subordinate value to the majority, or in fact, the very few, humans.

As Kant rightly recognized, only the ethics of duty have the form of universality proper to axiological and moral equality.[39] All have equal obligations to others, thus none are subordinate or superordinate in value. Equality is intrinsic to duty, which has the form of reciprocity. If each agent exercises their duty to each other, their relation is reciprocal and equal. If they do not, they ought to: a moral imperative. Universality as a formal requirement of Kant's ethic gives necessity to morals. We must do what is morally commanded as a consequence of universal reasoning: the conclusion is practically compelling. By contrast, hypothetical imperatives of the means to ends sort are contingent upon the conditional hypothetical premises: if x, then only y. Universality allows practical reasoning, which as a form of reason requires a universal premise.[40]

Despite Kant's inconsistent subordination of animals as means (*Lectures on Ethics*, 1775-80), the logic of his position is the best critique of pure speciesism in ethical theory. Species ethics are based on a purported or stipulated special human capacity, whether "reason," "language," "interests," or some other. These specific differences *violate the norm of universality in the Categorical Imperative*. As specific differences they are speciesist and ruled out by the requirement that maxims be universalizable: specific differences cannot be universalized by definition. As they cannot generate universality, they are not necessary, that is morally requisite or obligatory. Species differences are *morally* irrelevant, for they cannot be the basis of a Categorical, as opposed to a hypothetical imperative. They are also arbitrary: to argue that specifically or exclusively human capacities are the sole criterion of intrinsic value is as arbitrary as taking having a trunk as the sole criterion of intrinsic value and thereby limiting moral standing to elephants and trees.[41] Animals and plants are living things that depend on their environment as much as we do. In this respect there is no difference between living things.

Thus Tibor Machan's argument that humans alone are moral agents and thus may make occasional use of animals exhibits a paradigm case of violation of the first formula of the Categorical Imperative.[42] He argues both that humans are moral agents and that they can be immoral to animals—for treating humans the way animals are treated would obviously be immoral. Thus humans are moral agents who can behave immorally, a howler of a contradiction. Machan recognizes the necessity for universality in his critique of Utilitarianism, but is blinded by his own speciesism. He also argues that "rights involve the freedom to make moral decisions" without explaining why he thinks animals are any less free or moral than humans.

But isn't a species a universal of a type, for it applies to all individuals of that species? This is indeed the case. However, it is "universal" only in a reduced, specific or equivocal sense, that is, with qualifications. Life is a more comprehensive concept, with greater scope, and thereby more explicitly and precisely universal. In one respect, then, different degrees of universality are at issue. In another sense, they are not, for all of life is not an arbitrary division (compare my distinction of general and specific universals, above). As I noted, a species universal is universal for that species, but not for all. "Universal" implies everywhere in the (known) universe, and implies application to everything in the universe.

Critics of deontological ethics may argue that duty can be an excuse for evil. The critique of duty would point out subordinates who do their "duty" in respect to a "higher cause," for example, Nazis of all ranks, from generals and admirals to concentration camp guards. These individuals argued in exculpation that they were just doing their duty. Duty as empty form can be placed in the service of just such nefarious ends. The "duty" of a soldier to kill in war is another case, or at least a hard case, where it does not involve killing civilians ("collateral damage"). Kant himself made a related point about virtue at the beginning of the *Groundwork*, where he scored the "courage" of the brigand. The subordination of virtue to immoral ends shares with the subordination characteristic of soldiers in war in the name of duty the subordination of moral ends to non-moral ends. The lesson is not the immorality of duty or character, but the immorality of placing political, economic or other ends above moral duties, including duties consistent with a good character. Thus subordination of duty for immoral ends, particularly economic ends, is, perhaps obviously, immoral.

Reciprocity includes other species and their habitats, since ethics requires the consideration of the intrinsic value of the other in symbiotic association. Intrinsic value for a theory of creative actualization means the value of self-actualization, that is, of creative actualization of self through survival. Animals and plants creatively actualize themselves, and thus have intrinsic value.[43] The other in this case is wider than that of traditional anthropocentric ethics, since it includes other species. Since ethics requires universality and reciprocity, a general theory of duty should include other species. Moral progress lies in extending consideration universally: the *universal* other in association. Thus for cosmopolitan ethics, duty is universal.

Moral progress requires universalization, and reciprocity is the beginning of universalization.[44] But there is a distinction between a species universal and a general universal. A species universal is universal for that species, but not for all. Reciprocity has to do with the other, and ethics concerns the other in social relations; and traditional ethics extends moral consideration from individuals to universality among humans. However, moral progress is inherently melioristic, a judgment of better and worse. Extending ethical consideration from humans to other species is morally better, since a species universal is a limited universal: it is wrong in just the respect that racism is wrong, for it makes arbitrary exclusions. The environment is a requirement for all of life: we cannot survive without air, water, and soil. Moreover, humans require other species in a symbiotic relation. Since other species are required, they are practically necessary and thereby morally considerable.[45] In Kantian terms, other species are the "condition of possibility" of humans. Friedrich Nietzsche also saw that "the lower is the condition of the higher," but formulated an elitist ethic in response to this relation. At least he acknowledged that the lower is required. Nietzsche was worried about the dangers to the "higher type," from the masses of lower; in our age we no longer have this luxury: the "lower" is more endangered. Moral progress requires consideration of other species, and an expansion of moral considerability from humans alone to other species.

To whom is such a duty owed? Others in association are, as others, different. As noted, plurality of goods requires differentiation of goods. Monism of goods cannot result in reciprocity, for all differentiation would be subordinate to a highest good or even reduced or eliminated by the highest good. This would decrease the inventory of good in the world as incompatible with a supreme hierarchical first, the supposed source of good. Reciprocity as mutual differentiation, then, involves a duty to distinctive actualizations as intrinsic values. Does this mean their individual agents or bearers? The types of differentiation as such? Should it be extended to the conditions of such distinctive creative actualization? I will argue here for a view of obligations to the environment as a whole and secondarily to specific parts and species.

"Whole, holism and holistic" can have several meanings. One of these is a group taken as a whole, whether a species considered as a whole, the biosphere taken as a whole, a habitat taken as a whole, or even society taken as a whole. Another is some whole with parts in which the whole and parts are inseparable. This would include any organic body, such as the human body. We can remove the organs, but not without killing the organism. Culture could also be considered as a whole with parts. Holism in an environmental context generally involves an organic relation of parts to wholes, except where it refers to a whole group. Finally there is a distinct sort of value of a whole, which I call "holistic" value.

One issue this raises is the relation of intrinsic value to holistic value. Is holistic value intrinsic, instrumental, or some new relation of values: of parts to a larger whole? As I argued above, the relation of parts to whole is not quite that of instruments, but instead of functions.[46] The whole supports the parts and the parts support the whole in the human body. Similarly, there is a function of predators, of water, of soil, and of other parts in a healthy habitat, that involves an interrelation, instead of an instrumental relation. Since ecological "communities" are dynamic, with one species replacing another, often with the same function, it may be that no one species is essential to the health of the whole. But the function may be: water, for example, is essential regardless of its source. Similarly, bear could replace lynx, but the predator function is essential for the good of the whole. I referred to this as "systemic value." Systemic value is value to or for the whole. Thus I would argue that wild species have ecosystemic value: the value of a part to a working whole. Systemic value is likely to be confused with holistic value, but I would argue that they are distinct. The whole has a value distinct from such parts, holistic value, although inseparable from such parts: the good of the whole apart from any one part. Thus I am distinguishing holistic and systemic value.

Holistic value should not be confused with reciprocal relations, and other ethical norms.[47] Holistic value (part/whole; functions in ecosystem) is a part of value, not ethics. Holism is a valid ecological concept but more than just a moral concept. The whole has a value apart from the parts but a value among others and thus in a reciprocal relation with the parts. Indeed, the parts create the whole and the whole in turn aids the survival of the parts. Holism belongs to the environment, not to ethics, which is reciprocal, meliorist, and involves compossibility and participation, among other principles. Reciprocity is derived from the relation of intrinsic values, thus axiological—as opposed to *summum bonum*. Each element must be considered, but their relation is reciprocal qua inherent value. Reciprocity ties in to a community relation in the environment, a relation of random association. The moral relation is reciprocal interaction. Thus systemic value is also distinct from moral relations in a community. The addition of the holistic relation with its systemic value to value theory supports the view that value is plural, and that no one value can justify the rest. There are not only different kinds of value, but different value relations as well.[48]

The duty of humans is toward the environment as a whole and secondarily to species within it.[49] Why is the biosphere not only accorded its own intrinsic value but even due obligations on the part of moral agents? The biosphere as a whole is a condition of human and other life. The model of humans as self-sufficient subjects is ridiculous from an environmental perspective. Humans would be doomed even without bees (Freya Mathews, 2010), to speak nothing of air, water, soil, plants, and domesticated animals. Without them, we could not provide the food on which civil society is and historically was based. The species differences marking the human have done harm as well as good: setting "man against nature." The dependence of humans on a host of inter-related parts should be recognized. Human "control" of the environment is conditional on several environmental factors beyond human control. The human epoch does not so much set a species apart as a way of life, domestication. Domestication involves a relation to animals and plants that involves their close association with humans, the mutual domestication that brought all domesticated species out of the wild, including humans. As Leopold noted, this is a symbiotic relation, not one of master and slave. Value as creative actualization, grounding in the world, does not subordinate the world to a human subject. There is no distinction of "man" and "nature" in the relation of confrontation and conquest. Only in a Cartesian separation of the human subject from the biosphere can values be revalued as "subjective" and the world, the condition of survival of the subject, as somehow subordinate in value to that subject.

Humans are in the world, part of a community of plants, animals, soil, air, and other elements.[50] We humans are one species in the world among others. We live in and require the environment provided by the biosphere and interact with other species within it. Many of these, especially domesticated plants and animals, and not a few wild ones, are required for human survival as a species, continued actualization of humans as a distinct kind of value. Thus the value of the biosphere and the species required for human survival—whether a small number necessary for it or a larger number enhancing it—is reciprocal. Humans care for domesticated species and they provide their good to humans. There is no subordination of the biosphere to human good; the environment as a whole and different species stand in a reciprocal relation of equal worth.[51]

How can the relationship between humans and the biosphere be reciprocal if humans need the biosphere but the biosphere does not need humans?[52] The biosphere does not require humans as such, since it evolved independently of any one species. However, it is reciprocal with the many species that compose it. Organisms created the air and soil, which are the common elements for all life and essential elements of the biosphere. The relations of the species themselves in the web of life are an important element of the biosphere. All of these elements require mutual relations for the biosphere to function, symbiosis, although the biosphere is independent of any one of them. Thus if the biosphere as a whole is to be a continuing source of goods for humans, humans owe duties back to it. Humans are one species within the web of life and contribute to or detract from the whole. This is the ethical aspect of the relation, which constitutes a conditional reciprocity as the basis of an obligation.

However, its value as a condition of all of life gives the biosphere a special value apart from it being a unique creative actualization. It is necessary to humans thus obligatory for our life. Obligation to the biosphere as an instrument of human survival does not, however, entail that the biosphere merely has instrumental value. Unlike many instruments it is necessary for all value. The web of life is a unique creative actualization, both in comparison with the other planets and in terms of a whole in relation to all of the parts. The biosphere is a unique whole in which parts create the whole and the whole gives life to the parts. As a unique, novel self-actualizing value, it is intrinsically valuable apart from human use. Since the environment as a whole is a distinct good, there should be no subordination of the biosphere to human good: it should never be treated only as a means.[53]

Any distinct value can be used as an instrument, for example, prey can be used as a means by carnivores. The conditions of human life acquire value as instrumental to human life, surely. The distinction of intrinsic and instrumental value is not absolute, as Dewey has rightly argued.[54] Nevertheless, the biosphere is intrinsically valuable as the condition of distinct forms of actualization, reciprocal with human actualization. If value did not inhere in the environment, it would not be a suitable environment for life. The environment as a condition of life is an essential, necessary good for all of life, and therefore practically required. All subjective states have the biosphere as their necessary condition, as they are states or capacities of living organisms located in the living environment. The life of persons is the condition of valued subjective states and the biosphere the necessary condition of life. Life and the biosphere are more general goods than valued subjective states and capacities. If we destroy the environment we destroy ourselves including any chance of continued states and capabilities. Our environment is an essential part of ourselves. The continuation or survival of the biosphere is reciprocal with human survival.

Because the biosphere is the condition of all life, it has priority over any one species in terms of duty. Thus the duty to the environment outweighs the duty to any species—an important point in dealing with environmental decisions. However, duties to the biosphere as a whole may include obligations to species required for the health of the whole and their habitats, since these live in mutual relations. The biosphere has been actualized as the result of organic action, which involves many species in conjunction and interaction. Since the whole, the biosphere is good, and required, the parts are also good as required parts of a whole. Obligations to the biosphere may include such species, since they are the condition of its continuation and preservation. Duties may also include those to essential elements of the environment such as soil, air, and water, necessary for all species.

If the destruction of the environment as a condition of life is wrong, why is there an obligation to protect species of no value to humans? Other species may not "reciprocate" with benefits to humans and may even attack humans (AIDS virus).[55] Thus it might appear that there could be no duty to non-beneficial wild species. Although there may be no active reciprocation, there is surely passive reciprocity by non-beneficial species. These do not actively benefit humans but, with rare exceptions, do not harm or interfere with human survival either. Reciprocity involves difference: species survival does not require a relation to humans but is distinct and independent from human survival. It is autonomous as intrinsically valuable. Animals may not be able to reciprocate in the same respect as humans with ethical acts, but may do so in effect by ignoring humans (pragmatic reciprocity). Thus there is the basis for an obligation of mutual non-interference, and the duty toward a species as a distinctive form of creative actualization. There may also be an obligation toward them as intelligent, motivated creatures, a point which was argued in the first section. Most important, these species may be an essential part of the larger whole, the biosphere, essential to humans and non-humans alike. They may have systemic value apart from humans (Smith, 2010).

In the cases in which animals do pose a threat to humans, for example, disease or attacks, humans have as much of an instinct to self-preservation as any other species.[56] Reciprocity is not one-sided, like subordinating theories, but mutual. We, like other animals, have an instinct of self-defense. Our instinct to survive may involve self-defense is bound up with such an instinct. Humans attacked by animals should by all means defend themselves but this is distinct from exterminating grizzlies in California as potential threats. Reciprocal non-interference is the proper balance.

Meliorism as a norm requires us to increase the number and kind of goods. Thus there is an obligation even to wild species as distinct types of actualization.[57] Critics wonder whether species can be bearers of value, exposing their nominalist prejudices. I would argue that species are a type of a whole, and if any whole can bear value, species can also. Thus if the biosphere as a whole has value, so do species. Again, if collectivities like nations, corporations, clubs, cultures, and others can bear value, so can species.

Due to the instinct of survival, of self-defense among such wild species in their own relations, such obligations do not extend to every individual of a species. There is not and can be no absolute right to life if the conditions of survival and the right of self-defense justify some killing, some taking of life.[58] Indeed, the right to life is not absolute even among humans; the death penalty and the killing involved in war are examples. Wars like World War II are considered justified by most and the loss of life is the price for collective self-defense. What is ruled out is unjustified killing, deliberate extinction by human agency: genocide.[59] Preservation of species has priority over individuals. Species are the agents of the distinct form of actualization specific to a kind, which differentiates it as intrinsically valuable. A species consists in a collection of individuals, so some individuals must survive as the agents of the specific difference. But there is no obligation to any one individual as such, outside the human community, with qualifications, namely, rare individuals of a species with distinct characters or habitats. (Social obligations to other humans, based on intimacy, are distinct from the non-intimate relations to wild creatures.) We do not have an obligation to see that each acorn becomes an oak and to tend it such that it becomes one. This would constitute domestication and undermine the unique and autonomous value of the wild.

Do we have a duty to individuals as the reciprocal other in moral association? The inherent value of the environment is holistic and does not necessarily extend to individuals within it. Individuals should not, of course, be sacrificed on a whim or without just cause. But in fact, individuals are sacrificed every day as part of a larger whole (carnivore and prey) and this may contribute to the health of the whole.[60] Neither carnivorous species nor their prey are of superior value given the egalitarian provisions of reciprocity. While any agent that actualizes itself may have intrinsic value, value is not obligation. We may wish for the day when the lion lies with the lamb but this would be to devalue the actual world. Also, the relations other species establish to each other ought to receive consideration. Plants and animals must adapt to each other or they will perish. They must “fit in” to the world to survive. Too many deer would result in overgrazing in which the herbivores would kill off the herbs, one of their conditions of survival. This would diminish kinds and quality of creative actualization. Too many lions would result in starvation. If a rough “balance” or equilibrium is achieved wherein not too many prey are killed nor too few survive to feed the carnivores, and an equilibrium ultimately restored after a cycle, different types of creative actualization, values, live together as compossible and reciprocal. Different types of creatively actualized goods are then greater. If some go extinct, the types and quantities of good decrease, the forms of actualization are fewer. This trend would violate the norm of meliorism and would be anti-pluralist with regard to kinds and modes of value.

"The balance of nature" implies reciprocal relations, as each side of the balance is equal. Critics might argue that because habitats change over time, that the balance is continually shifting at best, and a myth at worst. But if the balance is restored, in the sense that each function remains: predator, water source, and so on—then a "rough balance" remains over time. Gorke argues that environmental systems, unlike mechanical ones, are "open" in the sense that new members appear and old disappear in the normal processes of evolution.[61] But if one function disappears, the habitat can be destroyed or radically altered, from thriving to severely damaged, which happened in Yellowstone before the reintroduction of wolves. The predator function was essential to keep the habitats healthy; several predators could fulfill this function. Similarly, water serves a function: without it the habitat becomes desert.

One issue this question raises is the relation of value to obligation.[62] For several figures in Western Philosophy, anything which has intrinsic value is entitled to moral consideration, and, more, a claim of obligation. Since individuals exhibit creative actualization in the form of self actualization, doesn't this entitle them to moral consideration? The essentially Utilitarian model in which all moral obligation is based on intrinsic value has been contested both in Western Philosophy and within Environmental Ethics.[63] J. Baird Callicott has argued that although individuals may have value, this does not necessarily entitle them to moral considerability, as only species can be said to have moral standing from a holistic, ecocentric point of view. Holmes Rolston, in his notion of "capturing natural value," has argued that the intrinsic value of an individual is what predators capture, and thus there is no direct relation between moral considerability and moral obligation. Individuals do indeed creatively actualize their own life, but obligations do not necessarily follow from intrinsic value. The separation of values and duties is required for the sake of morals, as I argued above. Not all creative actualizations are good; some would be bad, because they are not an improvement or may even be immoral. They would subordinate inherent values in a hierarchy (wealth) or other subordination relation. Further, not all values are duties; beauty may be of value, but it is hardly a duty, as opposed to a good, to pursue beauty.

The duty to individuals is mutual non-interference, but not an obligation to care for each valuable life. Indeed, wild individuals would not have value if they were cared for, as they would not creatively actualize their own life, but would be dependent on others, the relation of domestication.[64] The value of an individual is morally considerable, since it has intrinsic value. However, granting moral standing to individuals does not necessarily create obligations, since obligation is not based on intrinsic value, although related to it. Obligation is created by necessary conditions of creative actualization, what is required for it. Ignoring such conditions would have negative consequences for all of life. Individuals are not necessary to the health of the whole, nor, in the case of common species, to the health of the species. As individuals die in any case, their unique but finite good cannot be preserved. The good of the species is more morally considerable, since the survival of the species is a condition of individuals of the species and thus required for individuals: the unique good of individuals requires survival of the species.

Why should human individuals alone be exempted? In one sense they are not; we do not always exempt human individuals, for example, in war. In normal conditions, however, killing individual humans is considered murder, and in some cultures punished by execution of the murderer. I will leave to others whether this is a logically consistent policy; the question is why a premium is put on individual human life in many, if not all, circumstances. Why is there an obligation to individual humans? Isn't this speciesist? This places our theory on the horns of a dilemma, since to deny humans are exempt would go against the long tradition in the West of individual rights; while to give humans special status would be anthropocentric or speciesist. Like Callicott, I am going to argue for different levels of obligation but on different grounds. I argued above for a policy of letting alone with respect to animals. This does not differ in essentials from relations of casual association in human society. Within our own society, we have more intimate relations and more casual ones. Intimate association involves relations that include obligations greater than those of casual association, relations to those we do not know. We are no more allowed to murder strangers than family, but may have financial obligations to family that exceed those to strangers. The obligations to casual associates are to leave them alone, to preserve their persons and property. The obligations to the wild are similar, that is, to preserve the freedom of wild individuals, freedom from the restrictions of domestication; and the property of wild species, the autonomy of habitats. Leaving alone also involves not killing wild individuals without good cause. However, it does not involve the more intimate obligations of direct care. Thus the obligations to individuals in the wild is necessarily weaker, since it must preserve their autonomy as wild.

Matthew Hall has raised the issue of whether plants can be autonomous (Hall 2009). Critics of the idea of plant autonomy argue that they are neither moral agents nor capable of free choice and reasoning. Hall thinks that this is an anthropocentric projection and that other forms of autonomy include plants. I would agree with these critics that plants are not conscious in the same respect as humans, as far as we know, but with Hall that this is irrelevant. Self-actualization does not require consciousness. Moreover, the ethical other as other is different and thus human projection as the model is illicit extension of the same. Autonomy means that we do not direct the other, which applies to plants as much as other living things. Nor can we determine whether plants are conscious or not, since they have a different way of life. Indeed, even arguing for plant "intelligence," and "learning," as some of the authors cited by Hall do, may be a projection from human life.

Humans have further obligations to other humans, reflecting civil society. It is not speciesist to argue that humans cannot join avian or piscine societies: we are not anatomically correct. While the biosphere is all-inclusive, different societies are not. This creates obligations to the biosphere and to other humans that do not extend to individuals of other species. Our duties within and derivative from civil society are especial to human society, although they may include domesticated species. Civil obligations may exceed those of obligations to wild individuals. Such obligations do not derive from the biosphere, but only within human, civil society. These obligations may be reciprocal with civil rights as well, rights an individual may have in society which are not shared outside it. Freedom of religion is such a right, given the hypothesis that animals and plants do not have religions. Rights, then are confined to civil, human society and to other species under human care within it. "Civil rights" are granted within human society and are not relevant outside it. They are political.[65] They also include obligations to future generations, since we have a duty to those we have created.[66] Creative actualizations of societies are a distinct mode of value to which justice must be done. Societies must also be compossible in a pluralistic universe.

Duty to the biosphere may supercede the duty of mutual non-interference, even non-interference with fellow humans. Indeed, duty toward the whole may rule out actualization of individual goals, especially those incompatible with the survival of different goods. This includes individual desires, whether for profit or an excessive number of offspring. If preventing a madman from blowing up the world might justify killing him in collective self-defense, so destructive actualization may be ruled out by duties to the biosphere as a whole and to a plurality of goods within it. The duties to the biosphere are morally obligatory even if asserted against the (human) majority just like many other duties. The obligation is not dependent upon the whims of a majority, any more than those of ethnic, gender or other minorities are. The duty to the biosphere is extra-majoritarian, that is, an obligation over against the majority as it is morally based and a condition of the life of the majority.

Some have argued that Ecocentric Environmental Ethics violates the liberal tradition of "individual rights." Such rights require a duty to respect others' rights, however. Rights are reciprocal with duties, as my right requires your duty to allow exercise of it and your right involves my similar duty. Such rights involve a duty to respect the rights, that is, obligations to act such that the rights of the other can be exercised. The duty is identical for all or universal. In turn, duty or obligation involves necessity, since necessary actions are required, or obligatory. Necessity formally requires universality.

But universality cannot be derived from specific human differences by definition.[67] "Individual rights" require a universal ethics as their basis. Thus subordination relations of environment to human will ought to be avoided. I am not arguing against individual rights, but that subordination of all rights to individual rights might result in the end of life, including the life of individuals, and is thus a self-defeating, inconsistent view. In Kant's terms, it could not be consistently willed.

Moreover, individual rights are only one consideration in moral evaluation. Such rights are not absolute in any culture but require comparison with other rights, including the rights of the community, the rights of other individuals and competing rights.[68] The issue is not whether individuals have rights, but the problem that arises when two rights clash. The "right" of corporations to dump toxins in rivers interferes with the right of citizens to clean water—to speak nothing of the threat to wildlife of poisoned water. The right of individuals to use private property may violate the rights of others, especially where wrongs are committed on the property. The community has to protect itself in collective self-defense, and this may include overriding the wishes of individuals. Individuals are protected by certain rights they hold over against the majority. But individuals do not have any right to threaten the well-being of the community. Yet threats to the biosphere, to the air, water, and soil and other "commons," and to future generations and other species are such a threat.

The intrinsic value of the environment is, as intrinsic, categorically distinct: not subordinate to other categories of value such as economics or even aesthetics. In the latter case, the wilderness might be rearranged to suit human notions of beauty like a garden. But then the value of the environment would be subordinate. The wilderness is the category of value most autonomous from human desire; otherwise it would not constitute wilderness. Subordination of wilderness would be destructive of its distinct value. I should emphasize that this does not rule out all killing. Indeed there may sometimes be a duty to kill, for example, in eliminating invasive exotics such as eucalyptus in California. Whether the alien plant species were inadvertently or deliberately misplaced by human agency, their elimination may be required to protect rare or endangered species or threatened habitats. Human wisdom may be of value in such efforts as it can in some cases correct mistakes and imbalances.

Humans are intelligent enough to articulate, actualize, and exercise duties. This mode of actualization is a distinct form of creative actualization involving choice between the distinct modes. Actualization of duty follows from human recognition that we have a duty to protect the environment, which extends beyond other humans, its beneficiaries, to the environment as such. If this is not in fact practiced it ought to be: an obligation, that is, a deontological environmental ethic. Ecologically harmful technology is no more inevitable than the extinction of rare species and does not justify pollution or destruction of environments. We can actualize duty to the environment as an imperative of our continuation as a species.

Our obligations are, firstly, toward the biosphere as a whole and secondarily toward distinct species or strongly marked varieties. Thus the obligation extends to both enough individuals of a species to assure their genetic and environmental viability, and their distinctive habitats. In some cases, then, the agent of actualization of a distinct species type will be the same as the actualization itself, that is, in surviving the agent survives by actualizing itself over time. However, obligations do not extend to every individual of a species: we are under no *obligation* to see that each acorn germinates and then care for it. Instead, the duty to individuals is that of mutual non-interference. A borderline case might be a species going extinct but not due to human agency, as with "relict" species, and extinctions that took place prior to humans, such as the dinosaurs. The extinction of any species is unfortunate, but failure to adapt creatively to new conditions is a specific defect.

The obligation to the biosphere and to species is more positive, then, while that to individuals *per se* is more negative. In any conflict between duties to the biosphere and the duty of non-interference to individuals, the duty to the biosphere should prevail. Actualization of duty follows from recognition of the distinct value of the environment as of reciprocal moral importance. Different species surviving together constitute a world. As reproduction and survival are creative actualizations such species are valuable in themselves and worthy of preservation. The value of the habitat is not subordinate, but equal, even superior to any one individual, and as part of the biosphere, its preservation is an obligation. We have a duty to protect the environment that extends beyond other humans to the environment as such.

Since actual or general moral universality includes animals, plants, and the biosphere as a whole, the distinction between ethics and environmental ethics as genus and species is no longer tenable. Environmental ethics is the only possible framework a universal ethics can take: ethics properly understood is environmental ethics. Humane ethics is only a species of this more general form of ethics. Anthropocentric ethics is tribal: the valuation of the human tribe above other species, the ethics of the in-group, ignoring the general or universal other of reciprocal association. Anthropocentric rationalizations are ideological, the part masquerading as the whole. The interest of one species, as we argued above, cannot be universal by definition, and therefore cannot provide a universally valid premise of ethics. Starting from self-interest, whether of an individual or of one species, cannot provide an ethic. Properly interpreted, there can only be one ethic, an ethic of universal inclusion in the world that *includes* humans as part of environment. Environmental ethics is more inclusive; in its consideration of other cultures, species, and habitats it is cosmopolitan. Environmental ethics is the only general, universal ethics and therefore should be the basic approach for ethics. If ethics is a way of life, ecological ethics is the most universal way of life, the most moral.

The revaluation of ethics in radical axiology will have consequences beyond ethics proper, for it is a rejection of anthropocentrism and the problematic of anthropocentrism.[69] Cartesianism elevated problems of human knowledge to the forefront of philosophic concern. But human knowledge is centered on humans. Previous ethics has been derivative from metaphysics and epistemology; it must now form its own standards and even regulate these derivative studies. A revaluation of Cosmopolitan Ethics means that we must consider environmental consequences even in "first" philosophy. Ethics is first philosophy and anthropocentric ethics is a species of a much larger genus, the ethics of the wholly other, universal cosmopolitan ethics. The ecocentric revaluation of ethics follows from the expansion of duty to actual universality. But ethics is thereby transformed from partiality to the whole.

Two

RETHINKING COSMOPOLITAN ETHICS[1]

> Injustice anywhere is a threat to justice everywhere.
>
> Martin L. King Jr.

> The breakdown of the world's ecology is causing a shift in environmental sensibilities tantamount to a second Copernican revolution....In the second Copernican revolution we may be forced to abandon the even more self-aggrandizing belief that we are the center of the moral universe and have a special, privileged status in the biosphere—anthropocentrism.
>
> Lynton Caldwell[2]

> Anthropocentric ethics is neither capable of justifying general species protection, nor does it provide the social and psychological context that would be required for converting intuitively rooted moral concern for general species protection into deeds. Martin Gorke[3]

In this chapter I will propose that ecocentric cosmopolitanism is more in line with the notion implied and expressed by cosmo-politan (world citizen). The world as a place suitable for life includes the biosphere taken as a whole, and being an ecocentric cosmopolitan is being a citizen of the biosphere as a whole. We are residents of the world first, that is, in the world consisting of an all-encompassing environment. Being a citizen of the world implies an allegiance to the world, not simply to one particular culture. Indeed, living in harmony with the environment is a standard by which to judge cultures. I will argue that an ecocentric ethics is the only universal ethic, and thus the only one that can meet the test of universality required for a rational ethic.[4] Adopting an ecocentric cosmopolitan ethic would dissolve the distinction of speciesist/humanist ethics from environmental ethics, along with the problematic suggested by "moral pluralism."

I will first give a historical summary of ancient cosmopolitanism, and then the idea of moral progress. Next I will criticize the humanistic world-view that is at the basis of anthropocentric ethics.[5] I will defend ecocentrism and outline an ethic of ecocentric cosmopolitanism as a replacement for speciesist anthropocentrism. I will conclude with a brief look at the relation of environmental justice and social justice.

1. Cosmopolitanism

Cosmopolitanism is an ethic that emerged in Hellenic Greece when the intellectual and tribal barriers between the city-states were breaking down.

The cross-fertilization of ideas, which followed the mixing of people and thinkers from different regions, precipitated a crisis in belief. If different cities practiced different customs, which were best? Which of the varied customs of different city-state were "right" or in accord with nature? The problematic of human nature emerged from this reflection and in part as an answer to it. Human nature as *zoon logon ekhon,* the rational animal, was an abstraction from all cultures, which transcended the particular cities. Whatever the varieties, language as such distinguished all humans as human and thus as alike in nature. While each city had its distinctive customs, at a higher level humans are alike as linguistic and rational.

Once such a nature had been abstracted, it became a standard by which philosophers could judge particular cultural practices as natural or unnatural.[6] Since this standard was independent of any one culture, it led to a relative intellectual liberation from closed cultural ethics, tribal ethics. Cultural ethics are tribal since a tribe shares a culture, which excludes the foreign as alien and includes those who share customs, a language, and so on. Often it involves inclusive breeding, that is, everyone is related by blood, and often a common tribal identity. The denouement of the movement away from tribalism was the birth of the cosmopolitan philosopher, the "citizen of the world."[7] Cosmopolitan philosophers rejected identification with any narrow cultural ethic, since their allegiance was to reason, the standard of human nature as a whole. Nature for humans is human nature and this consists in rationality. Since human nature is the standard for all cultures, cultural allegiance is transcended. Thus cosmopolitan ethics was one form of philosophical ethics, as universal, trans-cultural, and critical.[8]

Cosmopolitan Ethics met the test of universality, for reason is a universal ability of humans in all cultures. Since reason is universal, the cosmopolitan is at home everywhere in the world: a "world-citizen." The cosmopolitan is also open to the world, since identity with a particular culture is rejected. Cultural exclusivity or tribalism is closed to the world, since what is alien to the culture is excluded as foreign. Cultural ethics is not cosmopolitan in either the literal sense of being a citizen of the world, since all mores come from the tribe; nor in the sense connoted by the spirit of cosmopolitanism. Cosmopolitans, by rejecting the closed ethic of the tribe, have opened themselves to the world: they were citizens of the world, not of any exclusive city, culture or tribe. Since they rejected the closed-mindedness of the tribal culture, they were open to practices of different cultures, from different parts of the world than their natal culture. The cosmopolitan is at home in the whole world. Thus a Cosmopolitan Ethic includes living in the world and being part of that world. Another aspect is the virtue of tolerance. The cosmopolitan tolerates ways of life and customs different that those s/he practices. Although the cosmopolitan may not personally endorse every cultural practice—for this would be impossible, if only because of variety—

s/he recognizes that such variety is itself good, if basic, universal ethical norms are not violated. Cultural variety means more kinds of goods.

A tension emerged between the ethical standard of nature and the tolerance of cultural forms and varieties, which are the essence of humanistic cosmopolitanism. The nature-law/custom distinction at the basis of Cosmopolitan Ethics favors nature as the standard, not law. The cosmopolitan is loyal more to nature than culture. This tension resulted in cynicism among the ancient cosmopolitans, for example Diogenes. Cosmopolitans became cynical of cultural practices not "in accord with nature." Thus Diogenes went about "clothed in nature," lived in a tub not a house, and acted as a gadfly toward his culture. The distinction of culture and nature and loyalty to nature is also at the root of the environmental movement. However, the emergence of the movement required an intervening historical sensibility.

2. Moral Progress

The idea of moral progress is the progressive expansion of moral considerability from white men to non-whites, women, sexual minorities, future generations, animals, and ultimately to all of non-human nature.[9] The theoretical development of ethics from the tribe to those beyond has as its logical outcome the extension of moral considerability to other species and the biosphere. The hope that humans can extend moral obligation from themselves to animals, other species, and the biosphere as a whole, just as they once extended it from the tribe to those outside the tribe, is the core of cosmopolitan environmental ethics. The goal is a universal ethic: all other living things are worthy of being treated justly, with mutual recognition in accordance with the principle of reciprocity. Why should duty be extended beyond the human species to include other species and ecosystems? Humans depend on the environment, implying a duty to other species based on this dependence, and the increase in the number and types of goods involved in speciation. However, the underlying rationale must involve a change of attitude and even more of orientation.

An anonymous critic has argued that this is the perspective of animal liberation, not the environmental movement as a whole. On one level I agree, since I would argue for *inclusion* of humans within the biosphere, not *extension* from humans. On another level, however, it is an extension, since human morals did not regard the non-human as morally considerable. Environmental ethics regards the extension of morals to include the non-human as its project, even if it does not start from an anthropocentric perspective. The perspectival question is distinct from the issue of moral extension, even if we argue that Western Ethics was wrong-headed to begin with by adopting the anthropocentric perspective. The critic was right in criticizing my original formulation, which only included animals.

"Brownlash" opponents have attempted to label environmentalism and the "green movement" as an ideology: as one more political movement. On the contrary, anthropocentric "ethics" which is ideological, not ecologically based environmental ethics, as anthropocentrism represents the interests of one species as if it were the whole, as if it were universal. The interest of one species cannot be universal by definition. Anthropocentric "ethics" is the rationale of totalitarian domination[10] by one species and in the interests of one species. The "domination of nature" means a world totally organized for human exploitation. The planet-wide organization of areas according to human use reveals the totalitarian reach of the management ethic: here a logging area, there a farm, here a suburban "development," there a "wilderness area" (usually so designated because it is on marginal land).[11] "Designated wilderness," implies only one species should decide what should be left as wilderness. The fallacious idea that forests or other lands have to be "managed" by humans despite the three and one-half billion years of life on earth before the rise of humans is the heart of the management ethic.

Human ethics up until the advent of ecocentrism and biocentrism has been tribal.[12] It has excluded the great majority in favor of a select minority, whether an ethnic, gender or other minority. Any reduction in the number and scope of living things deserving of moral consideration would be a regression. It would include axiologically false identification with a smaller group: the tribe, the nation, the saved, and so on. The identity is a false identity as a regression from humane ethics, cosmopolitan identification with humans. It is also a closed identity—bringing only some humans, not all, under mutual recognition, whether on grounds of race or class. This identity is a bad ground for action, as inconsistent, with a double standard, and grounded in axiologically false identities (race), among others. Small minded, bigoted, and intolerant, it values too little.

By contrast stands cosmopolitan ethics, the ethics of the "world citizen" articulated by Diogenes, against the notion of belonging to any narrow group.[13] The extension of duty beyond the tribe has a logical outcome in universalization of duty, since moral principles should include all other cultures, which constitutes cosmopolitanism as an ethics. In this ethic, a universal inclusion is contrasted with a narrow exclusion. The word "cosmopolitan" also connotes the rootedness of humans in the world (*cosmos*) as basic. As Chief Seattle was reputed to say, "man belong to the earth, not the earth to man." Humans are rooted in their environment and, as Leopold suggested, fellow-citizens in a world which includes other species.

Echoing cosmopolitan ethics, environmental ethics has attempted to extend moral considerability beyond the species to include animals, plants, and the biosphere as a whole. Environmental ethics is the most inclusive ethics, since it includes even landscapes, and is cosmopolitan in the sense of viewing humans as rooted in the world. Environmental ethics combines cosmopolitan ethics of openness to the world and the universality of moral principles with the thesis of extension of considerability: moral progress.[14] However, a change in what is cosmopolitan is connoted, since the logic of "world-citizen" is taken to include the biosphere as a whole, that is, literally the whole world. Environmental ethics interprets the world to include the biosphere, and all the living things in the world.

3. The Challenge from Humanism, Humanism as a Worldview: The Hierarchy of Nature as Human Chauvinism

One of the main issues debated in environmental ethics has been the issue of anthropocentric rights and values as opposed to non-anthropocentric rights and values.[15] Some have argued that only humans have rights or intrinsic value, and that the non-human world has value only for humans.[16] Others have argued for a more ecocentric view in which humans are just one species and the biosphere has a value of its own.[17] Still others have argued for a combination of the two views, for example an extension of human rights to higher animals.[18] To argue that all value emanates from humans, that the entire world is only of instrumental value for humans is a form of humanism. Anthropocentrism is the latest incarnation of classical humanism, since it would make of humans the great exception in the chain of life. I am going to address this issue from its justifying ground, then, namely, the idea that humans are the great exception and thereby entitled to use the world however they desire.

Humanism is any belief that involves the evaluation that humans are at the apex of creation or nature, the top of the hierarchy of beings. This explicit hierarchy is a structure and world-view that humanists adopted from medieval theism and never questioned but only transformed. The substitution of humans for God at the apex of the hierarchy of creation, the "great chain of being," began with the anthropocentric adaptation of Plato by Renaissance Neo-Platonists (Giovanni Pico della Mirandola) and reached its culmination in atheistic humanism. In this view, humans are supreme over all other species, as the apex of creation. Humanism involves an abstraction of humans from the environment, a utopian model also adopted by anthropocentric value theories.[19] It is utopian since in the actual world, humans depend on the environment, not vice versa.

"Humanism" is an ambiguous term, with many different meanings. Humanism as a literary movement, with an interest in human affairs as opposed to an otherworldly orientation, is distinct from philosophical humanism, which likewise views the study of man as central but differs in approach. Philosophical humanism is opposed to naturalism in philosophy, in that humans are placed at the apex of a hierarchy, replacing God. The view of humans as the center of the universe is the worldview of medieval humans. Humanism is only thinkable if attached to the notion of the earth as the center of the universe and of humans at the apex of the center.[20] Nicklas Kopernik (Copernicus) exploded this part of the humanist viewpoint long ago. The discovery of a universe with the sun in a remote arm of a huge galaxy among millions of other galaxies should have demolished this anthropocentric view. Nevertheless, just like astrology, another anthropocentric view, humanism survives. Humanistic thinking based on this worldview is an attempt to return to the archaic pre-modern perspective, a reactionary worldview of permanent hierarchy, with humans at the top of a biological pyramid.[21] Humanism is anti-progressive, based on its refusal to extend moral considerability to non-human nature; ideological, since it takes an arbitrary and questionable species difference and attempts to derive a moral absolute from it. Ideology is the part masquerading as the whole, that is, humans representing their "interests" as the interest of the whole. But the "whole" must include everything, that is, other species and the environment as morally considerable. Humanism places humans above the rest of creation/nature and refuses to see the whole out of a misplaced anthropocentrism. It is ironic that environmentalists could be accused of being ideological when humanism is so obviously partial to one species among millions.

Human chauvinism is more modern. René Descartes's belief that humans are special and animals, machines is a recent view. The ancient Greeks thought we were rational animals, a distinct species but a *zoion* nonetheless. For the medievals, God created nature along with humans. The exploitative form of humanism, of nature is a field for exploitation for humans is also modern. Descartes provided the rationale for human dominion surreptitiously adopted from medieval scholasticism, with biblical and pagan sources, at the same time as modern science was beginning to undermine this view in the heavens. It survived as metaphysical subjectivity and anthropocentrism.

Humanism has received metaphysical warrant from modern philosophy, the Cartesian view of a detached human subject confronting a world of objects, including other living things as objects. This view has been abandoned in all fields of philosophy except value theory! Led by the pragmatists and naturalists at the turn of the twentieth century, both analytic and continental philosophers have made metaphysical and epistemological critiques of a subject removed from the world. Humanism as a philosophical stance has been overturned and replaced by a more biological view in which humans are one species among others within the world. The humanists' claim that humans are special is untenable, given modern scientific findings, especially evolution.[22] If humans evolved from other species, why should humans be separated from other animals? They cannot be distinguished from other species as especial biologically, except as one species. Charles Darwin and even Aristotle believed humans were one species among others. Humanist philosophy cannot change the biological evidence.[23] Humans evolved from other species and biologically are fairly clever and successful primates. The view of humans as elevated has neither scientific nor moral warrant. A single species that is just as dependent upon the environment as any other cannot be privileged.

To be sure, humans differ from other species, just as any species differs in some respects. Horses also differ from elephants. Humans are also incapable of specific animal capacities, just as animals may be incapable of some human ones. The question is, are either of these facts morally relevant? What is at issue is whether such differences are morally significant: whether the differences which characterize one species provide moral warrant for that species to lord it over the others, destroy habitats, waste resources, overpopulate, and so on. Why is a species difference a unique ethical warrant? Why then in particular is a species difference morally relevant? It is speciesist to count only humans as morally considerable, to speak nothing of violating the basic premise of ethics: the golden rule or some variant.[24] "Do unto others" involves treating the other the same as one would like to be treated, implying reciprocity, which in rational terms is consistency. More formally, ethics requires universality or it cannot be rational; consistency or it cannot be ethical. Practical universality must include at least the majority of other species, or humans have put themselves in the position of the elite, an ideological stance. Since ethics must be universal and reciprocal, a species difference cannot provide such a warrant. It is not universal as distinct and unique to a species; not reciprocal as confined to a species. Anthropocentric "ethics" is irrational: there is no universal premise, since humans are an arbitrary part.

4. A Critical Evaluation of Anthropocentrism

Both the Western and Chinese traditions call for a humane ethic.[25] Humanistic ethics is specifically different, but also regulative for the rest of nature by extention, for example in mandates to be kind to animals. Humanist ethics, although an advance upon a narrowly tribal ethic, can itself be an exclusionary "ethic," and thereby the latest form of tribalism when taken as the source of ethics, or where value is confined to humans. Humanism—anthropocentric ethics—is elitist and exclusive, thereby tribal.

An ethic based on anthropocentric values is a species ethic, based on claimed distinctly human capability, whether "reason," "sentience," "language," "interests," or other purported capacities exclusive to humans. Thus one species is either the model for others or the source of value for others. Usually such posits are not exclusive to humans; in any case, they are axiologically arbitrary and morally irrelevant. As speciesist, they can never generate the universality required for rational duties, and thus are ruled out by reason.[26] Specific differences are not universalizable by definition. Since they are not universal, such posits are not necessary and cannot be obligatory, practically required.

A corollary of humanism is that human supremacy allegedly creates a moral superiority, which entitles humans to dominate all other species. Since humans are at the apex of value, they are the ultimate end or highest good in the hierarchy of "beings." The extreme formulation of this view is that humans alone have intrinsic value, and other species are only instruments of human good. All anthropocentric value theories, which confine intrinsic value to humans, are based on a hierarchy of value with humans at the apex. Humans give value to subordinate species in this view, which have value only on condition of their value to humans. The value of instruments is in relation to humans alone as intrinsically valuable, creating a hierarchy of goods in relation to their degree of worth to humans. Hay is good for cattle, cattle for humans, thus destruction of "worthless" land (habitats) to cultivate hay is "justified" because of its value to the ultimate, supreme end, subjective human states.[27] Such a hierarchy cannot be used as the basis for a universal ethic, however, without a fallacy in values. The values that differentiate humans are distinct, thus cannot be universal. Human value cannot be the universal value and thereby define value. The intrinsic value of human differences is intrinsic to humans, not to all value. Otherwise, they could not be distinctly human.

Like racism, speciesism argues from a dubious premise to an exclusive ethic. By way of example, one of the justifications for anthropocentric ethics has been that only humans have "consciousness." This judgment is both incorrect and morally irrelevant. Since the opposite of conscious is unconscious, animals obviously do have consciousness. When animals are awake they are conscious; when asleep unconscious, just as with humans. Consciousness is not especial to humans. But what of "self-consciousness?" This connotes consciousness of a self or of individuality. Animals exhibit this every time they attempt to avoid predation. Why else would an animal attempt to save its *own* life? Species differences are also morally arbitrary, just as tribal, racial, and class differences are.[28] The "reasoning" in human chauvinism is parallel in form to the reasoning used by the racist who argues one race is unique and therefore entitled to subdue/exterminate the other races. *Non sequitur* based on an arbitrary if not false claim. "Race" is not morally relevant any more than a supposed human moral capability (an ability which is counterindicated every time a human goes hunting for "sport," or the like).[29] It is false because humans are not as unique as is claimed, since animals practice many analogs of human behavior in their world, and humans are biologically just one species among others; arbitrary, as not morally relevant. Even if true, it cannot provide ethical warrant for extermination or subjugation of other races, since this is itself immoral. Why should an alleged superiority create a privilege? All alleged human superiority is arbitrary or irrelevant, as Paul and Ann Ehrlich have extensively argued.[30] "Superiority" connotes moral excellence and thus would exclude immoral acts like genocide, slavery, and others. These evils are neither based on consistent universal principles nor reciprocal.

Another version of this species difference premise is the argument that humans alone can reason, another case of arguing from a dubious premise to a non sequitur.[31] Reasoning is not confined to humans, as I argued in chapter one. More, many humans are not proficient in reasoning: but this provides no warrant to mistreat them. The lack of a capability others have, for example the ability to understand the mathematics of relativity theory, does not count as a reason among humans for moral difference, if only because of the norm of equality. Similarly, the lack of specific human capacities should not count morally. The arbitrariness of using such a species difference can be illustrated by substituting unique characteristics of other species for human ones and then basing moral considerability or an ethics upon it. Why not, for example, posit possession of trunks as the basis for moral considerability, and thus elephants and trees as superior? Or their superior size? Or the ability to make your own food as intrinsically valuable and thus green plants as morally superior? Human experiences may well be distinctive, but so are many animal experiences. We see the beauty in nature, but do not have the acute sense of smell of the canines or the long-distance vision of the birds of prey.

Humans do not live by reason in any case. In our society, "reason is the slave of the passions." Desire generates the insatiable demands of our consumer society, not reason, often in the form of artificial desires "created" by advertising. Reason would tell us to cut back to a sustainable level so as to preserve our world and prevent creation of greenhouse gases in the light of clear evidence of global warming. Reason would tell us to conserve the world in which we dwell, not to continue to pollute: to foul our own nest. Reason would tell us to preserve our species, not to be self-destructive.

Another argument for human speciesism is that only humans can be moral agents.[32] Clearly, moral ability, contra Carl Cohen, is not part of the essence of being human, a species difference, since humans are also capable of Auschwitz, Bosnia, and other "crimes against humanity."[33] After Auschwitz, the Gulag and the other events of the 20th century does this claim have any plausibility or validity? Humans may be capable of morality but it does not follow that they are essentially moral. Since it is not part of the human essence it can only be relevant if some humans actually achieve morality. I fail to see any great evidence for this, based on the twentieth century. For every Mother Theresa there is a Himmler. For every Martin Luther King there is a Hell's Angel. For every altruist there is at least one egoist, and likely more. For every person who treats animals kindly there are many who exploit, beat, or torture animals—to speak nothing of slaughtering them for food. Thus moral capability is not a specific difference on which privileges can be based anyway. Humans cannot be defined as "the moral creature." A certain amount of violence, particularly toward those outside the culture, is considered acceptable, indicating a moral defect or at least limitation of tribal ethics.

However, even if "human nature" could be taken as a starting point, then all manner of murders, tortures, perversions, rapes, and other crimes, which have been committed in all times and places, are statistically within the norm or "natural" to humans. Such crimes would have to be admitted into "humanistic" ethics and explained away.[34] Human "nature" includes the commission of crimes against humanity: genocide, rape, and so on—to speak nothing of crimes against the environment, and the horrendous treatment of animals. If human treatment of animals is a "humane" ethical stance and virtuous, then humans do not understand their own ethics. With virtues like this, who needs vices? Indeed, one could argue that humans as a whole have not yet attained the status of moral agents. Consider Albert Schweitzer's standard for morals:

> Ethics…consists in this, that I experience the necessity of practising the same reverence for life toward all will-to-live as toward my own…It is good to maintain and cherish life; it is evil to destroy and to check life….
> A man is really ethical only when he obeys the constraint laid on him to help all life which he is able to succour, and when he goes out of his way to avoid injuring anything living….[35]

By this standard, most humans are not ethical.

As for the exclusivity of moral agency, consider Lassie. "A man's best friend is his dog," as only a dog is always faithful and will always give her life to save her "master." While some humans are capable of such loyalty and self-sacrifice, these virtues are neither confined to humans nor universal among humans. What Cohen misses is the "ethic" of other species. An "ethic" is a way of life and of living together. All species have a way of life, and many live together, particularly herd animals. Moral agency by this definition is not confined to humans. Animals have their own "way," their own species ethic. Moreover, animal parents sometimes sacrifice themselves to preserve their offspring, altruistic behavior that Darwin noted.[36] Cohen also ignores the continuity between human and animal ethical behavior:

> Darwin believed there was a continuity between humans and other species, which led him to think of human morality as related to the sympathy seen among social animals. This long-disdained idea was resurrected only recently by researchers like the primatologist Frans de Waal. Darwin 'never felt that morality was our own invention, but was a product of evolution, a position we are now seeing grow in popularity under the influence of what we know about animal behavior,' Dr. de

Waal says. 'In fact, we've now returned to the original Darwinian position.'[37]

What is almost comic about such arguments as Cohen's is that the alleged exclusive moral agency of humans is used as a warrant to treat animals immorally, that is, as suitable subjects for experimentation! The basis of philosophical ethics is that for a human to act "inhumanely" is a contradiction, or in moral terms, hypocrisy. Moral superiority requires moral behavior at all times, and thus rules out immoral behavior toward animals, including useless medical experiments: useless because ultimately testing on human subjects is required anyway. As to the argument that humans are capable of moral goodness: most humans are quite ready to kill other species simply because it "tastes good," even when they know that the arguments from protein or nutrition are fallacious.[38] Only if humans actually act morally to other species would this be valid. But the record suggests just the opposite: other species are resources to be used for human convenience.

If humans alone are moral agents, as humanists claim, then why do they sometimes act "worse" than animals? Animals don't commit genocide; kill for the sake of killing. No animal would deliberately slaughter the bison to undermine the Native Americans. The idea of humans as moral subjects cannot be seriously maintained, especially given the tragedy such events symbolize. From this perspective, animals are more moral, since they kill only for food, not wealth, passion, "sport," "taste," and other crimes. Animals never kill for "fun." Animals do not inflict pain on other species, nor on their own species. Animals, with some exceptions, do not lie, steal, cheat or rape. Only predator species and a few insects kill their own kind. Humans are worse, not better than other animals.

It would be as much of a mistake to confuse moral capability with moral considerability as a moral ability with an essence. Non-human nature should receive moral considerability, *even if* animals are incapable of specific human capacities. Even if there is such a special human capability it does not create the privilege of raping nature but the responsibility not to, as a moral ability. A moral capability involves more responsibilities than privileges, including a responsibility to treat non-human nature morally. We should treat others morally whether or not they are virtuous, that is, extend them moral consideration, because it is morally right.[39] Only acting rightly is virtuous and in accord with duty, marks of morality.[40] This is precisely what humans fail to do with animals and plants. Similarly, from the alleged moral capability of humans, that only humans are or can be moral agents, it does not follow that only humans have intrinsic value. Intrinsic value and moral capability are distinct, as I argued in chapter one, following almost all moralists. Cohen has confused value and evaluation, in arguing that the moral evaluation of moral agents is equivalent to intrinsic value. If humans alone have a special moral ability they should treat other species with respect and care, not violence or exploitation. Otherwise this so-called ability is ideological: anthropo-fascism.[41] Human chauvinism provides a rationale for anthrofascism, a view without a conscience that lacks moral imagination. I will return to this point below.

What is lacking in humanism is the moral imagination to perceive a chimpanzee in a cage being used for "scientific research" the way we now perceive a five-year-old forced to work in a coal mine for a pittance in the 19th century. Some humanists argue that ecocentrism "puts humans on the level of the beasts" and thus is a devaluation of humans. In this view, it was precisely such a devaluation that resulted in the mass murders of the Twentieth Century. The victims of the Nazis, for example, were herded into cattle cars for transport to the crematoriums. They were treated like animals. But shouldn't the whole lesson be that neither humans nor animals should be treated in this way? The contrast of humans and beasts is thinkable precisely because of abusive human treatment of "beasts" that is, other species. If humanists were really worried about morals, they would be on the side of the environmentalists, who have a much larger moral imagination than the human chauvinists: Western culture needs a larger moral imagination. Exploitation and killing are wrong, whether of innocent members of other cultures or innocent animals and plants trying to survive. If killing someone out of desire is wrong, how can killing a forest out of desire to build a house be right? If killing is illicit, killing a woodland is also illicit.

Humanism is tribal, with a closed not an open ethic, with moral consideration confined to the in-group. It cannot serve as a basis for a universal ethic. The exclusionary principle, the notion that moral norms apply only to humans is human chauvinism masquerading as value theory. Humans can't have it both ways (privileged status). If we are moral agents, then treating animals as part of our food chain, as other predators do, is ruled out. For then we are only another predator species, contradicting the premise of moral difference.[42] If on the other hand we are just another predator species, then we cannot make the claim of moral superiority on which human chauvinism rests. *If we kill other animals we are behaving like animals, and so can't claim a special moral status. If we claim a special moral status above animals, we can't behave like animals and therefore must treat animals and the environment morally.* Otherwise we are morally hypocritical, which is much the same as particularism of the Nazis sort.

Humans cannot be both superior and dependent: this is the contradiction of anthropocentric ethics. Ecocentric cosmopolitanism is not anti-human, but expands ethics with a fuller understanding of human life in its environmental context. That humans require food, clothing, shelter, air, and water for survival is an obvious platitude; these are the condition of trivial "subjective" feelings, including happiness, and desires for consumer goods. The condition of moral agency is a healthy environment: we could not be moral agents without an environment. Food, water, air, and these other necessities are the environmental conditions of all life including human life, thus practically necessary. Practical necessities are required or obligatory as conditions of life. Polluting our air and water, destroying our soil is immoral and those who do so are morally wrong.

Finally, some have argued that humans have "transcended" their environment.[43] However, humans have not so much "transcended" their environment as transported it. Every breath we take, every drink of water reminds us of our place in the world, if "transcend" is taken to mean beyond the world. Humans still require housing and clothing in temperate and polar climates, for which they are ill-suited. We bring our originary hot climate with us, by adapting to novel habitats. Humans are still uncomfortable in frigid or excessively dry habitats. While human adaptability is remarkable, humans have no more transcended their environment than the lowly cockroach, which inhabits human dwellings in the same habitats, and could not survive without them. But if human values transcend mere biology, as humanists claim,[44] what is the place of environmental ethics? Human flourishing still has environmental condition. All the novel human values, marvelous though they may be, require an environment and are within it. Listening to Wolfgang Amadeus Mozart, reading William Shakespeare, and religious practices all take place within an environment. Humans have not transcended their environment in practice, and require one for the practice of such values. Transcendence is a humanist myth.

5. A Defense of Ecocentrism

Environmentalists and animal activists have been accused of hating humanity.[45] Just as the racist argues that someone in favor of civil rights is a "nigger-lover" so some make the argument against environmentalists or defenders of wilderness that they are "anti-people." Of course it does not follow from extension of moral considerability to include other species that one "hates humans." The extension of moral considerability involved in moral progress is just the opposite of hatred: it creates a larger field of moral regard, which is less exclusionary. Environmentalists and animal activists just do not think that the entire planet is a field for human exploitation with absolutely nothing left for the rest of life. They do not view the planet in economic terms as a field for fulfillment of human greed, self-interest, and enrichment. The idea that ecocentrism is somehow "anti-people" would be preposterous even considering the threat to human survival of the mindless destruction of different habitats and species involved in the model of individual consumption bred by consumer society.[46] Usually, such rationales are barely disguised greed, masquerading as concern for "people" for example in the dispute over logging old growth forests. The special, exclusively economic interests of a small minority, the loggers, whalers, and so on are presented as somehow "pro-people." Basing our ethics on human greed does not bring out the best in us. The most advanced humans were either non-violent, vegetarians, or both. They also rejected economic enrichment as a worthwhile life-style.[47]

One problem with any ecocentric ethics is that some authors treat environmental concerns as pre-moral. James Tufts, for example, analyzes the relation of societies to their environment, but does not treat this relation in moral terms.[48] By contrast, Aristotle and Charles-Louis Montesquieu emphasize the formation of the character of a people by their environment, a point echoed more recently by such authors as Darwin and Jared Diamond.[49] Such factors as climate and terrain influence the development of states, a more organic model. The many different forces that Aristotle and Montesquieu considered, which combine to make up political life, bear some similarity to the parts of a whole. Culture and political life are inseparable from a distinct habitat, however difficult to determine in some cases how and in what way the environment shapes culture. Ecocentrists argue that Tufts approach is an example of anthropocentrism, since it ignores the very environment that sustains culture. Tufts views are partial, since they do not consider the conditions of moral life fully.

Immanuel Kant connects obligation to free, autonomous, moral agents. Reciprocity is grounded in the autonomy of reciprocal moral agents, not in violence. As a moral limitation, reciprocity is obligatory even if it is not practiced. Morals as such are voluntary, since moral autonomy has freedom as a condition. Thus the charge of "ecofascism" leveled against ecocentric ethics is completely fallacious and misplaced.[50] Fascism is a political ideology based on violence, not a moral philosophy. On the contrary, environmental ethics tries to prevent violence against animals, habitats, and the biosphere. The documentation of such violence by Peter Singer, the Ehrlichs, and others reveals the horrendous use humans have made of animals and other living things, and shows where the preponderance of justification of violence is needed. This is to speak nothing of the wanton destruction of habitats and ecosystems by violent uprooting such as the deliberate burning of Amazonia. Violence is the tactic of anthropocentrism, not of ecocentrism: *anthrofacism*. Anthropological ethics is fascist, not environmental ethics. Indeed, given the definition of fascism, in which fascism "gains power by claiming to restore dignity, nobility, purpose and privilege to some unique people or race...,"[51] anthropocentrists are closer to "fascism" than ecocentrists. They wish to impose one culture, of one "people" on natural diversity by force. The genocide of other species and their habitats is justified ideologically by proponents for human *lebensraum*, whether capitalists or so-called "social justice" theorists.

Some have tried to connect environmentalism with fascist philosophies, since Martin Heidegger expressed reservations about the treatment of the environment and also endorsed, even if briefly or misguidedly, the Nazi regime. However, as I argued above, the modern environmental movement emerged along with the notion of moral *progress*, expansion of considerability from humans to animals, and the environment. The early figures in the movement in the United States, including John Muir, Aldo Leopold, and Rachel Carson, expressed concern over the impact of technology and the misuse of land on humans. Singer, among the first of those who articulated animal liberation, explicitly drew analogies to the civil rights movement. The existence of a right-wing environmental strain is as irrelevant as ultra right-wing forms of humanism. The environmental movement is not right wing, since it has adopted the idea of moral progress, that is, expansion of moral considerability.[52] Muir, Leopold and Carson never read Heidegger, so far as I know, and the environmental movement in America long predated him (John Muir among others). Nor would they have accepted the Heideggerian view of nature as secondary to some hidden, but higher principle, as the home of the mystical "Being of beings." On the contrary, the environment is dynamic; the land goes through cycles from pioneering to climax species.

Ethics harmonizes plural values. Ecocentric cosmopolitan ethics recognizes the place of humans within a larger whole. So long as there is no further degeneration of the environment at human hands and steps are taken to repair the damage that has already been done, the ethic is not "anti"-human. Humans have as much value as any other species. A universal ethics includes humans and would attempt to harmonize human life with other life, without overly privileging humans. A specific environment gave rise to life including human life. Humans are not at the center of the universe. Ecocentric cosmopolitanism is a revaluation involving the decentering of anthropocentric ethics.

The other danger of cosmopolitan ethics could be a feeling of disconnectedness, of not belonging to a place, people, time. Rootless anomie may result in excessive interiorization (Stoics) leading to ascetic attitudes or the attempt to replace culture with tribalism. Asceticism can lead to denial of the value of the exterior world. The cultivation of interior life can have value so long as does not result in neglect of the exterior world—rootedness—and referring everything back (*pros hen*) to the subject, in a devaluation of the world. Universality in ethics refers to external relation of duty. Subjectivization denies the value of external relations.

Cosmopolitans do not identify with one particular culture exclusively, but look for some value in each. The difficulty is not truly belonging anywhere, but everywhere. Cosmopolitanism is open- instead of closed-mindedness: seeing another culture, not a "bunch of savages." One identifies in part with each culture (*verstehen*), but is not bound by any one. One is not "rootless"; one's roots lie in several cultures. This is not anti-patriotic, for one lives where one does because one has access to more than one culture. One culture alone might be stifling; cosmopolitans—world-citizens—have a heart big enough to love more than one; a spirit big enough to encompass many; a vision of the importance of all. Cosmopolitan ethics is both pluralist in value and non-relativist: there are a plurality of cultures each with its own values. As for rootedness, ecocentrism arises from the insight that we are rooted in our environment, and require a specific habitat. Environmentalism can only increase awareness of our rootedness.

If everyone became cosmopolitans, would this amount to the death of culture and thereby of appreciation of all culture? Cosmopolitanism rejects tribal relativism, while recognizing the value within cultures. Cosmopolitanism is not mere relativism, but a larger field. The value of national forms is not denied, but classified along with similar but alien forms under encompassing rubrics. Cosmopolitans also appreciate non-national art under the category "art". The antithesis of cosmopolitanism is only cultural chauvanism—tribalism—not culture as such. On the contrary, cosmopolitanism celebrates the diversity of cultures. Nevertheless, a revalued cosmopolitanism realizes and responds to the neglected value of the environment and that culture depends on an environment as much as any other value.

Cosmopolitanism is one attempt to deal with the problem of cultural relativism. Cosmopolitanism includes the notion of the relation of the members of one culture to another. Cosmpolitanism acknowledges the value of distinct cultures while seeking a broader view. The obvious next step in moral development is an even broader view that acknowledges the place of humans in the larger world, just as one particular culture is a small part of a larger whole of all human cultures. Ecocentric cosmopolitanism involves the relation of humans, as one part, to other parts of the whole, that is, to other species, habitats, and the biosphere.

Another question that arises is, is environmental ethics a form of naturalism? Ethical naturalism is another way around cultural relativism. The ancient Greek distinction of nature from convention is one expression of this perspective. Moreover, most forms of naturalism share the view of humans as within nature, whether organic-process views like John Dewey's or the many contemporary physicalist-materialist views. However, there are several problems with naturalism. First, several distinct views are called "naturalism," from the Sophist standard of nature, through Aristotle's "naturalistic" ethics to "natural law" theories from the Stoics onward. "Naturalistic" ethics is equivocal: there are many kinds. Sorting out which form is most ecocentric would be problematic, since ecocentrism, a basically new approach, might become confused with some other form of naturalism that is not actually ecocentric and entangled with ancient and irrelevant disputes. A deeper problem is that the term "nature" is at best ambiguous and at worst contradictory. "Nature" is both the whole and a part of the whole, in excluding culture and human artifacts. Something that is natural in one sense—not supernatural—can be "unnatural" in the latter sense: a contradiction. This ambiguity is most acute in ethics where some behaviors that are physically possible are adjudged "unnatural," particularly some sex acts.[53] Some authors have carefully distinguished these two senses when approaching ethical issues. But because of the critical sense of nature-as-part used as a standard, the term would only be confusing when used in ethical debates. Finally, there is an influential school of thought that argues for the distinction of nature and value, and that value is a "non-natural" predicate.[54] Others have extended this view to a critique of natural law. While this view may be untenable, it would be unwise to entangle ecocentrism in such tangential, irrelevant disputes. The term "cosmopolitan" is also preferable on other grounds. Ecocentrists have no problem with "naturalism" taken as the whole, that is, more or less equivalent to "world" or cosmos. But "world" is more neutral with respect to the above distinctions and controversies, and thus better for purposes of a basic ethic.

However, cosmopolitanism is not uncritical, for it includes critical evaluations and moral judgments. Cannibalism is wrong. Sexism is wrong. Slavery is wrong. That ethics may not always get it right does not lend any support to tribal, cultural ethics or their theories of value. Cultural mores cannot be dismissed *a priori* by ethical theory, but must be individually examined. But the cultural justification of immoral tribal mores, for example slavery or cannibalism, is *itself* an evaluation: that culture justifies all. (Religion began as a transcendent justification of tribal ethics.) Such an ethic is contradictory, since it posits and denies a universal principle. Philosophical ethics provides a standard or norm for evaluation of cultures that is universal.[55] As Joseph Des Jardins put it,

> From the earliest days of Greek philosophy, philosophical ethics has not been satisfied simply to accept as right that which is customary. Ethics as a branch of philosophy seeks a reasoned examination of what custom tells us about how we ought to live.[56]

6. Humanist Critics of Ecocentrism[57]

Humanists and anthropocentrists have criticized ecocentrism on the grounds that ecocentric philosophies are formulated by humans, and thus they are implicitly anthropocentric. In this section, I will respond critically to this point.

Are deep ecology, the land ethic and other ecocentric philosophies, since they are formulated by humans, implicitly anthropocentric? George Bradford writes:

> Positing itself as a critique of 'humanism'…deep ecology claims to be a perspective taken from outside human discourse and politics, from the point of view of nature as a whole. Of course it is problematic to claim, to say the least, since deep ecologists have developed a viewpoint based on

> human, socially generated, and historically-evolved insights into nature, in order to design an orientation toward human society. At any rate, any vision of humanity's place in it that is the product of human discourse is by definition going to be to some degree 'anthropocentric,' imposing as it does a human, symbolic discourse on the non-human.[58]

On one level I would agree with Bradford, that our world-view is tinged with human norms, categories, concepts, interpretations, and other mediating symbols. However, it is obvious that we can either ignore the environment in our worldview or take it into account. Ecocentrists choose to take it into account. Moreover, we can interpret the place of humans in the world in terms of dominion—anthropocentrism—or in terms of Leopold's more holistic account, in which humans indeed have a place, but are symbiotes. Thus the issue is not whether human philosophers can escape human symbols, and interpretations—even human science cannot do that. It is whether the environment receives ethical consideration. The argument of ecocentrists is that anthropocentrists repeat the mistakes that led up to our environmental crises. At best, they pay lip service to the environment while acquiescing in human destruction of their own world

There is also the distinction of the actual world from our symbols of it. I do not think Bradford would argue that there is no distinction between the symbolic world of humans and the actual world. But ecocentrists are pointing to the human destruction of the environment, which is actually taking place, regardless of how we interpret it. The burning of Amazonia is actual, regardless of how we symbolize it or evaluate it. If Bradford is arguing for anthro-solipsism, that we can only discuss the human world because we use human terms, then how does he account for the so-called natural sciences, which formulate principles for the entire universe? Try anthro-solipsism when being chased by a bear!

Moreover, as Gorke has argued, there is a non-sequitur from human knowledge to ethics. What we know and what is right are categorially distinct. Gorke has also pointed out, along with Singer, Tom Regan, and others, that we take the interests of infants, those in a coma/catatonic state, senile, and mentally ill into account. They are considered intrinsically valuable like other humans—so verbal communication "is thus neither a sufficient nor necessary prerequisite for considering the moral claims of another being."[59]

Ecocentrists argue for the intrinsic value of other species, which is a critique of Bradford's main contention. Bradford would argue that the concept "intrinsic value" is a human creation, but this is exactly what is at issue in the debate between ecocentrists and anthropocentrists. Anthropocentrists argue that all value is projected from humans onto the non-human, but do not see that this notion is contradictory, since it denies intrinsic or inherent value. As with *summum bonum* theories, it confuses value and evaluation; humans become the *summum bonum* as the only intrinsically or inherently valuable species, giving value to all the rest. Our notion of creative actualization, particularly the self-actualization of organisms, encompasses all of life. An organism that actualizes its own life by nutrition, growth, and other activities is self-actualizing and thereby inherently creatively actualizing, or intrinsically valuable. Other species did so long before humans were around, and continue to do so—indeed humans depend upon such processes.

Again, the land ethic and other ecologically informed studies have introduced the notion of systemic value—the functional value of a species to an ecosystem. This value is independent from humans and thus cannot be encompassed by anthropocentric theories of value. The functions would be actual even if humans were to die off tomorrow, were around long before humans, and are independent of human symbols, however much our description is tinged or mediated by human categories and concepts. The world should not be confused with our language, the mistake made by some extreme forms of analytic philosophy and of idealism.

7. Environmental Ethics as Cosmopolitan: Ecocentric Cosmopolitanism

Human domination of the environment has come at too high a price, the loss of too many intrinsic goods in the form of species extinction and habitat destruction. Worse, it is comparable to a death wish since destruction of our environment is a form of self-destruction. Even human survival is threatened by the massive destruction of habitats by economic interests. It could well end in the extinction of the human species through destruction of as yet barely understood processes of ecological dependence which result in alteration of the climate or the chain of life to human detriment (global warming). On pragmatic grounds alone, then, it would be better to halt the rapacious destruction of the environment for the sake of the anthropocentric value "quality of life" or "standard of living."

The way that humans treat the environment is not balanced, weighed as it is almost totally towards anthropocentrism. I mean the balance between competing "interests," that is, different values: justice, but also the balance of nature.[60] Humans, from an environmental point of view, are parasitic overgrazers. Human ethics is tribal, with ideological justifications in the partisan sense: a part masquerading as the whole. Human chauvinism includes attitudes that think of nature as an instrument of one species. But humans' "mastery" of nature is a form of blindness, if not the height of arrogance. We need green plants, while they do not need us. Our human chauvinism is one more episode in the long history of moral blindness in human culture, which as an exclusive culture is tribal. The dismissal of global warming is indicative of this moral blindness, to speak nothing of the devastation of habitats in Amazonia and other tropical forests. Human culture has a chauvinist streak to which it is blind, due perhaps to its economic successes.

And yet. The idea of moral progress, that we can include larger and larger numbers and members in the moral community: women, non-whites, and so on, is peculiarly human, including the notion of universal human (political) rights. So culture is more complex. The best of it consists in a moral consciousness and a profound sense of the value of life. The worst of it is petty, small-minded, narrowly economic, and "one-dimensional." Culture can be an excuse for concern solely with gain at the expense of other humans and the environment. After Auschwitz and Wounded Knee, we can no longer ignore the moral dimensions of this destructive side as "the price of progress." The gas chambers have given us a new moral momentum, and consciousness. The lessons of nineteenth- and twentieth-century genocide, which can be traced back to ancient times (the Old Testament calls for genocide against the enemies of Israel) are the moral requirement for cosmopolitan tolerance of other cultures and ways of life. A merely tribal, anthropocentric reading of these events ignores its central lesson: that tolerance of different ways of life in the world ought to include ways of life of different species. Otherwise we are expanding the circle of moral considerability only to close it: a contradictory, hypocritical movement. Cosmopolitan ethics is not an option after these events but morally required. Our ethics should be cosmopolitan, and ought to be extended universally, even as we remain members of limited, overlapping communities: a nation, a family, a career, a religion, an ethnic tradition. Belonging does not exclude toleration, it demands it.

Anthropocentrism is a deliberate decision to exclude other species from ethical consideration, that is, an extension of tribal ethics. Our species or tribal identity is a decision to conform to tribal mores with their limiting norms and identities, and exclude a more cosmopolitan identity with the larger human community and the world. However, in terms of a critical philosophical ethics, an "ethics" based on either the tribe or the species is an oxymoron. Ethical duties are rules, principles, and norms that are obligatory. Such obligations occur in all ethics, tribal or not. But as Kant pointed out, practical obligation requires necessity: what is necessary is obligatory. What is necessary must, logically, be universal. If ethics requires universality, then it cannot be either based on the mores of one tribe among others nor one species. The environment is a necessity of all life, like food and water, which, indeed, include the latter. We are not devaluing humans by recognizing, with Leopold, that humans are not the "masters" of the environment but plain "citizens" within it. He called for a change in perspective comparable in some ways to the Copernican Revolution in astronomy. We are no longer the center of the universe, but remain within it.

Cosmopolitan environmental ethics recognizes the value of life and of species and habitats beyond the narrowly human. However, ecocentric ethics cannot only be an extension of "humaneness" to the ecology, as J. Baird Callicott, Paul Taylor, and Holmes Rolston, among others, have pointed out.[61] It is not enough to only "be kind to animals," to preserve the ecology for human society, or to conserve for "future generations." Anthropocentric ethics is based on specific differences, thus cannot serve as a model for preservation of the biosphere. The advent of environmental ethics marks a whole new orientation for philosophy, as figures at the beginning of the environmental movement tried to point out. This philosophy is ethical, concerned with the relations of social groups but is larger than anthropocentric ethics. "Deep ecology," ecocentrism, and other eco- and biocentric theories are more than just part of a new branch of philosophy.

The ultimate denouement of this movement is that environmental ethics *is* ethics. Thus the separation of ethics and environmental ethics as genus and species is no longer tenable. Ecocentric ethics, the ethics of cosmopolitan membership in the world is the only universal ethics and constitutes the new basic framework for ethics.[62] Environmental ethics also includes social contact between species, not only within human society. It covers social relations between different social types, humans, higher and lower animals, plants, and the biosphere as a whole. Again, this is not a devaluation of previous ethics, anthropocentrically based, but a fulfillment of the potentiality of ethics: its fullest realization.

The problem still remains of whether a "natural ethic" which includes killing by predators is somehow different than a civil ethic among humans. Which should the expansion of ethical considerability to include animals reflect? Does the positing of a distinct ethic in civil society negate the "land ethic" of Leopold, Callicott, and Rolston? Shall ecocentric and cosmopolitan ethics be one or different? I would argue that cosmopolitan ethics and ecocentric ethics are two aspects of the same movement toward true universalization and equality, equal moral considerability, between species. Such moral progress is a demand of meliorism. Ecocentric Cosmopolitanism is the ethics of the future. Cosmopolitan stresses that humans are part of a larger whole and, more particularly, *in* the world not its masters. We are not subjects detached from the world but live in our bodies in an environment in the world. It also includes the principle that the world includes different communities, including different cultures *and species*. The norm of reciprocity is the condition of compossible differentiation of cultures. Reciprocal difference preserves different values in mutually compossible relations in the world, including the cluster of values within different cultures. These communities may operate by different norms and still tolerate the values and ways of life of other communities, including animal and plant communities. Ecocentric stresses the dependence of humans on their environment as much as on technology and culture.

However, the differentiation of societies with their own values does justice to different forms of association. Humans can leave the wild alone and also interact in more intimate ways with domesticated species. Relations to other species are cosmopolitan, taking the form of different relations to each species in the world, which recognizes the distinctive value of the species.[63] These varieties of association are within cosmopolitan norms, since association with compatible species in the form of pets and cultivation is one distinct value. We may also avoid association with bears and lions when imprudent, just as we may avoid cultures that are hostile to outsiders, or even unpleasant individuals. Valuing of all such modes of association involves a more cosmopolitan outlook in which tribal ethics is rejected as a norm. A citizen of the world recognizes pluralistic sources of value in the world and rejects subordination of all value to the good of the human tribe. Tolerance of the non-human ways of life no more requires adoption of the non-human way of life than tolerance of the Islamic practice of covering females from head to foot or any other foreign culture's norms requires their adaptation.[64] True tolerance is not "moral pluralism," but a cosmopolitan ethic of universal inclusion, toleration of plural ways of life, and appreciation of diverse goods. Just as other cultures ought to be tolerated so ought the ways of life of other species to be tolerated: they should be left alone and unharmed. Plural ethics are not needed; instead, cosmopolitan sensitivity to different ways of life is requisite and the single norm of reciprocal letting alone.

A technical question is whether this is extensionist or inclusivist. Is a cosmopolitan, ecocentric ethic an extension of human ethics to the wild and the biosphere or an inclusion of the human within it? If an inclusion, how are duties derived from the biosphere? Is the biosphere the locus of universality? Obligation is grounded in consequences to the whole in an ecocentric ethic, the biosphere as the condition of all discrete values and thus its value to all of them. Value for all is universal and thus has universal moral implications for all intrinsically valuable agents. The universal consequences of acts on the value of the biosphere provide a universal reference, which grounds obligations and provides a priority principle in the case of conflicts. Kant argued that the Categorical Imperative could be stated as a norm or formula of a "universal law of nature." This norm is compatible with an ecocentric view, as it includes humans instead of extending moral standing from humans to the wild in some degree.

The direction of movement of both meliorism and reciprocity, the two conditions of value,[65] is toward greater and more universal inclusion. Cosmopolitan ethics is meliorist by comparison with tribal ethics, as it is greater in scope, egalitarian, inclusive, and aims at universality. Similarly, environmental ethics, or more specifically ecocentric ethics, includes the entire biosphere. As for reciprocity, greater universality connotes more general obligation, thus the moral force of obligation is increased by greater inclusiveness. The moral value of acts in accordance with duties considered under such universals increases, since the obligation is defined in terms of practical necessity and the latter requires universality. Ecocentric ethics is not confined to what Kant's critics, following Georg W. F. Hegel, have termed the "empty formalism" of the Categorical Imperative, the requirement for universality. Firstly, an ethic is a way of life, not only a list of rules. Even if ethics were to be confined to duty, a life of duty would be involved. Ecocentric ethics involves a way of life as well, as the philosophers of "deep green" and related philosophies have reminded us. Secondly, ecocentric ethics recognizes the content of duty in the reciprocal relation humans ought to have toward plants, animals, ecosystems, soil, air, water, and other non-human inhabitants of our common world. Humans are not the lords of the earth in this ethics in a hierarchy from man downward, but only one of a plurality of self-actualizing agents and necessary functions (air, water, soil). We also have multiple sources of obligation. Our loyalty cannot only be to our tribe or our species, but must include the environment, the source of the necessities of life. We are citizens of the world which we inhabit, not just our city, or culture.

8. Principles of Ethics

For ecocentric cosmopolitanism, reciprocity with the other in social relations is morally compelling, since we coinhabit our environment as self and other. The other is a generalized other, allowing for a universal ethic. Thus reciprocity involves actual universalization. Since extension of moral consideration beyond humans is moral progress, it is melioristic and we include meliorism as still another principle. As the world reflects our acts, we should act to improve it in the future as a consequence. Since we are in an environment as a living thing, we cannot formulate ethics without consideration of our habitat. I call this the principle of participation: that we are moral agents in all our acts.[66] These principles taken together form a larger whole, of which each is a consideration in moral evaluation.[67] The principles, then must be revalued in terms of consequences for the environment. The central problem of our time, the destruction of the environment must serve as a negative norm and standard against which we act. Philosophy should also include a way of life, not only idle speculation. If cosmopolitan ethics is superior, it should be practiced.

The extention of duty beyond the tribe, universalization is the heart of a cosmopolitan ethic and the basis of moral progress. Duty is to all others, including the biosphere taken as a whole. Acts of duty are *in* the world as relations to a universal other are external not internal. Responsibility affirms the value of the others. Thus morality as a cosmopolitan way of life is neither provincial nor "anti-life" (Friedrich Nietzsche); on the contrary, ecocentric cosmopolitan ethics affirms the value of the entire biosphere by moral recognition. The world is not an opposing nature/field for exploitation by egoistic subject, but a valued other toward which reciprocal justice is due: recognition of the intrinsic value of the other, not subordination to self-subject. The foremost duty in an ecocentric revaluation is to the living world as a whole that is the condition of all values. All our values require the world as the place and condition for such human values. Reciprocity with the world is required and thereby also in our self-interest. The revaluation of reciprocity means expansion of the other to include the whole, not just the human part. Reciprocity of duty applies equally well to an atomized individualistic society as to a more interdependent one. Both types of societies require an environment and living in harmony with the world.[68]

Another element of a cosmopolitan ethic is the principle of participation. A principle of participation involving the self is at the root of justice, and is another reason to be moral. The common expression of this would be the phrase: "what if what happened to others happened to you?"[69] We place ourselves in the place of the other and recognize their problems. The principle of participation means that we cannot exempt ourselves from ethics, since we are in society. If we are to make accusations of injustice, for example, then we must ourselves be moral. Our appeal to morals places us within the moral life. In ecocentric terms this means that we are in the world, and cannot deny living in the world and drawing upon its resources for our own lives and projects. Participation means the recognition that humans are in the environment. We can ask "what if what happened to the environment, or a particular habitat, happened to you?" or better, "to your habitat," as much as "what if what happened to other people happened to you?" Would you like to have garbage, poisons, and other effluvia dumped on you? Would you like your life ended and roots ripped out to make way for projects of dubious value? Environmental philosophy teaches us that the subject is not detached from the environment, but within the environment. Denial of this principle is environmental freeloading.

Cosmopolitan revaluation would include consequences for the environment as a prime consideration in moral evaluation. We are obligated to consider other species and their requisite habitats moral deliberation and act accordingly. If the consequences of our actions are destructive of other species and their requisite habitats, they are wrong.

Critics of consequentialism have argued that we cannot possibly predict all the effects an action, not even all the direct results. There may be "unintended consequences" of our action that we do not foresee. This argument against consequentialism, while partly valid, ignores the many consequences that we can foresee, particularly those having to do with the environment. We can clearly see the consequences of global warming, overpopulation, and the destruction of species and habitats. Since the environment is required for the survival of life, erring on the side of caution is called for: acting so as to avoid the worst consequences. However, we have duties to the others that are non-human regardless of consequences: ignorance is no excuse when it comes to duty.

Even if we could not know all the consequences of our actions, we can still speculate ethically: is something truly good or desirable—either in the long run (time-end) or in terms of some high standard of quality (evaluation)?

Moreover, the question of good in relation to other goods can be raised. This is the function of environmental ethics—for example should we continue to make plastic? Is it worth it? Thus an environmental ethic is not a segment of ethics. Rather, it is the ethics of the future, for it concerns the future of humans as part of a larger whole, which is essential to consider in any moral deliberation.[70]

In principle, we can distinguish moral values, moral limits, and consequences. Moral limits are limits to non-moral acts. Thus some economic acts fall within moral limits and some do not. Moral values ought to be creatively actualized, but moral limits only observed. Moral consequences should be evaluated.

Pragmatic ethics considers consequences as well as duty and virtue. Dewey argued for a holistic view in ethics.[71] He also argued that deontological ethics alone is not enough. Dewey argued that an ethic of duty tries to avoid responsibility for consequences; but that an undue stress on consequences is "one-sided." A complete ethic involves consideration of duty, virtue, and consequences.[72] Taking consequences into account is required for an environmental ethic, that is, ethics in general. An ecocentric view would evaluate consequences for the environment of any act as the foremost moral consideration. An ecocentric view would mean that humans recognized duties to the environment as "citizens" in it. Being a good citizen involves some ecocentric virtues as well, a good character. A moral agent makes moral decisions: a good person. Ecocentric ethics cannot be divided up into deontological, "virtue" or consequentialist ethics, since all elements are required in moral consideration of the environment. Surprisingly, Kant agrees on this point, since being virtuous is a duty and consideration of consequences of our acts is included in maxims as well as in moral deliberation (intent). His ethics does argue for the priority of duty in defining moral life and values: a hierarchical form of holism, in which the ultimate arbiter of ethics is duty.

We can distinguish the highest principles of moral evaluation from its various elements. Our concern for the future of ethics is to insure that environmental consequences receive the highest consideration. Consideration of consequences for others: the environment, humans, animals, is required also, not just principles and "intent." The limitations of Kant's ethic are that it is based on intent and intent is not enough.[73] Kant is right, however, to argue against subordination of morality to non-moral factors. Reciprocity is one element or moral principle along with participation, amelioration, and so on. However, *moral* evaluation means that morals cannot be subordinated to "interests," whether economic, political or any other human social category. Thus the purity of ethics is maintained while allowing for non-moral acts within moral limits. We must act within moral limits, as an immoral world can never be an improvement.

The limitation of Dewey's view is that there is no sense that some acts are immoral as such, regardless of circumstances. Clarence I. Lewis combined

Kantian and pragmatic elements in ethics. We must consider consequences within a framework of duty. Duty cannot be derived from what is given, but consequences must be considered in moral evaluation. However, our duty includes considering environmental consequences in evaluation of acts, as well as many other consequences. Thus pragmatic ethics is an approach to ecocentric cosmopolitan ethics

Another principle borrowed from pragmatism is meliorism. There is a value judgment in any conception of historical progress, including moral progress, namely, that we are better off than our ancestors.[74] Progress in turn, implies that we can improve our situation, even if we cannot perfect it. William James called this position meliorism. Thus meliorism is a historical outlook, since it connotes a relative improvement in the human condition. Ethics includes the consequences of our actions and the requisite moral duties and virtues. As our actions will be manifest as consequences—our present acts will make the world of tomorrow—our responsibility is to act for a better world, and we are blamable for a worse one. Thus ethics involves temporality and the attempt to improve the world of the future (meliorism). Meliorism is an ethic of future times: improve the future. Our future survival and that of the biosphere are bound together. Whichever potentialities I choose will be reflected in future time. Of the many actions and activities available, I must or should choose what will have best imaginable results, and thus work for continual world improvement. Consideration of the best consequences is essential to ecocentric ethics.

Moral progress and meliorism will occur when humans move beyond a narrowly human view and recognize their place in a much larger whole on which they are dependent. Humans are in the world: part of a community of soil, plants, and animals. As members of society already, we should be humane—and as within a habitat that we should be citizens of a world created by life and mutually dependent.

Western philosophy and religion have examined the question of duties fairly thoroughly. But the critical-evaluative question has received less stress. The question, what is the value of a value? has received only sporadic attention. There is some debate about this in alternative life styles, for example in "Buddhist economics" and the monastic life, with its rejection of wealth as the primary value. But the evaluation of "what good will it do?" is mostly left to individual desire by our atomized and thus unofficially relativistic society. The question of worth is neglected, that is, the value of alternative creative actualizations of alternative possible values.

I want to suggest that ecocentric ethics has often been about worth even more than about duty. Is cutting down an old growth forest worth it for a few more wood frame homes? Should we actualize atomic power or make more plastics? "What good would it do?" Would the world actually be improved by atomic power (development, more cattle, more wealth made by habitat destruction, and so on)? This is partly a question of meliorism, that is, the question of what actually will constitute an improvement. However, it also raises critical questions for evaluation, the value of what is valued. Ecocentric environmental ethics includes more than just duties as it extends to questions of what sort of world shall be actualized, a predominantly plastic world or a green one. While living according to rules or principles is integral to ethics, the comparative value of competing ways of life is also involved in ecocentric ethics. We should consider our way of life in the future, not only a list of rules.

One example of the contribution of environmental value theory to contemporary debates is the reintroduction of holism. As Callicott noted, Plato thought of values in terms of wholes: each human capability has its corresponding virtue, while justice is the virtue of the whole, when the parts are working together harmoniously.[75] Justice in a sense is the "one in many" of the virtues. Since that time, Western philosophy has underestimated holistic value and the relation of parts to a larger whole.[76] In a sense, categorial values, that is distinct kinds, are part of a larger whole: they have a reciprocal relation but are genera of a larger whole of value. In the background of the modes of actualization (imagination, value as creative actualization, and norms) is the *world as a whole*. Holistic value is the condition of all creative actualization, principles, and imaginings. The world is also the source of potency for creative actualization, that is, the sources of the ingredients and other elements of goods we creatively actualize and its locale or place. Indeed, no species can live without an environmental niche in the world—and there can be no environmental niche without the interrelation of parts in a larger whole, including rainfall, soil, atmosphere, other species, and so on. The world as a whole has holistic value greater than any part, since these latter require whole as their condition.

The world as the condition of all goods implies it is required. As required it is obligatory and we thereby have duties towards it. Its value as the whole, greater than any part means it has a unique moral value. Thus the world has *holistic* moral value: cosmopolitan ethics. Cosmopolitan ethics means living in the world. Justice must be done to the world/environment/habitats as parts of this larger whole. But is this a new framework for values, replacing intrinsic-instrumental, or another framework along side the intrinsic-instrumental—or can it be fit into the intrinsic-instrumental relation? Instrumental and intrinsic generally refer to individuals, individuals of a kind or collective groups. Thus they are parts of a larger whole. Holism in value includes the instrumental-intrinsic relation.

The question can be raised of good in relation to other goods, compossibility and inherent worth. This is the function of ecocentric ethics, the ethics of the future. For example, should we continue making plastics at the expense of habitats and wildlife? By comparison with not producing plastic, not actualizing it, would the world be better off in the long run? Is it worth it? In part this involves questions of reciprocity: we shouldn't pollute because it threatens the life of others (lead in fuels, the buildup of carbon dioxide, and global warming), and thus is ruled out by duty. But the question of worth also arises: should we continue to produce a product which won't biodegrade, is difficult to recycle, fills our already overflowing dumps, threatens wildlife, and other problems? Is it worth it? Wouldn't a slightly poorer but greener world, with more equitable distribution of necessities, actually be better, since it would be more sustainable and less likely to lead to ecological catastrophe?

One criterion for the answer to the question of the value and thus justification of a value is its compatibility with other life, that is, not so much individuals as the biosphere as a whole. The obvious because easy case is atomic war, which would probably wipe out all life. The compatibility of our creative actualizations with the biosphere may serve as a test for their worth; if they tend to degrade it, they are ruled out by duties to the biosphere. Environmental ethics is thus far more important than it might appear at first glance as it provides a test and justification for actions in all categories of value, intellectual, economic, social, and so on, not just relations to wildlife. If ethics is a way of life, environmental ethics is the most cosmopolitan way of life, the most universal morals.

Whether we ought to do our duty (others) or pursue our own pleasure (self) has been resolved by altruism in its view that helping others is a source of pleasures.[77] But our obligations to others cannot be unlimited in extent—if I gave all my food away, I myself would perish. Yet complete selfishness is amoral. Thus altruism and self-interest are not morally revealing. Again, should we defer our present interest for future generations? But then is our life worth living? Our present goals must dictate our ends as much as our future.[78] The question then arises: is ecocentrism altruistic? In one sense it is, since we are deferring gratification by acknowledging the other as morally considerable. But it is also in our self-interest to preserve the conditions of our life. Again, the contrast of altruism and self-interest do not appear to be morally relevant.[79] Moreover, ecocentric norms provide a value ranking in structure, independent of utility for humans. Wild flowers, insects, "weeds," and predators are all given a place regardless of their utility for humans.

The universality of moral duty implies moral equality and thereby justice. A cosmocentric revaluation would include justice to the environment in general and to other species in particular. Other virtues can similarly be revalued.[80] Doing justice to the universal other also involves self-restraint, the observation of moral limits in all our actions. From the obligation to other we can derive moral limits for application to intimate society. Duties to others (and recognition of my rights from others) follow from and involve doing justice to generalized others. These include the duty to support children and thereby the rights of future generations, as they are an obligation undertaken voluntarily. The inference is similar to that in keeping promises, which implies a form of truth telling. There is also the duty of mutual non-violence of family members and of benevolence. Such a way of life is possible in general conditions of non-violence. Atomic warfare has, paradoxically, made violence out of date. The "balance of terror" means we can no longer settle human disputes by violence. A new ethic of non-violence is thus called for by circumstances.

Justice is not only a negative principle, as a good way of life includes justice. Justice must be done to different pleasures in a complete life, no one predominant. All have a place: food appreciation, sexual ecstasy, the bliss of meditation, the refined pleasures of aesthetic enjoyment, and intellectual contemplation.

The issue of obligations to future generations is also raised by the consideration of the environment. Our descendants also will require the conditions of life: clean air and water, good soil to grow crops for food, and many other elements in the larger whole of the biosphere. In general, if we have children we have a duty to provide for them and from an ecocentric perspective this includes preserving the conditions of life. It also includes preservation of beautiful landscapes, rare habitats, and the possibility of experiences that should not be denied to our descendants.

Shared cultural values also involve receiving these from our ancestors and passing them on to descendents. This could provide a supplemental basis for responsibility to the past and especially the future. One ground for obligations to future generations is thus the responsibility to pass along our culture intact, and that requires passing along the conditions of culture. In some cases the culture may include a special place, considered sacred. In all cases it requires the larger whole of the biosphere, the *sine qua non* of any culture. It would be better yet to pass on an improved world, not just an intact one.

As an example of how human ethics could be reformulated in the light of ecocentric cosmopolitanism, the universal taboo or rules against murder and theft could be seen in wider terms. Morally, killing is wrong. If this is interpreted in terms of ecocentric cosmopolitanism and applied universally, it includes killing animals, which we do not need for food, trees, which we do not need for buildings, since we could use stone and synthetics. If murder and theft are universally wrong, then murder of animals and plants and theft of their habitat for "economic development" or other unnecessary human use is wrong. It is also unnecessary if a vegetarian diet is followed. Humans can survive fine on a vegetarian diet, which is healthier and cheaper. Moreover, raising vegetable crops is far more efficient in terms of costs than beef or pork, and soybeans and other crops do not cause, but instead alleviate global warming. Cattle, in particular, release methane in their digestive processes. A vegetarian diet will help with global warming by reducing methane, destruction of habitats by grazing, and among other reasons, requires fewer acres for production. On all these counts, a vegetarian diet should be strongly encouraged.

Another example is provided by the value of necessities of life. Air, water, soil, and space are universally required for life, including human life. What is required is practically necessary; what is universally required is obligatory. In turn, all theories of value require life as a condition, for example life as the condition of pleasurable states in hedonism or feelings of happiness in several theories. The conditions of life are required for any theory of value and are thus morally considerable. Thus elements of the environment that are universally necessary to all of life should receive moral priority.

Unlimited human expansion into all habitats, on the entire surface of the earth would be an obvious threat to all life. As in the Fascist concept of "total mobilization" for war (Ernst Jünger 1930), it would involve planetary-wide subordination of all life as instruments of human whim. The effect of such human expansion would be habitat destruction and species extinction. Biodiversity is also at risk. Since we do not know the long-term effects of fewer species, we are gambling with the fate of the planet. The interrelation of species and habitats involves symbiosis and holistic relations, for example, in the role of such predators as birds of prey and wolves in a healthy environment. The environment as a whole is a requirement for all of life, including human life. We cannot survive without air, water, and soil. We also require other species: our food as well as the condition of healthy air, water, and soil. Plants make food, anchor the soil, and help to create it.

Another result of human expansion is global warming; resulting in rising ocean levels. Global warming of itself may be a threat to all life, since we do not know the sustained effect of increased heat on current species. Forests may be destroyed by net warming as much or more than by acid rain. The ecological costs of cheap energy generation, including acid rain, global warming, nuclear waste, and many more are problems. There are the unforeseen consequences of releasing synthetic chemicals into the environment, of which ozone depletion is the most well-known case. Finally, the threat of overpopulation rivals the misuse of technology.[81]

9. Environmental Justice and Social Justice

Does environmental justice include social justice? In many cases, destruction of the environment has had a disproportionate impact on the poor, notably the location of polluting factories and refineries in poor neighborhoods. The poor have borne a disproportionate impact of pollution, including groundwater contamination, air pollution, and other ecocrimes.

While ecocentric cosmopolitanism is sympathetic to social justice, it cannot be sustained in the long term without environmental justice. Sustainability is the key to just economic distribution, since using up all resources will only defer the problem of scarcity of resources and scarcity will arise anew as an issue for our descendants. Putting off the day of reckoning can hardly be just in any moral schema. We are "externalizing" our costs over time, by passing our problems on to our descendants. By the standards of cosmopolitan ecocentrism, social justice does not mean increasing consumer goods for the dispossessed, since the whole model of consumer society is unsustainable. Consumer society is economically unfeasible, ecologically destructive, and thus irrational. We do not need more consumer goods for the "have-nots" but a sustainable life-style for all and the close intellectual connection of such a life-style to the place of humans in the biosphere as a whole. Similarly, "quality of life" issues ought to be considered in the context of ecocentric concerns regarding shrinking habitats, global warming, and other, larger environmental problems. This book advances a radically ecocentric view, in which any anthropocentric ethics, political and social philosophy, "identity" politics or even anthropocentric epistemology are adjudged out of date, replaced by ecocentrism as moral progress.[82]

As with feminist, gay, and other identity driven forms of environmentalism, social justice must be revalued from a cosmopolitan point of view. Social justice is a small part of a much larger whole, that is, justice to other species, environmental justice. Taking the part as representative of the whole, to speak nothing of identifying the part as the whole, fits the classic definition of ideology. The part is human life as privileged, anthropocentrism. No matter how much virtuous noise ecocentrists will encounter in protest, anthropocentrism is not actually friendly to the environment. For the program is rape of the environment, injustice to other species, including destruction of forests for housing; of habitats for food, and so on. Humans already have transformed over half the world's surface for their use, but are only one of millions of species. Just as there is something intrinsically unjust about a country with five per-cent of the world's human population consuming over thirty per cent of its energy, so is it environmentally unjust for one species to hog all the land. Justice for one species is a small part of justice for all species.

Critics of ecocentrism, such as Barry Commoner and Murray Bookchin, argue that social justice must have priority over justice to the environment.[83] Under their green surface, they are closet conservatives, for all their radical posturing. They cannot or will not face the reality that the price of human development has been the rape of the planet. Since they have come down on the side of human exploitation of the environment, they are "objectively" anthropocentric, and therefore willing to sacrifice other species' welfare to that of humans. Social justice theorists are blinded by an obsolete anthropocentrism. In this they are at one with their ostensive opponents, the so-called conservatives. In other words, like the capitalists they ostensibly oppose, they view the earth as a field for human exploitation, differing only in their distribution schemes. Like some conservatives, they in effect deny the animal in humans—particularly the urge to reproduce. They are still gripped by the discredited Judaeo-Christian view of human exceptionality. They have not absorbed Darwin, much less the lessons of environmentalism.

Human exceptionalism is not an environmental philosophy. Such pseudo-environmentalists cannot face the sad truth squarely, that unlimited human population growth and consumption mean our self-destruction: a future of famine, desertification, global warming, rising sea levels, deforestation, species extinction, and other disasters that will doom our descendants. Like Randians, they think that the world's goods are infinite. But this foolishness is so shortsighted that we only need to mention petroleum to refute it.

The vision of such authors is stuck in the fifties of the last century: the era of conspicuous consumption and the "affluent society." They cannot see or, like other humans who live on myths, will not face the prospect that the suburban wetdream they envision for the masses is both unrealistic—because unsustainable—and environmentally destructive. Rising expectations and affluence are contributing factors in the destruction of the environment, if not the direct cause. Human pollution and habitat destruction are for the sake of affluence. Such critics refuse to connect the dots between affluence and ecocrimes: between increasing yields to feed the world's poor and destruction of soils; between increasing productivity to increase the number of economic goods, pollution, and destruction of wildlife; between overpopulation and species and habitat loss. I am not arguing that we cannot have social justice. But from a cosmopolitan point of view, the threat to gorillas, rhinos, and many other endangered species is far more acute than the problem of turning the masses into consumers.

I am a pessimist about our future as a species because social justice issues will always receive priority over saving the environment and thus we are doomed. Ambitious politicians seeking votes will always cater to greed before considering the environmental impact of ambitious development schemes. Rising expectations that create too much demand will overwhelm the few remaining scruples. Only if we reduce population can we live in harmony with the environment. This is unlikely to occur. The social justice question should remain an intra-human problem in the polis: a part of ethics, not the foundation. Ecocentric cosmopolitan ethics can accept the priority of social justice over huge disparities in wealth only on the condition that sharing in prosperity does not come at the expense of the environment. The vision of miles of suburban tract houses to accommodate the masses from the "developing countries" can only be a nightmare to environmentalists. A smaller pie should be distributed more evenly, not a larger pie that allows "conspicuous consumption." The main principle should be that solutions to the problems of poverty and social injustice should not be at the expense of the environment.[84]

Let us be clear. Ecocentrists favor social justice: more economic equality. As I argued in chapter one, equal applicability of moral rules implies moral equality: justice and universal rights in political life are an extention of this equality. Morality is doing justice to others, to whom we have obligations as social agents and moral equals. Increasing social justice could even be helpful in achieving environmental goals, particularly in regard to overpopulation. For example, instituting social security in all nations might offset the tradition of having children just to take care of older parents. But social justice requires responsibility, not just redistribution. Control of population will do far more to raise standards of living than trying to carve up a diminishing pie. The poor share this responsibility. Any increase in the quality of life requires a reduction in population, including the number of children borne by the poor. Similarly, a sustainable economy is the prerequisite to long-term prosperity. Social justice has a place within a larger ecocentric philosophy, but should not be confused with the whole.

In this chapter I have argued for the revaluation of cosmopolitanism, and moral progress. Moral progress should include consideration of the environment in moral evaluation and deliberation and thus the place of humans in the larger world. In the past, cosmopolitanism has been interpreted as a humanist view. But humans constitute only a part and any attempt to make humans coextensive with the whole are unmasked as ideological. The world as a whole is much larger than the human world, and in the final analysis, cosmopolitanism implies a much larger whole of the world. Cosmopolitanism should be revalued from anthropocentric to ecocentric.

Only if humans become environmental ethicists are we moral agents. Otherwise, our ideology of humanism is a rationalization, a mask for self-interest. Moreover, the attempt to reduce environmental concerns to subsidiary importance is ideological, whether from the perspective of humanism, social justice or identity politics. I have argued that these approaches, important as they are, are anthropocentric parts of a much larger whole. Thus social justice and other anthropocentric themes have been revalued from the whole to the part. The future of ethics is ecocentric cosmopolitanism, in which intra-human relations are a part of a much larger whole, a revaluation of ethics as such.

Three

COSMOPOLITAN VIRTUE: A CRITICAL EVALUATION OF A GENERATION

> "Solitude, in the sense of being often alone, is essential to any depth of meditation of character; and solitude in the presence of natural beauty and grandeur, is the cradle of thought and aspirations which are not only good for the individual, but which society could ill do without."
>
> John Stuart Mill[1]

The revival of "virtue ethics," which many attribute to the work of Alasdair MacIntyre, presents a quandary for an ecocentric cosmopolitan ethic.[2] Virtue and excellence are bound up with human action, anthropocentric characteristics. Moreover, the label "virtue ethics" is unfortunate. Virtue is the ideal or optimum outcome of a process of development of character. The goal is a good character, whose highest development is measured as virtue.

I will argue in this chapter that the understanding of character in Greek philosophy and culture was in relation to the development of a moral agent, that is, a person with character. Secondly, I will argue that character is developed within a culture or more generally a society. Then I will argue for detaching the development of moral character from the larger system of Aristotle, especially metaphysical substance. In the second section, I will argue that my own generation has lost sight of its ideals. For the most part, this reflects the pull of economic values over all others. Finally, I will examine what cosmopolitan virtue might look like. I will argue that virtue, or excellence, is not confined to humans.

1. Free Will and the Moral Self: Development

Prof. Stuart Umphrey mentioned in a lecture that the Greek word *arete*, usually translated as "virtue," was less commonly translated as "excellence." A virtue or excellence meant that any power of humans could be exercised with degrees of attainment, from failure to accomplish some function or goal, through its achievement, to well-done or even exemplary actions. The elements of virtue include acting, the relation of means to ends, different human or animal capabilities, and the degree or worth or merit attaching to the way they are done, including the outcome. Virtue is exercised within action, and does not attach exclusively to feelings, the modern view. Excellence belongs to accomplishing some goal well or best: a form of creative actualization. Thus virtue belongs to the public space of actions, not the private, subjective world of feeling, with qualifications. Some virtues listed in Aristotle involved what we would call emotions. Thus there was a virtue involving anger in which righteous indignation was called for in select situations, whereas being angry all the time was a vice as was its opposite, being unmoved in even the most outrageous situations. However, anger is a peculiarly public emotion, we might say a behavior. Moreover, excellence was developed in the young, whether through teaching, as Plato's Socrates imagined, or through habituation, as Aristotle argued.

In *Academica*, Marcus Tullius Cicero writes as if "virtue" meant the excellence of any capability or function—thus the excellence of cooking would be a chef's virtue, of writing an author's, and so on. Virtue is possible in all activities then, as the phrase "Renaissance man" implies. Anyone can be virtuous in developing some talent, their work, home life, daily activities, or what not. Virtues are democratic taken in this wide sense: everyone is proficient at something. Virtue, the development of excellence, is possible in any activity and thus is democratic.

Development of character is the core of the ethics of virtue. Virtue is one element in a larger whole, a philosophy of animal capabilities. Animals, including humans as the speaking and reasoning animal (*zoon logon ekhon* in Aristotle), are normally born with specific capabilities. These differ from species to species: cheetahs have an incredible ability for speed, elephants for strength, dogs for smell, and humans for abstract reasoning. Because some individuals are born disabled in some respect, they fall outside the norm. A normative measure of such capabilities, then, is involved of a general, not universal sort. Species differences limit the function of each individual to specific capabilities. We no more expect a human to run as fast as the big cats than we expect insects to learn calculus. However, being born with a capability is not yet exercise of that ability. Such capabilities are potential and must be developed through education or training of some type. Whether by understanding or habituation, humans are transformed by such training from infants to adults. The process of maturation is at once a process of socialization and of the development of specific capabilities.[3] Of course some cultures may emphasize some abilities while others bring out alternatives. What they have in common is the view of humans, and many animals, as developable. Humans can be changed after birth from infants without language, talents, physical abilities, and other functions to adults with such capabilities. Virtue and morals are teachable because humans are capable of excellence—humans can change, by training and their own efforts. The function of education is not only to socialize or acculturate, but to develop capabilities to their highest.

Similarly, dogs can be trained to track criminals; birds can be trained to sing and even mimic human speech; horses, to race; oxen and elephants to do heavy lifting. Development of animal capabilities is not limited to humans.

Cats are trained to hunt by their own parents; other species may also require development of potentialities into actualization. Moreover, some may be better than others within a species: there may be a kind of excellence in the development of such animal and plant potentialities. This point would be considered a penetrating insight into the obvious by racehorse trainers, dog trainers, and others who work with animals, such as those who train animals for parts in film. Some breeds are better as seeing-eye dogs for the blind (shepherds and retrievers), some are better at following scents (hounds), some at racing (particularly greyhounds), and some at herding (the working breeds). An excellent sheepdog is one that excels at this function. Thus virtue, or excellence, is not confined to humans.

Education in the wide sense is imposition of culture. Education involves power relations of a sort, that is, parent-child, teacher-student, where the parent or trainer has some latitude in bringing out specific capabilities and perhaps thwarting others.[4] Some Sophists and Cynics argued that acculturation went against human "nature," since impulses and desires are thwarted by *nomos* (law, convention).[5] The counterargument, from Plato, Aristotle, and the Stoics, and separately in the East by Kong fu-tse, was either that "true" human nature is higher, for it involves rationality as the development of our species nature; or that living according to nature as it is presented by the Sophists is living like an animal. Thus some type of essentialism was contrasted with the relativism of individual impulses and desires. However, such training need not be coercive. Some people love learning, or sports, or music, or other activities and hardly need to be punished to partake in them. Development can indeed be a borderline case of wrong where culture is coercively imposed. It is better if it non-coercive.

Virtue proper, or excellence, consists in functioning well. Functioning well means not only avoiding the contrary, vice, but achieving the goal with some degree of skill, merit, grace or excellence. A house can be poorly built or built to function, but functioning is not functioning well. A home should have some beauty, some grace, and be well designed, not only keep out the elements. Similarly, actions can achieve the goal, but not well. To finish a race is not to win it; to be able to solve a mathematical problem is not to do it originally or quickly, with the proper steps. Excellence or virtue consists in the degree of value or worth that attaches to actions. Any good action may have a degree of excellence, but not excellence in the highest degree. A bad action falls short of the goal, but not all actions that achieve the goal are truly virtuous.

Nevertheless, the ancient Greeks' subtle treatment of virtue also includes aiming at and achieving the good apart from excellence. For a person may be good at one activity but excellent at another. As we age, our physical capabilities decline; older players in sports are not as quick or energetic as younger. But their other virtues may increase, for example their wisdom or their ability to be generous. Character could include a mixed combination of virtues at which a person is good and others at which she is excellent. Virtue, then is goodness of character, functioning well, and the attempt to improve oneself. Nevertheless, goodness in the highest degree also includes excellence, and the striving for even higher goals.

Aristotle distinguished actions that aim at some goal other than the action itself, from an action that is good in itself—which parallels the distinction of *praxis*, from *techne* or *poien* (in *Meta* VI). While excellence in achieving a goal that involves production was indeed virtuous, virtue attached primarily to actions as such that are well performed. Thus acts of liberality were themselves excellent although they included results as well, the act of using wealth to achieve results. Liberality at a wedding involved acts that had consequences, including a ceremony, a feast, and the like. Courage had consequences, whether in battle or in standing up for what is right. Thus the distinction of practice and production is not as sharp as some have implied.

Teleology includes a theory of action as aiming at some consequence. Action is not only behavior, since action is mediated both by social norms and by training. Instincts are moderated, and capabilities or functions of the individual of a species developed. Raw behavior is the sign of insufficient socialization or acculturation, a lack of "breeding." Although excellence is actualized in action, and thus includes behavior, behavior interpreted in some sort of mechanical sense is not enough.[6] For the action requires deliberation if the agent aims at achieving excellence in a given situation.

Virtues are species of moral goods in the widest sense of moral and are enacted at different times according to the situation that calls for the requisite kind of goodness. The situation determines to some degree the particular virtue that should be exercised or involved in the action. For example, courage is proper to battle, but hardly called for at a wedding, while generosity or liberality is appropriate to the latter. However, judging the situation involves either some degree of good judgment, itself an excellence, or practice at dealing with different situations.

Moreover, there are different kinds or species of virtue.[7] Although humans are not the fastest or strongest animal, many biological capabilities of our bodies can be developed. Achieving these in the highest degree is excellent, for example in the Olympics. Similarly, humans have unique capabilities that cannot be matched by animals. These include the development of talents, such as creation and performance of the arts: music, painting, sculpture, dance, poetry, theatre, and the like. Minor talents like cooking can also be developed. Humans excel at intellectual excellence and development of our language; mathematical and reasoning skills are another sort. Finally, there is moral excellence. "Moral virtue" is goodness or excellence of moral character: an added concern with being just or loving in our dealings with others. However, moral character is the genus of which these kinds are particulars, since development is toward moral character.

Character is the sum of such moral excellence; the more virtues, in kind or degree, the "better" the character. As Kong fu-tse (*Analects*) put it, virtues have "neighbors." I take him to mean that being excellent in one field, or one area of one's life or ability, does not make a virtuous person. As the different kinds of virtue indicate, a good person has attained excellence in many areas and has developed herself as highly as possible across the physical, in cultivating their talents, their intellectual and their moral life. Being courageous alone, but brutish in other respects does not make for a good character. Nor is a one-sided emphasis on one sort of development: developing the body at the expense of the mind, for example. Good persons, then, are many-sided and have tried to develop themselves to their full potential, including their body, mind, talents, and character.

All the species of virtue involve moral development: character. Intellectual excellence includes the virtue of truth telling and performance of arts or participation in sports include the virtues of persistence, endurance, and others that are required to develop the requisite excellence in such fields. Since a good person has developed their virtues in several capabilities, they are well rounded. The relation is of parts to a whole, an organic relationship. Character is holistic. Each virtue is called for in appropriate situations and the person can decide or deliberate as to which action is called for and to what degree. Consistently virtuous actions or behavior reveal a good character. Behavior and character are moral categories, which reflect social norms of acculturation as well as judgments of the worth or value of the actor. The ethics of character development involves a moral view of humans, in which they have developed virtues that have transformed them. Character is the development of moral habits, sensibilities, and actions in humans.

The article "Recent Work in Virtue Ethics,"[8] covered three "standard" objections to virtue ethics, namely, that virtue ethics cannot supply "guidance" for action, that it cannot establish a relation between virtue and flourishing or happiness—in other words, "why be virtuous?"—and that it cannot account for our "convictions" regarding rights and justice. I used to share these misgivings, but now realize that a deeper reading of the ethics of character raises the whole issue of the moral agent. On the first point, the virtue as a standard can evidently provide guidance qua standard.

Ethics concerns social, more than individual acts, insofar as we can distinguish these.[9] Ethics is social in the sense that it involves relations between people and others. The most obvious duty, *qua* duty, is to society, for every duty is a duty to someone. Stealing and murder harm *others*. Since duties are moral, the point is not so much about duty as that ethics is social. Duty is primarily social *as* duty. Thus decisions on having children, for example, are not entirely autonomous. The impact on or consequences for society are paramount from an ethical standard.

Critics might argue against social ethics on several grounds. One might be that individuals are the logical "contrary" of the social, in some sense. In other words, the interests of individuals and society do not always coincide. Moreover, society is nothing but a collection of individuals, and so individuals are primary.

Proponents of the ethics of character would argue that every person is already a member of society (or culture or community). There is no opposition of self and society, except as an abstraction. A difference in identity lies within a society; "anti-social" individuals reflect certain cultures that tolerate them or even encourage individual differences. Humans are born into a web of family relations that define their biological place and acquire other relations later in life: cultural identity, friendships, spouses, and others. Individuals emerge from this web of relations with a character: individuals are "made," not born, whether they like their socialization or not.[10] Starting from "individuals" is perspectival, and misses the larger whole. Society is not as "atomic" as critics claim or as liberal models posit. People share the world, their own species, nation, culture, language, and (often) variables such as gender, values, age, and so on, even prior to getting to know one another. Moreover, we need society and community. Association is manifestly to the good in a practical sense, since babies of most species could not long survive without parenting, especially humans. But more, society creates individuals of character, transforming them from an unformed infant into a person, with character, ability, and other virtues.

There could never be an individual ethic, suicide notwithstanding.[11] Individuals by themselves would not need an "ethic," since they would not have to take the consequences for others into account. There would be no limits on individual freedom for such individuals. Indeed, I cannot steal from myself, or commit adultery by myself! The limits provided by ethics are socio-moral ones. Harming another is wrong. Indeed, suicide is opposed by several societies on social grounds: as violating the sacredness of the body or life, or as bad for the morale of society.

Any ethics involves social norms or rules—obligations based the rights to human freedom of others. Such rules must at least hold in general, if not absolutely to constitute a universal ethic. Murder must be wrong for the most part, with only rare exceptions, such as for self-defense.[12] As universal, it should be binding on all: the powerful as well as the weak, institutions as well as individuals, and so on.[13] In other words, there must be some principle of justice that is reciprocal between each member of society. Since it is universally binding, it must precede individual happiness or utility (although even Utilitarians would agree that causing needless pain cannot be a good reason to act, no matter how much pleasure someone might get from it.) This is why the other in human association is morally compelling: also a member of our own society that we both require.

Some might say that this ethic is "objective," but the use of such a metaphysical-epistemological category in ethics is problematic. It is certainly not "objective" in the same respect that shape is. Moreover, the contrary is "subjective," that is, the framework of such an ethic would be Cartesian metaphysics, and the thrust of this chapter is against the Cartesian framework. As norms or obligations, they may be counterfactual—neither objective in the immediate sense nor subjective. They indicate what we should do in the future, but may not be inclined to do. However, if we creatively actualize the norm or virtue, that is, act on duty, we have in "fact" actualized the norm in the world. Thus it can be an "object" of perception in a sense if acted upon.

Some might argue that different senses of virtue in different cultures argue against any such universal ethic. However, students of anthropology have compiled lists of "universal virtues" that are valued by every culture. These include at a minimum: wisdom (knowledge), temperance, courage, justice, spirituality, and generosity, at least within the in-group. We might add trustworthiness, since this is the overall virtue that is aimed at by the special sorts. Thus the universality of virtue is not in doubt, and thereby the implicit view of human development that it entails.

The place of the ethics of character within a society can still be pondered. We discussed the necessity for specific rules above in terms of duty. Duty in the strict sense concerns right and wrong, not good and bad. We could argue that character types reflect values, not ethics in the stricter, more limited sense of right and wrong, moral rules as such or duties. Both Immanuel Kant and the Utilitarians deny that value and duty are identical. For Kant, we do not have a duty to actualize any but moral values and certainly no duty to instrumental values, but to universalize our own will and treat persons as ends. Similarly, the greatest good of the greatest number is not an obligation to individual pleasures or happiness or their conditions, instrumental values. We are required to act for the greatest good of the greatest number, not individuals.[14] Is character then aesthetic, a perfection of society, but one not strictly required from a more narrowly moral view?

Proponents of character might argue that the virtues of moral excellence may not be necessary for a society to function.[15] But it tends to promote a good society—of virtuous persons.[16] For association with the virtuous is surely better than with the nefarious. Character may not be a necessity for society to function, but a society of persons with character is a superior society. Another point that can be urged in defense of the development of character by society is that it results in a person as such, that is, a person is someone with character—or who has developed his or her capabilities. Society permits and actually constitutes persons, since persons have moral character. Since we all require social relations in childhood, and all develop socially, the case for the practical necessity of virtues is stronger.

The counter-example is the ideal of self-sufficiency, but in any case, complete self-sufficiency is impossible. Only after character has been formed can people thrive and pursue a many-sided development. Moral autonomy does not mean economic self-sufficiency.[17] The ethic of self-sufficiency in Stoicism and Cynicism was a prelude to the ethic of the isolated subject in modern times: from the "Meditations" of Marcus Aurelius to the *Meditations on First Philosophy* of Rene Descartes. From the self-sufficiency within flux metaphysics the historical turn was to the subjective values of nominalist empiricism, to the internal relations of self to feeling and desire in value theory. In the history of philosophy, character declined from public actions to inner feelings. But such values do not appear by magic: pursuit of individual values is made possible, indeed constituted by socially derived values.

Absolute relativism is implicit egoism. The subject legislates according to individual ends. Duty is left to individual discretion. The denouement of the subjective turn is selfishness, anti-social values, and individuals sacrificing the common good for individual ends. What egoists do not see is that egoism is only possible within a peculiar social milieu, and that it requires toleration in and by a society.

However, a stronger argument is that society does require its members to have a character. For persons with character think of duty to others in decisions. Those with moral character are more likely to resist temptations and the blandishments of desire. Doing duty may be enough from a moral point of view, but who is likely to do their duty? People trained in moral virtue are more likely to behave morally, putting principle before inclination.[18] Character is helpful, if not required for upholding and sticking to moral principles in difficult situations. Also, there is the issue of moral agency, that is, who is a moral agent? I will cover this issue in the next section.

Another criticism is deeper: will inculcation of character or virtue achieve the goal of moral behavior? Critics of character ethics could argue that no one type of characters necessarily commits bad acts or good acts. We cannot predict that an individual raised a specific way will always act virtuously or morally. There may be a connection between a virtuous habit or disposition and a suitable action in some circumstances, but there is no logical necessity. A person could have a good character and sometimes act badly—or a bad character, and sometimes act virtuously. How can we possibly know what anyone will do, even if they have been trustworthy in the past?

The case for the character development ("virtue") ethics in terms of social consequences depends upon trustworthiness—itself a virtue. We can trust virtuous persons to be just, courageous, and so on. But this begs the question of whether we can trust them to be trustworthy. Even if they have proven trustworthy in the past, or for the most part, there is no guarantee they will do so in the future.[19] If for no other reason, they might change. Thus marriage vows only work some of the time. Also what if people have other virtues but are not trustworthy? While virtues as such may involve generic trust, character is instilled. Proponents of character might argue that those who deviate do not have character. They could also make the case that all of ethics face the problem of immoral behavior and thus that this issue is not confined to the ethics of character. However, opponents might point to this as even more crucial for the ethic of character, since it is supposed to anchor moral behavior in the person. Moreover, character training works in general, but will not invariably result in moral behavior. It is general in another sense as well, since it is not absolute, but not relative either. Instead, moral agency involves moral deliberation and may result in actions that go against the norm. The model is that of generality and exception, not moral absolutism as exceptionless.[20] Similarly, we cannot know whether a given individual has a conscience, the notion of which probably originated in the religious ethics of the Middle Ages, much less whether s/he will act on it in a requisite situation.

One issue this raises is whether, in terms of social consequences, virtues reduce to trust? Or is trust generic with different forms, for example, a person who is trustworthy in battle is courageous? I will discuss trust later.

Another problem is if virtues are the qualities required for success in a specific society, as MacIntyre claims,[21] they are duties, as per Kant. For they are required, the condition of successful associations. Thus they are practical necessities, duties. We could argue against this view that they have hypothetical, not categorical form—to borrow Kant's language—that is, they follow only *if* a person seeks success. This brings consequences into the relation, as virtues would then be required for optimum social success. Duties are defined within a consequentialist ethic of social success as a value. Virtues in this case are not done, contra Aristotle, "for their own sake," or autonomously from their social impact. Proponents could argue both that ethics is always in a social milieu, as we argued above; and that the ethics of character does indeed involve teleological elements. It is not totally removed from consequences to others, but, on the contrary, takes these into account. In this case, the notion of actions done for their own sake is problematic. Aristotelians might argue that these are not virtues, since he distinguished actions done for their own sake from actions done for a further end, *techne*. But he recognized the social element in character.

Such goods or excellences are categorical in that each social category makes possible an action or field of action not possible in the jungle, which allows humans to develop their capabilities to full measure: virtue. Thus not all virtues are strictly moral; they are good or of value, but not "right"; Adolf Hitler's troops might have been virtuous soldiers but not moral. Virtue makes for a better life, then, but *only under a moral condition*—an excellent sadist is hardly worthy of esteem. Virtue is a spice added to a moral life to increase its value yet further. But it will not enhance an evil character or act.

Character ethics could be defended on the ground that the Nazis hardly exhibited moral virtue. Kant's example of the resolve of the criminal ignores the holism of the ethics of character: that having one virtue is not enough to make a good person. Those who cannot control their impulses, for example criminals, and thus wrong someone else, have shown themselves to be underdeveloped adults. The moral self that cultures develop is in general normal for that culture or for society. Criminals are the exception.

However, the criticism that virtue ethics cannot always tell us what it is right to do in a situation may still have some validity. Perhaps this is why most classical authors do not want to separate the ethics of duty and the ethics of virtue, namely, Plato, Kong fu-tse, and Kant, echoed more recently by John Dewey and James Tufts. All these authors consider it a duty to be virtuous and virtuous to do one's duty, rejecting a one-sided emphasis on either virtue or duty. Although Kant is considered the supreme voice of deontological ethics in modern times, such a view ignores his writings on virtue in the *Metaphysics of Morals*. Some writers treat these classifications as fixed, but they tend to blur in the actual texts.[22]

Another issue is whether society is responsible for transgressions of individuals? Since character reflects social influences on development, is a particular society responsible when character does not "take" and an individual does bad or evil things? Society is partly responsible, but the individual bears some of the responsibility, having had character training. But can they be blamed if they have not achieved autonomy? To some degree, this logical point raises issues of the limitations of character ethics that require a relation to duty. Agents with defective characters still have an instilled sense of duty. They know what is right and wrong, from their moral training. Even if they have not achieved the moral virtue that is a mark of a good character, they are required to do their duty.

The same point applies to the issue of suffering. The problem of suffering is, what if an individual suffers despite practicing virtue and doing duty?[23] Why should they be moral? One aspect that both character and duty share is that they are done for their own sake. We act virtuously because it is virtuous, not for any reward. We do our duty because it is right—for its own sake—if we are moral agents, doing what is morally worthwhile. Rewards, even social approval and commendation, are beside the point morally, as Kong fu-tse (Confucius), Plato, Aristotle, Kant, and Dewey and Tufts realized. In terms of an ecocentric cosmopolitan ethics, doing what is right and excellent for the environment as a whole may require risk of social disapprobation. But then it is even more virtuous, just as a person without means who lives the life of virtue is even more commendable than one who is well off, and does not have to make sacrifices.

What is the relation of the ethics of character development to creative actualization as a theory of value? Intrinsically valuable agents are self-actualizing, whether human or other species. Character development aims at some type of emergent autonomy, where agents can govern their own actions without adult or public interference or guidance. At this point the agent is self-actualizing in the moral sense. They are capable of making moral decisions, considering the moral consequences of actions on others and for the environment. Self-actualization, or the intrinsic value of a self-actualizing agent, creatively actualizing itself, in one sense involves both inner character and regard for consequences for others, for as inner resolve to actualize itself in the world it is private, but also involves a space for activity. We actualize our character in actions, which are public and have consequences. Character development involves the cultivation of traits that will enable individuals to act well on their own. They can then act autonomously to some degree as self-actualizing agents. In turn, acts reflecting character are creatively actualized by such agents, validating them by bringing them into the world.

Prof. Simon Critchley has argued against "virtue ethics" on the ground that the Aristotelian philosophy or worldview is obsolete.[24] However, the ethics of character contains multiple elements and not all of these are obsolete. Certainly, modern thinkers have rejected the notion of natural *telos*. The Aristotelian model of human nature as fulfilling some function defined by nature is archaic. However, the notion of the function of organs and other parts as well as that of the function of mental capabilities has made a comeback.[25] Moreover, developmental psychology is a recognized subject of study by academic and practicing psychologists.[26] Thus function and development, which underlay ancient virtue ethics, are very much in play. Finally, in the notion of motive is a modern refinement of acting for an end or goal: teleology.

As far as the metaphysical underpinnings of virtue goes, the substance-attribute metaphysics, I would note that the ethics of character is not limited to Aristotle or his ontological "paradigm." Plato, Kong fu-tse, Kant, Shaftsbury, and others articulated ethics involving character and virtue but rejected the metaphysics of substance and attributes, if not the whole idea of an ontological "first philosophy." Thus it is unclear in what sense the ethics of character is obsolete, since it depends neither on the Aristotelian metaphysics nor its view of human nature. However, Critchley may be criticizing MacIntyre's attempt to revise the ethics of character in its Aristotelian form, and thus Aristotelian philosophy as a whole, instead of the ethics of character as such.

Aristotle also argued that action aimed at some end. In other words, he could be interpreted as having a teleological theory of action. Action was distinct from motion in that an end, or at least a function, is the goal. (Plants have functions, but not ends.)[27] As goal-governed, we can characterize it as achieving the end or falling short, and thus of relatively good and bad actions. Moreover we can evaluate degrees of excellence in achieving outcomes. We can distinguish the distinct capabilities of different species and their proper functioning and corresponding excellences. Finally, we can note the development of (human) capabilities and functions as well as that of animals.

Actions can also be separated in terms of kinds and situations aiming at distinct ends. Economic activities are distinct from intellectual for the most part. Human capabilities come in different generic kinds: physical, talented, intellectual, and moral. Humans can develop their physique, and have different capabilities than that of other species. Humans can also develop their talents and intellectual capabilities. Finally, cultures develop moral character.

2. The Self and Free Will

The point of the ethics of character is cultivation and development of the moral person. Moral persons are free as they are in control of their desires and immoral or anti-social impulses. A person of character has moral virtues they can exercise appropriately to a situation. In this view, humans are "made," they are not simply "existing" in the sense of full blown without any social development—thus the ethics of character does not have a subjective starting point.[28] Cultural values inform all individuals, and autonomy, as Virginia Held (1990) has argued, is created as a result. Cultures attempt to instill virtues, creating moral individuals, moral selves. Character is developed or made in the process of maturation. Individuals with character do indeed act independently within society, since they have character and thus can be trusted to act virtuously. In this view, a person cannot be an "individual" except as a moral self.[29] The formation (or perfection) of character *is* the formation of a self. The "subject" is formed, refined, and developed by culture and social values: social virtue. This development of character is the development of moral agency. The "subject" also includes the willing subject: Augustine's notion of the will, insofar as we accept the validity of this notion.

The value of development of character is that we can creatively actualize moral agency, self-actualizing agents who lead a moral way of life. Individual autonomy and moral personality, which develop out of socially instilled habits, belong to self-governing moral agents. They self-actualize virtues and duties in appropriate situations. A "virtuous" agent has developed to the point at which she is capable of behaving morally.[30] Autonomy is a consequence of character formation in youthful development, and represents excellence and full actualization of potentiality.

The moral self is trained—habituated—to consider others within a culture, and often beyond. The project aims at a moral self with a character, so we do not behave poorly, we do not behave only out of self-interest, pleasure, or individually regarding motives (egoism). Conscience is culturally created—cultivated in moral development. Education, training, and habituation combine to constitute the moral self, involving the development of the plastic imagination.[31] The moral self is developed out of the imagination of the individual, which is somewhat plastic, into self-consciousness. "Self-consciousness" means self-aware, thereby responsible. The moral self created by culture evaluates by the categorial standards of the culture, including religion as a source of ethics, and normative polarities like truth and falsity, right and wrong, real and unreal, beautiful and ugly, and so forth. The judgment of truth also requires a moral self,[32] as do judgments of beauty and even economic worth (price). The moral self will not deceive (although compare the occasional conflict of politeness and honesty), but those with a less than sterling character might. Similarly with beauty and even economic worth. These are choices and thus require a moral self.

Plato counterpoises virtue in general to a type of slavery: to desire. A moral self in control of his or her desires actualizes virtue, that is, acts with excellence in moral situations. The moral self was the ancient equivalent of the "will," the chooser, but with a moral core, since the moral agent chooses responsibly. The self is in one sense a series of choices.[33] However, an infant is no more capable of making mature decisions than they are of providing their own food or shelter. The ability to make responsible choices, that is, to act with a good character, is formed. The moral self, the person with character, is developed as part of socialization during maturation. Morality involves choosing well between impulses, that is, precisely *between* determining motives, and thus freely.

Decisions *project* futures, and therefore, are not physical. "Mind" moves the body (not the brain alone) toward future goals. A projected future cannot be "present" atoms, "facts," or other physical kinds. "Mental" causation is clearly needed—rocks do not jump out of the way in anticipation of danger. I make this point in anticipation of physicalist objections to the whole notion of mind, and making free decisions. On another point, inaction is distinct from indecision—either not choosing (indecision), not being able to choose—from not acting on a choice, and from inaction of restraint. These distinctions are tied to ethics, although many have argued that inaction is a choice.

The moral agent is self-actualizing, that is, intrinsically valuable. By choosing morally, the individual self-actualizes as a moral agent. Moral choices over time reinforce the moral character of the agent, in both senses of "character." Such choices affect whom we become by the choice, thus the whole person, not just the "will"—which is desire plus some "x" (reason, or some other element). Persons make choices as moral agents, not "wills." Otherwise, desires determine choice, since will is desire, thus they are not choices, but enacting desires. The latter require evaluation by moral agency and intelligence: critical evaluation. As Dewey noted, acting impulsively on desires, without critical evaluation, is the mark of a child or a spoiled person, not one of character. The standard for such evaluations is: would it improve the world in general to creatively actualize such a goal, including improving society, and the environment?

Since the ancients did not discuss free will, but did discuss character as a function of cultural implantation, this view is consistent with ancient philosophy. This complex view was lost by the time of the late Roman Empire, which created a moral crisis. Briefly, the Hellenistic period was the beginning of the interiorization of value, in the rationalism of the Stoics and the hedonism of the Epicureans. Both these theories of value are connected with internal psychological capabilities, respectively, reason and feeling. Roughly, these theories became the basis for modern theories of value, the Stoics of Kant and the other rationalists, the Epicureans of Thomas Hobbes,

John Stuart Mill, and the other empiricists. Thus value was connected to interior states, not actions, undermining the idea of the moral agent as the whole person. Aurelius Augustine's answer was the doctrine of free will. I will examine this notion in the next section.

3. The Actuality of Character

The "being" of character addresses the issue of who or what we "are," not only what we do.[34] Virtue persists through training of some sort and thus makes us a specific kind of person, different from whom we were before. But this is a becoming, a change, not being. It does involve dispositions and potentialities of habit. Thus it implies potency and actuality of an evolving moral self, not being, persistence. Moreover, the exercise of different virtues at different times argues against their "being."

For Aristotle, virtue involves bringing potentialities of being into being: of powers into full, perfect operation. However, virtuous actions as the actualization of "being" is a strange doctrine. Courage is not present, not observable, not hidden under or within—it is a mere "disposition" or even just a potency. A person "is" courageous without courage being "present," whether as a substance or attribute. Such a virtue does not have "being" in the sense of "existing." It may be visible in actions, but this involves interpretation, mediated by cultural or social norms of courage. If courage then comes into "being" at points of courageous action, it is a becoming or a change, not a being—it cannot be present if the future is involved, if we predict courage in the face of circumstances. For what if cowardice is exhibited instead? Moreover, duty in Kant is not ontic but regulative. Moral values are logically independent of what "is."

The change from some other behavior to courageous behavior is a change, and a challenge to the "being" of a virtue. Instead of the clumsy application of "potency" and "attribute" to bring virtues in line with ontological metaphysics, why not describe character in terms of the pursuit of different values at different times by moral agents through action? Thus the goal is not perfection as such but achievement of the goal—itself a form of excellence. We are not perfecting a blank slate or potentialities, but pursuing valued goals with moral deliberation.

Although the pursuit of excellence as a good in the highest degree is a form of value, the worth of values is distinct from perfectionism, as I argue in the next section. At the same time, action reveals the worth of values, including cultural values such as knowledge. Historically, this raises the question of the "metaphysics" of values. Compare powers (disposition, potency) and operations (actuality) with rights (positive freedom as power). Action involves change and changing, not being, and action is the agency of creative actualization: value. I am not arguing against the potency-actuality relation but its ontologization.

Ontologists might object that a power or disposition is ontic, as it endures. However, powers and dispositions are not present except as actions, which come and go. An agent finishes one action and goes on to another. Similarly, habits remain only as potentialities, not actualities. They do not endure in the same respect, since they come and go: they become. Moreover, habits may change over a lifetime in the process of perfection of character; the world also changes, however slightly, as the result of virtuous and non-virtuous actions. Character habits are normative in the general sense, that is, in general others can rely on a person of character to do the right thing, or act virtuously in the appropriate situation. They are general norms of an individual character, not substances, entities or other metaphysical fictions. They are principles whose actualization comes and goes. Critics might argue that reduction of virtue to actions is incompatible with potency and actuality, but these are required for good habits. I would respond that the "moral self" has the requisite potentiality and must be capable of creatively actualizing appropriate virtues at different times. Thus the potentiality belongs to the character of the moral self.

The ontic fallacy has produced more bizarre creatures in ethics than anywhere else: "will," a "sense of responsibility," the "moral sense," and "conscience"—all *logos* requirements for embodiment of norms and values—determining the differentia of ethics.[35] They are required by the metaphysical view of character, in which the good is an attribute of being, and so must have being.[36] The standard treatment of virtues abstracts from their temporal actualization at different times and treats them as "beings," for example in the persistent exhibition of temperance. Temperance is a "state of the soul," in Plato's words. The underlying principle is that virtue must have actuality as an entity in the agent. Virtuous acts issue from embodied parts of the soul that retain virtue as duration over time.

Their temporality in truth points to their basis in the freedom of the moral agent as such.[37] For we can do different acts at different times: freedom of time, space, means, and ends. That is, we can act freely. The agent is free, not the "will." Similarly, the modern age, following the inward turn of the Stoics, Neo-Platonists, and Medievals, treated virtues as feelings, attitudes, and other mental projections and posits. These creative posits follow the inward turn of Neo-Platonism in method and thus its surrender of public space—in which it abandoned the spirit of Plato.[38] "Will" is in any case ambiguous: it can mean desire or the choice between desires. Thus it is subject to infinite regress. But the notion of a will is not needed, since the moral self involves the development of character. Augustine in effect separated the self into a will and a soul. Kant added rationality to the human will, which acts according to principles. Later figures argued for the irrationality of the will, as in the case of Arthur Schopenhauer's "blind will." Such ambiguity indicates a problem, which in my view is that the "will" is a makeshift, brought in to internalize moral agency, but inadequate to the task

The evidence for character is as a history of actions—behavior, if you will, in the original sense. Such actions proceed from a moral self, a person disposed by culture, habit or inclination to perform such actions. One agent can perform different actions and with different degrees of finesse. The unity is in the agent, but this agent is not a substance, much less a being. For their virtue has been developed and they change in the process of development. Moreover, the action cannot be separated from the agent at the time of the action, although the agent remains after the action. Creative actualization of qualities of a creatively actualized whole often aims at a "perfection." It can be a quality of a part in some cases, however quality is defined, for example, fineness of a performance in music. Thus its qualities or perfections constitute a further good: better. Virtue is putting the good into action: creatively actualizing good qualities or excellence in an act. In this view, character is revealed in action, not an essence. The character is manifest in different situations in any case—there is no common element or essence to the actions as such in different situations.

Characteristics are a convention, not a nature, about pragmatic qualities on which we are willing to act. For all characters—human, chemical, taxonomic, and so on—are revisable in principle, that is, in knowledge; and in practice: reform. We are willing to act on believed characteristics that are yet unrefuted: including those of people (trust). Diverse characters of each kind are categorially distinct: a diamond has economic as well as scientific characteristics (valuable vs. carbon crystal). None are supreme unless one is judged categorially supreme: intellect (positivism), economics (Marxism), and so on. This binds society through imposition of a special kind of association on all other kinds.

A moral agent reveals their moral character by avoiding bad actions, or acting in accord with and based on moral principles, not some essence or substance of personality. The idea of an essence of a character is based on the substance and attribute principle—that there must be an underlying substratum to the different acts—but there is no such basis in actuality. "Character" is not a reflection of the metaphysics of substance and attribute, but of hierarchical evaluations by a culture.[39] Some features are evaluated as subordinate to others in a culture for social reasons.

I am not arguing against the central idea of development, that education and practice transform humans in the course of maturation. In some sense, they "have" a good character, which is part of them. We are in broad agreement with the notion of potentiality to actuality—the dynamic part of Aristotle's metaphysics, which was also broadly accepted by Dewey and the other pragmatists. However, "operations" (*energeia*) can be replaced by actions, activity, work, and functioning (as a verb) in moral contexts. There is no reason to think that moral agents are anything like non-moral actualities. Action aims at an end, but non-agents do not.

Nor am I arguing against an inner self that deliberates. Since virtue is exhibited in action, no internal states need to be posited; the action and end are enough. But this is not a denial of any mental element in moral deliberation. On the contrary, wisdom involves a certain amount of deliberation. Habit does not have the field to itself. However, character involves habits that are formed that have some influence on actions but do not absolutely determine them. For the moral agent is capable of acting otherwise and must do so in some cases. Moral prohibitions on killing are put aside in time of war. Moreover, situations determine the application of the power, not mere habit. Otherwise, we would constantly be exhibiting inappropriate virtues in every situation. We must be able to control when we exercise the capability as well as controlling its exercise (discipline).

The moral self is the agency of such decisions, with or without a good character. The moral self is not only one element of the self but is the self, at least with regards to practical affairs, that is, life. The point of ancient ethics is that we are a moral self with a character that has developed well or poorly. There is not a self apart from our character, since the two are inseparable, given that our self has been formed by our society. My argument for development of character, then, is that the ethics of character is one with the development of autonomous, trustworthy moral agents. Such agents are free from strong determination by desire and feeling, since they are in control of their desires and feelings.

Virtue is in the public sphere, while "free will" is internal. Virtue is governed by the end and creatively actualized in action. The inward turn is Neo-platonic, not Platonic: a prelude to the philosophy of the subject. Virtue requires training (Plato, Kong fu-tse) and habituation (Aristotle and Dewey) but involves choice of the mean based on deliberation regarding circumstances. Thus virtue involves a free agent, but is end governed, not psychological: creatively actualizing some goal.

Perfectionism: another criticism of the ethics of character is that the social molding of persons is a misuse of power. Such criticisms vary from the anarchistic view that any attempt to enforce a type of character is detrimental and political, to the contemporary view of Michel Foucault, with his analysis of the self-policing agent who reflects the installation of "biopower" in the individual. Foucault concentrated on the "classical age," that preceded and included the Enlightenment understanding of humans. In particular, it could be seen as a historical critique of the model of humans as perfectible: the perfectibility of humans.[40]

Perfectionism can be a theory of history, as part of or independently of a progressive ideology. In the former case, perfectionism is made historical and tied to intellectualist values. Perfection is in proportion to the advance of knowledge over error in history, of reason over irrationalism. This is the world-view of many of the more prominent figures of the Western Enlightenment.[41] In the political version, perfection is the measure of progress, and progressivism is an ideology linked to perfectionism, its *sine qua non*. The perfectibility of humans is taken as a universal principle, whose value is self-evident.

Oddly, a corollary to perfectionism can consist in blaming the social environment for every social problem including lack of character. Bad character is due to im*perfect* circumstances. Since we are products of our social environment, our faults are faults of our environment. This type of blame is ultimately a denial of the moral perspective: that we are all free to develop ourselves, act responsibly in spite of circumstances, and do as we wish: free choice, autonomy, personal integrity, and dignity. It involves a revaluation of intellectualist values over moral values (despite the association of the latter with perfectionism) as a determinate material cause is sought: satisfying intellectual cognitive interest. This environmentalist-mechanist model, humans as *tabula rasa*, is then imposed institutionally—conditioned responses, "adjustment," reward and punishment, and so on. The model is self-fulfilling, since it rewards behavior that conforms to the model, encouraging it.

This view includes the medicalization of evil: Hitler was "crazy" not evil. The principle provides a blank check to underdogs: victims of environment. In whose interest? The Medical profession and psychiatric technocracy who prescribe sanity. Social maladjustment is a curable illness. Economic circumstances are responsible for all social ills. But development of character includes the virtue of self-improvement.[42]

Kong fu-tse recognized the element of fate in our lives—that we may be born poor, or in humble circumstances. How much more virtuous is the person who has a good character in spite of such circumstances! I am not arguing against the role of the social environment in shaping or failing to shape character, but that individual choices cannot be laid entirely at the feet of society.

I distinguished excellence from character development above, since no one is capable of excellence in all his or her potentialities. Virtue ethics can be tied into a perfectionist theory, but it need not be. Virtue as "well done" and as "excellent" is a matter of degree. Both are commendable, but not everyone can attain excellence in all situations, or at every time in their lives. Tone-deaf individuals do not become opera divas. Moreover, moral character can be separated from the virtues in the abstract, and as a whole distinct from its parts. Character is more important than virtues in the sense of goods or excellences, for it implies the development of a moral agent. Excellence in non-moral virtues is good, but not *morally* required.[43] Developing our talents is good, even virtuous, and those who do not develop them when they could may exhibit one form of vice. But in a pinch, moral character is more important than non-moral virtues, as Kant's example of the courage of the thief indicates.

Virtues may also be a matter of having a moral disposition of a kind, whether courage, justice, moderation; or not having it, instead of having it in the highest degree. "Virtues" can mean the perfections of character, but having a good character is morally sufficient. Indeed, perfections are only relative to human capabilities, since old records are continually being surpassed, and thus perfection is only a standard, not constitutive. Improvement is a more realistic position regarding human capabilities. Meliorism only requires moral character, not perfection; bettering or improvement, not excellence. But excellence is not ruled out. Excellence means achieving development in the highest degree, but it is virtuous to act morally, even if we do not do so with distinction.

Excellence cannot mean perfection for finite creatures; doing justice to difference must mean that excellence takes different forms in different individuals. The development of these must take the participation of the individual into consideration: it must involve interaction. This precludes perfectionism, as this standard involves a single, universal model. Participation requires autonomy: self-reliance or maturity as a goal, not passivity or dependence. A moral agent creatively actualizes virtue in appropriate circumstances. The ability of the moral agent to pursue what is right and good autonomously is the mark of moral character, not perfection of every capability. Self-improvement, not pursuit of the goals of others is autonomous.

Perfectionism can become obsessive, and we should not lose sight of the proper goals of the development of character. The important thing is a person of good moral character, a moral agent, not a person with excellent development of talents and capabilities but no moral character. Proper excellence is a good character in the highest degree and is a perfectly good goal, so long as it does not become obsessive. For it can breed an anti-social kind of competitiveness that defeats the purpose of creative actualization of a good character, a person who can act with autonomy in society, who can be depended upon to do the right thing. I am not arguing against excellence, but only that we do not miss the forest for the trees. A one-sided development of intellectual, or athletic excellence, or that involved in musical or artistic talent, may work to the detriment of overall character. I noted that Kong fu-tse mentions that virtue has "neighbors." Development of the person as a whole, with emphasis on moral character is of greater importance than specialized talents, although the latter should be developed within the whole.

The virtues, though social, are not arbitrary. They can be evaluated: is such and such a virtue worthwhile? Does it add to life? Is such and such a model actually the good life? The question, then, is what constitutes a good life, worth living, not what is human nature.[44] The two issues have been confused. This lends itself to a variety of answers, no one model is universal—good for everyone. Individual capabilities are as much of an issue as social norms. Some do not have the talents for sports but may excel in the arts and vice versa. Part of moral agency is coming to a decision about what is possible with regard to self-development. Sometimes we cannot change what we are. However, several norms can be formulated, for example, alcoholism is bad and achieving a well-rounded life, good. The good life is a guideline, constituting a useful or valuable standard, although not a duty: not required, only good. In general, the positive part of ethics is the good life, the negative part, forming negative limits, is norms for social living: duties.

Perfection of human nature presumes a norm of human nature: the tyranny of "nature" as a universal standard for everyone. But such norms are erroneous if natural and arbitrary if cultural. No ethic involving perfectionism can be truly naturalistic for virtues are artificially cultivated. Hobbes recognized this: the Leviathan is artificial.[45] Aristotle thought he could bring this under potency and act, but the conditions of virtue are highly artificial (education, training) unlike say, an acorn becoming an oak. Thus the question of the ethics of character involves the development of artificial character, as Plato implied.[46] But the question remains: is the possibility of perfecting or developing humans their "nature?"[47] Is it in character or habit for most animals to invariably behave in such and such a way, given specific circumstances? Character ethics has both a naturalistic element and a developmental element, since we cannot expect development of capabilities foreign to a species. The origin of the state of nature as model or norm lies in materialism, as it looks to past determinations.[48] Fully developed nature, that is, perfectionism, is more idealist: an ideal striven for in future.[49] Natural capabilities are shaped, developed according to "non-natural" standards in this model.

One problem with perfectibility is that it poses a static model, in which a specific standard is reached. Both Kong fu-tse and Dewey saw problems with such a stasis. Kong fu-tse preferred to stress the striving for virtue as an ongoing process, rather than a fixed goal. Times change and character may change with it. At all times we strive to do what is right and good, regardless of circumstances. Similarly, Dewey argued for growth as an ongoing process. Moreover, thinking of the achievement of character as a goal may be detrimental to further growth, since, once reached, growth of character may cease. What is ultimately virtuous in this view is continual development of character, the continuous striving for good.

One counterargument to our critical evaluation of virtue and character is that consideration of others involves sympathy, and that this is a feeling.[50] However, we could consider others from a moral point of view without feelings. Indeed, only if it is moral would or should we be sympathetic. Sympathy for the Nazis executed after the war seems misplaced.[51]

It is interesting and important that some virtues have to do directly with feelings, such as the control of anger. We could argue that the example of control of anger as a moral imperative provides a counterargument to the "moral sentiments" view of virtues. Moral virtue sometimes consists in *overcoming* bad feelings—thus a struggle to subordinate feeling to moral standards, for example equality and justice used as standards to overcome hatred, anger, and other emotions that might lead to immoral acts. Martin Luther King Jr. and Mahatma Gandhi both argue that we must purge ourselves of hatred for our opponents in exercising non-violence.[52] Some modern psychologists argue that this model is unhealthy. They raise the issue, can we help what we feel? Can we or should we control our feelings? Indeed, we may even ask if this is morally required? Or only that we not act on such emotions? Are virtuous actions, and not virtuous feelings, all that can be expected from morals? If suppression of strong feelings is not healthy, then the arguments against affective theories of virtue acquire even more force. Virtues may control feelings, but they are not themselves feelings. Virtue is exhibited in actions, as characteristics of actions, not of feelings. The subjective view is undermined.

Control should be redefined as not acting on feelings, then, and sometimes overcoming them. Being angry is not a vice, especially righteous indignation at outrages (feeling it). Acting it out can be. George Washington was reputed to have a temper, but also to have it under control. Aristotle would describe this as virtuous, adding that righteous indignation is appropriate in some situations, for example, if someone overhears another bragging about raping his sister, or daughter. Control of anger is a virtue along with anger in the right circumstances, at the right time, and so on, as per Aristotle, but in the sense of not acting on it, instead of suppression. We cannot always help what we feel, but we can control our actions.

Similarly, all humans have a desire to eat, but some cannot control their appetite. Those who can exhibit moderation, a most underrated virtue in our time. Desire and feeling are not excluded in a good character but channeled and developed into excellences. Thus desire is transformed by moral development, not extinguished. But Kant may be too rigorous if he is interpreted to mean that moral sentiments ought to be excluded from moral deliberation: there is nothing wrong with moral sentiments for others, particularly family and friends, on pragmatic grounds, however instinctive ("empirical").

4. Ways of Life

Ethics also includes and involves a way of life.[53] A genuine philosophy is a way of life that pursues what is most worthwhile, and not spurious values. That the worthwhile is not necessary but free gives it a range which different temperaments and personalities can utilize to personal advantage. But this does not mean indifference or license—a genuine philosophy should issue in practical activity, and give a life meaning. Indeed, "doing philosophy" is a way of life. Philosophic Ethics examines the basic possible ways of life and evaluates them. It also derives social norms for interactions of persons.[54]

Indeed, Aristotle concluded his famous lectures on the ethics of character, the *Nicomachean Ethics*, with a comparison of different ways of life and an evaluation of which constituted the best life.[55] Philosophy, the best life as the contemplative life, is a way of life in Aristotle. The ethics of development of character involves instilling a way of life in youth that is socially responsible. Aristotle modified but retained the essence of Platonic character ethics. The ethics of character was also prominent in the Stoics and even treated by Epicurus. Indeed, ancient ethics might be called study of the ethics of character as a way of life.

Morality is a way of life in several respects.[56] One is that a moral way of life contrasts with an amoral life. Another is the contrast of the moral life and other categories within life, such as the "religious" life or the "esthetic" life (Søren Kierkegaard). Morality also constitutes our life as autonomous agents, in control of our own destiny and not ruled by whim, desire, circumstances and other external forces. Above all, morality as a way of life means that morality is practiced in our whole life and is not only a list of rules or a technical study. The "spirit" or defining character of morality permeates all of our life, not only obvious moral situations, like telling the truth.

Similarly, moral limits are required for the religious life, the esthetic life, and any other life devoted to one or another category of value. These kinds of life are the gifts of society, made possible as the fruits of moral peace over the ages. Spiritual contemplation is learned from others and requires a peaceful haven, free from social conflict. Kierkegaard's "esthetic" life may be taken as either the hedonistic life or the creative life of the artist. In both cases, creative actualization is required: either of suitable pleasures or of art works. Moral limits are the condition of all kinds of values, including pleasures and art works. New goods must be compossible with previous goods; hedonists and artists should observe the same moral limits as other members of society. In a sense the artist evaluates in creative actualization of the art work, in selection and treatment of different tonal, verbal or visual possibilities incorporated into the work of art. These possibilities are compossible with the other elements of the work and make up a whole, a work. Artistic creation involves as much an evaluation as any other creative actualization. Indeed, the artist is in a sense the model of the creative life since creative actualization of works of art is done for its own sake.[57]

Cosmopolitan Ethics is a Way of Life. Creative actualization as a theory of value involves a creative life. The creative life aims at something good in itself in this respect. Creativity is a norm for all areas of life but since morality is a way of life, the creative life must in some sense be moral.

Aristotle argued that the contemplative life is self-sufficient, but since it aims at knowledge, it is, like desire, empty and requires satisfaction from the external world. However, there is wisdom in the pursuit of self-sufficiency, even if this is an impossible goal for humans. There is the story of the philosopher who when asked what he had lost in the sack of his city, replied that he saw no one carrying off wisdom or knowledge. Since they cannot be lost, they are more durable goods. Moreover, their value is not dependent upon others, or outer conditions, but self-development and industriousness. Lives that depend on others or external goods are incomplete. Of course, no human is totally self-sufficient and the goods of intellectual life are possible in a culture with language and a store of knowledge, to speak nothing of knowledge of how to obtain food and other necessities. Moreover, the contemplative model as Aristotle presents it has contributed to the interiorization of values and detachment from the world.

Cosmopolitanism is such a way of life. Cosmopolitan virtue gets around problem of the cultural origin of virtues mentioned above.[58] Cosmopolitanism assesses cultural practices and celebrates the best of them as the best way of life. Thus it builds on practices to formulate a universal ethic of character. Character includes living by the principles and virtues we preach. Philosophy at this stage of its history does not separate theory and practice, but lives what it preaches. Moreover, it attempts to practice such an ethic as the life of the sage. Philosophy is trans-cultural: philosophical ethics is the attempt to describe or articulate *universal* norms.

Ethics as a way of life means ethics is a way of philosophic, wise life, involving knowledge. (Again, moral values are the ground of knowledge, which grounds all acts within a life.) It is not so much that there is no value to the life of a prostitute or alcoholic: the drunk may claim that he or she prefers this life on one level; the "sex-worker" that sex is easier than other kinds of work. In such a case it will not do to argue that environment or social conditions are responsible for such judgements, as these beg the question of value. Although the truth of the alcoholic's or prostitute's justification is recognized, their own recognition that there is a better life is sufficient argument for social values that deplore such lives. No argument for social utility is needed, which does no justice to individual preferences that are victimless. The alcoholic may always prefer to return to the bottle, but at some level recognizes the harm done to health and other values in life by this life-style.

The Greeks thought that the best life is possible only among intimates. Thus they looked to the city-state as the model for the best life for humans, and looked down on the large empires of their day as ruling through force, not virtue, and founded upon commerce (desire) not human good or reason.

The modern view is that casual association is possible and even desirable if economic improvements are to benefit all. The condition of this is the extention of morals to humans as such—humanism—and not merely to kin; and the recognition of the other as fundamentally human and alike, and thus as having rights like oneself. Whether this has truly happened or is truly possible is another question. Large scale, non-intimate, casual association is the rule in modern life in every institution; intimacy survives only in a *part* of life (family, relatives, social life in the sense of friendships) except perhaps in small towns.

Behind Roman law lies the recognition of it as a law for all nations. Cicero saw the limitations of the *polis*; he tried to enlarge the Greek perspective with a mixed state representing all classes making laws for all peoples. Roman law embodies the morals of casual association. We can also see this in the Stoic conceptions of natural law, and world citizenship in Marcus Aurelius, and in Polybius, Justinian (Flavius Justinianus), and other Roman authors.

"Professional" norms have been articulated and interpreted to include the principle that we should not be judgmental, and that professionals only analyze problems. The creeping of such norms into philosophy has meant that philosophers censor themselves, and many do not make judgments about their society. One problem with this "professional" view is that philosophers impose a kind of self-censorship on their work, with the effect of blunting any serious confrontation with their society. They castrate their moral insights.[59] If philosophy absorbs professionalism as a goal, which has been the agenda of a host of APA presidents going back to Arthur Lovejoy, the greater will have been absorbed by the lesser. Professionalism is basically an economic category, not a philosophical life. The triumph of "professionalism" will mean that economic norms will then hinder philosophical judgment: economic values will regulate moral values. This is similar to the problem in implementing farsighted ecological projects and standards, for example, those having to do with Global Warming. Our political leaders have been so reluctant to propose measures that might have an impact on the economy that they have allowed Global Warming, habitat destruction, and overpopulation to proceed apace. Morality has been put aside to cater to economics. Similarly, anthropocentric issues have dominated philosophy, especially human learning and epistemology, when the crisis represented by destruction of the environment has been marginalized. So-called "professionalism" in philosophy, the economization of philosophy, has encouraged this tendency.

We can no longer play this self-indulgent game. Preventing global catastrophe, in the form of rising sea levels, has priority over every other problem, no matter how pressing. For if we do not prevent the melting of the Arctic, Greenland, and Antarctic ice-sheets, many of the island countries, and the coastal areas where most humans now dwell, will disappear.

There is also the tradition of the gadfly in philosophy going back to Socrates. Philosophers should, like Cato the Censor, excoriate their contemporaries for their moral failings. Philosophers are moral critics who speak truth to power. Philosophers should return to their role as gadflies with a vengeance. If this means being judgmental, so be it. Either we believe in moral principles or we do not. Those in philosophy who do not are not really philosophers. There is no longer room for self-indulgence when it comes to environmental issues. Human folly, the destruction of the conditions of human life, should be challenged at its core. Philosophy, the love of wisdom, is the office of exposing folly. Raising the question of the value of a value—critical evaluation—which has included a critique of the excesses of technology (pollution), should now include a critique of the excesses of affluence, as a way of life. With this in mind, I will examine how my own generation has dealt with the looming catastrophe in terms of the ethic of character.

5. Cosmopolitan Virtue

Why do virtues need an ecocentric transformation? What is the place of human character in an ecocentric, cosmopolitan ethics? Some might argue that an ethics of character appears superfluous, that "virtue ethics" is a secondary part of ethics, and not required for a consistent ethic, especially with respect to the environment. Others might argue it is dated and irrelevant to modern conditions; an ecocentric cosmopolitan ethics should proceed on another basis. Such an ethic does not need any notion of character or virtue to derive duties to the environment. I have tried to address this last point above, by contrasting the model of the moral agent with the more modern "will." The modern notion remains seriously defective as a model for the moral agent, both in theory and practice. While there is no guarantee that a person with moral character will do the right thing in all situations, there is far more reason to hope if the person is a moral agent, trained and habituated to do what is right and virtuous, than if we are to depend upon the autonomous "will," even presuming that we could define the latter.

The irrelevance of virtue to an ecocentric, cosmopolitan ethic could also be argued on historical grounds. What is considered virtuous may change with changing circumstances. For example, specific ancient virtues like bravery or courage have no meaning in the face of contemporary technology. We would hardly judge it courageous to march, sheep-like, into the slaughterhouse created by machine gun fire or a tank, which happened in the two World Wars. The courage of a soldier is even more superfluous in the face of atomic weapons, which have made one on one combat obsolete. Since ecocentrism emphasizes the relation of parts to the whole, individual courage appears out of place.

Moreover, since virtue changes over time, what is virtuous is ambiguous. Virtues can no longer be taken as having a nature (essence). For courage may be standing our ground in one situation, for example against Hitler; and walking away from a fight in another, namely, atomic war. Bertrand Russell argued that walking away from a fight might exhibit far more courage than engaging in one. If virtue varies according to the situation, it can have no common essence. Again the "excellence" of something may vary by context—a horse may excel at both running and pulling and still be a horse.

There is some truth to these criticisms. Previous accounts of virtue have been primarily anthropocentric and tied to a distinctive culture.[60] The changes in the definitions of virtue and the metaphysics of virtue over the centuries has not been treated as problematic. Thus the ethics of character needs a critical evaluation and a reworking in the context of an ecocentric cosmopolitan ethics. The virtues should be redefined to accommodate the insights of the environmental ethicists. We are free to choose what our virtue will mean, as all history attests.

In sum, the ethics of character is incomplete. Perfectionism, development, "self-realizationism," and "virtue ethics" have ultimately been a tribal ethic, involving realization of Greek virtue or English, or Italian (*virtu*) or Chinese (Kong fu-tse). While there has been refinement in the form of perfection, as tribal the ethics of character is still the virtue of a part, and not universal. By contrast the ethics of duty, in the form of the Categorical Imperative applies to all (Kant). The Categorical Imperative asks whether our maxim can be universalized, and even be made consistent with a universal law of nature. Indeed, Kant contrasted the universality of the Categorical Imperative with the self-centered and egoistic quality of hedonism and affectivism.[61] Agapism or the ethics of love also applies to all (Christianity, Moism), as does the ethics of non-violence (Gandhi, M.L. King Jr.). The lack of universal scope is a defect of the ethics of character. Thus the ethics of character has not yet become cosmopolitan: the virtues are not yet cosmopolitan virtues. However, some virtues are universal and the socialization or acculturation of children is also universal. Thus there is not in principle a ground to prevent a universal, a cosmopolitan ethic of character.

Another sense in which the ethics of character has been tribal is its anthropocentrism. To argue that humans alone are virtuous is blind.[62] All the evils of human civilization: genocide, murder, rape, social injustice, and cruelty, are absent from most animal species. Gregarious animals tend, on the contrary, to behave in a manner that would be considered exemplary among humans, for example, protecting each other and their young from predators. Very few species make war or kill within their own species. Animals do not have concentration camps, atomic bombs or chemical weapons. Domesticated animals are sometimes far more loyal than humans.[63] Moreover, the contrast of acting "humanely" as opposed to acting "like an animal" implies a false dichotomy. Darwin, among others, noted many cases of altruism among animals, especially parents.[64] Indeed, he was so struck by this behavior that he thought he needed to explain it in terms of its function in the struggle for existence. Humans can be virtuous and animals can be virtuous. The case for human exclusivity is dubious. Moreover, humans are animals, as both the ancients and modern science recognize.

What would constitute ecocentric ethics of character? Rising expectations and affluence are the direct cause of threats to all of life. Overpopulation and the resultant attempt to feed too many people are contributing factors in the destruction of the environment that is the home of all life. New technologies also threaten our environment, (ozone depletion) and the relentless spread of older ones, like the omnipresent automobile. Another is energy generation: the ecological costs of cheap electricity, including destruction of forests through acid rain, Global Warming, and nuclear waste. Their combined effect has been the increase of Global Warming, resulting in rising ocean levels. As I have noted, the environment is a requirement for all of life, including ourselves: we cannot survive without air, water, and soil. We also require other species as our food as well as contributing or creatively actualizing healthy air, water, and soil.

By contrast, the habits cultivated by our consumer culture work to the detriment of our environment, and will also be a threat to future generations, our children and grandchildren. The threats to future generations include salinization of soils and erosion, and the effects of Global Warming, resulting in rising ocean levels. They also include destruction of beautiful landscapes, whether for commercial exploitation, or to feed burgeoning populations or provide them with water and other necessities. Our descendants may not be able to have the experiences of wild and scenic places that we enjoy, due to our malfeasance. Finally, there is the disproportionate effect of pollution and environmental degradation on the poor: issues of environmental justice.

In raising the issue of a revaluation of character we must ask: to what extent should humans be willing to make sacrifices to preserve the environment? To what extent should humans be willing to go out of their way to consider other species, for example, in preserving rare species? What virtues are suitable to an ecocentric ethics of character?[65]

Some philosophers have already attempted to refine an ecocentric "virtue ethics" relevant to environmental issues. The "deep ecology" movement also includes a way of life that tries to live in harmony with the environment, a type of ecocentric character ethic, but as defined by Arne Naess is larger than an ethics of character.[66] Thus I make no claim to originality in what follows, but will only attempt to revalue traditional virtues, retaining their original spirit while universalizing them. The ethics of virtue as it stands is still anthropocentric. The question is, can it be revalued to constitute ecocentric ethics of character? Can virtue be ecocentric?

Character and Holism: Kong fu-tse (*Analects*) remarked that virtue has "neighbors." Confucianism always stressed development of the whole person, with an overlapping and interconnected set of moral characteristics. A harmony between these characteristics was the ideal. Plato and Aristotle echoed this model with analyses of the different sorts of virtues, and their harmonious relation in the virtuous person, as opposed to the person with a virtue. Virtue involves a harmony of the development of different sides of the character.[67] This harmony is greater than any part: it is holistic. Thus the ethic of character ties neatly into a holistic ethic. The parts are the various virtues; the whole is the person of character, who exhibits appropriate virtues in different situations and has the wisdom and intelligence to evaluate the situation and choose which virtue in what proportion is appropriate, and the discipline to actualize it. Just as a habitat involves the relation of different plants and animals to a whole, persons of character have achieved the maturity, have developed their character sufficiently, to apply different parts of themselves to different circumstances over time. Character must be a whole, for a person with but one virtue is virtuous in that respect, but not yet a moral agent. It would be unwise, unjust to cultivate one need at the expense of others. Such is a narrow life, not a whole life, just as destruction of many species of plants to cultivate those of use to humans is narrowly shortsighted.

Although I argued that being a moral agent is more important than excellence at any one or even set of virtues, this should not be taken as an argument against virtue. Rather, character is a whole that is greater than its parts, as the philosophers of virtue recognize. A person without the wisdom to judge when the virtue is appropriate is not virtuous in the greater sense. Individuals without discipline can know what is virtuous, but are less likely to act virtuously. On the other hand, we sometimes see examples of virtue from unexpected sources: young children and those who are uneducated but pure of heart. Thus while the virtues may be a sign of character, we should not miss the forest for the trees. Character is more than just the sum of its parts. Character must be whole, must not be fractured, or the person has not achieved maturity, the condition of autonomy. Those without character are not moral agents. However, I am not arguing that the virtues themselves are not required, for they are the parts of character.

Universality is morally required for its lack would come at the expense of hypocrisy. Applying one set of ethical standards at work and another at home is fraudulent and hypocritical. Moreover, I mean ethical standards in the wide sense, that includes both the person and their acts and a wide scope for what counts as ethical situations; and ideally, the best in each circumstance, that is, excellence. A "whole" life will apply similar standards in the distinct ways of being a human as a social creature, an economic, psychological, intellectual, and creative one. We must value things, material and economic, by the same standards by which we value, for example, art—if not as intensively (a matter of taste and choice). As long as we apply a consistent set of standards we are "whole"—we have integrity.

The problem has been that these standards and virtues have not been applied consistently. We have not applied them to other species, much less the environment as a whole. On the contrary, humans as a species would achieve moral maturity when they actually behave like most other species and do not destroy for the sake of sport, fun, and other immature rationalizations. No other species "hunts" for the sake of "sport"; nor tortures animals for juvenile "fun," nor exterminates plants and habitats for self-indulgent nesting: "suburbia."

In terms of the relation of parts to wholes, we are not whole without an environment. Even the beauty of a beautiful character requires an environment in which the individual can thrive. The larger environment is the condition of human life, not vice versa, and thus we are only part of a larger whole. Thus the ethics of character needs reworking to reflect the actual relation of humans to their world, a cosmopolitan reform.

What duties and virtues have in common is a concern with the other. Duty is to others, an external, not an internal relation, like that of the self-regarding or self-interested subject. Philosophical ethics is the attempt to universalize the other as beyond the tribe—ethics is not confined to the in group. Philosophical virtue is cosmopolitan. But anthropocentric notions of the other should be expanded. An environment is the condition of any society. In an ecocentric cosmopolitan ethic, the other is revalued as the habitat, the biosphere and the non-human world, because these elements are required for society, which in turn is required for human life and flourishing. Moral character develops within an environment and requires such an environment. Virtues, like duties, occur in the context of an environment and virtue is primarily ecocentric. For societies come and go, but their need for a habitat is a required principle. Ecocentric cosmopolitanism includes the biosphere as a whole and its various elements as the larger "other." We will revalue the virtues in this light.

Duty can follow from virtue. A virtue is a state of character: in instances where courage is called for the individual who is courageous can generally be counted upon to be courageous.

Justice, a virtue, entails specific duties, as does courage, for those who possess it or aim for it. Thus, as Dewey argued, following Kant, Plato, and others, the separation of duty from virtue is at best artificial and at worst a moral travesty.[68] It is our duty to be virtuous and a virtue to do our duty. Virtue and duty are properly seen as reciprocal elements in moral agency. A moral agent has a good character and is therefore inclined to do his or her duty. Likewise, a moral agent is dutiful and thereby disinclined to avoid self-improvement. As Kant argued, we have a duty to develop our talents and exercise virtue. Above all, a moral agent will not shirk their duty to their own environment out of laziness, greed, and other false idols of contemporary consumer society. They will see the greater good that includes the larger whole and expand their outlook accordingly.

6. The Revaluation of Virtues from an Ecocentric Standpoint

What would constitute ecocentric virtues? We could say redefining virtues should reflect the sense of them as being parts of a whole. In this sense, the virtues have been too narrowly defined, since they have been primarily cultural, not cosmopolitan. They should be revalued as parts of a much larger whole. However, there is also the sense of redefinition as an expansion, in accord with moral progress. Redefining the virtues with a cosmopolitan purpose is at the same time an expansion, for it includes a wider, meta-cultural world, and includes consideration of the wild. Thus whether ecocentric cosmopolitan ethics is a revaluation or an amelioration in the sense of moral progress, they will be broader in scope.

I noted above that some virtues are universal, considered parts of a good character in all cultures. Thus they are a good starting point, since they already have a kind of universality, at least within the "in-group." Another argument for universal virtues is deriving them from different kinds of practices. Most virtues are such in relation to a category. Bravery is virtuous given the end of a state to be militarily defended; liberality in view of economic prosperity; wisdom in view of truth, and so on. Thus a theory of practical categories could provide a principle, in the sense of origin, for a set of the kinds of virtues, whose cultivation would tend towards worthwhile goals by instilling character patterns. Each kind of activity has a corresponding virtue. Meanwhile, evaluation would provide a hierarchy for the virtues—for example, moral virtue over economic.

The universal virtues include wisdom (knowledge), temperance, courage, justice, truthfulness, spirituality or piety, and generosity. To this I will add trustworthiness, as a condition of all the virtues, a point I argued above, and respect. Two virtues are most in need for an environmental ethics and thereby of particular importance, moderation and wisdom. What consequences would such revalued virtues or do such virtues have on such vital environmental issues as preservation of habitats, of species, overpopulation, and Global Warming?

Ecocentric cosmopolitan moderation, or temperance, requires a major revaluation. Moderation was such an important virtue to the ancient sages that for two of the most famous, Aristotle and Kong fu-tse, it was the key to the virtues. Virtue could be defined as being moderate or following the mean in all actions. Redefining or expanding moderation must include self-imposed restraints of all kinds. These include especially moderation in consumption of goods, since production of all goods has an impact on the environment. Even non-organic materials may involve habitat destruction, notably mining of metals, coal, and others. Thus the consumer life style, in which purchasers buy all sorts of unneeded goods, with little thought to their environmental impact, is the opposite of virtue. Nor are such goods preserved, but are tossed out even before they are old, creating mountains of trash. Such a life-style involves the cultivation of desire, not character. Virtue has, in other words, been sacrificed to consumption; moral, to economic values. Virtue in general is counterpoised to slavery in the form of slavery to desire, by Plato. Virtue is the quality of a moral self in *control* of desires, not their slave, a person who is self-disciplined. What this says about our society I will point out below.

Moderation cannot mean an automobile for every member of a household, especially in view of the known contribution of autos to global warming, pollution and other environmentally destructive consequences. Nor will virtuous citizens seek to have a huge house occupied by only a few people. Such extravagances will rightly be seen as the sign of deficiencies of character: vulgar, tasteless, tawdry, and as compensation for personal deficiencies: showing off of wealth out of insecurity. Since a vegetarian diet has a lower environmental impact, it is also virtuous. Estimates are that it takes eight pounds of grain to produce one pound of beef. If humans consumed that pound, instead of feeding it to cattle, they could take a great deal of land out of cultivation and return it to wilderness. Cattle also produce methane during rumination, a major contributor to Global Warming.

One objection to anti-pollution arguments could be that other species dirty their own habitat, so it is "natural" to do so. But this cannot be a norm for humans, unless they are to live like animals. General Electric's polluting the Hudson is not comparable to pigeons fouling everywhere in any case since pigeon droppings are recyclable.

Moderation must also include restraints on the number of children each couple produces. Having too many children has a deleterious impact on society.[69] Worse, it exploits the environment that conditions the lives of such children. Bringing too many children into the world is to condemn them to a life of misery, if overpopulation continues. Nor will people of character seek to use dangerous means to increase the number of people in the world: cloning and the like. Their temperance will keep the desire for children in check. Having too many children exhibits the vice of immoderation.

Overpopulation and the resultant attempt to feed too many people are major contributing factors in the destruction of other species and their habitats. A person of character would not need to be persuaded of the inherent value of other species. A virtuous person would realize the intrinsic good of other self-actualizing agents as well as the distinctive intelligence of animals. They would recognize that humans are one species among others, not especial.[70] They would be cognizant of the threats to other species humans represent as well as the holistic interrelation of species and habitats in every environment, symbiosis, and equally the importance of biodiversity. They will shun products that torture or even use animals for testing cosmetics and human medicines, knowing that tests for humans on animals are scientifically worthless. They will recognize the importance of preserving the wilderness for its own sake—just so that there will be some unspoiled wilderness.[71]

Virtue would involve choosing goods that have less of an impact on the environment. This might include stone and metal dwellings instead of wood, goods made of recyclable materials, sustainable energy sources, and many other selections. Virtue (excellence) would include not wasting despite a prosperous condition. For although a prosperous society should be able to afford some waste, it would nevertheless be better, more virtuous not to waste. A person good at putting seeming waste to good use is acting even more virtuously: thrift.[72]

Moderation is vital in counteracting the pervasive effects of the consumer "society" or instead the way of life promoted by a misplaced emphasis on "economic growth," or more accurately, greed. The consumer way of life does not promote virtue, especially temperance, but cultivates infinite desires. Advertising is used to create artificial "needs" and promote a wasteful and extravagant life. The virtue is not indulging ourselves, but controlling our appetites, whether these be a desire for wealth or for any other good. Only a corrupt, decadent society is ruled by appetites, since the normal, civilizing influences have not been sufficiently instilled.[73]

For example, many chief executives earn as much as 400 times as much as the average worker.[74] Their job consists in squeezing even more productivity out of their staff while reducing such benefits as health care and eliminating retirement pensions that were paid in good faith by their employees. Such salaries are hardly a sign of moderation or generosity. The executives are rewarded for moving jobs out of the country, with the motive of paying even smaller wages. Worse, such hatchet jobs are admired in several circles and these individuals are made into celebrities! Such an outcome is only possible in an age that valorizes thugs on television, in movies, and other media.[75]

The weakness of character in consumer society is revealed in the addictions so common in modern society: gambling, cigarettes, alcohol, and the like. Some citizens complain about taxes and then proceed to throw their money away at casinos for the thrill of it, lubricating their vice with alcohol. The casinos never lack for business, while the schools have to hold bake sales to come up with enough money. Many use leisure time, which our ancestors had to fight so hard for, in decadent dissipation, not self-improvement. These vices reveal the underlying values of consumer society: consumption of anything, so long as there is consumption. Thus the worth of life comes from outside the individual, in the form of consumables that make life "worthwhile." Boredom and apathy are the hallmarks of this way of life. Life itself holds no worth in this model, without outside crutches.

The desire for a lot of money traditionally meant wanting to control necessity.[76] We can judge it an inferior life, ruled by necessity, not freedom. It is not the pursuit of future possibilities and novel creations, but a reaction to want, thereby determined by the past. With the advent of the "affluent society," the desire for money has meant controlling the means required to pursue a life of vice. Many in our society complain about their income and then use disposable income to buy intoxicants, and other forms of vice. In other words, they want more money to pursue their vices. This sea change is hardly in pursuit of virtue.

Another area of their lives where self-indulgence has won out over virtue has been in sex. Even priests and other clergy, who should set an example for others, ignore their vows and prey on the young, sometimes even children. Fundamentalist ministers rant against fornication and adultery and then some indulge in sex with prostitutes. They thereby contribute to destroying the moral fabric of society, since their actions defy their words, undermining moral principles. Such self-indulgence is not moderation, and exhibits neither character nor virtue.[77]

In sum, desire and greed are cultivated by consumer society, not moderation. Moderation in the consumption of goods is both more virtuous and an environmental necessity. The revaluation of moderation in an ecocentric cosmopolitan ethic will mean radical changes in contemporary life-styles among both rich and poor.

Ecocentric cosmopolitan wisdom would involve critically evaluating of all actions with respect to their environmental impact. All actions would be subject to wise moral restraints for a person of character. Wisdom would thereby come into its own, since it would be refined by its revaluation as well as by expanding its role.[78] Wisdom in the relation of humans to their environment would include wise use, but also the wisdom to know the limitations of human understanding, and thereby avoid the vice of excessive pride and stubbornness. The idea that we know best in "managing" the wild is preposterous, since the wild got along fine for 99.9% of the time that life has been around without any human intervention whatsoever, much less management (Ian Smith, 2010).

Despite the fact that they are the best educated generation in history, and that their society put huge amounts of money into providing them with a virtually free education, contemporary consumers often prefer drugs and self-indulgence to self improvement, to development of virtue. They have been too lazy to do serious intellectual work, so they let others do their thinking for them, following crackpot television personalities who echo corporations and promote the consumer life-style.[79] Indeed, their slavish watching of television has meant the virtual end of culture and its replacement with entertainment: "consumer" culture, an oxymoron. A genuine culture uplifts; "prole-feed" is not a replacement for culture.

Often, contemporary consumers have not used their education as insights or tools for use in public affairs, or indeed, become involved in politics or civic affairs at all, preferring partying to civic virtue. They allow evil to flourish unchecked in the political process with their political indifference, and by not working for public campaign financing. They ignore the evil of open bribes, in the form of massive campaign contributions, as the engine of the political process. They know about these evils, but despite their cynicism, they have done little to change the status quo. Activism would be more bother than sitting in front of a television. Civic virtue is carried on by the few activists left from what was at one time a huge movement of reform. Aristotle argued that good men enter politics to avoid being ruled by bad men. Unfortunately, despite the wisdom of this observation, it has been lost to television sound bites.

Wisdom is needed as never before to guide the changes that will be necessary in life-styles. Only a fool could believe that we could go on creating more greenhouse gases, by driving bigger and bigger cars and buying larger and larger houses. But the issue for an ethic of character is, what does consumption say about the mentality of the consumer? The underlying anxiety of the pursuit of status has been remarked upon many times, as has the urge to conform by "keeping up with the Joneses." The mentality does not reveal character, but its lack: self-indulgence and lack of autonomy.

Generosity is redefined to acknowledge the environmental impact of excessive consumption. The movement should be away from consumer society and the type of person it has generated. If, indeed, we can call a generic unit of consumption a person: someone who is completely defined by others. The conversations of such persons are not cultured but consist of evaluations of different consumer products that the "person" could do without and would be better off without. My grandmother once said, "ownership is a responsibility." Owning excessive goods requires care of such goods and thereby diverts time and attention from more worthwhile pursuits, including instilling character in children.

Generosity is a virtue only in a hierarchical world. If moral equality and equal rights were taken seriously, there would be little need for generosity, since everyone would have an equal share. Thus perhaps generosity would disappear in a virtuous world, where persons of character would replace freeloaders and exploiters. One can only hope. Perhaps generosity would mean taking the time to educate well-meaning consumers on how humans got along for centuries without all the economic toys that they feel they "need," and thus how little such trifles are really needed.[80] Indeed, the focus on generosity in all the religions of the West stands in contrast to the greed of many ostensive believers in our generation, who give possession of consumer goods priority over generosity, while attending church: confusing the ritual with the spirit of their faith. Another form it could take is redistribution of incomes so that equality is the norm, not the exception.

If "charity begins in the home," then concern for future generations should take precedence over self-indulgence. Gas-guzzlers and mansions should be sacrificed for our descendents. However, we should not limit charity to our own species; generosity to other species is also a virtue, exhibiting true generosity of spirit. True generosity is preserving species and their habitats from human use, exploitation, and above all destruction, even if this involves fewer goods. This would extend generosity beyond one species to where it is truly needed. Sharing affluence within a responsible environmentalism is still a mark of character. However, generosity to those less well off should be limited to those who have shown some regard for limiting births. Generosity to freeloaders is hardly a virtue.

Ecocentric cosmopolitan justice would be revalued by redefining and expanding norms of justice to include justice to the wild including other species. This would involve distributive, rectificatory, and other kinds of justice. The unequal distribution of land and territory between humans and other species would have to be corrected: redistributed. Species whose habitats have been taken would be due justice, requiring rectification; restoring habitat is a duty of rectificatory justice.[81] Individuals of other species would be treated with reciprocal justice. Major projects could be initiated, such as restoration of the tall grass prairie on the Great Plains of North America, and of the jungles of Amazonia. Since neither is suitable for agriculture in the long term—the plains lacking sufficient rain, and the rainforest a deep soil cover—these rectifications would come at little expense to humans. Justice means doing justice to the demands of the environment as a whole.

Another issue is the tradeoff of higher employment with a reduced standard of living, by contrast with lower rates of employment with an increased standard for some. Previous generations were willing to accept a lower standard of living in order to pursue other values, particularly the raising of children. Now "latchkey" children are sacrificed to mommy and daddy's ego, so both can work and live in a wasteful suburban eyesore, with far more space than is needed, and whose heating and air-conditioning contributes disproportionately to Global Warming. Conspicuous consumption has triumphed over virtue. This tradeoff is not confined to market economies: it is a moral issue.

Selfish individuals of the middle class put the burden of fighting wars on the poor, who often have few options. The baby boomers tried to claim the moral high ground, proclaiming that the Vietnam conflict was wrong, the war, "immoral." But they were far more concerned with saving their own hides than moral principles. After the war was over, they did not attempt to carry forward the moral crusade, put an end to militarism, and reigning in the military-industrial alliance. Instead, they joined the very bourgeoisie they had heretofore criticized. The vice corresponding to lack of justice is injustice and its corresponding lack of principle: hypocrisy.

Social justice means fighting for more equality in society.[82] Instead of attempting to realize the ideals of "harmony and understanding," to speak nothing of equality and cooperation, democratic virtues, contemporary consumers suck up to hierarchies in corporate society, so they can pursue status symbols such as "supersize" houses and autos. With the naivete of children, they suck up to celebrities—including royalty! —instead of pursuing virtue and self-improvement. Movie star celebrities are their heroes, not reformers and social activists. They have in effect endorsed social hierarchy, not the autonomy that is the goal of a more equalitarian society. They have embraced evil, by allowing species genocide.

"Freedom" has meant the freedom to pursue our self-interest at the expense of others, whether in a consumer life-style or in so-called identity politics. Some women, who rightly rebelled against a second place role in society, have, like many other groups in our history that have achieved success, left reform behind to live in suburban affluence. African Americans rightly put an end to discrimination but some fail to see that their social conservatism on issues of gay rights is of a piece with discrimination against any other minority. In turn, gays have often pursued self-indulgent affluence, ignoring social justice to speak nothing of the environment.[84]

The rationalization that they do it for their children is absurd: if they were really worried about their children, they would work to prevent Global Warming, destruction of wild habitats, and loss of biodiversity. On the contrary, their children and grandchildren will suffer the consequences of their self-indulgence. Consumers have let the environment go to hell, using air-conditioning, which has resulted in holes in the protective ozone layer; V-8 engines, which wastes fuel to a ridiculous degree and thereby denies it to succeeding generations; and suburban sprawl, eliminating habitats and living space for other species. They have chosen careers over children; self indulgence over their moral responsibility to instill character in the next generation. Generally, such careers have the goal of more affluence and thereby more status: bigger houses, bigger cars with a bigger negative impact on the environment. Again generally, they watch sports while letting a generation grow up physically unfit and overweight with a combination of lack of exercise and bad diet. Justice to future generations would require much more exercise of virtue.

Similarly, so-called "greens" whose agenda is really social justice, not environmental justice, are actually allies of capitalism. They wish to divide the pie differently, but the same pie. They are blind to the fact that if we do not prevent Global Warming and habitat destruction, there will be little pie left to divide. They do not deny the economic ethic of greed/self-interest, but only add the caveat that it should be spread out to include more of the human species, excluding all other species. In other words, they have little interest in the environment as such. However, if we do not expand the concept of justice to include justice to the environment—to other species and their habitats—then we are not progressing morally and are not really interested in universal justice, only anthropocentric justice.

The philosophers of virtue adjudged courage a moral virtue. Although courage appears most appropriate in combat, this application is dated. This moral genus stands apart from any one situation like battle. *Moral* bravery, as opposed to military bravery, consists of doing what is right in the face of social pressures to conform, or to go along with what is evil. Cowards will look the other way, excuse, equivocate, and accommodate evil. To ignore hostility, sarcasm, threats, ostracism, and the anger or even violence of others for the sake of what is right is the true mark of courage. Courage in this sense involves overcoming fear of the group and its pressures on oneself. Indeed, we should treat esteem and popularity critically on moral grounds. If good character is not popular so be it. Then unpopularity is a badge of moral distinction.

It will take courage to confront the necessity for limiting human desires, both for too many children, for excessive affluence, and for the use of environmentally pernicious technologies. The courage must be exhibited to others—confronting their wrong-headed affluence with the crisis of Global Warming and overpopulation—and oneself. Heroes in the future will focus on saving the environment, not humans, who in their billions hardly need saving. It will be preserving the wild in the face of the endless expansion of human domestication for more food to feed more human mouths, the ultimate in self-indulgence: having too many children.

Ecocentric cosmopolitan courage revalues and thereby expands courage to include standing up for the right in any context. It includes courageous acts of defending the underdog, especially helpless plants and animals. It would also include endurance in the face of adversity, the courage of our convictions as exhibited in practice, that we do not abandon the cause of the environment in the face of hostility, or the indifference of an apathetic majority. The courage to continue to stress environmental issues in the face of indifference and hostility, and to do so even to friends and family is virtue in the best sense (compare diligence and persistence in Kong fu-tse). Moreover, courage will be called for in confronting those from other cultures who insist upon their right to overpopulate. Objectively, they are wrong, for the earth does not have an infinite carrying capacity, to speak nothing of the effect of such freeloaders upon responsible moral agents, who voluntarily limit births. People with character should not have to pay the price of irresponsible breeding, regardless of cultural and religious rationalizations.

Consumer culture promotes the vice of conformity, not the virtue of standing up for what is right. Many of our institutions actively promote this vice, where they do not cheer it on or acquiesce in it. This is hardly virtuous.

One hard case for ethics is the status of truth telling. A classic case is provided by medical ethics: should we tell a "white" lie to a dying person in order to possibly save their life, or, as a kindness, to ease their death?[83]

The issue for an ecocentric Cosmopolitan Ethic is the place of truth-telling in this ethic, the issue of whether truth telling is absolute or general; a duty or a virtue.[85] Deceit about the environment comes in two forms in contemporary society. One is outright lies and deceit. An example is the attempt, funded by the oil companies and drummed up by the right, to discredit Global Warming as a "hoax." The other is the oblivion and blindness of the average citizen on issues of the environment. The first is outright falsehood, but the second, like medical professionals evading telling the truth, involves a kind of deception, whether that of the media or perhaps a kind of self-deception.[87]

The issue of truth arises for an ecocentric Cosmopolitan Ethic, because lies and evasions harm the environment. Truth telling is a duty, since we should tell the truth about Global Warming despite the fact that it might be "inconvenient" to do anything about it. But it is also a virtue to courageously fight for the environment despite the power and wealth of the brownlash, reflected in media blackout and public ignorance and indifference.

But can we as ecocentrists exaggerate, for example, the effects of Global Warming to induce people to abandon their lethargy and indifference and start to reverse Global Warming?[86] I think that such shaving of the truth about the environment will have bad long-term consequences, as well as being intrinsically wrong. It is wrong as a misstatement, a half-truth. In the long term, it will harm the cause of the environment, as people realize that the movement exaggerated, and the consequences they envisioned have not come about. The public might come to mistrust statements by environmentalists and ignore future calls to action based on future truths. Moreover, environmentalists can hardly appeal to the truth of their statements and lie or exaggerate at the same time—moral hypocrisy. I am aware that predicting consequences is difficult, and therefore that predicting the exact consequences that will occur due to Global Warming is also difficult. People may actually believe that a specific consequence will occur that does not. Such qualifications should then be incorporated into environmental literature. We should openly admit that our projections are estimates and uncertain, even where they are based on the soundest data, analyzed by the best experts. It is as virtuous to admit intellectual limitations as to have the courage to speak the truth. While such nuances may make the battle for public opinion more difficult, as brownlash opponents exploit every opening, they will be better in the long run.

Thus I do not generally believe that subordination of truth to other values or virtues is morally right. There may be desperate circumstances where lying is justified—hiding Jews from the Nazis, and denying that we are doing so comes to mind.[88] But normally lying is counterproductive in the long run and wrong as such. Subordination of the truth to "higher" values like life is based on a *summum bonum* model. But reciprocity of values means that truth has as much value as other values.

Moreover, enough consequences of environmental degradation are already known that environmentalists have plenty of valid information for their case as it now stands. There is no serious scientific criticism of the fact of human warming of the atmosphere, nor of the long-term consequences of rising sea levels and changes in weather patterns, even if the exact amount the oceans will rise is nebulous. Two of the United Nations are already close to being underwater and some of their inhabitants have had to be evacuated from the lowest lying areas. While we do not know the precise threat to all life that Global Warming will cause, we do know that it will be considerable. The evidence for Global Warming is so overwhelming that only a fool would deny it. Moreover, our descendants will pay the price for our malfeasance, to speak nothing of threats to other species and the loss of biodiversity. Clearly, some species will go extinct due explicitly to Global Warming, for example the threatened polar bear and the various penguin species.

Similarly, the effects of acid rain in killing all life in some lakes of northeastern North America and other areas of the world, are beyond dispute.[89] The cause is equally indisputable: energy generation, smelting, and other human activities. The pollution of rivers, lakes, and even the oceans through the attempt of corporations and governments to externalize costs by pouring pollutants into the water is well documented, as is ozone depletion. We know that spent nuclear fuels will be "hot" for thousands of years and thus a threat to our descendants. Similarly, salinization of soils and erosion are threats to future generations, while destruction of beautiful landscapes will decrease their quality of life.

Truth telling is not only a virtue in intellectual life (as it might be in a love affair), but absolutely required for all intellectual life—a moral duty. Truth is a moral value, not only an intellectual quality, the condition of all intellectual life, science, and scholarship.[90] Universalized lying would destroy society not just intellectual life. However, the revaluation of such a virtue would be in terms of telling the truth about environmental damage. Truth telling is the refusal to compromise with humanist rationalizations of the human destruction of the environment. Deliberate distortions, as with Exxon's deliberate attempt to mislead the public about Global Warming, are wrong. Denying Global Warming on grounds of a political or economic ideology—libertarianism or the like—is comparable to denying the Holocaust, or the denial of any wrongdoing by the government of the Soviet Union by some Marxists. Cosmopolitan virtue is telling the truth despite attempts by selfish interests to distort, to defame the character of such witnesses to the truth as Rachel Carson, and generally to get away with evil.

Ecocentric cosmopolitan piety would be revalued by defining and thereby expanding the notion of piety in terms of reverence for the environment. The "holiness" of life in some religions could be a model for such a revaluation. Children could be trained early to revere life, and the disgracefulness and immaturity of harming helpless creatures. Moreover, piety could be the basis for human stewardship over domesticated animals and plants. However, "human" stewardship is tied to an anthropocentric view of human dominion. Ecocentric piety is precisely reverence for the environment as the condition of human life, and thus as larger than us all.

Ecocentric cosmopolitan respect could be revalued in terms of leaving the wild alone: respect for the wild as the condition for our own: Paul Taylor's "respect for nature." Such respect would include human habitations as well as other habitats, that is, habitats in general. The diversity of such habitats and of species would be respected even though we might never come in contact with such species.

The key to the virtues in Aristotle and Kong fu-tse was moderation, since they both advanced the doctrine of the mean. By contrast, the key to virtues in

Plato is wisdom, which is required to know what is virtue. To these I would add trustworthiness, as a condition of all the virtues. The virtue of the whole is a trustworthy character. For a person with character can be trusted to act rightly on his or her own, out of public sight. Can the person be trusted to act well regardless of the situation? If so, they have a good character. Can we trust them to tell the truth, to have the courage to stand up for what is right, to be moderate in their desires, to be reverent and respectful? If so, they are moral agents, and not freeloaders, criminals, exploiters, and others who use other persons, as well as the environment egoistically, for their own selfish ends.

What of cultural virtues that are not universal? These may well be good, even if they are not universal. They may still have cosmopolitan value, within a culture. Thus cosmopolitanism does not exclude particular virtues, even if they are not universal virtues, since cosmopolitanism looks for the best in each culture. Kung fu-tze, for example, includes the virtue of perseverance. Such humane virtues can also be revalued in accord with ecocentrism. Perseverance could be reworked as standing up for what is right in the face of human destruction of the environment, despite setbacks. Another is benevolence, which is roughly good will towards fellow humans—echoed in Kung fu-tze's notion of humaneness, which includes benevolence. Benevolence could be revalued as benevolence toward all life, not just human life.[91] There is also the virtue of thrift, which is ecofriendly.[92]

Some would add caring, as in caring for children (Held). However, although this is universally required for human societies, caring for wild individuals is not an obligation as such, and would result in domestication, ending wild value as different, as I argued in chapter one. Caring for belongs primarily within families, not others, whether strangers or wild animals and plants. However, caring for injured wild animals and then releasing them is generally virtuous.

On the other hand, not all so-called virtues are worthy of cultivation. One such is the supposed virtue of loyalty. Despite its philosophical defenders, loyalty is the virtue of concentration camp guards, organized criminals, and other fanatics.[93] Even where loyalty is to good, it can easily turn bad. Only if loyalty can be tempered by other virtues and by duty can it be benign. But then it is not good in itself, much less the soul of virtue, as Josiah Royce would have it. We might leave such "virtues" to history except loyalty to environmental ethics.

In this chapter, we have revalued the virtues in an ecocentric direction. Moral agents act responsibly, and responsibilities are recognized to the environment as the condition of moral agents. Character is relocated from the self to agents in relation with an environment. Such characters have ways of life that involve interaction with their environment: a cosmopolitan ethic. We have thereby expanded the ethics of character development to include its hidden sources, a more holistic view.

Four

THE PROBLEM OF "RIGHTS" IN THE WILD

1. Wild Rights

In Aldo Leopold's famous book, *A Sand County Almanac and Sketches Here and There*, he examines two approaches to rights. Leopold mentions these two approaches without actually resolving the differences between them. The first is the notion of the extension of rights, based on moral progress. He notes that Ulysses killed his slave girls after his delayed return from the Trojan War. Slaves at that time had no rights. Since his time rights have been extended to larger and larger groups of people, from the nobility (Magna Carta) and property owners first to all white males, then to all males and later to women. One influential group within the environmental ethicists has adopted this model for the debate over animal rights and argued for the extension of rights to animals.[1] They have argued that rights for animals are similar to other civil rights issues and the struggle for animal rights resembles the fight for women's and minority rights. Leopold himself states that "The Land Ethic simply enlarges the boundaries of community to include soils, waters, plants and animals, or collectively: the Land."[2]

Leopold also outlines an alternative view, which he derives from his studies of the interaction of parts in an environment. In this philosophy, humans are one element in a larger whole that includes them. Humans are less the lord and master of nature than coinhabitants of an environment with complex and little understood interrelations of the parts within a whole. The parts include such non-organic elements as air, water, and soil, in which, combined with organic elements, each part contributes some essential function to make the whole work. Plants convert the sun's energy into organically available form. Insects pollinate the plants; herbivores feed off the plants. Predators keep the herbivores in check so that they do not overgraze. Microscopic organisms serve diverse functions from soil enrichment to aiding decay.

The place of humans is as symbiotes. Humans care for prized species and clear land to cultivate valued plants that in turn feed our species, whether directly or indirectly. Humans require the elements of the environment as much as any other species. Humans breathe air, drink water, and eat the food grown in soil, or animals that in turn do so. Human survival is therefore linked to the healthy condition of these parts and their interaction that forms and functions as a whole. Our survival is also linked to the survival of the plant and animal species that we use as food. Thus maintaining the land in a healthy condition is an imperative of human survival. Leopold thought that an ethical formula could be derived from these considerations. "Quit thinking about decent land-use as solely an economic problem. Examine each question in terms of what is ethically and esthetically right, as well as what is what is economically expedient. A thing is right when it tends to preserve the integrity, stability and beauty of the biotic community. It is wrong when it tends otherwise."[3] This principle alters humans' relation to the rest of the environment. "In short, a land ethic changes the role of *Homo sapiens* from conqueror of the land-community to plain member and citizen of it."[4]

Did Leopold realize that the Land Ethic contained two disparate ethical or rights theories, hierarchical and holistic?[5] The first, the extension of rights from humans to the land, is anthropocentric. Rights are grounded in human rights and extended from humans to the rest of nature. The latter, the Land Ethic proper, what is most novel about Leopold's approach, reconfigures both the human relation to the environment and the corresponding norms and values. If humans are just "plain citizens" and members of the land-community, then they cannot be the source of value. For they are part of a larger whole, not masters and thereby creators of values within the whole nor the source of values. The holistic model more accurately describes the human condition, since humans are both within and dependent upon an environment, however much they alter it. Humans depend on air, water, soil, and other species, and live within selected places, not beyond or above them.

John Muir stated that a spider has as much "right" to be in the wilderness as any human and thus for respect for the rights of wild creatures. Despite Muir's contention that a spider has as much "right" to be in the wilderness as any human, the extension of "rights," which is basically a legalistic concept, to the wild is untenable given predator-prey relations. The spider lives in the wild, but does not have civil rights precisely as wild. As J. Baird Callicott rightly maintains, the extension of the notion of rights, which is political and legal although based on ethics, to the wild, would be "ludicrous."[6] While we can give wild species moral considerability and even recognize an obligation to preserve each species, the notion of "rights" within the natural world is implausible, given the food chain and the necessity of predation. If a deer has a "right to life," then its predator must respect that right. It is unlikely that a lion or a bear could be made to follow such human norms. For were it to do so, it would starve. Thus the principle would be self-defeating and is thereby impossible.[7] A right to bear arms or choose a religion is even more absurd.

Does it follow that animals have no rights? I would argue, contrary to the position of Tom Regan, for the separation of the language of rights from morals proper. Although Regan and others have argued that the "rights view," as they call it, is, echoing John Rawls, an ethical view, rights are actually as much a political and legal concept as an ethical one. The notion of having rights evolved in political, not ethical philosophy. To be sure, rights are *based* on ethics: politics ought to be grounded systematically in ethical principles. Politics can be judged by higher, ethical standards, by a "higher law," or moral duties. Political idealists argued that politics should be regulated by ethical principles. Thus the main ethical theories: divine command, deontology, the ethics of character, consequentialism, pragmatic ethics—do not argue directly for "rights" but duty, value, the good life, and the like. Rights are based on such ethical notions: they follow from them. Legitimate rights should be based on moral principles, not only the law. The alternative would be a politics dominated by the realist standards delineated in the Machiavellian tradition, force and fraud. But civil rights are legal; the rights of a citizen were within a determinate *polis*, that is, a political order. They did not apply to a creature in the wild—despite the oddity of the attempt in medieval courts to haul animals into court in violation of some statute! "Universal human rights" are indeed confined to human communities. The mistreatment of other species is an ethical, not a narrowly political or even legal issue.

The notion of "rights" as political is complicated by the different senses of "right" in English. "Right" also has an ethical sense, that is, the opposite of wrong. We do our duty because it is the right thing to do. However, this ethical sense of right is distinct in meaning from the civil or legalistic sense of "rights", however much they have been confused or conflated. "Rights" were legal guarantees, not universal ethical precepts, values, or norms. "Rights" evolved out of the law and were guaranteed by the power of the state. A "right" is at first a correct action: "he was right to do" some action. Only later do rights become political and legal freedoms and claims, "I have a right to…" based on the propriety of that freedom. Thus there is a moral basis for legal and political rights.[8]

The model of the animal rights theorists is that rights follow from intrinsic value, not duties. Duties follow from rights. The question is why do rights follow from intrinsic value? For Cosmopolitan Ethics, intrinsic value is the basis of duty, or moral considerability, not of rights. Rights are derivative from ethics as applied to politics and thereby limited to the political realm. Our duty is to regard or morally consider a bearer of intrinsic value in our moral evaluations; from this value, duties are derived.

A problem here is that if we have a duty the other has a right to expect it. It could still be maintained that a duty implies some type of right, so if we have duties to the biosphere, then rights are implied.[9] Another problem is that duties and political rights are reciprocal, but duties and intrinsic value are not. The duty of non-interference with an intrinsically valuable agent is not reciprocal since intrinsic value is not a relation, that is, extrinsic, but intrinsically valuable, that is, by itself or in itself. John Dewey argued that intrinsic value is not out of all relations. However, the point is that the value is not derivative from any relation. This is why a better norm to cover this aspect of our relation to the world is Callicott's "moral considerability." Individuals of wild species should not be killed or harmed without good cause, because they are morally considerable. For a cosmopolitan ethic, moral considerability attaches to the other as other, for the other has a value of its own. Given a cosmopolitan environmental perspective, moral considerability can be extended to the wild, as a functional part of a whole. However, moral considerability does not mean "rights" in the political sense.

Each instance of a kind, a species, actualizes a kind and is thus inherently valuable, but not necessarily a bearer of rights. We don't have an obligation to it as valuable, although it remains morally considerable. Self-actualizing agents—living organisms—are morally considerable, but not rights bearers. They are intrinsically valuable, but do not thereby always generate duties, only moral consideration.

As Callicott has convincingly argued, talk of "rights" in the wild is an illicit extension. Consideration of the wildness of the wild rules out extension of political and social rights from civilization to the wild. Otherwise, the wild is domesticated and ceases to be wild.[10] We do not establish rights for wild qua wild for the wild is not subject to anthropological institutions. The wild is different—alien—other.[11] This is why a cosmopolitan relation is requisite for our relation to the wild. Just as we recognize the value of cultures that are not our own, we can recognize the value of parts of the same whole as our own that are not our own parts. We do not live in swamps but can recognize the role they play in the environment as a whole. We are not bees, but can recognize their value. Moreover, the extension of rights would constitute a politicization of the wild, bringing it under political norms. Politics would be totalistic, extending even to wild creatures with their own agenda. The life of animals would be treated under political rubrics, and the wild thereby dominated. But what is distinctive about the wild is that its flora and fauna do not behave in ways controlled or managed by humans.

Some might argue against this view that political philosophers have universalized rights, beginning with Thomas Hobbes's *Leviathan* and continuing with the Social Contract tradition up to Rawls. Rights are not tribal or ideological but apply to all humans as humans. Thus there is the notion of "universal human rights" which is a standard for judging particular governments, to determine if they are ethical. An ethical standard is applied to politics, namely human rights. I would not deny this standard, only its validity outside of the political sphere. No rights theorist has argued that such rights are more than a special application of ethics to one area, the state or the political order. We do not have rights when hiking in the wilderness and confronted by a bear. Instead, we have rights that are, or ought to be, guaranteed in our human associations, rights under the law. Indeed, the term "rights" is probably legal in origin and connotes a legal order, a government that makes and enforces laws, as Hobbes implied. If rights are universal, they are universal to political animals, humans. If we argue for human rights for persons outside our own country, we recognize them as members of another state whose political rights are either being violated or not being protected. While I would agree that rights should be ethically based, this basis is especially applied to the political sphere. Since the environment taken as a whole is outside the political sphere, even if affected by political decisions, rights do not apply to it. However, the environment is morally considerable, as the condition of all association, political and otherwise.

Moral progress is measured by the extension of rights in the first approach. Why does this constitute progress? One reason is that moral considerability is extended to more and more individuals, an increase in scope and quantity. Secondly, the extension of rights is of greater universality and thereby approaches equality: more consistent. However, this model is more appropriate to civil society than relations with other species. Rights require the environment, not vice versa. Rights are possible within the biosphere as a whole, and thus part of a larger whole, that predated it.

Value in the biosphere does not proceed from a first, from a hierarchical supreme value, as I argued in chapter one. The value of non-humans is not derivative from that of humans, whether the use value or any other projection or emanation. This point reinforces our critical evaluation of the extension of rights from humans to the wild. Wild creatures, whether animals or plants, have a value of their own, intrinsic value. This value is attached to their life, their membership in a species and to their continued survival (self-actualization). Since life is of a finite duration, it has a finite value, not an infinite value. Intrinsic value is not absolute but is not relative either, whether to humans or some other source of value.[12] Instead, life is finite not infinite as it requires self-actualization. Thus life has intrinsic, finite value. Moreover, its value is dependent upon a habitat, whose value may exceed that of any one individual as their condition.

We should develop a more holistic, cosmopolitan environmental ethic. Rights should be respected within civilization, but a more comprehensive ethic must govern our relations to other species and the larger whole of the earth. The issue of rights must be examined within an overall environmental ethic, as a special case. Humans are within a larger environment. Rights require reciprocity with symbiotes, that is, other species, and forces that make life flourish. Environmental ethics should include moral consideration of all the parts and elements that make up the biotic whole, including air, water, and soil, as well as other species. Individuals of other species are self-actualizing agents with intrinsic value as well as inherent value as members of that species.

The function of the species within the whole is relational and value in this systemic sense is functional. Rare species thereby increase in value, since their function within the whole is threatened. Were the species to disappear, its function might go unfulfilled and thereby threaten the whole. This has happened with some pollinators in Hawaii, where flowering plants grow that are hand pollinated by scientists, since their pollinator is presumed extinct. Individuals of more common species have less functional value, although no less intrinsic value as individuals, and thereby less overall value from a holistic point of view. Since their value requires their own effort to maintain themselves, self-actualization, it is finite, for it is in proportion to their actualization of their own lives. Death ends this intrinsic value, and in the wild, predation ends it. To put an end to the system of predation in the wild, the food chain, would be to put an end to the wild. Since humans depend upon the diverse elements that make up the whole in the wild, imposing human norms upon the wild would threaten both human and non-humans as a whole.

Joseph Des Jardins (2001) pointed out that we do not know the "function" of every species. But just because humans cannot figure out the function of a particular species does not mean that they do not have one (Ian Smith, 2010). Moreover, they can be classified in a rough schema, for example, predator, prey, pollinator, and so forth.

An ethic that considers the whole world not only takes the relation of parts to a whole as the model, but also considers all the elements that some twentieth-century ethicists tried to keep apart, namely, duty, character, and consequences. The ethic of duty is still seen by many as the contrary of the ethic of consequences and they are seen as incompatible. However, this was not the view of either ancient or modern philosophy. Plato's Socrates argued in the *Crito* for duty to obey the law, the virtue of doing so, and the evil consequences of not doing so, for example, the lesson or message it would send to his children.[13] Similarly, Immanuel Kant argued that we have a duty to be virtuous, in a part of his moral philosophy many commentators have neglected.[14] Maxims include consequences of actions as a consideration; the content of duties are ends. John Stuart Mill argued that several virtues were necessary for a happy life. Finally, Dewey argued that all these parts belong together in a larger ethical whole.[15]

No environmental ethic can ignore consequences to the environment. Nor can duties to the environment be dispensed with, and the moral self that is responsible for enacting them. A moral self is a self with character, who is morally cognizant. All of these factors are involved in any environmental ethic.

One problem with a limitation of the notion of rights comes from the tradition of arguments for a "right to self-defense." I argued that rights must be confined to human societies, following Callicott. However, several rights have been called "human" or "natural," for example, freedom and equality, and thus are pre-political, not positive.[16] If such rights are universal and cosmopolitan they are not political. But then the question of the "rights of animals" is reopened, since rights are not only civil, but pre-civil or supra civil. Again, I would argue that human rights are confined to humans, but that animals in a wild state are morally considerable as inherently valuable. Animals have intrinsic value and an instinct for self-defense but no rights in the human sense, since they are not politically organized, and must defend their own lives in the wild.

Against a "natural right" to self-defense I would argue that plants generally do not exhibit self-defense behavior. Of course it could be argued that plants often carry chemicals that act as deterrents to predation, concentrating heavy metals in the leaves of some herbs or even poisonous plants.[17] Some species of plants are even carnivorous, although the ability to trap and digest insects is not explicitly a deterrent so much as nutritional in function. However, the notion of self-defense requires, I would think, at least some form of consciousness or teleology. Either a creature must be aware of danger and try to avoid it, or must aim at its own preservation. Clams, hardly high in the scale of animal intelligence, tend to dig themselves deeper in the sand if they sense a predator. Whether this is due to consciousness at some level or mere "instinct" is moot. What is clear is that plants are not conscious in the same sense as many animals nor do they actively avoid predation, and therefore cannot be said to deliberately defend themselves. Do they have a "right to self-defense"—or only by analogy? For example, bark functions to "protect" trees from predators. But a function is not always identical to a purpose, although they can overlap.[18] The argument of animal liberation is that animals, who exhibit self-defense behavior, should be defended but not plants. But this does not create a right. So the question of rights may be gratuitous in the wild, and not universal. Moreover, animals fleeing from a predator are not, strictly speaking, defending themselves.

The hard case is domesticated animals, which humans use instrumentally. Since all animals and plants have intrinsic value as self-actualizing, domesticated animals are morally considerable. The harder question is whether they have rights: should they be protected by virtue of their being within civilization? These are questions for individual cultures, but at the least, cruelty is ruled out by moral considerability. Thus keeping animals penned up their whole life and other forms of mistreatment are morally wrong. However, in another respect, bringing animals within human society instills human rights of a sort to domesticated animals. Domestication involves human association with such animals and plants and thereby close social relations. Since ethics regulates social relations, it includes relations with animals. We have at least some duties to animals including, minimally, the duty not to be cruel or use torture. But then domesticated animals have at least some rights, since rights within civilization follow from duties.

Some would argue that animals are property and thereby can be treated however we wish. However, the kosher laws of Judaism indicate that the treatment of animals was an ethical concern for that quite ancient text, the Old Testament. The animals were to be slaughtered in a way that minimized pain, for example. Thus the ethical treatment of animals has an ancient pedigree. This principle is valid *a fortiori* for Hinduism and Buddhism, with their proclivity to vegetarianism.

What of cosmetic and medical testing? We can dismiss cosmetic testing easily. We hardly need another perfume with thousands already on the market. Torturing animals to test such products violates the norm of avoiding pain. The question we should ask here is: would we be willing to have such products tested on ourselves? This acid test of the golden rule should provide us with a compelling reason to outlaw all such testing: for the profiteers do not test it on themselves first, the only ethical approach, but use helpless rabbits and other species. But there is a pragmatic test as well, which applies to medical testing. The tests are almost all useless, despite the braying of the very profitable drug industry to the contrary. The results for other species will tell us almost nothing about the effects of such products on humans. Since the cosmetics, chemicals, drugs, and other products will have to be tested on humans anyway, there is no point in torturing animals by holding their eyes open and spraying the chemical in their eyes to test the products. The only way to test the effects of such products on humans is to use humans.

Another point here is that since the notorious experiments at Tuskegee, Auschwitz, and Manchuria, informed consent has become an ethical requirement for all research on human subjects. But animals cannot be informed and cannot consent. Thus their use violates such norms. The evidence we have from the animals' behavior is that animals do not like to be tested, since if their cage is left open they will attempt to escape. In other words, they are prisoners. They often cower in fear at the sight of their human masters and try to avoid the tests, and the pain of the tests. If animals are domesticated, and brought into association with humans in civil society, they have some rights, and we have duties toward them. Since their behavior makes it obvious that the animals do not want to be held captive or to be experimented upon, anymore than a human would, there is a strong case to be made against any such experimentation on rights grounds.

Another problem is the lack of a clear border between wild and domesticated areas; and of predator intrusion into ranches and farms in West. The reintroduction of grizzlies and wolves into Yellowstone National Park has come at a price. The grizzlies and wolves neither recognize nor respect the unmarked borders of the Park and have often wandered off Parklands into the surrounding areas. These areas are used as ranches by humans and contain valuable livestock. The predators, always on the lookout for a meal, have killed livestock, threatening the living and viability of the ranches. These intrusions complicate the issues. Do endangered species like grizzlies and wolves have "rights" when they enter areas appropriated for human use? I will examine the duty to preserve species below, but as for the separation of wild and civil land, I should note that I am a pessimist. Humans are unlikely to live with dangerous predators about, endangering livestock and children. No matter how large the areas set aside for animals such as bears and wolves, they would inevitably wander out of them in search of food or as an expansion of population. Calls for their elimination would then drown out voices demonstrating the essential role such predators play in keeping the land healthy.

However, this situation brings up another argument for vegetarianism. If humans did not have to commandeer large areas for livestock grazing, the problem of predator intrusion would not arise, or at least not so acutely. One advantage of poultry is that it requires far less grazing land than beef, and far less grain input per unit of meat. Human carnivores could take a step toward sustainability and preservation of species by eliminating beef and lamb from their diet. They could be even more virtuous by becoming vegetarian.

The examples of intrusions support the holistic environmental view, since they provide evidence for the artificiality of the wild-domestic distinction. The distinction of human, domestic, and wild is not absolute, since some wild species live in human dominated areas; while humans looking for scenery often build homes in basically wild areas. Wild birds, insects, trees, and many other families of wild species persist within human settlement areas and remain wild. Human settlements are not completely free of wild species. Moreover, the wild intrudes in air, water, soil, many soil organisms, and so on. Since the wild and the domestic interpenetrate, human expropriation does not affect all species equally, and may even benefit a few such as pigeons and coyotes, by eliminating or restricting their predators. But other species require wild habitat, including, tigers, spotted owls, and many plant endemics.

Moreover, I agree with Callicott (1989) and others that humans are part of "nature," that is, we dwell in the same world as other species and interact with them even in the least wild areas. Further, very few areas of the earth have not been impacted or occupied by humans (Antarctica is the main exception). Humans have penetrated almost all of the "wild" areas. Yet there is still a difference of degree between urban and "wilderness" areas: the extent of human penetration and destruction. "Desolation wilderness" in the Sierra Nevada has much less human impact than the Los Angeles basin.

Thus the distinction of the wild from human civilization, or domestic areas is somewhat arbitrary (Daniel Simberloff, 2012). Human settlements are a transformation of part of the wild that does not transform it entirely. It is not as if humans eliminate wild species and elements entirely in civilization: they do so selectively or inadvertently. Wild species may also migrate into domesticated areas, especially the more mobile species like birds and winged insects. They may also carry less mobile species in the form of seeds, microbes, and others. The domestic is a special case of the wild, then, and still a part of a larger whole. Value does not proceed from humans to the wild, but includes humans as part of a larger whole. Instead of projecting rights from humans to the wild, humans should recognize and respect the intrinsic value of the irreducible other: other species as wild.

This argument also answers the critics of the notion of "wilderness," who argue that if we designate it, set it aside, manage it, and otherwise interact with it, it is not "pristine." But humans interact with the environment all the time and are part of a larger whole, the biosphere. Wild land, as Scott Friskics put it, is primarily but not exclusively occupied by wild species.[19] Wilderness was there long before humans; setting aside preserves prevents human encroachment on land that wild species occupy. If one human traipses on wilderness, does this mean that area is no longer wilderness? To argue for an absolute standard of "pristine," would mean that there is not one square inch of wilderness, for there is virtually none that has not been affected by humans, as Bill McKibben argued. This is the view of those who favor development: since there is no wild land, we may as well develop it all. I would respond that "wildness" is a matter of degrees.

What of our distinction of rights and moral considerability, then? If the wild intrudes into the domestic and human settlements are a "part" of the wild, a special case, how can the area where rights are in force be distinguished in practice? Since the wild intrudes into the domestic, the distinction is abstract, and the area that can be considered "human settlement" is an abstract and in a sense arbitrary border, just as the borders of nations do not always follow geographic or environmental borders. Rights are attached to these abstractly created areas, namely the political realm. From another angle, can we claim that species have rights since they transcend the civil, domestic, wild distinction? Since humans and their domesticated animals are species and have rights, why not wild animals? And what of the symbiotic relations between humans and some wild species? As argued above, this would constitute quasi-domestication and thus cancel the distinct value of the wild. Humans would thereby interfere in the relation that other species have to each other. Moreover, it would be impossible, as Callicott argued. However, wild species have intrinsic value as well as inherent and systemic value as part of a larger whole, their habitat. While they do not have rights in the wild, they are morally considerable.

What we can say is that the best solution to human-wild predator confrontation is some expansion of the range for such species. Ranches and other human habitations in the vicinity of these ranges could be purchased at public expense to provide more habitats for endangered predators, both to preserve them and the habitat they protect. This will not solve the problem of human-predator confrontation, but it will diminish it from a holistic point of view. Enough individuals of the predator species will survive in the wild to continue the species. Another approach would be to capture rogue predators and return them to remote parts of the wild. This approach is already being used at Yellowstone and other areas. If too many predators wander off wild preserves, banding could be used to set off alarms on the borders and allow capture. Ranchers on the borders of wild lands could use trained dogs to keep the predators away from herds.

The problem of conflicts of intrinsically valuable agents is similar in form to conflict of rights. Such conflicts can arise even in an anthropological context. Whether we use rights or value, conflicts of kinds can arise. Moral pluralism does not solve the problem of conflict.

2. Moral Pluralism

Christopher Stone, in his book *Earth and Other Ethics* (1987), argued the case for moral pluralism in environmental ethics.[20] Stone argues that any notion of "rights" in nature means "rights" for trees, forests, and other non-traditional candidates. He also notes that an environmental ethic would involve conflicts of rights. However, as I argued above, and as others have argued before me, notably Callicott, the extension of rights to the wild is impossible, without destroying the wild. Moreover, conflicts of rights are no greater in environmental ethics than in anthropocentric ethics, for example in the conflict between life rights and liberty rights in the abortion debate, or life and quality of life in the literature of euthanasia. Both have hard cases and yet there is no reason to suppose from the problem of hard cases either that there should be a pluralistic ethic or that it could solve such hard cases. Thus we can agree with Stone that applying a single ethical theory to every case is difficult and that some cases are hard to decide.[21] But such problems are hardly confined to "moral monism," even if we grant that there is such a thing.

Stone also notes the conflicts between the ethics of self-interest and that of duty.[22] But this is hardly an issue for ethics, since, as I will argue in chapter six, an "ethics" of self-interest is an oxymoron. Starting from an "ethics" of self-interest begs the question, since in regard to this "ethic" we can raise the critical question formulated by Kant: whether actions taken from self-interest have any ethical content. Kant's point is still valid: moral principles cannot be derived from desire, self-interest or any other egoistic source. If self-interest were the basis for action, anything in my self-interest would be "right," presuming I could get away with it if I so wished—including Auschwitz, the ripping off of investors by the management of Enron, drive-by shootings, dragging African Americans through the back roads tied to a pick-up truck, acts of terrorism against innocent civilians, torturing animals, and so on. Doing what is right may, and often does involve going against our inclinations. Overcoming desires and feelings are often what give ethical value to an action. We must put aside our hatreds and enmities and resist the urge to murder, steal, and rape the land.

The realist response is that this is how humans really behave. But how humans behave tells us little about moral value. Humans massacred innocent Native Americans and kidnapped black slaves, but both were violations of moral laws that these Europeans paid lip service to. Ethics reminds us what we should do even if we do not. Ethics is a restraint on self-interest and desire, not its rationalization. The moral issue is where we draw the line between individual interests and rights, or moral duties to others.

The ethics of interests is realist: Callicles and Glaucon arguing for justice as the interest of the stronger in Plato's *Gorgias* and *Republic* I, respectively.

But this is hardly ethics: instead, they argue for power or economic strategy (compare Bryan Norton's discussion of "cost-benefit analysis" applied to issues in environmental ethics). As one author put it, "economics has dictated to ethics for too long already." Moralists from George E. Moore and John Dewey through contemporary authors have argued against different versions of John Stuart Mill's mistake, that from what people do we can derive a principle of what they should do. I am not completely persuaded by such arguments, but in the case of self-interest and desire, no strictly moral principles can be derived.

Stone starts from the law, since that is where he received his training. But moralists since World War II have almost universally rejected any basis for moral principles in the law, since the law may violate moral principles. Martin Luther King Jr. pointed out in the nineteen-fifties that while racial segregation was legal in the South, it could hardly be considered moral. Similarly, he noted that turning in escaped slave during the period before the Civil War was legal, as was turning in Jews in lands occupied by the Nazis. But this could hardly be called moral.[23] On the contrary, the law should be based in morals.

The theoretical issue is whether all moral problems can be dealt with by one set of principles. Theories unify a set of events, phenomena, or other content at different levels. Some theories cover and explain a small set, others cover a larger set, and comprehensive theories cover a trans-generic set.[24] Theories that do not succeed in explaining a set with a uniform explanation are revised not supplemented, unless the supplement covers a discrete set of contents; or they are replaced. Thus Sir Issac Newton's theories were replaced by Albert Einstein's; no one argues for pluralism in physics. Theories in ethics should be as revisable and improvable as in any other field, for this is one measure of moral progress. Indeed, Dewey called for "Reconstruction" in ethics as well as in philosophy, and thus the notion that ethical theories can be improved: meliorism. My own work includes an attempt to expand Kantianism in a pragmatic and ecocentric direction.

The main theoretical content and concern of ethics consists in duties, values, and their relation. All ethical theories agree on this model, differing mainly in their theories of good and right, and the relation of the good to what is right. Ethical theories compete: they do not supplement one another. Like physical theories, moral theories are universal, not covering particular content, but every evaluation, act, consequence and for rigorists, even emotions and intentions. Moral theories are trans-generic, and not only applicable to one society but all. Morals do not simply cover "conflicts," as Stone would have it, but are a view of how life should be lived, a way of life.

Universality is required in order to deal with the problem of culture, and the many different values that different cultures have pursued. The claim of philosophical ethics is that there can be a universal ethic that applies to all members of every culture as humans, that is, at the specific, not the cultural level. Despite the charge of some that ancient philosophy is rooted in one culture, at its best ancient philosophers attempted to encompass all groups. The notion of a "nature" of man, however naïve it may appear today, appealed to the universality of human species membership above that of culture. Thus the United Nations "Universal Declaration of Human Rights" links universality of rights to humans as a species, bypassing membership in a particular culture. In other words, as humans, individuals have such rights even against a culture. Such a view is consistent with a cosmopolitan ethic that recognizes cultural membership, but that we are citizens of the world first, since the world is the condition of culture.

Stone argues that hedonism does not apply to collectivities without realizing that this point constitutes a good argument against pleasure as a supreme value or hedonism as an adequate theory of value.[25] Stone's argument may be valid with respect to rights, since rights would be difficult to extend to the wild, but I argued above that rights do not constitute a universal ethic, but a more narrowly legal and political application of ethics. Again, he argues that different principles may apply in different circumstances, or to a different set, human and non-human.[26] But while we have what Robert Merton has called "middle-level" theories to deal with smaller sets, an ethical theory as a whole must provide a universal principle of some type to decide cases, as Callicott has argued.[27] While we might argue that both killing and lying are wrong, we might argue that killing is a greater wrong.[28] But we could still apply one principle, such as the Categorical Imperative, to different circumstances or contexts, as Kant attempted to do. We might say that both murder and theft violate the Categorical Imperative (or alternatively, the Greatest Good of the Greatest Number, and so on). The alternative is contradicting ourselves.[29] Applying duties to animals and other species is theoretically feasible: don't murder is equally applicable to other species as to humans. Don't steal can be applied to habitats of other species as much as to human property: don't steal the habitats of wild species, especially rare or endangered ones.[30]

The thesis of environmental ethics is that we can extend moral principles to non-human agents. However, this is not an entirely new extention, since it has always been considered virtuous to be kind to domesticated animals, to kill them in as painless a way as possible, and the like. Moreover, the universality of ethical theories is consistent with dealing with non-human species ethically. The problem raised by Stone and Callicott of ethical or moral "pluralism,"[31] then, collapses. Properly interpreted, there can only be one ethic, an ethic of universal inclusion. An "ethic" that excludes is ideological, not ethical,[32] as it cannot be universalized.

As I will argue in detail in chapter five, the distinction between "consequentialist" or teleological ethics, and "non-consequentialist" ethics, which has been overdrawn, is not a matter of circumstances, as these are factored into both moral theories by the universality of the major in moral deliberation. Calculating in terms of consequences as a rule *is* a principle as is its universal claim: what everyone ought to do in the same circumstances; the appropriate virtue under the circumstances. The difference lies in the moral framework: whether duty or virtue.

3. Situation ethics

Stone implies that we have to treat each situation as it comes and according to distinct principles. Thus he implies that he favors adopting a situation ethic. Situation ethics are radically nominalistic. Different situations differ by time, history and space (place), circumstances, relevant type, and so on.[33] In this view, ethical situations are never of a kind—thus no principles or norms can govern all situations. The view is a form of historicism applied to ethics, since each moral situation differs in some respect from preceding ones—radical difference.

With respect to the thesis of radical difference, we can argue that each situation is an instance of a concept, thus not individual or a unique situation. The situation is brought under a concept and thereby a universal.[34] In this view, pure situation ethics is impossible. The recognition of moral aspects in a situation: that something is "bad" or "wrong," implicitly recognizes a rule. It subsumes the situation (in the relevant respects) under an abstraction from past moral cases, or an *a priori* rule in its cognizance of the wrongful or bad aspects of the situation.[35] Moreover, if the situation governs ethics then situation ethics is itself a norm. If we must look to the situation to tell how to act, then a universal principle is invoked: in all situations we must follow the dictates of the situation. But then situation ethics defeats itself, since it offers a universal principle that transcends the particulars of the situation.[36]

All cultures have rules. Relativism does not fail on this score, although it does on others; it either does or does not subsume the situation: relativism's argument is that the case fits for some and not for others under recognized moral rules (or that the rules themselves are relative). It does not deny the rules as such. That the rules are culturally relative does not make them any less categorial: Western science may have arisen in one culture, but is nevertheless categorially necessary to achieve its fruits: public truth, and a semiotics for technology.

What is the role of principles in ethics? Dealing with a situation morally requires rules and principles. First, what sort of situation is it?[37] Classifying the situation requires rules and norms: rules of classification and norms or conceptual rubrics under which to identify them. Then there is the matter of moral classification: what is the right or wrong involved in this situation? There is the issue of what moral principle to invoke: categorical, Utilitarian, or some other approach. Finally, we have the principles of practical reason. Principles in ethics are practical: practical reason, whether of the means-end or rule-instance type. Since valid reasoning requires universals, it requires principles.

Are there universal moral principles? Kant's ethics is correct as a theory especially insofar as it recognizes the role of principles, universal norms, in moral reasoning, over "intuitions," or "feelings." For Kant reason is the faculty of principles. Feelings and desires reflect social conditioning: they are learned from our social circumstances. As such they cannot attain universality nor transcend the society: they are uncritical. Thus they cannot be used critically to analyze inferior social practices.

Kant's moral problematic consists in deriving rational ends, which are not empirically mediated. Nor are they based on natural law, and understood by reason, as in the Stoics.[38] Instead, he derives the basis of morality from the principles of reason *per se*, in the form of consistency. The Categorical Imperative uses the form of rationality as the basis for a formal moral principle, namely universality. Universality as a formal requirement of Kant's ethic gives necessity. We *must* do what is morally commanded as a consequence of universal reasoning: the conclusion is practically compelling (as opposed to the contingency of hypothetical imperatives). Since reasoning requires at least one universal, situations cannot be dealt with rationally without a universal element. Self-interest alone cannot provide one, based on a singular. Self-interest can only end in relativism.

Morality consists in rules governing unique situations with a tenuous logic: action must be consistent with moral principles. We invoke rules as covering a situation according to the situation. The rule must fit the situation, but must also govern it in some sense, treating ends as an instance of a universal. The rules may only limit action but may also define actions appropriate to a situation. In some cases, actions are not inconsistent with moral principles, and so are allowable. In other cases, moral principles require us to act against our self-interest. Reasoning of the means to end type, the "practical syllogism," is governed by such moral principles. Duty is following such principles—and their creative actualization as actions. Voluntary principles are recognized as obligatory.[39] Obligation is necessary (the ground of possibility) for other values, for example, environmental ethics is necessary for all human values.

John Dewey emphasized the elements of the situation as having an important bearing on moral evaluation.[40] Dewey would like situations to "rule" our actions, solutions, decisions, and means.[41] Above all, Dewey criticized the traditional understanding of the relation of theory to practice. In practice, we have to deal with concrete situations, for which theories are only an instrument. However, Dewey also recognizes the role of previous rules, cultural norms, and overall goods as aids in moral evaluation. He is not truly a situation ethicist. What he at times ignores is that situations and problems have a *likeness* as well as a difference—categorial and cultural connectedness. This is not only convention: the case has its own "*logos.*"

Cases come under rules against stealing, lying, murder, and so on. Nevertheless, we can agree with Dewey that ethical practice interacts much more with theory, and affects it. The recognition of the rights and value of previously marginalized groups, moral progress, involves dealing with the historical situation we are confronted with in creative ways. Theory also evolves to some degree as we progress from naïve to more comprehensive ethics, and take impact on the environment into account.

One issue is Dewey's method in relation to his principles. Is every situation to be treated as an experiment? Would this be practical? There is the related issue that morals are learned as rules, not in terms of the uniqueness of the situation.[42] The situation comes under the rule; exceptions are a matter for deliberation. Exceptions can even be described as coming under another rule. Indeed, Aristotle's long list of virtues can be seen as situational: virtue is appropriate to a situation. Courage is required in battle, not in pecuniary matters, for example. Deliberation is contextual in fitting appropriate ends and means to a context or situation. However, the situation is brought under an appropriate virtue, or general principle.

As such rules are often reliable, why bother with the experimental method? If the reliability of rules has been taken for granted, Dewey's emphasis on the uniqueness and novelty of the situation is overstated, although in justice to him, he does not claim every situation is unique. On the contrary, he argues that many circumstances can be handled using routine, habit and established "folkways": they do not require any experimental ethic. Using such situations as a starting point is Dewey's own "given," as he takes for granted that the situation is unique and basic. However, situations can be described using generalized concepts, as an instance of some kind.[43] While his emphasis on the novelty of the situation rightly incorporates change and close attention to the actual circumstance, the absolute uniqueness of any circumstance is implausible. There is also a tension between his valuing the present meaning of activity and his emphasis on consequences. Consequences flow from present activity and are arrived at in a later stage, the relative future or a later present. Thus an emphasis on consequences must detract from attention paid to present activity, or vice versa.

Nevertheless, we can still raise the question of whether any ethic is a "moral monism," whether "moral monism" is a valid notion. I would argue that in some respects "moral monism" is valid and in some, not. For all ethical theories have multiple elements in a relation. Utilitarianism has a notion of right and wrong, and of the value of virtues like justice. But these derive their value from the supreme value, happiness. Mill, for example, argues that the feeling or sentiment of justice is vital for a good society and thereby the happiness of the greatest number. The ethics of character development (so-called virtue ethics) and deontological ethics also include more than one element or one kind of moral consideration. However, they unify these under a supreme principle, whether consequences or right. No ethical theory that I know of would argue that there is only one element in a valid ethic, or even one principle. But there may well be a supreme principle to decide between different lower level principles or between conflicting values or the like.

Moreover, Kant argued that non-humans are not moral agents and therefore do not warrant moral considerability. Does this make Kant a "moral pluralist," as Stone's explication of the criteria of moral pluralism would suggest? But Stone argues explicitly that Kantianism is an example of moral monism.[44] We must wonder if moral monism is a straw man as he defines it. Again, Dewey has argued against the separation of duty, virtue, and consequences, and thus for a number of elements in any moral decision. Does consideration of all the elements of an ethical whole make of Dewey a "moral pluralist?"[45] But every ethic considers these elements, even if one element has priority.

I argued above for rights within civilization but only moral considerability for relations in the wild.[46] Is this a call for ethical "pluralism" then, with no one ethical principle valid for all situations, places or times? Is there one ethic for humans within civilization and another in the wild and still another governing the relations of humans to the wild? Or are some principles universal? I already argued for collapsing the distinction of "anthropocentric" and environmental ethics, in chapter one and two. In terms of values and duties, there is only one ethic; human affairs are a subset of universal, cosmopolitan ethics. The question of rights only arises within human civilization and cannot be extended to the wild. Rights are political, although they have a moral justification. Rights imply specific duties, as some have argued. However, rights are still a part of a much larger whole. Also, morals should not be confused with axiology: plural values do not make a plural ethic. Humans have plural values of different kinds that are insubordinate to a hierarchical first, a *summum bonum*. But moral principles are universal and regulative over creative actualization, if only as a limit. An ecocentric, cosmopolitan ethic is one ethic for wild and domestic, with priority given to moral evaluation, not "self-interest."

The limitation of rights to human civilization indicates how these questions can be approached. Political and legal rights are confined to human affairs but not value and duty. Indeed, I argued for a "deontological" environmental ethic in chapter one. Duty should override inclination, self-interest, desires, "attitudes," feelings, and pleasures, where there is a conflict with moral principles. The "deontological" ethic in chapter one follows Kant in rejecting non-moral sources of moral value. Duties to the environment override some other moral considerations. However, a deontological ethic is not only an ethics of duty, despite misinterpretations of Kant. Kant also recognized the value of virtues and happiness, but duty was adjudged first, since it alone involved consideration of moral value. But reasoning in maxims includes ends, that is, potential consequences, since achieving ends as goals of action will bring about specific consequences. Our view is that consequences for the environment must be a primary consideration for an ecocentric cosmopolitan ethic. But there is certainly a role for virtues and character in such an ethic, as I argued in chapter three. Kant believed it our duty to be virtuous and therefore that the moral agent is a moral self, a person with character. In this combination of duty, virtue, and consequences, Kant, like Dewey, outlined plural elements of an ethic. An ethic is not simply one or the other, duty, character, or consequences. All the great classical ethicists attempted to formulate an ethic containing elements of each. Epicurus argued that virtues were likely to make for a happy, pleasant life and that doing wrong would result in feelings of guilt that disturbed tranquility. For a cosmopolitan ethic, the best of each of these elements must be considered. Ethical theories are themselves plural elements in some sense or contain multiple factors for deliberation, since all the morally relevant elements must be considered.

An ecocentric cosmopolitan ethic recognizes that no ethic can ignore its own conditions, namely, the holistic relations of parts to the whole in a habitat and our place as moral agents within a habitat. There can be no happiness or virtue without taking breaths of air, drinks of water, and portions of food grown in soil. Since these are required they are practically necessary elements of a cosmopolitan ethic. Thus while rights are confined to humans and perhaps domesticated animals, value is not. The elements of any habitat have a niche or functional value of their own independent of their value to humans. Moreover, each individual of any species has inherent, if finite value as a representative or "token" of a kind and intrinsic value as self-actualizing. Thus the individual is morally considerable without having rights. We have a duty to leave wild areas alone and to consider the environmental consequences of all of our actions, individual and collective. It is virtuous to do so. But such duties arise within a larger whole, not out of human notions of right and wrong "extended" to the wild. Human civilization is within a larger world (cosmos) and requires the relation to this world for its own actualization, its own values. Although the other parts of this world are not identical with human civilization and its norms, they require ethical treatment, since they are practically required.

The treatment that is required is holistic in the sense that the good of the whole is greater than the sum of its parts. Thus the good of the environment as a whole, the biosphere, has greater moral weight than any of its parts, including the human part. The reason is obvious: if the whole were to disappear, so would the human part. If there were no atmosphere, all human values would cease. All the greedy little dreams of Wall Street and the dittoheads would disappear as they choked to death, asphyxiated like fish out of water. Limits to the expansion of human civilization are not only virtuous, they are required.

Nor is property a fundamental right. John Locke argued property is as fundamental as life and liberty, incorporating it into his contractual model in political philosophy. Property rights are one of the rights we retain from nature in signing the contract. His argument, however, is invalid, since by signing a contract, the previous "natural rights" are superceded or modified by the needs of the civil society.[47] The contract itself must spell out the accepted rights, which may or may not include property. Thus the whole notion of "natural rights" is confused and mistaken. The contract establishes rights, as Hobbes argued, it does not modify them. Before there is a sovereign, there are no rights. Rights that we have independent of government are human or moral rights, not "natural" rights. Nature does not establish rights.

Also, Locke ignores the "natural rights" of other species, to speak nothing of native inhabitants. Why do humans alone have "natural rights?" Other species have at least as much right as humans, since they are "in" nature even more than humans. Jefferson was right to modify Locke's list of natural rights and exclude property. While everyone has biological requirements for life, such as food, these are not "rights" but must be procured. The "right" to property is limited by the requirements of the health of the whole, and thus not absolute but relative to other goods of greater value.

For an ecocentric cosmopolitanism, then, any notion of absolute rights is untenable. But this principle applies equally within any theory of rights. If the right to life were absolute, no criminals could be apprehended, no just war conducted, no right of self-defense exercised, since the rights of offenders would be absolute. Some majoritarian elements—the "good of society," the public good, the "common good," must be evaluated higher than absolute rights. This is not a denial of the universality of rights in principle. But conflicts of rights require evaluation of priorities and of fundamentals, for example in law enforcement. Similarly, the right of other nations to defend themselves against aggressors like the Nazis follows from the right to self-defense. Those who appease the violent deny the right of self-defense. The value of one's own life in effect denied: conditioned on the whims of assertions of the violent. "The other person's point of view" can be taken too far: respect for others should include reciprocity, and should not be automatic—we should not give a blank check to *every* point of view. Killing soldiers of the Nazi war machine is not a violation of their absolute rights. Rights are part of a larger whole, including property rights. Prof. Frank Derringh (personal communication) believes that even Locke saw limits on property rights. John Cronin and Robert F. Kennedy Jr. have noted that courts have consistently ruled that use of private property that adversely affects the larger society, for example, pollution, can be regulated. Similarly, ground or running water contamination by a single private property owner can be restricted. The state has the power of eminent domain, which preempts private property rights. They also note Rachel Carson's arguments for a right not to be a victim of pollution.[48]

Moral limits derive from duties to others. Such duties can be applicable to intimate society in the sense of family, especially future generations of family: our descendants. Duties to others, and the recognition of rights by others, center on the norms of justice. The duty to support children, and thereby the rights of future generations, are an obligation undertaken voluntarily. In turn, since duties and rights are reciprocal, parents can and should expect a limited right to obedience from them. The limits are violence against family members, including children. The model is the Confucian norm of humane benevolence toward the young and especially kin.

The view of some "natural rights" theorists, for whom children are the property of the parents, to do with as the parents see fit, is economistic. Children are not "property" of the parents: they are *kin*, as intrinsically valuable moral agents. The relation is social not economic. To subordinate intimate society to economics is not only a category mistake, but also cold and calculating, absurdly self-interested, and not in the interests of the child. It turns children into mere commodities, undermines the family, denies the children any rights, and imposes a false model—an economic contract—on an entirely different relation. As usual with economistic thinking, egregious evil is rationalized by profit.

If we have children, then we have obligations to them, whether of support and nourishment, or of other forms of providing for their welfare.[49] The universality of such obligations for all cultures is beyond dispute.[50] Such obligations imply rights, for an obligation to another, in this case children, implies that the children normally have a right to expect fulfillment of such obligations on the part of the parents. Moreover, if we have children, consistency argues that our children may have children and thereby subsequent generations will also have rights. The act of having children implies the rights of future generations.[51] If we have children, we have an obligation to future generations.

From the perspective of duty, if we have children we have a duty to them. From the perspective of the ethics of character, we are virtuous to provide for them. Even from a Utilitarian perspective, the greatest good of the greatest number, the greater number will be in future generations taken as a whole, and thus their happiness requires ethical consideration. From a cultural relativity perspective, the culture cannot continue if its future members die out. From an agapistic perspective, if we love our children it will hardly do to make their lives miserable. Thus there is no ethical warrant from any theory to deny the goods of the world to our descendants and every reason to make it our duty to see that they have as good a life as we do.

Thus there is a contingent right of future generations to intergenerational equity, although more binding on those who do have children. We cannot argue for any unconditional rights for those who are not yet alive, and may never be. Only those who actually have children—and this is not something that should be encouraged, on other grounds—have a strict obligation to them. The right is also contingent on historical conditions like the actual availability of resources and other goods. This right is thereby of an odd sort, not as binding on everyone, and perhaps more contingent than other rights.

Some might wish to extend the right universally, arguing that even those who do not have children have some sort of obligation to future generations. I will not argue against this point of view, but only defend a more limited view of duties to future generations that I think is the "minimally decent" duty. At the least, then, those who have children have such an obligation. I would not argue for any duty to have children and thus for a universal duty to them on that ground either. If some humans prefer a good quality of life without children to the joys and sorrows of family life, that is their right. But since most people have children, the duty to and rights of future generations extend to the great majority of humans. However, even those who do not have children may value the continuation of our species, and therefore may recognize a right of future generations. Yet the case could be made on other grounds that even those who do not have children should share the goods of the world with them for the sake of the whole.

William Blackstone included the right to a livable environment among the rights of future generations that includes environmental conditions. Blackstone agrees that there is "a correlative duty or obligation" on the part of other people to either act or refrain from acting in some ways.[52] These would thereby include recognition of "interests" not recognized before, namely, environmental factors including an interest in clean air and water and to wilderness preservation. However, we should note that our duties are not limited to those with rights, since we may have duties to intrinsically valuable agents. Duty to others is wider than to those with rights.

Short-term solutions that seem attractive are actually not benevolent, if they will have such disastrous long-term results that they leave out the mass of future humans in order to give a few more trinkets to our own generation, their ancestors. Since I am addressing those influenced by Marxism, which despite all the attempts to turn it into an environmental philosophy is as anthropocentric as can be, I will express this point in Marxist language. Having unlimited numbers of children—creating life—and then making it harder, if not impossible in some cases to sustain that life is a "historical contradiction." Destruction of the environment is a threat to future generations. Having children and not providing for them is the height of irresponsibility. If we have children, we have an obligation to them, to provide for them. To their credit, the Marxists of China recognized this problem and dealt with it, however heavy-handedly. If we want affluence we must control births. If on the other hand we want more people, or the right to unlimited reproduction, we must kiss social justice and affluence goodbye, to speak nothing of wild species and habitats. By limiting births we recognize the rights of future generations, a right implicit in obligations to our children.

The rights of children should include a decent standard of living, not hopeless poverty. Such a standard argues against having too many children to support. Fewer children will mean an increased number of goods available if only in virtue of less demand. Also, child labor should be outlawed; countries that tolerate it should be boycotted by more enlightened countries. Some might argue that child labor has been used for centuries by agrarian peoples as supplementary farm labor in many cultures. But in poor, overpopulated countries abuse of such parental prerogatives also occurs, including selling small children into prostitution or full time jobs.

The right of support is only the most basic of rights. Treatment of children should be within moral limits. The child should have some right to protection from exploitation by either their family or outsiders. Children are not mere clay to be molded in whatever way the parents wish. Developmental and perfectionistic models of human growth could be misused to justify pain, deprivation, and the like in the name of creating a superior human. Children's rights should protect them against absolute perfectionism as a social right—there is no right of a society to absolute perfectionism as social right. Nevertheless, proper development is a normal expectation. Children have a right to an education, for their own development. Obtaining an education is inconsistent with a full time job. If a country does not provide one, it is mismanaging its assets and should be boycotted. Children are the future of our species and endangering our species is wrong.

4. Wild Value

Since rights do not extend to the wild, animals and plants in the wild have no individual rights, not that there ever could be! However, individuals of wild species have a value of their own, intrinsic value, as self-actualizing agents, responsible for their own survival. The issue is the bearer of intrinsic and inherent value. An ecocentric cosmopolitan ethic places value in agents, actors, and thereby organisms, not persons.[53] Value as creative actualization includes respect for self-actualizing agents as intrinsically valuable, and thus argues for non-interference with actor-organisms. However, their value is finite and since we should not interfere with the wild, it persists only as long as that individual continues to self-actualize. The value is limited to the ability of the agent to actualize its own life through survival.

With species, we have a different story. Species play an essential role in a habitat, and some of their functions are not known. Moreover, species have value as a unique kind that individuals do not have. Thus from a holistic point of view, as well as an axiological one, preservation of species is practically and morally required.[54] This is to speak nothing of the symbiotic value of species, both to humans and to other species. We also require other species for our food as well as the condition of healthy air, water, and soil.

Several problems arise connected with preservation of species.[55] One is if there is a duty to preserve species, that a duty to other species implies rights, if duties and rights are reciprocal. In one sense, we can claim that some species do have rights, since humans and domesticated animals are species to which the notion of rights correctly applies. The problem here is wild species, since I argued that the notion of rights does not apply in the wild. Does the notion of rights extend to all species, since the concept "species" transcends the domestic-wild distinction? This quandary is made even more complex since a clear line between domestic and wild species is not geographically or biologically possible. Wild species invade "civilization" and interact with it. Wild birds, insects, rodents, plants, and other species live in "civilized" areas; while grazing animals are turned loose in the "wild." I would argue that since the notion of rights would not be practical in the wild, and if pressed would end any "wild," that rights do not apply to wild species. Nevertheless, we still have a duty to them, and must consider them morally. The language of rights does not apply, since we should restrict it to predominantly human and domestic areas. Duty follows from the intrinsic value of other species and the duty includes moral considerability of individuals of other species and preservation of all species. The normal relation of duty to rights does not apply, since it is complicated by domestic/wild distinction. Rights are reciprocal with duties in civilization, but not in the wild. Indeed, we cannot expect reciprocity from wild animals. Some duties are not reciprocal with rights, notably, those of professionals, and our duties toward infants.[56] For this reason, my colleague Prof. Frank Derringh argues that animals do have rights (personal communication), since, as we have duties toward them, it follows that they have rights. Again, the problem is that there are no rights in the wild—no "natural rights."

The caveat is, ironically, that rare and endangered species increase the value of individuals of that kind, for they carry the concurrent genes that delimit the species.[57] As Callicott argued, an individual whooping crane has far more value than an individual pigeon, since the former is rarer. Should humans save species at risk due to competition, predation, habitat changes, and threats from invasive species? Species have always gone extinct, apart from human intervention (Darwin). Should humans intervene to save a species that is at risk due to "natural" factors?

Another problem is to separate out extinction due solely to competition and predation from that induced by humans. Some dangers from pests that threaten species may come from changes ultimately induced by humans. For example, thousands of conifers in Western North America are at risk due to global warming. The bark beetles that used to die in the frigid winters of the Western mountains are now overwintering, leaving thousands of acres of dead and dying trees, particularly pines and firs. Human hunting resulted in the extinction of the wooly mammoth and wooly rhino, the great auk, the passenger pigeon, and others; and the near-extinction of the bison, the whooping crane, and many others. Such threats may be direct, as in loss of habitat, extirpation, or overharvesting; or indirect, including global warming, the introduction of invasive exotics, which has reduced the numbers of the bluebird in Eastern North America, the extinction of pollinators, and many other causes due to humans. In the case where humans are directly or indirectly responsible, humans should indeed attempt to preserve the species.

Should hunting be allowed? After all, there is hunting and predation by other species. Why shouldn't human animals hunt? If hunting is for food, there is no alternative due to poverty, and does not involve rare species, it is morally permissible. However, it is not virtuous, since humans do not require meat in their diet and therefore killing animals is gratuitous. Hunting for "sport" is morally depraved: it provides evidence that humans are worse than animals, since no animal kills for "pleasure" or "fun." Such hunters are not superior to the "beasts," since they kill them solely out of self-interest, not on moral grounds. In other words, they are not moral, but selfish. Hunting rare species is immoral, and genocidal.[58] However, hunting of individual members of a species that is not rare or threatened is not. I mentioned above that I am pessimistic about humans permitting other predators to survive and thrive. In the absence of predators, humans may find it necessary to cull herds of herbivores, to prevent ecosystemic collapse. However, reintroduction of predators would be a superior solution, since it would also increase the numbers of such predators and thereby their chances of survival as a species.

I would also argue that we should preserve rare, threatened, and endangered species if there is no compelling reason against its preservation, even if its rarity is not the result of human activity. Why not? The species represents a unique genetic combination, a unique kind of good, and thereby has inherent value. Although it may be losing its intrinsic value, its ability to compete and self-actualize, it still has a value as unique. Moreover, it may have a function in the whole of which we are unaware, and its preservation may serve some potential holistic purpose. Our descendants may be grateful for this wise decision, for they may realize the unique value and understand the function that we cannot grasp. Moreover, they may be happy that they too can witness the ongoing actuality of the preserved species.

To what extent should humans be willing to make sacrifices to respect such species, for example, preserving rare species and biodiversity? Minimally, preservation of rare species must also include preservation of their habitat, since no species can live without a habitat.[59] Generally rare species are endemic to a particular habitat that they require.[60] This means that economic development must be circumvented in such areas, and more generally where it will have a negative impact upon the survival of such species. We are not asking too much of a species that has taken over half the land for itself, and another large chunk for grazing, frequently extirpating competing species, to preserve the few remaining wild habitats for other species. This may well require limiting our population so that we do not appropriate such habitats.[61]

Moreover, humans should also prevent air, water, and soil pollution, which knows no boundaries. Burning fossil fuels in the Midwestern United States has caused dieback of trees and dead lakes from acid rain in the Appalachian Mountains and into Canada. Burying industrial wastes affects underwater aquifers that may extend for thousands of miles. The radioactive fallout from Chernobyl poisoned thousands of square kilometers. We cannot pollute locally and ignore globally. The consequences of such actions must be considered if there is any ethical merit to such decisions. In a holistic view, all parts are interconnected and the impact on the wild and the domestic may be far-reaching both in space and time. Obligations to other species require actions to cease pollution as much as obligations to fellow humans, including future generations require the end of pollution. The culture of pollution must be terminated.

I would argue that a species that has incorporated over half the land for itself has violated norms of every ethical theory, save self-interest (which is hardly an "ethical" theory). If rights follow from obligations, it cannot be our duty to subdue over half the territory on earth for ourselves—especially the best land. Any genuine sense of duty would prohibit human hegemony on grounds of justice to other species. The same argument goes for virtue, or the greatest good of the greatest number. Even divine command ethics, following Noah, would have to concede the validity of species preservation. But if only self-interest argues for human hegemony, and as I will argue, that an "ethics" of self-interest is an oxymoron, then our duty as moral agents is evident.

As with rare species, the value of unique landscapes increases with their beauty, uniqueness, or other values of rarity.[62] The Grand Canyon is not economically productive, but has far more beauty than many other areas, notably boring suburban tracts and subdivisions, monotonous scenery produced by monocrop agriculture, and urban slums. Such beautiful areas should be preserved as public parks or wildlands, with limited access in case of more fragile ecosystems. This would be to continue with the policies established by President Theodore Roosevelt over one hundred years ago, of preserving beautiful or unique landscapes.

Subordination of the beauty of landscapes to their economic value is hierarchical and category mistake.[63] Beauty is valuable in itself, indeed may provide one standard of what is valuable in itself.[64] Putting economic development first would be a bad evaluation: economics, the means, replaces beauty the end, by its subordination. Some values are "above all price," as Kant put it. Surely, preserving a few acres here and there regardless of their economic value, purely on grounds of their beauty is not asking too much. All that money is a mainly a means to pay for quality of life, and this includes beauty as an essential element. What human would prefer an ugly domicile to a beautiful one, *ceteris paribus*? Almost no one prefers a large bank account with none of the goods it will buy to quality of life (Midas).[65] Imagine being rich but stuck on a desert island and unable to spend a penny. While such preferences do not prove the intrinsic value of beauty, they do indicate general human evaluations as between economics and aesthetics. Economics is a means and any attempt to make it an end is axiologically fallacious and morally corrupt. Commercial exploitation of beautiful landscapes is also an extension of economics into what should be aesthetic. Nothing is more depressing than miles of billboards on the way to a unique landscape.

Parts of Florida, California, and other unique habitats contain flora and fauna that occur nowhere else on earth.[66] Such "botanical islands" should be preserved as having unique value in themselves. If preserving rare species has value and is morally required for a true Cosmopolitan Ethics, then ecocentrism also requires preservation of habitats. Preservation should be for its own sake, in accord with strict duty; but also for the sake of future generations, that they may enjoy the experience of such species. Among other consequences, preservation of unique ecosystems, including large botanical islands, will have value just so that there will be some unspoiled wildlands.

Some argue that habitats change, and so preservation of a habitat would interrupt habitat dynamics. I would respond by pointing out that "dynamic" ecology can be exaggerated over the middle term. New inhabitants of the forest or desert are still in a forest or a desert. True, the Sahara and Mohave basins were originally seabeds. But this is over eons, not centuries.

What of previous human actions that have harmed either beautiful or unique landscapes or both? As with rare species and their habitats, humans should act to restore such areas. Given that humans have appropriated well over half the land for themselves, we are not asking too much for humans to give back such land for all to enjoy. For example, the City of San Francisco flooded the Hetch-Hetchy Valley, by all accounts an area as beautiful as neighboring Yosemite, to assure its water supply. But the State of California has plenty of water from other sources, if it stopped wasting it with poor agricultural practices, urban plantings, and other attempts to turn a basically dry state, at least in the south, into a wetter land. Thus Californians, along with the burghers of Southern Nevada, Arizona, and other nearby states, establish lawns in areas with little rainfall, extirpating unique natives. They also draw water from the Colorado River, with the result that little of the river reaches the Gulf of California, and bones of species once common at the mouth of the river litter the former delta. Farmers spray water in orchards haphazardly, instead of using drip or other water-conserving methods; and agriculture uses well over eighty per-cent of stored water in the state. Similarly, Californians destroyed two large lakes, Tule and Owen, with their unique fauna, in order to water lawns in the desert. Other ridiculous agricultural practices from the southwest include growing water intensive crops in the desert including citrus in Arizona, cotton in the San Joaquin

Valley of California, and rice in the Sacramento Valley. In the California cases, only massive government subsidies keep the farmers afloat. In other words, the produce doesn't even make economic sense, much less ecological sense.

In sum, both Hetch-Hetchy Valley and the area drowned on the Colorado to make Lake Powell could be restored, and the consequences would be minimal in terms of actual human hardship. Get rid of the "damn," and rainfall along with the again free-flowing river will do the rest. Restoration of these beautiful valleys would have the added benefit of increasing areas for wildlife. Some might argue that we cannot restore these valleys to their original pristine state. Scott Friskics in turn argues that this argument is a red herring, since no one claimed that anyone could restore a habitat to a "pristine" condition.[67] But in any case, all ecosystems undergo changes, so the changes caused by these areas having been underwater for a period is comparable to such ecosystemic changes. True, they will not be identical to the original, but they would not be like they originally were so many years later even if there were no flooding. In any case, it will be a start. Perhaps time will heal these valleys as well. Tule Lake and Owens Lake could be restored for the unique fauna they support, for example the endangered Tule Deer. The case for restoration of the Florida Everglades and Tall-grass prairies of the American Midwest is even stronger. The Everglades are home to thousands of species, and the attempt to destroy them to benefit the handful of families who controlled the production of sugar cane in Florida is incomprehensible even in economic terms. As with California rice, the sugar costs more to produce than the world price, so tariffs and price supports keep the plantations a going concern.

The obligation of humans is to preserve wild lands as part of a larger whole, including biodiversity. This cannot be accomplished by its domestication, that is, by "management," or a "management ethic."[68] For this would be domestication of the wild. The wild must be preserved as wild, and therefore interfered with as little as possible, and preferably not at all. Management of wild lands would interfere with their development apart from human norms, plans, values, and the like. In one sense, then, a revised cosmopolitan ethic is "pluralistic," since it calls for different approaches to domestic and wild areas. In another sense our ethic is not pluralistic, since both humans and individuals of other species have intrinsic value, while our duties remain the same to both in principle. However, these are different notions of rights. Humans have rights in the political sphere that individuals of other species do not have. Moreover even among humans we have two senses of rights, civil and cosmopolitan, or universal. Civil rights obtain within our own society. Cosmopolitan rights extend to other societies. Universal human rights are cosmopolitan. The extension of rights to other societies has the relation of parts to a whole: the recognition of the membership of individuals of other cultures in a larger whole of one species.

However, an ecocentric cosmopolitanism is not pluralist in the precise, contemporary sense of multi-cultural, which means that different cultural norms within our society may require us to come to a consensus between different cultural factions for public purposes. Such a multi-cultural approach may apply to human affairs, such as abortion, but is of limited applicability to human-non-human relationships. Cosmopolitanism takes the best of many cultures, or is cognizant of them. But reformulation of cosmopolitan ethics in the direction of ecocentrism is required by the practical requirement that human cultures are part of a much greater whole upon which they are dependent. Thus ecocentric cosmopolitanism is pluralistic in one sense, but not another. We do not live in the wild and do not follow its mandates. But our ethic applies to both: the whole. We are above all citizens of the world, and a world much larger than human civilization. Human rights are distinct from the relations in the wild, and thus an ecocentric cosmopolitanism may involve different practical principles for different parts of the whole. However, values are the same for individuals of different species; and our duty to leave the wild alone is similar to that of other bearers of intrinsic value. Just as persons and cultures have the right to autonomy and to be left alone, so does the wild.

Cosmopolitan ethics includes the holistic relation of one part to other parts making up the whole. Although we are members of a culture, other cultures are appreciated. Cosmopolitan ethics is inclusive. By contrast is any tribal or exclusive ethic, which does not recognize the other as morally considerable. In extreme forms, such tribal ethics are genocidal, as in the Nazi ideology.[69] A reformulated cosmopolitan ethics takes being a "citizen of the world" more seriously and comprehensively. Humans are both within a culture and also a larger whole. The larger whole includes other cultures but these in turn are in the biosphere, whose elements make our world possible. Thus the relation of humans as parts of a larger whole replaces the anthropocentric relation of humans dominating nature: dominion, manifest destiny, "management," and other models that ignore actual relations of humans and their dependence upon the biosphere, including non-living elements such as air, water, and soil.

A cosmopolitan ethic is pluralistic in the other sense since we have different cultures, and the survival of such cultures requires the survival of each. The alternative is a war of all against all, and the potential for cultural genocide, as happened with many of the Native American cultures. Thus cultures also require reciprocal relations, the basic relation of duty that I examined in chapter one. Pluralism in the sense of many different cultures living together must be reciprocal. Cultures cannot be subordinated or the ethic is not truly plural. A pluralistic community is more just, as relations between them are reciprocal, not hierarchical. Each agent, including cultures acting as wholes, has powers ("rights") but also obligations in a correlative relation. Ecocentric cosmopolitanism extends this model to the world as a whole: the non-human as the morally considerable other. Non-human parts can thrive in reciprocal relations with civilization if they are left alone.

In this chapter we have attempted to revalue both the notion of human rights and pluralist ethics. “Rights” are part of a much larger whole of greater value, the biosphere. Moral considerability should expand to encompass non-human agents, following from the actuality of plural values; but the notion of rights cannot be expanded, since this would undercut the distinctive value of the wild, of other species, and of landscapes. Rights are not absolute, as parts of a much larger whole. However, several universal principles of ethics remain valid. The rights of future generations are only potential, but require duties if we have children.

Five

VALUE AND OBLIGATION

> There is no concrete act or specific way of acting which could be determined as right or wrong without reference to the anticipatable consequences of it as good or bad. And there is nothing relevant to any concrete act or specific way of acting by which what the categorical principle dictates could be determined except such predictable value-consequences of this way of acting. Clarence I. Lewis[1]

Cosmopolitan Ethics was formulated as a deontological ethic in chapter one. Modern moral theory tends to sharply distinguish deontological ethics from teleological ethics and thereby from consideration of consequences. Recent deontologists, such as W. David Ross,[2] have emphasized this point as well as some so-called meta-ethicists, who, working within the analytic tradition, favor sharp separations, divisions, and distinctions. For an ecocentric cosmopolitan ethics, consideration of consequences for the environment is paramount. Thus the issue is raised of whether deontological ethics, especially that of Immanuel Kant, can evaluate consequences; and what is the relation of value and obligation in general.

The usual division made in ethical theory, between "deontological" and "teleological" systems, is apt to be misleading, since all ethical theories try to make room for the diverse elements of ethics, including duties, virtues, consequences, and rights. A more precise division might be made between teleological systems that define moral good in terms of some other good or value (eudaemonism, hedonism) and those that see moral good as overriding (Stoicism and Kantianism).[3] The first tend to stress difference over identity, and internal goods (my pleasure); whereas the second tends to the identity of norms (rules, justice, reciprocity, moral laws, and principles) and external goods (social relations, altruism, the public sphere, and so on). Epicurus, John Stuart Mill and Friedrich Nietzsche are in the first group; Plato, the Stoics, Kant, and Jean-Jaques Rousseau—and even Aristotle in a qualified way—in the second. However, as is often the case in ethics, the split is not precise. The distinction between "consequentialist" or "teleological" ethics and "non-consequentialist" ethics is not a matter of circumstances, as these are factored into both moral theories by the universality of the major in moral deliberation. Examples of universal ethical queries are: what everyone ought to do in the same circumstances; or the appropriate virtue under the circumstances. The difference lies in the moral framework: whether duty, virtue or consequences form a first principle that is decisive in deciding cases.

I will argue in this chapter for consideration of consequences, especially environmental consequences, in moral evaluation and thus for an explicit expansion of deontological ethics to include consequences. Since for environmental ethics consideration of consequences of an act on the environment is paramount, evaluation of consequences must be revalued to a reciprocal relation with obligations. Thus the relation of value to obligation cannot be foundational. I will consider Kant's cosmopolitan ethics and whether the revaluation involved in incorporating the consideration of consequences is consistent with the Critical Philosophy. Our revaluation of Cosmopolitanism will include consideration of consequences within a moral framework and thus more than an exclusive emphasis on duty. I will also examine the role of principles in ethics and creative actualization of moral values.

1. Consequentialism

The distinction between "consequentialist" and "non-consequentialist" ethics is not a matter of circumstances, as these are factored into both moral theories by the universality of the major in moral deliberation. (Consequentialism is also distinct from the ethics of character, the unfortunately named "virtue ethics.") For deontological ethics this means what everyone ought to do in the same circumstances; for the ethics of character it is the appropriate virtue under the circumstances, which would be chosen by a virtuous person. The difference lies in the moral framework: duty/virtue, not circumstances. However, "deontological" ethics encompasses both consequences (for example, the pursuit of pleasure) and character in the form of self-realization, for in no case is pleasure or development ruled out, but only subordinated to duty.[4] Where they do not conflict with duty they are morally permissible. Indeed, in "hard" deontology (Kant) self-realization becomes a duty.[5] (In "soft" deontology self-realization is morally permissible and valuable, but not obligatory).

Since an ethical framework is the spirit of deontological ethics, the latter is the best humanist theory. Room is allowed for development of talents, and even for the pleasures of social interaction. Deontological ethics is also morally rigorous, putting moral values before other values, as supreme. Evaluatively, it encompasses more: allowing—even demanding on a strict reading—both development and pleasure. This does not work in reverse: strict hedonism may result in dereliction of duty or lack of full development. It could be argued that just as duty rules out some pleasures, hedonism rules out some duties or virtues. Duty is social: duty is to others. Pleasure, by contrast, is private. Virtuous acts of character are both public and private, since they assert character in the world: the decision to actualize the virtue in public space. However, all three, representing the main moral theories—deontological, character, and consequentialist—involve activities that require forbearance from others, duties from others. Duty, ethics in the narrow sense, requires a reciprocal structure of mutual differences between agents. Pleasures also require mutual forbearance. Thus the three main moral types themselves have a reciprocal structure: they are mutually involved, not contrary types. This point is consistent with John Dewey's argument that the division of ethics into narrow types misses the connections between them (Dewey 1930).

However, the ethics of duty comes with drawbacks. In Kant's critique of the ethics of virtue at the beginning of the *Groundwork of the Metaphysics of Morals* he argues that the courage of a brigand is morally reprehensible. The principle is that the subordination of morals to non-moral grounds is immoral or can lead to immoral acts. Similarly, religions that preach love and turning the other cheek are used to justify burning at the stake, and violent jihad. Kant's critique of virtue can be turned and used to criticize the ethics of duty: subordinates do their duty in respect to a "higher cause," the excuse used by Nazis, and soldiers in war, among others. Duty as empty form can be placed in the service of just such nefarious ends. Similarly, the duty of a corporate officer to obey superiors can be used to justify polluting the commons. Thus the subordination of duty to other ends is itself immoral.

Kant has also been interpreted to mean that consideration of consequences does not belong to purely moral decisions as such. I will contest just this point, since Kant's ethics also contains teleological elements, for example in the notion of persons as ends in themselves and the kingdom of ends. Maxims also require consideration of consequences. The contrast of the ethics of principles (deontological) and of consequences may be overdrawn, which may reflect the influence of Ross.

Be this as it may, the consequences for the environment are an essential for any environmental ethic. Fortunately, there is a later version of Kantian ethics in which consideration of consequences plays a major role. I am referring to the work of Clarence I. Lewis although his position differs from Kant in many respects. Lewis, as a pragmatist, argues that the right is based on the good, or good consequences. I would argue that Kant takes both the right and consequences seriously in that the second statement of the Categorical Imperative enjoins consideration of persons as ends in themselves. I examine this point below.[6]

2. Pragmatic Consequentialism

All rational ethics is consequentialist to some degree, insofar as the form of rationality is ground and consequent: responsibility implies that we are responsible for the consequences of our acts. A reason for an act is the ground of a specific consequence that is aimed at by the action. Practical reasoning aims at an end that will result from specific means. The end in this type of practical reasoning is the effect or consequence, whose ground is the means as the cause. Considered in another sense, teleologically, the end is the ground and acting on the means is the consequent, in order to achieve the end. If all ethics is slightly consequentialist, then all ethics is also partly rational; ethical reasoning has the structure of ground and consequent.[7]

Consideration of consequences is compatible and consistent with a deontological ethics that stresses action, instead of intent. Moral consideration of consequences should not be taken as a species or variety of Utilitarianism, which is only one species of consequentialism.[8] The ethics of good intent is not enough. Indeed, our duty includes consideration of consequences in moral evaluation of actions, since our action will have consequences. Deontological ethics can encompass consideration of consequences, then; consequences are morally considerable, if not the only consideration.[9] To be sure, we should also do our duty because it is right: we should do what is morally right. Our duty is doing what is right. Duty and consequences are not two distinct ethics, but one: a duty is to consider bad consequences of our possible actions on the world, on others and ourselves. Kant's moral teleology, which was overlooked in the hasty interpretation that criticized the "empty formality" of the Categorical Imperative (Georg W. Hegel), includes persons as ends in themselves. This point is evidence that moral practice should include both consideration of others and consequences for them. Intents are in any case unknowable and multiple. The multiple possible intents of any act make the specific intent unknowable unless we make it public. The act is both knowable and unified, and responsibility for acts follows regardless of intent although the latter may mitigate the degree of responsibility.[10]

Such an ethics was the project of pragmatic consequentialism, particularly that of C. I. Lewis.[11] Lewis agreed with Kant that ethics, as opposed to values, could never be completely grounded empirically, that obligation for creatures with instincts and impulses comes in the form of imperatives, and that moral values have priority over other values. However, his statement of obligations takes consequences into account more explicitly than Kant's does. Dewey argued against a one-sided emphasis upon duty or consequences. I agree with Dewey that in raising the moral issue we are evaluating the consequences of our acts: on the world, on others, and even on ourselves. We should not try to "evade responsibility for consequences"; they are one factor in evaluation of actions.

I would also argue against Kant that consequences are not empirically derived or heteronomous as a projection into the future. Kant would probably respond that many of our actions are habitual and based on inductions from experience: that some acts will have specific results. This account of practice leaves out the novel aspects of creative actualization as practice. Creation of novelties cannot be empirical by definition, but habitual acts have such novelties as their condition: they follow initial successes. We must calculate or predict an outcome or consequence based on norms of past creative actualizations or on imagination. Pre-valuation of consequences requires imagination, even in habitual actions, for no one knows better than Kant how actions can go awry. Indeed, the connection of ends and means in creative actualization of novelties is an office of imagination, and not of reason, as Kant would have it; for there can be no connection in reason without induction from sense data, if the first Critique is accurate. But creative actualization of novelties cannot be based on induction. Habit is the generalization of creative actualization; predictability is common to habit and science. Moral rules may indeed be independent of nature, as Kant argues, but their consequences cannot be if they are creatively actualized in the same world as creatures. We can derive an is from an ought if we creatively actualize the duty in the world.

One standard objection to consequentialism is that we cannot possibly consider all the future consequences, all the remote effects and ramifications of an act.[12] Indeed, even some of the direct results may escape our consideration. Here intent, cognition, and moral obligation enter. Although our plans may indeed go awry, we can try to do as well as possible, to carefully consider consequences, and plan for as many contingencies as we can. If things still go wrong, we at least know that we were conscientious and tried our best. We may still be responsible for the results, but did as well as we could. We may be able to rectify our mistakes. Fortunately, the world is such that for the most part our endeavors are successful and life thrives. Pragmatism would argue that we are considering as many consequences as we practically can.

What if "pure" duty and consequences conflict? Telling the truth may hurt someone, but be required as a moral norm. Here we are required to be as courteous as possible; virtue is also a duty. In any case, we are sometimes required to creatively actualize moral norms as a consequence of our actions regardless of other consequences. The conflict here is not, then, between duty and consequences, but the negative consequences of doing our duty by contrast with those of not doing it.

Duties are derived from the reciprocal structure of social norms. By this I mean that moral duties are generally duties to or for others.[13] Thus they arise primarily in a social context. The general statement of these is some variant of the Golden Rule, which I take Kant's first statement of the Categorical Imperative to be, although he pointed out the differences. The golden rule is a paradigm of following rules with respect to the other. They are not exercised or undertaken for their own sake but for sake of society. What then of doing what is right for its own sake? But doing what is right because it is right arises in a social context of obligations to others. "Right in itself" might appear in conflict with consequences as the ultimate rationale, as Ross, among others, argued. But there is no reason not to consider consequences as well as inherent right. Indeed, as C. I. Lewis argued, it would be irrational not to consider consequences. Ross might argue that in any conflict of values or ends, where doing what is right is in conflict with some valued consequences, doing what is right has priority. But evaluation of what is right is inseparable from consideration of consequences in moral evaluation. The moral principle involved is putting morals ahead of interest, pleasure or other non-moral values. Moral duties are also co-extensive with rights/freedoms, as my right is your duty and your duty is my right. Of course we may have non-moral duties arising from our situation, such as the duty of parents to children, of soldiers, and so on.

Consequences for others are not the only concern; the consequences of our acts in the world are also morally considerable. How will each act affect the environment, our own habitat, required for our own survival? "Intending" to do no harm to the environment, or be "virtuous" is not enough. The act must at the least not harm the environment; it is better if it improves it, even if minimally, since small but widespread actions may have a great collective impact. Consequences for the environment are paramount for living things, who are totally dependent upon their environment and, for that matter, live entirely within their environment. The ultimate standard is ecocentric, and this standard is cosmopolitan, transcending cultural loyalties.

Another problem involves value conflicts, which can give rise to hard cases. What of situations where there is a conflict of duty and consequences? Are there some acts we should (not) do regardless of consequences? "Do unto others…" or consistency involves reciprocity and implicitly universality. Universality must include animals and plants, as I argued in chapter one and thus this principle is universal not only with other humans, but with all self-actualizing agents. In moral evaluation our duty includes considering the effects of our choices on other living things. As Lewis pointed out, not considering consequences is irrational. But is murdering one individual to prevent genocide wrong? Because murder is wrong? This is a hard case—causing one death to prevent the death of a kind. Would we murder Hitler to prevent the 25 million or more deaths in the European theater in WW II? I do not see how this can be wrong if so many deaths would have been avoided. Is this a surreptitious appeal to utility, to the Greatest Good of the Greatest Number? I would argue that our appeal is to the good of the whole, a type of holism, not to the greatest good of the greatest number. But on moral grounds, surely we also have a duty to prevent genocide as a *per se* wrong and as evil. Other species have a value above and beyond any individual as a unique kind. Thus killing one to save a kind is defendable on deontological grounds. Moreover, killing to save a kind could be defended on self-defense grounds: the peoples who fought the Nazis acted in self-defense. Within a deontological framework, there may be higher and lower duties, as Ross argued.

Also, as I argued in chapter four (and will argue in chapter nine), the principle of duty applies only to species, not individuals of species. There is a slight disanalogy here, since rights occur among humans as political creatures that do not arise in the wild (see chapter 4). Thus Adolf Hitler violated rights, but a predator does not. We can violate rights by harming one individual human that a lion does not violate in the wild. Intrinsic value alone does not necessarily entail moral obligations to individuals, although it does entail moral consideration. But the analogy holds in another respect. Genocide is a wrong done to whole group, a universal wrong. Duty arises for species where the wrong would be universal and terminal. Killing all the Jews is a much greater wrong than killing one individual Jew even though killing just one is also a wrong.

Another consideration is the distinction of moral evaluation—what is my duty in this case—and performing our duty. Moral evaluation includes consideration of consequences and other relevant factors, since a person's act will have consequences that are morally significant.

It is also our duty to achieve the best outcome. This is implied in Kant's argument that we have a duty to develop our talents (compare virtue and Kant's formula of ends). Thus achieving the best outcome is one factor in moral evaluation. Duty has primacy, but that does not mean it is not our duty to achieve the best outcome, in choosing among different possible actions. Acting to creatively actualize morals has consequences, so we must consider consequences of our acts. How will each act impact the environment? This is the question raised by ecocentric Cosmopolitan Ethics as the primary one for moral evaluation. Acts should at the least not harm the environment and preferably should improve the environment. Thus the pragmatists stress meliorism as a goal in all actions. Meliorism includes improving the future as a consequence of our acts and this also includes moral improvement or moral progress. There is no incompatibility between meliorism and moral progress, since the one is a species of the other. Moral progress is the measure of meliorism. Thus acting for the good of the whole is melioristic. Meliorism is consequentialist and constitutes the pragmatic alternative to Utilitarianism.

The formula of the "law of nature," in *Groundwork of the Metaphysics of Morals* implicitly includes the wild.[14] Namely, holism in the environment is the condition of possibility of humans. Since we would "will" our decision to be a "universal law of nature," the consequences of our moral decision would be a criterion of its rightness. Could we will it universally, that is, as a universal law of nature, with all that would entail for us as a consequence?

Thus the conflict of deontological and teleological ethics is overdrawn, as the distinction of ethics of duty and of consequences (or ends) is misleading. We have a duty to try to improve the world with each act. We are morally required to creatively actualize duty and our acts will thereby have consequences in the world. Whenever two values conflict hard cases can arise, but consideration of consequences is part of moral evaluation. Also, since we can consider consequences, we can distinguish long-term and short-term consequences. As Garrett Hardin suggests (1993, p. 169), we can evaluate the increase in suffering that will be the long-term consequences of feeding people today without tackling the population problem in our moral deliberation.

3. Kant

Kant called for a kind of Cosmopolitan Ethics, but rejected strict consequentialism in ethics. Our own project is an ecocentric Cosmopolitan Ethics, in which consideration of consequences for the environment is central. How can we reconcile the consideration of consequences for the environment that is the guiding principle of this project with the interpretation of Kant in which he rejected consequentialism, especially since our opening chapter used an approach that is deontological?

In defense of Kant, we should again point out that he did not reject consideration of consequences in moral evaluation. On the contrary, in the formula of persons as ends in themselves and of the kingdom of ends, Kant explicitly requires consideration of consequences for persons in order to achieve moral worth. What are the consequences for persons of my maxims? What of the community as a kingdom of such persons? In his examples of moral evaluation, the example of whether to be charitable to those in need explicitly considers the consequences of not being charitable.

> For a will which decided in this way would be in conflict with itself, since many a situation might arise in which the man needed love and sympathy from others, and in which by such a law of nature spring from his own will, he would rob himself of all hope of the help he wants for himself.[15]

The consequences of willing such a law would be to rule out charity for oneself. Kant, then, does not reject consideration of consequences in moral evaluation. Instead, he distinguishes consideration of consequences from consequentialism as such in moral evaluation. Actions are duty-governed, not end-governed if they have moral worth. But it does not follow that there is no consideration of consequences. Indeed, as C. I. Lewis has pointed out, rationality means the ability of an individuals to understand the consequences of their own acts.[16]

Kant does not fit or conform to many of the marks of Gerald Gans's definition of deontology, which follows Hastings Rashdall (Gans 2001). Contrary to Gans and Rashdall, Kant does include consideration of consequences as ends as the "matter" of the Categorical Imperative, both in the statement of the "universal law of nature," of people as ends in themselves, and in the kingdom of ends. Kant also rejects "intuition" (Harold A. Pritchard) as the source of moral norms in favor of practical reason as the source of practical principles. Intuition for Kant is tied to sensibility and pure reason. Kant's moral problematic is deriving rational ends, not empirically mediated, nor based on natural law, as in the Stoics; and understood by reason. He derives the Categorical Imperative from the structure of reason and of moral agency *per se*. Finally, most rules are not absolute: there can be a hierarchy of principles. In Kant this takes the form of the conformity of maxims to the Categorical Imperative. Maxims are subject to consistency with the Categorical Imperative.

Kant continued in the tradition of the ethics of intent, that is, that moral worth is contained in intentions, not in actions. There were several reasons for this. The first was metaphysical: the "pure" or free will was not in the phenomenal world and therefore required a separate noumenal dimension for moral deliberation. By contrast, action is in the world of nature and subject to determination by prior causes. Free will could not be subject to determination by the world of nature, unlike action, and remain free. Another ground involved the moral purity of the will. Conforming to the moral law is not enough for moral worth to accrue. We must act for the sake of law. Our intent, and not our acts are the measure of moral value. The other ground was practical. While we may intend to do good, our intent may not always be translatable into action. We can be morally right but empirically wrong; intend well but act badly. Since action is subject to the vicissitudes of nature, it is not entirely in our control, unlike the will.

The problem with the ethics of intent is, as Dewey put it, that intent is "one sided."[17] The ethic of intent "evades responsibility for consequences" by leaving them out of moral deliberation. The result is that moral agents could hypothetically live in an amoral world, ignoring practical consequences out of a misplaced sense of moral purity. Contra Kant, value, *including moral value*, is established in practice through creative actualization of what is right. Unless we practice what we preach, no moral *value* accrues. Intending to do what is right is not enough; we must strive to do what is right as well. Using intent as the sole standard is a diminishment of responsibility, not an increase. Morally inferior, this standard evades responsibilities in the actual world. Bringing about moral acts is as important as intending to. Moreover, an intentional theory would imply an intentional theory of wrong: we are wrong if we intend to do some wrong act, even if we do not succeed. If we are wrong if we intend, what is the difference between a temptation, acted out in our fantasies, and intent? To be tempted is to commit wrong in this view.

An ecocentric cosmopolitan ethic requires a somewhat critical evaluation of Kant, despite his greatness as a moral philosopher. In praise of Kant we may recognize his personalism, his consistency in maintaining a moral theory of value, his rigor in defending duty, and his ongoing critique of any attempt to ground ethics by any inductive, empirical method, or any hypothetical foundation. Kant's ethics is correct as a theory insofar as it recognizes the role of principles, that is, norms in moral reasoning, over "intuitions," or "feelings."[18] Intuitions and feelings reflect social conditioning: they are learned from social circumstances. As such they cannot attain universality nor transcend the society, they are uncritical. Norms derived purely from society cannot be used to criticize inferior social practices—a problem for experimental philosophy. However, consistency can be used to criticize norms within a society, whether a society applies the norms justly, to all individuals, and are they living up to them. Kant's technical precision in showing the role of necessity in transcendental ethical justification—the arguments from conditions of possibility—has provided a model for all subsequent ethics. Kant also made the autonomy of the person a central focus of his ethics. His critique of any ethics based on self-interest in its different guises should also be mentioned. (Autonomy should not be interpreted in terms of self-interest, which it often implicitly is in the literature.) He also realized that universal duty is an egalitarian norm, as universal, and therefore created an ethical theory that enshrined equality and freedom, in the form of autonomy, as the content of ethics.

(A question that arises in Kantian ethics is whether he reduced a prior cause to a necessary ground. Given his treatment of cause in the first Critique, causality requires the logical relation, not the reverse. Be this as it may, necessity is a modal category, in which conditions of a conditioned are in a relation. Conditions of experience may require a conditioned to meet certain principles. As not all causes are necessary, the two were not conflated.)

I have already examined the issue of consideration of consequences as an ethical requirement. Recognizing the validity of moral ideals and principles or intending them is not sufficient. Unless they are put into practice as moral values, they have no effect. Obligation *per se* is to others, not to principles: it is social. It therefore requires consideration of the effect of our actions on others, that is, more than just intent. Although the rubrics for contemporary ethics sharply divide deontological and consequential ethical theories, we may at least raise the question of whether Kant rejected consequentialism? That is, did he argue that we should disregard consequences? A Kantian would argue that the value of persons is included as the content in the second statement of the Categorical Imperative; society, in a sense, in the third. But if morals must include consideration of the effects of actions on other persons, it includes consideration of consequences. True, the derivation of the moral law cannot be from empirical elements since obligation requires necessity. But this is distinct from consideration of consequences in moral deliberation. The second statement of the Categorical Imperative makes consideration of persons morally obligatory; this may imply that consideration of consequences for persons is morally considerable. The sharp division of deontological and consequentialist ethics may hide a more complex relation.

Since actions have such effects, as the cause of them, control of intent is insufficient for moral worth, since well-intentioned but harmful actions may violate moral principles. We may intend good by violence but achieve harm. Intent is more important in cases of moral error, or unforeseen consequences. However, part of moral agency should include responsibility for consequences, as Dewey noted. Moral agents attempt to creatively actualize moral outcomes, and to correct their errors and shortcomings.

Admirable as Kant's ethics is, he overemphasized intent without responsibility for action. This emphasis was at least partly motivated by his theory of nature as deterministic, and of action as within the world of nature. Kant was a pessimist about our ability to achieve moral goals "for the most part," since he thought the laws of nature largely determined our actions in the external world. We cannot be held completely responsible for our actions in such a world. We could always intend the best, at least in principle, but not always achieve good results in the world. These presuppositions are now dated. The "billiard ball" model of determinism in the universe is factually mistaken (indeterminism in physics) and logically implausible: Kurt Gödel's theorem argues that a formal system cannot be both complete and consistent. Thus it is unlikely that any formal deduction, no matter how complete, could actually determine the subsequent place of every particle in the universe. The model is also contrary to experience of altering the world in accordance with projected values: novelty and *telos*, creative actualization of imagined alterations.

Consistency, Kant's ultimate criterion of moral life, is not enough. I am not arguing against moral consistency: consistency is one condition of moral life, that is, non-hypocrisy in our principles. But consistency is ultimately a logical principle, instead of a moral one. While it may be unfair or unjust to treat two people inconsistently, this wrong is wrong as unjust, not as inconsistent. Morals give consistency its ethical value. Obligation is based on consistency in Kant's view, not directly to consequences. Consistency is tied in a sense to reciprocity, since an identical principle for each situation is invoked as a norm in the former and for each inherent value in the particular situation.

A problem with Kant's ethics is not its distinction of the purely moral from and empirical, although that also can be a problem, but its hidden logicism: it sacrifices right to the formal requirements of universality. Given Kant's conception of morality, this could hardly be otherwise, but it may well (and ironically) be right to lie if this will save a life; in other words, if life is the fundamental value. Such reasoning is hypothetical, but the conflict of values in many moral situations is not. Kant's absolutism and rigorism appear all too dogmatic to a pragmatist. Where logic conflicts with right, logic must give way, as contingent upon right. The logic of genocide dictated that the most efficient means be used, gassing, since gassing was more cost effective and quicker: more lives could be taken. Similarly, destruction of the bison was the logical condition of destruction of the Native Americans of the plains, their culture and the ecology of the long grass prairie. Logic without ethical limits lends itself to gigantic crimes.[19]

However, Kant's obeisance to a distinctive reading of science along mechanist lines caused him to avoid worldly commitments in ethics and he attempted to separate the ethical and natural realms. Kant separated the realm of freedom or ethics from the realm of nature by arguing that ethics concerns what we ought to do, a normative question, not strictly with what is practiced. The problem here is that moral practice is left in limbo. Moral agents could not always ensure results, since the consequences were too dependent upon practical-empirical principles, and thus morality was separated from the actual world of practice, a strange result for a "practical philosophy."

The separation of theory and practice in ethics would here reach its climax. We can be moral in theory without being moral in practice, so long as our intent is moral. In turn the problem of hypocrisy arises, namely, the separation of moral theory from actual practices. Further, theory no longer serves as a theory of practice, since what is true in theory may not be practiced.[20] Responsibility is indefinitely postponed by this model in which "right" is assigned to practice by theory (Aristotle, who first explicitly separated theory from practice as separate realms of "being"), and to theory by practice (Kant). Blame is assigned to theory by practice and to practice by theory in an endless cycle in which ethics is decentered and ungrounded in the world. The so-called "is-ought" distinction is the latest version of this dichotomy; the distinction ignores the connection of the two in practice, creative actualization of what we should do as acts in the world. Kantian ethics in this sense has unwittingly resulted in an imaginary "realm" completely removed from the actual world, where ought and is may never coincide. For creative actualization, practicing moral values is required, since moral values regulate practice. Value is the synthesis of evaluation of the best course of action with actual practice in creative actualization of the goal in the world. The "descriptive" is the result of the "prescriptive"; they are not in separate universes. A moral command that is "prescribed," if acted upon, is creatively actualized in the world. The distinction is not between "is" and "ought," then, so much as potential and actual.[21]

Kant distinguished prudential and technical reasoning from moral, and the former two were also treated under "practical reason." The actual world of practice was distinguished from morals on technical grounds. However, the actualization of duty, and thus the consequences of moral action were neglected by Kant. The result was later errors in theory, in which value was first reduced to ethical norms, to "oughts," by several schools of ethics in the twentieth century and then the claim was made that ethics, which concerns only the ought, is "unnatural" or has no relation to actuality. The connection of values and ethics to practice was lost.

Are ethical propositions descriptive or prescriptive? The difference is one of degrees, from purely descriptive—ethology—to unactualized standards and principles. Creative actualization thus attenuates this distinction, by bringing norms into the world as acts. I would also argue that the descriptive-prescriptive distinction involves a false dichotomy, as descriptions use normative identities to form judgments like all language. Thus descriptions always have a prescriptive element as normative, in the form of linguistic judgments: bringing a case under a concept. More, things can be as they ought.

Kant has confused value and evaluation in his moral philosophy like much of the tradition; intent is the product of evaluation and choice, but not practice. We can procrastinate till the end of time, claiming we intend to make good, but not do it. Value requires practice—if it does not indeed constitute practice. Moral value requires creative actualization in the world. Ethics concerns actions, not merely commands or the intent to carry out moral duties. Now a Kantian would immediately object that Kant never intended to argue against putting moral decisions and principles into action in the world. However, he was so concerned to show that moral reasoning is free of natural determination that he neglected philosophy of practice, which brings ideas into the world. He thereby neglected to treat the value of moral practice, of actual results, as opposed to moral intent. Moreover, melioristic consequences are also a valid moral goal: creative actualization of an increase of good in the world, including new and different kinds of value. We should leave the world a better place than we found it. While we may not succeed in creative actualization, and not achieve the goal, our responsibility is to try again, not retreat to quietism.

Another issue raised by Kantian ethics is that of the role of limits. Practical reasoning in Kant implies both the distinction of non-moral goods from moral goods and moral limits on non-moral goods, because potential actions have a moral condition if they are to have moral worth. We must test our actions against the standard of the Categorical Imperative in a moral evaluation. The implication is that actions that pass the moral test are within moral limits, even if they are not moral actions, but of another kind. However, Kant failed to make this point explicit, resulting in confusions of all sorts, from the mistaken idea of the "emptiness" of the Categorical Imperative to a moralistic reading of non-moral values. Moral principles do not "prescribe" or "command" all actions, but in some cases only limit creative actualization of non-moral values. Since this limiting role involves evaluation, especially moral evaluation, moral deliberation should be clearly separated from value, which Kant noticeably fails to do. It appears to incautious readers as if value were prescribed by the Categorical Imperative, although Kant does not actually argue for moral principles as the source of non-moral values. This point applies to both cultural and personal values as well. Moral values do not "command" the content of these other sources of value.

Kant's treatment of the virtues suffers from his emphasis on duty. Kant argues that virtues in general are duties, while duties are moral. However, this was not the predominant view in ancient thought. Plato, Aristotle and Cicero wrote as if the "excellence" of any function is a virtue.[22] Thus there is the "excellence" of a good race horse, of a loyal or useful dog, and so on. Excellence in cooking would be a chef's virtue; of writing, an author's. Such virtues are of different kinds of values in that different kinds, or social categories of value make specific actions or different fields of action possible, which would not be possible in the "jungle," that is, if there were only autonomy and no society. Such kinds allow agents to develop their capacities to full measure: virtue of various kinds. Thus many virtues are not strictly moral: they are good, or of value, but not "right" or required. Virtue in such cases may make for a better life, then, even if it requires a moral condition or limit.[23] Virtue is like a spice added to a moral life to increase its value yet further.

Virtue is possible in all activities, then, since we can increase our excellence in anything. Anyone can be virtuous in developing some talent, their work, their society, home life, daily activities and so on. Virtue is equalitarian taken in this wide sense, since everyone can be proficient at something, even domesticated animals (seeing eye dogs). Only "moral virtue" implies excellence of character. "Character" implies social reliability of some type, that is, that an individual can be trusted to act virtuously in requisite situations. Character is the sum of such moral excellence; the more virtue is creatively actualized, whether in or by an individual, the "better." However, it would be difficult to argue that perfection or excellence of non-moral virtue is a duty. We can no more expect a tone-deaf person to excel in singing than a child to explain physics. Kant did argue that we should develop our talents. Presumably, he would have argued that a tone-deaf person does not have any potential talent for music to develop. But the distinction of some virtues from duty is not clear.

Another problem with Kant's ethics is that he was writing in an age before the development of the modern social and anthropological studies. In his stress on autonomy, Kant also missed the role of culture in forming the values of the person, including the value of individual autonomy in Protestant culture. Many actions are culturally conditioned, even as the individual believes that such actions are a result of personal choice. Others are culturally commanded. (Whether culture is an *a priori* condition as a universal or "empirical" I will leave to others to determine.) "Autonomy" is itself socially conditioned as are duties defined by a culture. The individual can decide whether or not to follow such social rules, but this may itself reflect a loose society, which allows deviant behavior, by contrast with a strict one which does not, for example, fundamentalist Islamic regimes which literally carry out such sentences as stoning of adulterers, execution of gays and cutting off of the hands of thieves. More relaxed regimes in the same countries winked at such actions (the Shah versus the mullahs in Iran). Again, reforms of society may be the result of the pressure of social movements, for example, for civil, women's, gay and other rights. Individuals within such movements of course contribute to the consensus. However, they are in turn shaped by the possibility of reform.

Indeed, morals grew out of cultures and may even be characterized as the identity principle of different cultures that emerged from history. The way of life of a culture, however, is the differentiating element. Different cultures represent creative actualization of different values in a unique, hence inherently valuable way of life. Even cultures intolerant of individual differences may be the source of some good values. In turn, individuals acquire value within the culture as members and bearers of the culture. The distinction of bad cultural values, such as violence or intolerance, supplies grounds for the distinction of morals from values. Morals provide grounds for evaluation of some practices as bad or evil. This is the task of philosophy. Kant did not stress the wider view of ethics as a way of life in his analysis of the role of moral "oughts" or norms.

Moreover, morality is social not individual: the individual is inculcated with social and cultural values in education and acculturation. Individuals are expected to creatively actualize cultural values in appropriate situations: courtesy, honesty, generosity, and the other virtues. In other situations, moral norms limit what ought to be creatively actualized by individuals, in the form of rules, taboos, and other principles. But as I have stressed, in such situations moral values *per se* may not be the goal of acts, for example, in art, economic life and other categories of social activity. Moral values are creatively actualized where they conflict with other possibilities or where requisite as virtue. Again, character is a limit, which does not delimit the content of acts that do not violate norms of virtue.

Nor can there be a duty to one's self, given the norm of duty. A duty is due from one agent to another. We do not have two selves of whom one owes a duty to another, but one self. Kant acknowledges this point with the use of the term "imperfect" duty, but I doubt whether any duty to self can be derived. What is the value of moral rules prohibiting acts that do not affect others? No good is served, or right. Duty is primarily social: duty to the other, in Emmanuel Levinas's language. Cultural taboos affecting the person, for example the norm against suicide, are based on cultural values, and therefore not moral duty in the strict sense, since the norm is not universal. We may indeed argue for the value of staying alive to the individual, but this places this value in the area of personal values or else of virtues. Personal values are those individual projects that a person creatively actualizes; while within moral limits, such values are not moral values.

Kant also ignored the historical development of ethics: moral progress and improvement. In Kantian terms, morals are the unchanging conditions under which new and different situations must be brought as the ground of moral worth. For Kant, events change, moral principles do not. However, such a view does not do justice to novel circumstances, giving circumstances proper evaluation in terms of moral consideration. New issues also arise for ethics, for example, whether we can ethically clone humans. The central issue of our time is one that Kant ignored, namely, the consequences for the environment of human actions.

Finally, meliorism and moral progress argue against absolutism in value theory. Moral progress is the nemesis of any final completion of morals, meliorism, any absolutism of values. Compossibility implies plural, finite inherent goods and thus limits on their value as finite. Categories of value constitute different kinds that reciprocally limit one another if their differences have been creatively actualized to constitute different parts of the world. Their worth lies in their own distinct value and not any reference to some external absolute. Absolute values would limit new creative actualizations to accord with the absolute value, whether according to a political, religious or other standard. Different kinds of value would be contingent upon the absolute value alone, their worth derivative, not inherent. But the value of a good plane is not contingent on anyone's will; it will fly even if all the Kantians will it not to. Kant's arguments for the absolute value of a good will have the technical difficulty that a good will is contingent upon willing in accord with the moral law, since Kant rejects voluntarism as whimsy. But then a good will is not absolutely good, the moral law is absolutely good. A will becomes good by willing duty, not of itself.

Time is the enemy of all inherent good as death, decay and oblivion. Excellent ploughs from the nineteenth century will find little use on today's mechanized farms; the body as the "temple" of the soul wears out and dies after even the longest lifetime. Even inherent or intrinsic value is not infinite, then. Moreover, improvements in quality and kind leave the most perfect goods as quaint historical artifacts. Historical progress in moral evaluation has meant an increase in the scope of moral considerability; the value of humans has been expanded in only the last fifty years to include some non-human values, whether some animals or the environment. The norm of meliorism cannot be absolutist; there can be no absolute standard if improvement or betterment is a constant challenge. However, creative actualization is not relativist, whether in terms of culture or egoism, since such standards are rejected for moral evaluation. Instead, creative actualization is pluralist, since a variety of agents may creatively actualize, as I argued in chapter one, and a variety of kinds may be good. Intrinsic value is not confined to one over-arching good, a *summum bonum*. Properly understood, intrinsic value is incompatible with a *summum bonum*. It is also incompatible with a speciesist definition of intrinsic value.

What of moral standards in evaluation? Are these, as supreme, absolute? But if such standards improve over historical time, they are not final, but finite in value. They are good relative to the moral progress of the time, but not as refined as those of later, better ages, with even higher moral standards. Moral progress is melioristic and moral improvement lies in the scope of morals as well as perfection of its practice, that is, improvements in creative actualization of moral values. Moral progress has reference to history, not a *summum bonum*. Preservation of life is also a moral value. However, preservation of life is not an absolute since there is a time to let go of life, to "let die." In a beautiful view of death, Chester Mahooty, a Zuni, remarked on the death of his wife that it was "her time." The value of life is not absolute in war, in self-defense, or in the wild. In the case of any one individual, the value of life is finite.

4. The Relation of Value and Obligation

A central topic among the environmental ethicists has been the relation of value and obligation. Indeed, since many have adopted foundational models, they have based their entire approach on this relation. They attempt to justify moral treatment of animals, living things, species or holistic relations, like that of the elements of a habitat to the habitat as a functioning whole by arguing for their inherent or intrinsic value. The argument then proceeds that if something has inherent or intrinsic value, it is entitled to moral considerability. This is the hidden minor in many of the intrinsic value arguments in environmental ethics. The conclusion, that moral agents have a duty to protect bearers of intrinsic value, does not follow directly from the "presence" of intrinsic value. The minor is required to connect value to obligation. This creates a warrant, ground, reason or justification for the protection of non-human nature, however we define it. The problem with this approach is that obligation cannot be logically derived from value. This problem arises due to the foundational model that is widespread in ethics, that is, that obligation must be grounded in some justifying value.

In this section I will discuss this type of foundationalism in ethics and the problems with foundationalism. What is the significance of the relation of value and obligation to the topic of the first section, consequentialism? The connection may be foundationalism. For example, utilitarianism argues for a foundation in happiness and this should be what we aim at; happiness is a foundational value that is also a consequence. I will discuss the latter as well.

What then, is the relation of value and obligation? One answer is that all goods, including moral goods, receive value from a highest good, a *summum bonum*, including good actions. The good of the many is contingent on their compatibility or consistency with a highest good. Thus for vitalism, goods are good because they contribute to life. Another answer is in terms of the relation of intrinsic value to moral considerability: intrinsic value stipulates precisely what will be morally considerable. Obligation is defined in terms of or in relation to intrinsic value. What is right increases intrinsic value as a consequent. This is the tactic of consequentialist ethics, including both Utilitarianism and some forms of Pragmatism. This formula of practical reasoning was adopted by many environmental ethicists, who argue from intrinsic value to moral considerability or even moral obligation, following a consequentialist model. (If intrinsically valuable, then morally considerable; if morally considerable then morally obligatory (McDonald, 2003, ch. 1). They apply it to living things in general.

Another theory is that moral values are intrinsically valuable or a special class of intrinsic values: the original contribution of David Prall's version of the interest theory of value. Moral values are defined as intrinsically valuable in this view and thus no conflict between intrinsic value and moral considerability can arise. Kant defended another view. Kant argued that obligation is the condition of possibility of intrinsic value: if obligatory, then intrinsic value. Intrinsic value is not the condition of moral norms, but on the contrary, moral norms are the condition of the goodness of value. A good will is a moral will, that considers moral principles and acts for the sake of duty. Obligation may even be said to define intrinsic value: the value of persons as moral agents in the second statement of the Categorical Imperative.

Finally there is the holistic view best represented by Dewey.[24] Dewey was a consequentialist, but differed from simple consequentialism in that a number of factors in interrelation should be considered in contemplating action. While consequences are one factor and they provide the warrant for the means, they are not the only one. Others include the particular situation, past solutions and traditional moral values, overall values, the means, and so on. Intrinsic value is only one factor among others, although, contrary to early readings, Dewey did not exclude consideration of intrinsic value. What is right or obligatory is what is practically necessary, but this is a reflection of a multitude of relevant factors.

The question I want to examine here is whether (intrinsic) value provides a sufficient justification for obligation. Why is a person obliged in terms of an intrinsic value? How does intrinsic value logically entail an obligation? Why is there an obligation to respect the rights of what has intrinsic value?

First, the distinction of stipulative and inferential obligation should be made. A theory of intrinsic value may stipulate what is morally considerable or morally obligatory or both. Utilitarianism and Prall's conativism stipulate what is moral by defining solely in terms of value. However, other theories argue that obligation follows from value as if by logical inference. Here the problem arises. Similarly, how can an obligation be derived from what "is" good? (This problem is parallel to the "naturalistic fallacy" argument that an "is" cannot be derived from an ought.) Let us say that everyone agrees that happiness is a good, however many conditions we place upon it and however we ultimately define it. The question then becomes why anyone has an obligation to be happy, increase the happiness of him/herself or others or regard happiness in moral terms? Good can include other kinds, such as economic and aesthetic goods. Surely no one would argue that we have an *obligation* to increase the number of consumer goods or art works in the world? It may well make for a better world, but no obligation is logically entailed. Nor is it axiologically entailed, since goods other than consumer goods might be even better, such as world peace, social justice, environmental or wilderness preservation. Similarly, it may well be that happiness is a good condition. But does it follow inferentially that we have an obligation to it?

One way in which some believe that intrinsic value justifies obligations is that intrinsic values are valuable as ends not as means, for their own sake, not as instruments. While we have no obligation to preserve or increase instruments like consumer goods, we may have such an obligation toward intrinsic goods. The question is why this follows. Why does anyone have an obligation to preserve another person's happiness? Another way of putting this is to consider the following. If intrinsic values necessarily entailed moral obligations, we would be required to preserve and increase every intrinsic good.

Ethics is the condition of social values, values that require society, and those required for life, that is, almost all values.[25] A condition of values does not have a foundation in values nor in metaphysical postulates such as being, or the subject. A condition is the basis for what is conditioned and the conditioned cannot thereby provide its foundation. As for the more metaphysical candidates for such a foundation, I argued in earlier chapters that value is independent of such a foundation as their basis. Goods in the world are actualized as a result of evaluations of worth, and thereby have evaluation as their initial ground. The actuality of such goods follows from evaluation as their basis and the decision to creatively actualize them as a worthwhile project.

One issue is whether the condition of a value is a value? In one sense a moral value is a norm or a moral principle of evaluation and thus is not yet a value. However, if the moral principle is acted upon it becomes instantiated as a moral value. As a condition for evaluation it is an "ought" but as creatively actualized it becomes a good in the world. Moral values are a required condition of other values and if they are creatively actualized, they are values as well.

Morals and values have a reciprocal relation, not a foundational relation, as I argued above. Moral scrutiny is a check upon creative actualization, since what is immoral cannot be justified. The morality of values is a condition of their worth, and is creatively actualized in the reciprocal relation of intrinsically valuable agents. Inherent values require the moral relation of reciprocity if they are not to be subordinated as instruments. But moral norms, in the form of principles and obligations require creative actualization as moral values in acts of duty, virtue and other distinctly moral kinds. Also, the value of morals must be manifest. Neither is foundational; both are grounds or conditions in different respects. Basically teleology has a double direction, in that the logical ground, the end, precedes its realization as result. The consequence of the act grounds the end as result.

To ground in the subject would be a denial of the world: it would be solipsistic and egoistic to argue that the subject comes first.[26] The subject is in the world and formed by culture, preconditions. "Ethical egoism" is an oxymoron, since the ego cannot be universalized as distinctive. The subject cannot generate duty, for the basis of morals lies in other-regarding duty. The philosophy of the subject represents an inward turn; duty is to others, an external not internal relation. The inward turn began with Socrates and his linking of virtue with wisdom and wisdom with reason. It was developed by Aristotle (*eudaimonia*), the Stoics (reason) and Epicurus (feelings of pleasure). The more relational notion of value in Plato, the order and harmony of the parts—an external relation—was eclipsed by psychological accounts of value and their use as a foundation for morals. The inward turn was accelerated by the Neo-Platonists and given a "foundation" in Cartesian subjectivity. Kant tied value in agents to the pure will but also rejected psychological accounts of value and first articulated the transcendental arguments in ethics: that moral principles are the condition of moral worth.

Ethics is manifest in action, which is public, not private, and creatively actualized. Duty governs actions that affect others, social actions, not private worlds. Moral rules are taught and culturally sanctioned and enforced by public criteria with public constraints. Virtues are exhibited in public actions, not private whims. Intent, conscience, guilt, anxiety and other internal aspects of ethical life are either byproducts of public rules and virtues, or religious in origin. Guilt follows from breaking social rules not as the cause of breaking the rules. Morals do not have a foundation in character but rather form the character.[27] Ethics is the originary ground, in one sense of that term, the condition, and thereby does not require a foundation, especially in a subject that, on the contrary, is formed in terms of the ethics of a culture.

In another sense, foundation means something either indubitable, self-evident or "given" from which reasoning can proceed with certainty. Value is not self-evident, since it must be practically attained, and complete success is uncertain, even when the same goal has been accomplished many times in the past (agriculture). Nor is the value of results always self-evident since the actual value of the novel may not always be apparent, and may be subject to improvement: meliorism.

Since value has to do with creative actualization of a goal, which generally lies in the future, the achievement is not "given" but made. Presuming we could agree on what is "given," a dubious prospect judging by the variety of hypotheses on every topic in the history of philosophy, it would have no bearing on ethics. The "given" is often negated or altered by moral actions, since they stand in need of correction (slavery). What "is," like the treatment of the Native Americans, may be morally reprehensible and stand in need of reform. If "being" is what is "given," or what now exists, then it cannot provide a foundation either. Ethics is creatively actualized as a novel action in the world and its consequences, and is thereby a source of change. Change upsets what "is." The mark of moral value is improvement of what "is" and reform or replacement of the "given," not rationalization of what "exists" and its accommodation. The origin of philosophical ethics was a critical evaluation of the defects of the "given" and of what "is." The question for ethics is not what "is," but whether what is "given" can withstand critical moral evaluation. What good or purpose does it serve? Does it meet moral standards?

5. Creative Actualization of Moral Value

What is the role of principles in actualization of moral values? I critically evaluated situation ethics in the last chapter. To summarize: situation ethics are radically nominalistic. Ethical situations are never of a kind—thus in this view there are no principles or norms. Yet if the situation invariably governs what to do in morals, situations are themselves a norm. In this sense, situation ethics is logically problematic, since it appeals to principles in making of situations themselves a standard. In this respect, a pure situation ethics is impossible. Situations and problems have a *likeness* as well as a difference—categorial and cultural connectedness. This is not only convention: the case has its own "*logos*". Certainly, problematic situations have unique aspects, and these differences must be considered in evaluation. However, situation ethicists have not made the case that such situations differ morally in the relevant respects. In other words, the moral principles can remain valid in different situations. Is the validity of moral principles justified?

Recognition or discovery of moral aspects in a situation—that something is "bad" or "wrong"—implicitly recognizes a rule. It subsumes the situation (in the relevant respects) under an abstraction from past moral cases, or a rule in its cognizance of the wrongful or bad aspects of the situation.[28] Moral principles unite situations of a kind, which then come under a rule. Situations are morally classified as kinds, identities. In this respect, morality provides a tenuous "*logos*" for rules governing unique situations: they limit action but do not define actions appropriate to a situation. Different situations differ by time, history and space (place), circumstances, relevant type, and other factors. Rules are invoked as covering a situation according to the type of situation. Similarly, calculating in terms of consequences as a rule *is* a principle as is its universal claim. The normative character of moral concepts is an argument for their generality: the appeal to duty is an appeal to a principle—the identity element. Duty is following such principles, and their creative actualization in actions. Voluntary principles are recognized as obligatory. Obligation as necessary (the ground of possibility) for other values, includes the environment, necessary for all human values. Ultimately, principles are a moral interpretation of situations or moral instances that bring them under moral evaluation and judge their moral significance. The moral aspect of situations and circumstances is thereby revealed.

Relativism does not fail on this score, although it does on others; it either does or does not subsume the situation. Relativism's argument is that the case fits for some and not for others under recognized moral rules (or that the rules themselves are relative). It does not deny the rules as such—although if it did it would fail. "Act" theories also fail on this ground. They are too atomistic to be moral. To define an act as moral is to appeal to *a priori* or *a posteriori* rules (respectively, Kant and Mill). However, the distinction of the ethics of acts from that of rules is probably a false dichotomy, since we frequently act according to rules; if so, acts embody rules as instances. (I am indebted to William Frankena (1963) for his discussion of the distinction of the ethics of acts and rules, in the context of Utilitarianism.)

Universality as a formal requirement of Kant's ethic gives necessity. We *must* do what is morally commanded as a consequence of universal reasoning: the conclusion is practically compelling (as opposed to the contingency of hypothetical imperatives).

However, the application of rules to cases is pragmatic, a matter for decision, as Kant implicitly recognized in his doctrines of free will and autonomy. In one sense this means that rules are not provisional so much as their application to cases is a matter of deliberation. We must decide which rule to apply in a situation, and in some circumstances we may decide not to apply the principle. The decision to apply a principle or not to apply one means that rules are not exceptionless, for they do not cover every situation. Rules against lying do not cover murder directly. In this respect, moral situations are relevant, for the relevant moral principle to the case is one factor in moral evaluation: moral judgment. Judgment is (1) the capability to apply principles to situations (compare "moral sense") but also (2) the willingness to apply principles to situations.[29] In another respect, consideration of exceptions and hard cases may be pragmatic in tone.

Moral virtues are similar to scientific predictions, and act as a predictor of character. So in a sense moral rules are similar to mathematical axioms: in either case if violation takes place, a person goes beyond the limits of the axiom (immoral; incorrect); and the axiom, which is a rule, defines what is included in the realm (Euclid's fifth postulate and parallelograms; rule against stealing and honesty). Just as math can be applied, so can ethics, although both can also be viewed at the abstract level. The violation of ethical rules in society leads, or could potentially lead, to such and such results, just as violation of mathematical rules in engineering will cause of result in the collapse of a bridge. That some rules are culturally relative does not make them any less categorical. Western science may have originated in one culture, but nevertheless be categorically necessary to achieve its fruits: a semiotics for technology. However, the analogy of ethical rules to math can be overdrawn. Ethics has many parts that are not analogous to math, even applied math.

Moral principles are only one element, albeit an important one, for evaluation of moral problems and situations. Morality is more than simply applying rules to circumstances, which taken alone would be excessively rationalist—although it includes moral principles. The role of universals and reason in deontological ethics strongly support underlying intellectualist values, despite the attention paid to values of the person, among others. As I argued above, circumstances and consequences should be taken into consideration. Moreover, Dewey's critique of the relation of theory to practice raises the question, does ethical practice interact with theory much more? Does theory reflect longstanding practice to an extent that Kant either ignored or did not consider important? Thus we can argue against prescriptivism that we use examples of good or right actions in teaching and explaining ethics. Ethics is in part abstract or general class descriptions of actions, not mere commands or imperatives.[30] Creative actualization includes bringing principles to life in the world through action.

The main issue that arises is whether there can be exceptions to general principles or whether moral principles are inviolable, or absolute, as Kant implied. In one respect, Kant is correct, for obligation overall cannot be evaded and moral worth accrue. Moral agents have obligations or they are not moral agents.[31]

However, because we are free to evaluate, we can, as moral agents, make exceptions to rules for the sake of morals. Moral limits are not violated by saving one or another when only one can be saved.[32] The choice of which can legitimately rest upon other factors, whether kinship, intimacy, species identity, and others. Thus if the issue is saving a human vs. an animal, it does not violate moral principles to save the human, just as bird parents save their own young. Moral evaluation is general.[33]

A rule specifically to handle exceptions is needed. Such a rule was implied by Mill: a higher value or principle justifies violation of rule. Only ethical exceptions override general ethical rules. Even religions that hold their moral principles inviolable recognize exceptions, for example Orthodox Judaism, Protestantism and Islam, all of which make an exception to their strict rule against abortion to save the mother's life. This is because her life is an instance of life; life is the value the rule is trying to uphold. Again, in medicine, white lies are used strategically where these contribute to patient survival. Another example of justified lying might be hiding refugees from oppressive governments, for example, from the Gestapo. The casuistics of truth telling are general, no absolute rule suffices, even though many ethicists argue against such lies, and even against evading telling the truth.

Is this the essence of problem of hard cases? Evaluation of hard cases involves a choice when two important values or principles are in dispute. I would agree with Ross that a higher duty may supercede a *prima facie* one, as this is consistent with Kant's view of the relation of maxims to the Categorical Imperative, namely, a hierarchy of rules. The problem is: will morality break down if exceptions become built into it? But aren't there always exceptions? Relative to an end, desire, context, means? The fear of absolutists is that allowing exceptions will also allow wiggle room. Decisions will fall short morally and agents will rationalize them away as exceptions. One exception will provide grounds for many. Moral rigor and purity will be lost. But moral purity can be added as a principle to moral evaluation, without bringing in absolutism. One additional principle would then be to be as morally rigorous as possible in moral evaluation. We should only make exceptions on moral grounds, not because it is easier to do so.

Morality is only a limit on some acts, and not constitutive of them. Often ambiguous value situations are not decided on moral grounds. Two opposed actions may both be morally acceptable even if there is a value conflict involved. Morality only limits such acts as a border. It does not decide for us. Morality is a limit on actions done in a social context and thus limits all actions in social situations, but not those outside social situations. Thus factories that pollute but are profitable are immoral, because they have a bad impact on others. Moral limits concern the relation of value and obligation, in the creative actualization of non-moral values.

I am not arguing that all moral situations involve limits. On the contrary, in some situations duty should prevail and we should creatively actualize moral obligations. I should emphasize the point that morals can be creatively actualized as such and thus moral principles are not only a limit. However, other situations are morally neutral, and can be decided on other grounds so long as moral principles are observed as a limit. Moral values ought to be creatively actualized, but moral limits only observed. In all such situations, then the moral consequences should be evaluated. We may legitimately distinguish moral values, moral limits and consequences.

John Dewey's and James Tuft's arguments regarding character may constitute a counterargument to my moral limits principle, since agents may actualize duty because of their character. As they put it, "all actions which call out lively esteem or disfavor are perceived as acts of persons: we do not make a distinction in such cases between the doer and the deed." (1932, p. 297). Moreover, Dewey and James Tufts, David Hume and others argue that all these are "moral" if "morals" is taken in some generic sense that includes all values. However, this does not change the principle that "non-moral" acts, in the sense of actualizing non-moral values must accord with moral limits. As for character, virtue is creatively actualized in virtuous actions.

Another problem with Kant's account is that the moral status of non-moral values is unclear. If we must act on maxims for the sake of duty, then what of beauty and other types of value? If we bring about more beauty, whether through gardening, art, or preservation of beautiful landscapes, are we acting for the sake of duty? But how can our motive be beauty if it is duty? If acting for the sake of beauty is our motive, are we only in the realm of what is consistent or permitted by the moral law? But then how can we be acting for the sake of duty?

I think that the solution to this riddle is that in our deliberation our maxim must be subjected to a test and we must always act for the sake of the law. If the maxim of our action would not be in conformity with the law, then we must act for the sake of the law, not against it. But acting for the sake of the law means that we must bring about an act of duty, that is, that morals require creative actualization. Acting for the sake of duty or morals means acting in the phenomenal world and instantiating duty at a time and place. Kant argues that intent is what is morally decisive. But surely his argument is not that we should intend one thing and enact another, that we are not responsible for actions at all? The question is the degree to which we are responsible. We are at least responsible to try our best to enact moral values where these conflict with other values. To resist bringing about evil is to act morally. To act virtuously is to creatively actualize moral values. Morality governs action more than intent and actions aim at an end or consequence. Moral evaluation concerns the evaluation of such actions in deliberation for right and wrong consequences.

Morals require creative actualization, as moral principles must be brought about in the world through the practice of duty. I argued, "intending" to do something is not morally sufficient. We are required to act morally if there is a conflict of morals and other values, although values that are morally acceptable are not done for the sake of morals. But to achieve moral value, the action must be successfully completed as a duty. Value is deliberated upon in *evaluation* under a moral condition. We must evaluate if an action would be right or wrong. But bringing about a moral action is practically required: we are obliged to do it. Otherwise, morals are Utopian. Either we achieve the goal of creatively actualizing the moral value or not. Success in enacting a moral value marks a virtuous or good action, failure a bad or defective one.

Ethics does not only concern the ought, it also critically describes what is actual, and even predicts what will occur. Because value is creative actualization, what humans do describes their values, their idea of the ought. A scientist values truth or physics or sub-atomic particle explanation. Scientists believe they ought to be engaged in such research. Moreover, projects involve creative actualization of plans in the future. Thus the separation of ethics and "fact" along the lines of the normative and the existential is untenable. Creative actualization achieves valued goals, including normative goals, the creative actualization of what "should" or "ought" to be.

The relation between moral obligations and values is reciprocal, not hierarchical, since moral values require creative actualization. But morals are the justification and condition of creative actualization of non-moral values. Since their relation is reciprocal, they are conjointly required. Reciprocity is the relation of intrinsically valuable agents as well as the condition of compossible values. Morals and values are independent variables then, but related. Moral values are exhibited in acts of duty and virtue that are undertaken as choices that are alternatives to creative actualization of non-moral possibilities. Actions should be morally evaluated, since they have consequences for others and for the biosphere as a whole. Morals are not self-enforcing; limits are freely observed. Freedom and duty are reciprocal then. This internal reciprocity of moral values is distinct from external reciprocal relations of moral agents.[34] Moral limits to actions which actualize values constitute a reciprocal relation, reciprocal values in a just relation; their enactment morally limited.

Although morals may be an independent variable in evaluation of what should be actualized, the *value* issue is independent of moral evaluation. Its inherent value does not tell us if we should creatively actualize a good. As I noted in the first chapter, we are not required to creatively actualize every possible inherent value, or for that matter novel intrinsic values, but only moral values. Whether an H-bomb is good is in one respect independent of its moral valuation (McDonald, 2011, ch. 3). The bomb is "good" as a bomb if it has been actualized without defects. But this does not tell us if we should *use* it (Hiroshima, India and Pakistan) or even *build* it. These are questions for moral evaluation, not value theory as such.

Inherent value as such does not have a moral condition. But actions do, since they must be evaluated and their consequences for society, the environment, and other factors must be deliberated. We are only required to creatively actualize moral values. I agree with Kant that moral considerations are independent of value, that is, some actions are right apart from valued ends. But this belongs to evaluation, since we deliberate about morals, evaluating different courses of action prior to enacting them. (Moral evaluation should not be confused with moral value. Moral deliberation is a form of evaluation, not to be confused with value as such. But moral values are a species of creative actualization.) My point here is that morals have a value condition as well. We are required to successfully achieve moral ends in the world, as part of our responsibility. If we succeed our act is good, if we fail, bad. Moral responsibility depends upon such creative actualization, that is, upon value, for unless we achieve the moral good, our actions are defective, and moral ends are not achieved. Thus morals depend upon value in *practice*.

Similarly, norms require creative actualization. "Oughts" are not *a priori* but teleological—they involve future time if they are practical and truly valued, not spurious. Genuine oughts predict what *will be* but *are not*, unless an "is" corresponds with an "ought" already, for example, in bravery. If we ought to we will and making the "ought" actual confirms it. We creatively actualize what we ought to do in principled action. "Ought" implies can, as Kant noted, but it also implies that we can creatively actualize oughts in the world. Thus the world will be altered, however slightly, by our norms, attenuating the distinction of is and ought.

Reflection on and commitment to values generates ought, not a metaphysic of morals. Values imply duty or they are not valued. Values have a moral condition and morals have a value condition. The moral condition is consistency with duty, although not necessarily for the sake of duty. The value condition is creative actualization, successful realization in the world. Unless we try as hard as possible to act dutifully where this is appropriate, our morals are feckless. In other words we are responsible for the consequences of our acts in the world and this is one factor that should be evaluated in moral deliberation. Intending to do good but achieving harm is not enough.

The issue I raised above was how intrinsic value can justify morals. By making morals intrinsically valuable, which we are required to creatively actualize, the issue is redefined. Morals are not justified in terms of extrinsic consequences, in the sense of non-moral intrinsic values like happiness, the foundational model. Creative actualization of morals is itself a consequence of acting for the sake of duty. Moral values are intrinsic values and therefore do not require further justification as values. Moral values are the creative actualization of morals as virtues or moral actions, duties, done for their own sake. We are required to act morally and by doing so we realize moral consequences. Moral values are then actualized in the world, not confined to the intent of a will. Enacting them, acting for moral ends where these are in conflict with non-moral ends, has superior consequences in the world, amelioration.

For an ecocentric cosmopolitan ethic, justification of a value judgment is in terms of cosmopolitan moral considerations, that is, all the relevant elements of culture and the environment taken as a whole.[35] Each element must be considered, but their relation is reciprocal qua inherent value. Actions are undertaken if we judge that they would be better than alternatives (meliorism), in terms of environmental consequences, and within moral limits. Ends regulate means (action) but are only justified if they are within moral limits. They are not justified by intrinsic value as a *summum bonum*. Intrinsic value is morally considerable, but does not warrant obligation. Only practically necessary conditions generate obligations, especially threats to the biosphere as a whole or species as a whole. There can be no positive obligation to intrinsically valuable individuals in the wild or they would be domesticated, undermining the whole.

Ethical justification has multiple elements, since ethics has plural principles and values. Since ethics actualizes moral values, one part of Ethical justification is axiological. The value is worthwhile in itself, but may also be instrumental to a good life (way of life), or a part of overall moral excellence or character, a pluralism of values. Ecological virtue means acts that are better adapted to habitat, less destructive of habitat: constructive not destructive. (2) Normative justification concerns others and our relation to them: do unto others as a universal principle actualized as moral relations (reciprocity, justice and the principle of participation). Is the act within moral limits? (3) Teleological justification concerns the conditions of a good society, non-violent and categorial (meliorism, moral progress). To these, ecocentric cosmopolitanism would add an environmental condition. (4) Does a possible or proposed action help or harm the biosphere as a whole, species, including our own, and local habitats? (compossibility, consequences).

Since we can consider consequences in moral deliberation, we can consider environmental consequences. Since we can consider consequences for the environment as morally regulative principle, doing so is our duty. But then the separation of duty and consequences is overcome—it is our duty to consider environmental consequences in every decision and care for it in action.[36] This duty arises upon consideration of our own condition of possibility, the conditions of all life as self-actualizing (intrinsic value). However, The duty to consider consequences for the environment transcends the egoism-altruism distinction. The consequences for the self in environmental impact cannot be separated from the consequences to others.

The outer world is not excluded from moral consideration: considering consequences for the world is required. Mindless destruction or harmful alteration of habitats raises a moral issue. We can act to prevent such consequences and are required to since the environment is the condition of all agents, all intrinsically valuable persons. Indeed, we are obligated to avoid harmful actions, which might hurt the environment. Acting to preserve the environment is acting for moral consequences, realizing morals as goods in the world. By doing so we are taking responsibility for our actions, not simply intending to act rightly.

In sum, the relation of duties and consequences should be revalued in light of the importance of consequences for the environment. For an ecocentric cosmopolitan ethic, our duty includes consideration of consequences in moral evaluation. Bad consequences of our possible actions are an element that must be evaluated in moral situations. Duty and consequences have an internal, reciprocal relation, since we cannot fulfill our duty without consideration of consequences and also in many cases, creatively actualizing our duty as a consequence of our actions. In turn, consequences have a moral condition, whose principle is manifested as a duty. The duty arises in avoiding more nefarious consequences, namely, to what extent should humans be willing to make sacrifices to preserve the environment, given the threat to all life of human expansion? Our duty is redefined in relation to our world, in which we are at home. The consequences of our actions on the world are a primary consideration.

Six

RATIONAL SELF-INTEREST, AN OXYMORON

> The fundamental source of moral conceptions of right and justice is that men and women live in society, in a community of some sort.
> John Dewey and James Tufts[1]

It is unfortunate that Andrew Light, Erin McKenna, and other pragmatically minded environmental ethicists have adopted the phrase "animal interests," probably under the influence of Peter Singer.[2] Bryan Norton, for example, has argued for value within the tradition of "interests" and finds it hard to clarify what the notion of the interest of a species would be. He notes that analogies to individual interests are not helpful in a holistic context.[3] He is echoed by Joel Feinberg and Nicholas Rescher, who, starting from the individualistic premise that morality concerns the interests of different actors, argue that species cannot have interests.[4] The term "interests" implies self-interest and its origin as an "ethical" term lies in the adoption—one might say adaptation—of Utilitarian ethics by economists late in the nineteenth century. "Interest" is a banking term that has no place in ethics proper; its use by contemporary Utilitarians as a substitute for the classical "happiness" is unfortunate.[5] Indeed, John Stuart Mill, one of the fathers of Utilitarianism, proclaimed that only a "moral blank" could live their life on self-interest alone. It is not for nothing that Aristotle and other moral philosophers condemned the loaning of money for interest on moral grounds. As Immanuel Kant argued, an ethic based on self-interest is an oxymoron, since ethics is concerned above all with the other, despite Kant's concern with autonomy. Indeed, Clarence I. Lewis (1969, p. 34 et al.), who was influenced by Kant, even separates moral from prudential and other concerns by whether others are involved or would be significantly affected by an action. Worse, "(self-) interest" introduces a morally dubious concept from economics[6] into ethics as a theory of value, contributing to the economization of ethics that has brought on the problems this book is addressing.

In this chapter I will argue against any notion that ethics can be derived from self-interest. I will first consider the relation of reason, desire, and self-interest. I will then consider a critic of environmental ethics who argues from the point of view of self-interest, and also critically evaluate the "ethic" of self-interest. I will then critically evaluate Utilitarianism. I will examine their connection in this substitution of "satisfaction of interests" for happiness by contemporary Utilitarians. Finally, I will critically examine Economic Utility.

1. Rationality

The assumption or premise behind the phrase "rational self-interest" is that self-interest can be rational.[7] This could be interpreted in a number of ways. The first is that self-interest can either be rational or irrational, the most prominent meaning. In this sense, what is rational can generally be distinguished from what is not and self-interested actions can be judged in terms of their rationality. Some self-interested actions are rational while others are not. The task of reason is instrumental, providing a schema for actions that will be rational. Reason in this respect is a standard independent of self-interest by which the rationality of any case of self-interest can be critically evaluated.

A second interpretation would be that only self-interested actions can be rational. Actions that are not self-interested are foolish, since they cannot be justified rationally. This is the view of the cynic and the freeloader. The cynic argues that anyone who believes that they are acting on altruistic motives is fooling herself, since her real motives lie in self-interest. And reasons can be shown for this judgment, since ultimately, only self-interest can be a motive and thus self-interest alone is a rational explanation.[8] The freeloader is a practical cynic, who, knowing that some will be fooled into giving more than they take, or following discretionary rules, exploits such altruism for his own advantage. This interpretation makes a stronger connection between rationality and self-interest. Anyone who does not act on self-interest is behaving irrationally.

A third interpretation is that rationality is itself self-interested, that is, that they are closely connected, or, less plausibly, identical. (Only) self-interest is or can be rational and rationality is self-interested.

Rational self-interest differs from psychological egoism, although they are sometimes treated as one theory.[9] The premise of psychological egoism is that a person cannot act in any way but out of self-regarding motives. Rational self-interest implies that reason and self-interest are related but distinct. Reason is a means, a test for whether self-interest is rational or not in a particular case. Presumably, a person could act on self-interested but irrational grounds. In principle, then, these are distinct theories, although some authors treat egoism and self-interest as synonyms, for example Thomas Hobbes. Some authors jump back and forth between the two theories. I will examine self-interest as an "ethical" theory in the next section.

In this section I want to make the case that these connections between reason and self-interest are fundamentally in error. First, I will examine the diverse uses of the word "reason."[10] Brand Blanshard (1961) sees multiple roles for reason: reasons as seeking causes (p. 345), corresponding to our #3 below (explanation, ground); and reason as inference, corresponding to our #1. There is also reason as calculating how much good each activity will bring and examining alternative goods and consequences echoing our notion of comparison, critical evaluation, and foresight of consequences. He adds the notion of reason as examining identities ("likenesses") and differences (ibid.); and also reason as seeking the universal in the particular.[11] He also adds reasons as causes in emotions, (p. 245); and reason as composite with desire in practical matters (346 ff.). Finally, he cites reason as consistency, for example, of ends (349)—therefore incompatible with desires (350)—but this contradicts his compositism (He provides a counterargument p. 365). Blanshard redefines impulse as rational in a vague compositism of reason, feeling, and desire.

Steven Toulmin, who argues that reason is finite in scope (1970, p. 202 ff.), covers reason as grounds, as cause, and as evidence. However, he adds that reason may also mean valid inference, as worthy (p. 71 ff.) and as categorially distinct according to subject and circumstances (114 ff.; compare Aristotle on the difference between math and rhetoric in *NE* I). Practical Reason includes determining consequences (what we discuss as means-ends) (p. 147-8); following the rule (compare p. 174; rules and ends p. 145-6) (our rule-instance: norms); evaluation and choosing from alternatives ("Decision theory", p. 147). However, this is "practical reason" only as weighing consequences, not evaluation as such.

"Reason" and "rationality" are equivocal, since "reason" can mean the same thing as formal logic but can also be a ground or condition, a human faculty, and even a standard of procedure, among others. "Reasoning" in practical philosophy and ethics can mean, firstly, using logic, a normative science. "Reason" means consistency with specific logical principles, or deduction from them or in accord with them using norms of formal inference. This form of reason presumes these principles as well as consistency and the other so-called "laws of logic." (Mathematical reasoning also is formal.) Informal reasoning can follow deduction, as in an enthymeme; or induction as in informal generalizations from common sense observations, from a few facts or other sources. Reason in this sense includes valid inferences from premises.

Second, In "practical reason," means are related to ends as suitable for achieving them; or instances fulfill the rule.[12] Thus "practical reason" can include either the means to end model or that of rules and instances.[13] Practical reason connects ends, means, and action: this means is appropriate for achieving x end. The role of reason in practical reasoning is contested ground in philosophy. David Hume and Aristotle thought we do not "deliberate about ends," but that "reason is a slave of the passions." But other authors argue that the ends also require rational justification. Moreover, the end or goal can be the "reason" for an action (John Searle), although this may mean "ground." A third candidate is decision theory and evaluation of consequences, although these may be forms of the first two.[14]

Another issue is whether practical reason is one form of reasoning from ground to consequent. Are ends a consequent, or, alternatively, a ground? In one sense, the end is the ground that leads us to seek means to achieve the end. In another sense, as the effect of using means successfully, it is the consequent. John Dewey distinguishes the end as the goal with the term "end in view," in order to avoid this ambiguity. Dewey has noted that means are causes of ends in situations where we bring about some effect. Thus the means-end relation can be causal. (Arthur Schopenhauer argued that both cause and effect, and means to ends are forms of the "principle of sufficient reason.") However, means to ends are different from cause-effect relations in having a goal, although Dewey is correct to tie means to causes. The same factor separates means-ends from ground and consequent.

There is also a distinction between reasons as "causes" from motives—motives are future oriented, but causes are present oriented. Future grounds can include desire and feeling as motives, but these are not rational *per se*. They can, however, pass rational tests, especially critical evaluation.

Ground and consequence: a third meaning is giving explanations, that is, the reason for an event, product, change, and so on. These explanations can be causal, or give other grounds, including motives, "rational explanations." Reason as the final cause in Aristotle can mean goal, plan or *explanation*. Thus he runs together our two and three. Note that whether means are the "cause" and ends the result (Dewey) or means the result and ends the cause (Aristotle in practical syllogism) is contested. One is first in order of events the other in the order of reasons.

Seeking a rational explanation, understanding, is distinct from irrational, mystical or far-fetched "explanations." Fundamental causes, explanations, and so on are the ultimate ground of ethics in some theories, whether within ethics (eudaemonism) or not ("meta-ethics"). Also included in this sense is giving evidence as a form of giving reasons or grounds. Evidence provides "grounds" to adopt a theory in science, so facts can be part of practical reasoning. This meaning of reason can also include desire or other non-rational elements as the "reason" for an act, that is, the ground or explanation. However, this latter sense is problematic, as I will argue below.

Fourth is reason as a standard, which includes rational origin (that is, not empirical, mystical, passionate, and so on in origin), the use of rational principles, or that some thought or judgment "accords with reason" that is, is reasonable. This meaning may be equivalent to (1) but may mean "reasonable sounding" that is, feasible, in accord with common sense, and so on, especially in practical contexts involving the new or untried, as in what is "reasonable to believe…" The former sense, reason as a standard, may include rational procedures (rational norms in Vilfredo Pareto and Max Weber). In the last guise it serves as a standard for actions (for example, in John Rawls) and decisions (decision and game theory). This sense of reason, unlike the first, attributes some sort of content to reason.

Reason as the warrant or justification is a fifth meaning. This can be equivalent to both (3) and (4) but is stronger, since the implicit claim is that the reason justifies, and does not simply explain. (Toulmin's discussion of giving reasons that are worthy probably comes under this use of reason: good reasons.) Thus it can include any legitimating normative notion. The reason as the justification is not only evidence either, but a warrant. It can also involve giving "sufficient" reasons for a belief. As with (3), this meaning of reason might include desire or other non-rational elements as the "reason" for an act, that is, the warrant or justification. However, some have argued that desire alone cannot be a warrant.[15]

Sixth is reason as a faculty, capability, or power of those who reason. Plato contrasted the "rational" part of the soul from the appetitive and the "spirited" part. For Aristotle, reason as a capability can becomes a virtue that ought to be cultivated or inculcated: an intellectual virtue. Kant defined reason as the "faculty of principles," that is, the faculty which, as understanding universals, was also the origin of universal principles. Thus reason, unlike formal logic, has a content in the form of principles.[16] I believe this also covers or includes the sense of reason as distinct from understanding in the German idealists (Kant, Georg W. F. Hegel).

Also, rationality may be required to relate or connect intrinsic value to moral considerability, an issue examined in chapter five. But although the argument for the inherent value of non-rational parts of environment is transcendental (thus Kantian): the condition of (possibility of) all living things, the relation is holistic, including such non-living elements as air, water, and soil. Thus their rationality is not required for necessary value and thereby a moral condition or duty.

In the first meaning, logic, the value of reason is more limited, since reason as logic is purely formal and has no content. It would have only instrumental value as a tool limited to restricted circumstances. Ends extrinsic to reason regulate the use of reason (Hume); non-formal goals are the ground of its use. The other grounds may also be thought of as a "reason," that is, a ground. However, non-rational grounds may also be causes, for example, feelings as a ground for action (ground and consequent). Since a feeling or desire can be a ground or cause, but do not derive from reason *per se*, reason as a concept is ambiguous, as inference, or even rational origin, are not coextensive with ground or cause.[17]

Causes, grounds, and reasons are distinct. Causes determine, while reason is more inferential. But ground, in the sense of ground and consequent, can have content, including irrational content: feelings. The problem here is that the three terms overlap in normal usage. A present danger can be the cause, the ground, and the reason for action. We could separate the reasoning involved in acting from the perceived danger to the inference that we should act to prevent harm from the danger itself as the cause or ground. But this is an abstract separation.

To say a feeling is a "reason" as opposed to a ground is distinct from reason in the sense of the rational capability, which is different from feeling;

or from reason as a standard. Thus reason as a "ground" must be separated from other grounds. The question is what is a "rational" ground if reason in the formal sense has no content? We may certainly infer logically from other grounds, but these grounds themselves cannot be logical.

One answer might be the meaning of reason in (4): reason as a standard. Reason as a standard, which includes rational origin, presumes the judgment that what is reasonable is better than alternatives. We ought to reason, be reasonable, reason from specific premises or principles, and so on. This would be superior to alternatives like feeling or intuition. Our contention is that by doing so it becomes a norm or value, as it cannot of itself qua formal logic provide such norms without falling back on extraneous factors. Reason as normative means we ought to have grounds for our statements, for knowledge, or for action, and that inference should follow norms of logic. In all these cases reason is normative and thereby involves a hidden evaluation. Reason, like nature, has become a hidden norm, which mediates the is and the ought. "Reasonable" is established as a standard and then used as a norm, for example, "reasonable" in law (compare "natural law"), and as a criterion of legitimate belief.[18] Again, in comparison, we evaluate some reasons (grounds) as higher that others. Finding a "good" reason or explanation is selective: picking out and emphasizing particular factors, causes, events, and so on as more important or valuable than others.

The main problem in analyzing "reason" is in the notion of "grounds" for action, that is, the sense of reason connected with ground and consequent. C.I. Lewis has argued that practical reason includes consideration of consequences: to act without considering consequences is irrational. In his view rationality is "the capacity of the individual to understand the consequences of his own acts, and hence to govern them by reference to what is good…" (1969, p. 146). This view ties together such different senses of "reason" as connecting ground and consequences; of reason as a standard—or perhaps warrant—and as a capacity. However, consideration of consequences is what Kant would have called the "content" or "matter" of rational consideration, not the rational element as such. The reasoning is the inference that connects ground and consequent validly; consequences are variable.

A ground may include a great many things that are not rational in the sense of warranted, logical, or even reasonable, to speak nothing of rational origin. Rational origin in the strict sense is, however, problematic. For René Descartes reason is more of a critical faculty that analyzes the data of the senses. Similarly, for Kant, universality was a mark of rational origin, but in form. Sensuality (pure reason) or desire (practical reason) provide content. Thus there is no form of knowledge originating in reason as such. Even intellectual intuition (some Scholastics) requires "intuition." Reason is formal but empty.

The overlap in the term "ground" between rational and non-rational has been hidden by usage. A desire can be a motive, and classical philosophical psychology has argued that desire can influence the will, leading to action. Thus desire can be the ground of action, and a "reason" in this sense, and thereby, some argue, a warrant. But desire does not originate from reason, whether logic, reason as a standard, as a warrant, or the faculty or power to reason. Pure logic has no content, and this excludes desire and interests. A desire to rape is not ethically warranted, and so if desire or interest is a "ground" its origin is not the same as rational "justification," or "warrant." A desire or interest tends to be confused with a legitimate "reason."

The same arguments apply to feeling, in the sense of emotion or Hume's "sentiments." The distinction of reason and passion goes back at least to Plato, but was continued by Hume, for whom reason is the slave of the passions, and thus distinct from it in some sense—and Mill, since feeling sets ends or grounds. So ground means more than "reason" for these figures as it does for Kant. Thus the justifying ground is now a contested point. However, for Blanshard, reason can form a composite with desire—which also occurs in Kant as a synthesis.

The problem with "moral sentiments," "caring," and so on is that these have nothing to do with ethics. Ethics is concerned with right and wrong, a good and bad life, or good and bad actions, not feelings. But good feelings, as Kant noted, will not make a wrong act right. Moreover, feeling is morally irrelevant—a throwback to Cartesian metaphysics in which mind and matter are in separate universes as subject and object, and mind cannot influence reality (reductive materialism). But value as creative actualization is tied to action and consequences, not feelings. Even moral control of feeling is not a feeling but control over feelings. Ecocentrism decenters ethics from a subject who sets value—the egoism at the center of modern ethics. I will critically examine these points in more detail below.

As I noted, practical reason connects ends, means, and action: this means is appropriate for achieving x end. Reason at the least connects ends and means.[19] However, reason also has a critical sense that may limit means in terms of propriety. This brings in reason as a standard and as a warrant. A desire may ground, but is it warranted? Is it sufficient to act on? Thus Arthur Murphy argues that "when the question of reason does arise, a want is not the proper answer to it unless it is understood not merely as a want, but one whose satisfaction, in these circumstances, provides a sufficient ground for action." (1964, p. 46). As Dewey argued in his critique of conativism, ends and means ought to be evaluated. Which end is worthwhile? What is the best means to the end? Practical reasoning includes decisions about alternative courses of action, choices, if you like. We must evaluate the best means available.[20] Moreover, critical evaluation also includes moral evaluation. Are the ends and means within moral limits? Reason in the sense of "justification" involves an appeal to moral principles, such as "murder is wrong" or "it is unjust."[21] Reasoning in the moral sense brings a case under a rule that is required to justify it.[22]

Critics of this view of reason might argue that desire, feeling, and self-interest are frequently warrants or justifications for action. Thus while conceding that reason can be useful in connecting means and ends, the end is decided by feelings, whether sentiments (Hume, some feminists), pleasure (Epicureans and Utilitarians), desire (conativists and others). Mill believed that ethics provides the end, while science provides the means. For his classical Utilitarianism, the theory of value is hedonism, which argues for the supreme value of happiness, a feeling. Thus the highest good is a feeling and means are ultimately warranted by this feeling. Practical reasoning, then, is conceived as ultimately justified by feelings as their ground. Feelings tend to be confused with reasons. In sum, what counts as a warrant for actions is contested ground. I will critically examine this point in more detail below also.

A further problem is the notion in Aristotle that reasoning comes in different kinds, or is distinct according to subject and circumstances.[23] What counts as a reason on practical grounds may not meet the formal standards of proof in mathematics or formal logic. Moreover, a "good reason" for buying some commodity in economics may not count as a good reason in music or intellectual life. While acting on desire and for self-interest may make perfect sense economically, it may be to the detriment of morals.

We could deepen the Greek notion of reason with their notion of it as involving good judgment, or wisdom, as well as learning and experience.[24] This view of reason is tied to what has been called "intellectualist" values, with reason as a standard. Reason as the standard is normative for practice. Although reason may indeed lend itself to good judgment, the source of knowledge is not reason: again reason has no content. Wisdom is not equivalent to reason, but its virtue in Plato, who calls them "akin," but not identical. Even if we were born with reason, or wisdom, occasions of experience bring this capacity into play. A related problem is that if reason has no content, rationality is limited. Thus the warrant must be supra-rational, but not based on desire or feeling. Reason should be subordinated to moral values like justice, right, and duty. In any conflict, we should choose a higher value, which means a moral value. But is this rational?[25]

A possible caveat is that Kant argued that the Categorical Imperative is based on the formal structure of reason itself, particularly consistency.[26] He thereby established a purely rational principle for morals—a "pure" practical reason. From the notion of moral agency he tied this to the statement of the person as an end in himself or herself, and claimed that they are "equivalent." Thus from moral agency and rationality a content of practical reason can be derived. Many commentators have failed to be persuaded regarding this equivalence. But be this as it may, the content is based on reason, not desire or self-interest.

To show why reason cannot be included in any account of self-interest, I will begin by examining the Pythagorean distinction of opposite principles, or contraries.[27] Two of these are the limited and unlimited; two others are good and bad. The significance of the first distinction, between the limited and the unlimited, is that reason cannot, in the Pythagorean view, comprehend everything, since it can make no sense of the "unlimited" or "boundless." An example is pi, which can only be expressed as a ratio, not a finite number. Reason is concerned with the finite, the limited, what can be de-fined. Reason cannot comprehend the endless, what is unbounded, such as endless desire. Desire is boundless because endless: the desire for wealth knows no bounds and will not cease with the accumulation of a vast fortune.[28] Since happiness and self-interest can "justify" anything, they are infinite and irrational in that sense—incomprehensible. Also, they are uncritical since as "justifying" anything they are thus irrational in critical sense. The critical sense is required in practical reason: will the means achieve end? well or badly? Reason can only understand desire, passion, and willfulness by imposing limits on them. Rational limits are imposed either by allowing only so much of desires and no more (moderation), or by allowing only "rational" desires, and feelings, that is, by regulation of content. In this case, reason gives form to desire by limiting it. Desire is subordinated, in such cases, to reason, that is, with the rationally conceivable, achievable or compatible.

A hierarchy is thereby created with reason at the apex, regulating subordinate capabilities such as desire, and commanding them with rational regulations and direction. The ends of desire and passion are limited to what is "rational." Their ends are subordinated to those of reason. Only desires and passions that are "reasonable" can stand or be justified as a reason. Any "Dionysian" celebration of excess is ruled out. A hierarchy grounds a scale of values, with reason as the highest value, intellectualism.

But is this a valid interpretation of Pythagoreanism, and its model of reason? The best evidence for its validity is that Plato, who was influenced by the Pythagoreans, argued in several texts against any identification of reason and desire. On the contrary, Plato distinguished limited reason from unlimited desire: reason is separated from desire in distinct "parts" of the *psyche*. More, desire is portrayed as out of control, and only if it is subordinated to reason is it worthwhile.[29] I argued above that reason in a practical setting is a value, not only a neutral calculus. Reason as the origin of value is intellectualism, the theory of value associated with Plato and other rationalists. Virtue is equated with knowledge, an intellectualist formulation at odds with non-intellectualist accounts of virtue in Aristotle, the hedonists, and other moralists.

If reason is the supreme value, equality cannot be far behind. The link here is the ancient Greek notion of humans as the rational animal, the specific difference between humans and non-humans. Humans speak: they put sounds into words, in an order that make sense. This universal attribute divides humans from nonhumans essentially. While we share desire with the beasts, they cannot reason, speak or conceive, the preliminary to valid inference.[30] Above all, other species cannot grasp universals, the higher order abstractions that are requisite to pure reason, science, and mathematics. Reason as the human essence marks human nature, which as a universal itself is equally valid for each bearer of this nature. As all humans are capable of reason, they are equal by nature or in nature, essentially equal. This essence is both finite, restricted to one species, and universal for each member of the species.

Reasoning, the primary sense of reason, requires universality in the sense that any valid syllogism must have one universal premise. *Prima facie*, this requirement excludes calculations of self-interest.[31] Self-interest is inherently singular, not universal, since it involves a self, a singular one. A self is an individual, not a universal. Calculations of self-interest consider only one person, not any universal interests. They do not dwell in the strictly rational, the realm of universals required for reasoning. Only if my interests coincide with those of others can we obtain a universal. But if we are concerned with the universal or "general" interest, as Rousseau put it, then my own self-interest qua self may be sacrificed for the universal. Moreover, there can be no guarantee that my self-interest will coincide with that of others. Qua self, it places a premium on what is different from that of others, what is distinctive to oneself. Altruism is a contrary of self-interest for altruism takes the other into consideration. Through this consideration, altruism derives duties to the other. These duties are expressed in universal form, that is, as general rules or principles. Altruism dwells in the universal, the requirement for rationality. Alternatively, universal rules or principles, such as Kant's Categorical Imperative or the like, provide the form for specific duties.

Prof. Thomas Avery, a friendly critic, has stated that my arguments against any specific connections between rationality and self-interest are in error, since "there is nothing in formal logic that excludes application to individual things."[32] I would agree, since reasoning can have particular premises, as in AIO syllogisms, so long as one premise is universal. My point is that self-interest alone is not universal by definition, since it involves the interest of one individual. To bring in universality we must go outside of and beyond self-interest. Self-interest cannot generate universals without reasoning from like cases or extending an individual fact to universality: a fallacy. The ubiquity of the phrase "rational self-interest" has blinded some to the fact that the phrase is synthetic, combining two quite distinct notions. While we can indeed apply logic to self-interest, the two are thereby not identical. My argument is directed against the synthesis, but first requires analysis of both reason and self-interest, to show that they are not identical.

Calculations of self-interest are based on desire or passion as the source of the good of the individual, not rationality. I am pursuing my desires or my feelings, and using reason only as an instrument. The warrant of self-interest is ultimately desire, not reason. As I noted above, this is one meaning of "rational," in which desire can be a ground for an action. However, such a warrant is not rational in the strict sense of rationality as a standard or ground of the action, much less its warrant or origin. Moreover, it is not rational in the strict sense of inference and cannot be, since it does not contain a universal premise. As singular, it can only yield fallacies. In Kantian language, we could not "will it to be a universal law." Similarly, we cannot derive universal obligations from a theory of value based on (individual) feelings either.

Desire can only be interpreted as rational by incorporating the fallacy of ambiguity. If a "ground of action" is a reason, and desire is a ground, then it may be a reason for action in the sense of ground. But then any ground whatsoever is a reason, even what is not strictly rational, including the irrational. With such contradictions, reason loses its logical and normative force. Astrology can be as much a "ground" for action as science; theft that is in my self-interest as "reasonable" as a virtuous character. Otherwise, desires or "self-interest" as a "ground" of action can never be rational, for they are unlimited, cannot provide universality, are opposed to reason as a standard and cannot provide any valid inferences. Nor can self-interest alone provide a warrant or justification for action. I will argue this point in the next section, but the basic point is that what is in my self-interest may not be morally justified, for example, rape.

Reason in the modern sense is normative.[33] Reason does not describe how people think; in the words of Professor Reuben Abel, logic is not "psychology."[34] Instead, logic is a norm of how people should think, or more precisely, what conclusions may be validly inferred from specific premises. Self-interest, on the contrary, is a concept covering actual thinking and behavior, not a normative concept. The argument for self-interest is realistic: people actually behave out of self-interest, not high ideals, willy-nilly. Be this as it may, the concept is only rarely critical, and is generally descriptive. Advocates of this position might argue that in economic modeling self-interest is used as a norm and a premise from which specific inferences follow.[35] But such models presume that people act out of self-interest in some sense, and any critical sense only follows from reasoning, not self-interest itself. We might argue critically that an action is not actually in someone's self-interest, but this would be impossible to argue if the person argued that it is indeed in her self-interest. The point would have to be conceded. Thus self-interest is categorially distinct from reasoning in this sense as well. Inference is exact and universally valid, but self-interest is vague and relative. Rationality is normative, but self-interest can never be, since self-interests of different parties often conflict. Different people often have different, sometimes *conflicting* interests. Self-interest alone can only lead to relativism. No universal can be logically induced from conflicting interests; nor can practical reason in a *situation* with different motives of different individuals be formalized.

Advocates of self-interest might argue that self-interest could be valid in means-ends type of practical reasoning. In this type, the end can be singular, a particular end. Achieving my self-interest is the goal; using the best means to achieve this goal is rational. In this view, rationality is being used as a means, or connects means with ends, as in Hume's and Mill's model. However, the rationality of the self-interest as an end has not been established, only the validity of the use of the best means in view of this end. Here self-interest is accepted axiomatically as self-evident, tying it to the second meaning of self-interest examined above, where only self-interest can be a valid motive.[36] Only those who act out of self-interest are acting rationally. However, this begs the question. How do we know self-interest is actually rational? Since self-interest is not and *qua* self can never be universal, it cannot be the ground of any valid inference. Self-interest only establishes a singular, not a universal, and is of itself particular. In logic, I-I-I inferences are invalid.

On the contrary, we must go beyond self-interest to the practical syllogism to obtain a connection to means, invalidating the theory. If it is rational to calculate the best means to some end, then this principle must be added to self-interest to make self-interest a ground. But what if this is not in my self-interest? Only by appealing to a principle beyond self-interest, especially a rational one, can we make self-interest "rational." Nor is self-interest equivalent to any faculty or power of reasoning, or any other meaning of reason. As for self-interest as a "ground," as I argued above, unless we are to accept any ground whatsoever as valid, self-interest is no more reasonable than desire, feeling or other capabilities different than the capacity of reasoning. But then anything whatsoever would count as a reason and nothing would be irrational. As John Searle points out, reason only makes sense by contrast with the irrational.

Advocates of self-interest might counter that the issue is not whether rationality has content, but whether self-interest can provide this content. Since reason in the narrow sense is tied to valid inference, and has no content, the issue is, can self-interest provide such content? In the sense of a "ground" for acting, self-interest can be such an overall ground. However, a ground is not a reason, unless we accept anything whatsoever as a ground, including irrational grounds, which would invalidate the distinction of the rational and irrational. If an insane person thought that the doctors and nurses in the asylum were actually space aliens who were trying to take over, and thus thought it in her self-interest to resist their therapy, she would have a "ground" to act. But although the reasoning may be valid, the premises remain false, and thus the overall argument fails. As the notion of "good reasons" implies, giving good reasons includes critical evaluation of grounds and consequences.[37] Are self-interested reasons good reasons? Do they meet critical normative standards, whether worthwhile as ends, morally right, or "good and sufficient reasons?"[38] To assume that they are without critical evaluation of them begs the question. Moreover, they do not meet the sense of reason as a standard, which is more relevant in a practical context. Self-interest does not have its origin in reason, but in the irrational.

A "universal" ethics involves norms or rules—obligations based the equal rights of others. "Right" as a norm (normative identity or concept) is a universal. Thus it cannot be relative to individuals. As universal, such a rule must be binding on all and invariable to constitute a valid ethic: involving obligations on the powerful, institutions, individuals, and so on. Such a rule must then be derived from the principle of reciprocal justice, and it must precede happiness, utility, or self-interest.

2. A Critique of Environmental Ethics

One challenge to the environmental movement has come from within its ranks. Janna Thompson has argued "environmental ethics is not properly ethics at all." Her arguments are that environmental ethics "fails to satisfy the requirements which any ethical system must satisfy…or they fail to give us a reason to suppose that the values they promote are intrinsic values."[39] Thompson argues that expanding the circle of value to include living things and the biosphere is arbitrary: why stop at life? Why not include rocks? Why not organs like the kidney, which may have a good of their own apart from the good of the organism? Thompson also argues that the method of determining value in environmental ethics, by analogy or in terms of a good of its own, is inadequate. Another problem is determining what will count as an individual, whether an organism or its component parts. Thompson argues that one requirement environmental ethics fails to satisfy is determining what has value and what does not (non-vacuity), since if the analogies are pushed far enough, virtually everything has value. Limiting intrinsic value to living things would be as arbitrary as limiting ethics to one species.

Similar problems arise when alternative proposals are considered, for example, Richard Routley's multifactored account of evaluation of natural systems. The problem is "determining and limiting the scope of the application of the value criterion."[40] Nor is the problem limited to intrinsic value. Another problem is application of the value to ecological systems, determining what policies and activities ought to be promoted. Why preserve an old growth forest at the expense of a monocrop tree plantation? Why are human depredations bad but uprooting of trees by storms not? Thompson argues that although a criterion of value may some day be formulated which is not arbitrary, that this is unlikely, since what we judge good or bad is too dependent upon point of view, interests, and convenience. An ethic based on intrinsically valuable individuals with a point of view, which supports the satisfaction of interests, needs, and preferences of such individuals, is an ethic that satisfies the criteria required for any ethic, unlike environmental ethics.[41]

The most immediate criticism of Thompson's view is its inconsistency. Thompson argues that it is arbitrary to assign value to individuals, but immediately does so in her own ethics. If individuals can be determined for the "ethics" of self-interest, then they can be determined for environmental ethics on the same line, namely, an individual organism. If anything the ethics of self-interest is far more individualistic than the more holistic approach of ecocentrism, since the interest of selves is precisely the interest of individuals. Thus Thompson's own arguments could be turned back against her: why ignore the "self-interest" of the kidney? Why then do organisms have value but not organs? It is unlikely that anyone would argue against the idea that organs do have a good of their own, that caring for one's heart requires specific actions which contribute to the health of the whole. However, the question is not the value of the heart but whether it has intrinsic value. The answer for all ethical systems is that the heart has instrumental value not intrinsic value—or alternatively, functional value within a whole, the organism.[42] Like a tool, it must be cared for, but does not have intrinsic value. For its value is as a part to a whole. The heart could not survive outside the organism, and requires interaction with other organs. However, in our revaluation we can take the insights of environmental ethics and apply it to this issue: the heart has systemic value as a part of a larger system, the organism.[43] It cannot survive without the other parts.

One approach to environmental ethics is that living things have intrinsic value but that non-living things can only have instrumental value. Since Thompson has failed to establish why such a longstanding, widespread, and widely accepted distinction is acceptable for anthropocentric ethics but not for environmental ethics, she has not made her case. The same considerations apply to her arguments that rocks should have a value of their own and that an ethic must be selective by specifying what has value and what does not. Rocks may indeed have value, since agents may use them, but their value is instrumental. As I argued in chapter one, an organism is self-actualizing and thus has intrinsic value. A rock is not self-actualizing. One has intrinsic value, the other instrumental value.[44]

As to the idea that including everything in the dimension of value somehow violates a core principle of ethics, I would respond that for the "ethics" of self-interest, everything can potentially have value if turned into instrumental value, and thus her own theory fails the same test. In the ethics of self-interest, everything has value, since everything is turned into an instrument of self-interest. Nor is the universality of value equivalent to a null set, given the instrumental-intrinsic distinction. Some instances of value have instrumental value and others intrinsic value. Every individual in the world has potential value of one sort or another: good is a universal term. That different proposals for what has intrinsic value have been made within the environmental movement is hardly a problem confined to the movement. Yes, environmentalists disagree. But as one commentator has succinctly put it, "diversity of philosophic opinion has been a scandal since ancient times." Again, Thompson is inconsistent, failing to note the dozens of anthropocentric theories of value in the history of philosophy—and even different versions of self-interest!

Environmental ethics, taken as a whole, does not argue that everything has value. Environmental ethicists agree on one common strain in theory of value, even if the details vary. This is that environmental degradation is bad. Some theories argue it is bad for humans, others that it is bad for animals and still others that it is bad for the land, the biosphere, or the environment taken as a whole. However, in common they argue that it is bad. Since this is a common strain, it both shows that environmental ethics does have a common strain and also that it has a valid theory of good and bad, however variegated in the details. Environmental ethical theories advance dynamic theories of good and bad, consistent with developments in the environmental history as exhibiting change. Bad is realized as an alternative to good in the process of human and environmental interaction. But by no stretch of the imagination is it a proposal that nothing is bad. On the contrary, environmental ethics arose in response to the very bad rape of the environment and continues to chronicle the sorry state of the relations of human to their world. Environmental ethics also raises the critical question of the value of "goods" like plastic in consumer society: it also has a critical and evaluative aspect.

Again, the self-interest theory is subjective. Thus it has all the problems of subjective or psychological theories. First, we can raise the question of the value of self-interest itself (the critical question). Is acting in our self-interest actually for the good?[45] There is also the category mistake of confusing all the goods in the world, all the many kinds of goods with self-interest (Aaron Simmons, 2010). The egoism of such theories results in a kind of worldlessness. I will address these issues in the next section.

Thompson's argument that environmental ethics cannot distinguish intrinsic value actually characterizes her own self-interest view. In terms of self-interest, everyone else is an instrument with which to serve my self-interest. Since everyone else is an instrument for me, how can we ascribe intrinsic value to other self-interested agents? The theory is inconsistent on this point, for my self-interest argues in favor of using others as instruments, while in principle they have intrinsic, not instrumental value. If my self-interest includes kidnapping innocent Africans and selling them as slaves, my "property," then I am justified on grounds of self-interest. If it is in my self-interest to steal the land of Native Americans and use it to enrich myself, then I am justified on grounds of self-interest. If polluting a river used by others as a source of drinking water is in my self-interest, then I am justified according to this theory.

Thompson's theory argues that value can be encompassed by the notion of self-interest. The issue for value theory is why self-interest is the only good? Even if the intrinsic value of self-interest were to be granted, the reduction of all value to self-interest is dubious. The theory confuses value and evaluation, and denies the validity of non-self-interested value judgments. While I may evaluate whether something is in my self-interest, this evaluation is not equivalent to value. For I may find that what I thought was in my self-interest actually is not. Moreover, the value of what I deem in my self-interest should not be confused with its actual value.

It is difficult to see what value the notion of "self-interest" adds to value theory even if it could be determined what the value of self-interest itself is or the legitimacy of this norm. If water is superior to poison for quenching thirst this has to do with the inherent value of water, not self-interest. Water does not lose value when our self-interest changes. It may indeed be in our self-interest to use water for quenching thirst based on its inherent value, but then self-interest cannot be the ground of value. Again, beauty has little to do with self-interest, since beautiful objects remain so regardless of our self-interest.

No matter how much it may be in our self-interest to go somewhere in a plane that will not fly, we would not attempt to travel in it: it's no good, it won't fly. The theory is defective as a theory of value, then, since it cannot tell us what is valuable about things other than what is in our self-interest. The *pros hen* to self-interest conflates self-interest with value, but many kinds of value cannot be explained in terms of self-interest. A tool is valuable whether or not it is in our self-interest to use it: the self-interest we get from a tool is irrelevant with regard to how well the tool functions, its usefulness. The theory is reductive, but fallaciously. The value of a tool must be inherent or it has no value as a tool, but according to self-interest, only as an instrument of a self. Like other theories with a *pros hen* to the subject, self-interest is the denial of inherent value, that is, the value which inheres in "objects." But self-interest alone will not make a plane fly.

Above all, even if self-interest has potential value, only the creative actualization of that which fulfills or meets our interest has actual value. Creative actualization of economic goods and arrangements is also required in order for them to be in our self-interest. At best, obtaining objects deemed in our self-interest is a species of value, and at worst a devaluation of values. Moreover, self-interest as such does not have value, but fulfilling our self-interest. Unfulfilled self-interest would mean a life of frustration, according to the theory. But then self-interest as such does not have intrinsic value, but other goods beyond self-interest that fulfill it.

The reduced scope of intrinsic value in the theory of self-interest means that "self-interest" is the source of the value of the person, the human, and the body as a whole; or even gives value to them in such theories. Self-interest alone constitutes them as a valuable being, that is, as a person, a human, or as of value in this view. Humans are valuable as the distinct locus of self-interest as intrinsic value. The specific difference of humans is a value difference: self-interest as a locus of reduced value, an evaluation of what is important in humans, uniquely valuable. "Self-interest" defines value for the whole person. The person is an instrument of such self-interest, even in a "life of" satisfactions of interest, desire, liking, pleasure, and so on. Put this way, the theory is a perversion, for it treats the means as the end. The self-interest of persons or of humans would place intrinsic value in the person or human. But this defeats the theory. Self-interest as such does not have intrinsic value, but the person or human who acts in self-interest.

Self-interest requires the living person as a whole, as the beneficiary of the life over the long term. All such valued subjective states as self-interest depend upon a person's life since they take place during the course of a life, whether satisfactions of desire, calculations of interest, liking, and so on. If the life of the person is required for the presence or occurrence of valued self-interest, they are their condition, and necessarily valuable. Since the person and life are required, they necessarily have moral value as such, that is, the life of the person is intrinsically valuable and this entails duties from others. Further, if the life of the person is necessary for self-interest, but not vice versa, the value of self-interest must derive from the person, as valuable to persons. This constitutes an argument against the ultimate value of self-interest as instrumental to living persons. The living person is a higher value still. Yet the self-interest theories do not acknowledge the overall value of life and the locus of value in the person. A worthwhile life contains many other elements than mere self-interest, including life itself, social relations, and the world.

The fallaciousness of the reductionism of the theory is again evident if self-interest is considered in relation to life. If "conscious beings" have "intrinsic value" because they can calculate their self-interest, a modern sentience view,[46] does value lie in the self-interest, the calculation of self-interest or in the "conscious being," that is, the life? The self-interest or calculation of self-interest cannot be separated from the life in which such desires arise. Life and agency is the context in which self-interest arises and the latter would be inconceivable without the former.

Like other theories that attempt to reduce all of value to one category of value, "self-interest" theories are in the class of value theories called "*summum bonum*" theories. *Summum bonum* theories argue for the primary legitimating value of a single standard whose value justifies the rest. The value of instruments derives from their actual or potential value in fulfilling my self-interest. The question arises of why "self-interests" is to be adjudged the highest good, the value that justifies the rest? The argument that interests are tied to individuals, and that no one can determine another's interest, argues against a trans-personal notion of self-interest. What is in my self-interest is not in your self-interest and that is the end of it. What is in my self-interest is completely subjective, and thus not a specific kind. Even if self-interest is a distinct kind of value, a distinct category of good, this cannot justify the arguments which would make of it a *summum bonum*. Thus self-interest as such cannot be used to condemn those with amoral proclivities, such as torturers and rapists.

The theory of self-interest reduces all intrinsic value based on relative self-interest creating hierarchies of value in the process. Agents who rate or evaluate other people as "useful" to themselves place them in a subordinate position in a hierarchical evaluation, which presumes instrumental values in relation to intrinsic value. Since others are subordinate in value such evaluations are unjust, for they do not recognize the equal intrinsic value of the other. To be an end-in-itself one cannot be only of instrumental value—this is a contradiction to speak nothing of unjust and bad grounds for acting. If other agents have intrinsic value, they should be included in the moral stage of evaluation that is the condition of good grounds for action. Agents must thereby make a different evaluation, in which they are not mere instruments.

Self-interest is the locus of highest value in this theory, instead of the rest of human consciousness, life, or society. If drugs are perceived to be in my self-interest, the life of an addict is as good as DaVinci's. Pursuing one's self-interest is more important than health, overall well-being, and the effect or consequences of such satisfactions on society. If consistent, satisfaction of interests is seriously defective as a theory of what is good or valuable. Thus we may wonder if calculating self-interests, satisfying our interests, or any other version of the theory tells us much about value—whether they are standards, criteria or measures of value at all. Moreover, as Plato knew, we do not always know what is in our self-interest and thus calculating what is in our self-interest is problematic. If we do not know what is in our self-interest, how can we evaluate any action or good?

The defects the "self-interest" theory shares with other kinds of conativism are that values must be independent of self-interests, and regulate them, even if we concede any validity to self-interest. Self-interest does not create value: value creates self-interest. Self-interest is formulated from evaluation of desires in which creative actualization of various goods or options enables an agent to rank and choose them. No option no choice. Value may be creatively actualized despite my self-interest: goods produced against my self-interest. A car is in good condition not because of my self-interest but because it meets mechanical standards. I may choose to try to sell cars that are inferior in value because of my self-interest, but this fraud will ultimately alienate, not attract buyers.

Such a theory rules out the intrinsic value of other goods that are not in my self-interest. If the theory is consistent, only landscapes, plants, and animals that serve my self-interest have value, since they derive their value from their contribution to my self-interest. Less esthetic landscapes, less cuddly animals, and less ornamental plants cannot have a value of their own in this view. Yet such landscapes may have a value for the whole that is important if not vital. Unattractive landscapes may harbor species essential for environmental balance; ordinary looking plants and animals may serve a vital role in the balance of nature or the food chain. Yet this theory of value not only cannot justify the value of such environmental goods, it may well justify their destruction. If killing "worthless" animals and plants is in someone's self-interest, then self-interest justifies it.[47] If cyclists calculate that it is more in their self-interest to decimate fragile desert landscapes than to preserve them, then their self-interest is a good reason, according to this view.

A self-interest theory of value would have a hard job avoiding relativism. If one person gets pleasure from a landscape but the majority wants to turn it into a shopping mall, how can a justification for preserving the landscape be established? If the Nazis get pleasure from torturing minorities, and the majority does not lift a finger to save them, how can anyone object on grounds of self-interest? Because it is not in my interest if I am in the minority? But then self-interest cannot be a highest good, since there is no one interest that forms the standard.

If I determine my self-interest without ever experiencing art, the beauty of landscapes, intellectual achievements, the development of talents or other goods then these are without value. Only my self-interests have value. My self-interest in being a "couch potato" would be equally valuable with preservation of the Grand Canyon and all other external goods of the world for such a purely subjective theory. But if art, the life of the mind, natural landscapes, and other goods do have value, then self-interest must be erroneous. Ultimately, a "self-interest" theory is a devaluation of all value, since collectively everything could have value to someone, and the distinction of what has value from what does not would vanish.

We may wonder whether self-interest can be the basis of a society at all, since a consistent self-interest theory would result in a "society" of individuals pursuing their self-interests. This "society" would be more accurately described as a collection of individuals than a society with several common social goals. Some economic theories, including economic forms of utility, envision just such a life a "rational." Indeed, pursuing my own self-interest is identified with pursuing my own happiness. The self-interest of corporate raiders who put people out of work, corporate executives who rape a company's assets and other criminals who put their self-interest ahead of society is not simply justified but promoted by the pursuit of self-interest.

Clearly, this view is not how most humans live. As Sam Harris has put it,

> Nor are most of us resolutely selfish, in the narrowest sense of the term. Our selfishness extends to those with whom we are morally identified: to friends and family, to coworkers and teammates, and—if we are in an expansive mood—to humans and animals in general. As Jonathan Glover writes: "Our entanglements with people close to us erode simple self-interest. Husbands, wives lovers, parents, children and friends all blur the boundaries of selfish concern...Inescapably, other forms of friendship and love hold us hostage too…Narrow self-interest is destabilized."[48]

Whether we ought to do our duty (others) or pursue our own self-interest has been resolved by altruism in its view that helping others is a source of pleasures and satisfactions (William Frankena, 1963). But our obligations to others cannot be unlimited in extent—if we gave all our food away, our self would perish. Yet complete selfishness is amoral. Thus altruism and self-interest do not appear to be morally revealing. Again, should we defer our present interest for future generations? But then is our life worth living? Our present requirements must dictate our ends as much as our future duties. Time impinges upon our choice of ends.

More important, the reduction to self-interest means that many goods in the world are reduced to means. They are only good if they serve self-interest; thus their misuse is justified if they serve self-interest. They have no value of their own. This view devalues the world, for the world and the other kinds that make up the world have no intrinsic or inherent value. The surreptitious attempt at appropriation of ethics by advocates of self-interest is the origin of the devaluation of the natural world, the origin of the instrumentalization of nature and its exploitation. Thus the self-interest theory, by denying inherent, systemic, and holistic value, justifies misuse of goods of the world. In this respect, the theory can hardly be a basis for environmental ethics.

Is self-interest alone valuable or the sole source of value? A completely consistent self-interest theory would be impossible. If self-interest alone confers value, then the theory would not require real objects upon which to confer value, nor an external world. However, any such theory requires both: objects that are instruments of my self-interest as well as an outer world. The self-sufficiency of the solipsistic subject again fails.

Self-interest is self-centered. The "self-interest" theory, whether narrowly economic or as a more psychological model for all human behavior, is actually an "ethical" egoism theory.[49] As such it devalues the whole for the extremely reduced part. All the thousands of species, unique landscapes, environments, other people indeed the entire cosmos are reduced to their value for one paltry ego! All the many wonders of the world, its flora, fauna, landscapes, and resources are reduced to instruments of the whim of a subject who may or may not take an interest in them, may or may not realize their value to her or him. This view belongs to comedy, not ethics, to hubris taken seriously by tragi-comic actors, blind to their relation to the world around them. Without such a world-wide environment, there could be no selves and thus no self-interest. The environment predated self-interested humans and would continue without them.

Grounding the intrinsic value of species or environments in subjective theories is impossible. These theories confine value to human subjects, as only humans are considered subjects. To ground value in the subject is ultimately solipsistic: an absurd denial of world. Yet it is at least odd to argue that the world is only of value for the subject, in that the subject could not exist without world: the subject is in the world. The world is "objectively" necessary to the existence of subjects. The subject actually depends on the world for a life in which such interests and self-interests are possible.

Subjective value requires a different relation: the value of interests requires values in the world without which no subjective valuings could occur. The grounding of value in the subject is ultimately circular in this view, for the world is only of value to subjects, but subjects require the value of the world as their precondition. Again, what is required for an end has intrinsic moral value, as a condition of other intrinsic values.

"Representing" animal "interests" presumes representationalism, Cartesian epistemology. The question is whether we can "represent" the interests of another species adequately. Could another species "represent" ours? Judging by human history we cannot, for whenever humans have encountered other species they have acted on self-interest, including the extinction of other species (mastodons, great auk, passenger pigeons, and so on.)

Another issue is why "self-interest" is always confined to human interests? Why not the interests of other species? The theory reveals its true colors in confining self-interest to that of human agents: the rape of nature. In fairness, some environmental philosophers working with the concept of interests argue that the interests of animals should also be taken into account. But again, this defeats the theory.

3. The Lack of Ethics in the Self-Interest Theory

Thompson's "ethics," far from succeeding as an ethic, fails completely. An "ethic" based on self-interest is an oxymoron. If everyone consistently pursued their self-interest, the result would be the grossest sorts of immorality, as Hobbes argued: a *bellum omnium contra omnes*. It may be in my self-interest to enslave others, to steal land from the natives, to exterminate species (elephants for their tusks), and to rape the land. Self-interest involves calculating what each situation holds for oneself and thus reduces all value to random episodes in a relation that must refer back to the subject who calculates her self-interest. Subjectivity as the inward turn cannot generate duty. Duty is to others is part of its meaning, an external, not an internal relation. A critical evaluation of duties to self ends in the realization that there cannot be any strict duties to oneself, only duties to others. Thus self-interest cannot be source of duties. Emphasis on the subjective self is Cartesian, that is, it assumes a universe of subjects and objects. Since ethics primarily involves relations to others, it cannot be based on an individual or perceived individual interests.

Any "ethics" that begins with the individual, and not the social—especially the "ethics" of self-interest—confuse economics, or other non-moral categories with moral values. "Self-interest" starts from some egoistic model of humans and thus begs the question. Such a model ignores the notion of community, which Peirce correctly argued was the source of norms—not the "autonomous subject."[50] The subject is the product of social relations that constitute morality, not the source of morals. Ethics is primarily social. Ethics is relational and includes consideration of systemic and holistic values: the relation to others constitutes the moral aspect of the situation as well as the good of the whole (compare Kant's second and third formulation of the Categorical Imperative.)

Using "interest" or "self-interest" is a bad revaluation of both ethical values and the original Utilitarian theory of value. Balancing "interests" is a political calculation not a moral one. The use of "interests" in ethics has been unethical. The ethical issue is control of self-interest, not its gratification. The latter represents greed, not morality: a pig philosophy. Self-interest is at best of significance for the realm of value, not ethics. Moral values should function to keep self-interest in check, not rationalize it. Ethics is often hard, not easy, for we must put aside self-interest to do what is right. Kant rightly argued that acting out of self-interest, out of inclination, has no moral worth, regardless of its value to a self.

Ethics does not mean any rationalization of self-interest or desire, but constitutes a challenge to them. In terms of consequences, ethics concerns consequences for *others*. Ethics is primarily social, involving relations to "other" persons, not the self.[51] This is consistent with the Principle of Participation, involving the self at the root of justice: "what if what happened to others happened to you?" The "Golden Rule" makes no reference to "self-interest," but is reciprocal. The self is not excluded from ethical consideration, but on the contrary, becomes a measure of the worth of others through generalization. Others should be treated as we would like to be treated. Thus justice to one's self is included, if only an abstract self, since everyone, including oneself is treated equally.

The Golden Rule could be reformulated in terms of self-interest, but this would violate its spirit. It could be interpreted as looking out for the "interests" of the other. Even here, it goes beyond self-interest as such, by universalizing one's own interest to others, a logical fallacy. Since the interest of the other is considered, self-interest alone is not sufficient as an ethic. The Golden Rule as such states "do unto others as you would have them do unto you." The point concerns regard for the other and reciprocity of action. I would argue that their good should be considered, not their interest, and thus for their long-term good, including moral evaluation. However, as I noted in the first chapter, Utilitarianism, egoism, and elitism are all ultimately anti-egalitarian. Some, not all, benefit. Thus they cannot be universal in this respect. Only egalitarianism can constitute an ethic: the universality of duty requires moral equality.[52] Since self-interest involves treating others as means, it is seriously deficient, for others are never equals as means to selfish ends.

We cannot derive an ethic from self-interest alone. On the contrary, as Kant and Hobbes both showed, an additional principle must be added to obtain ethical life. It may well be "in our self-interest" to put self-interest aside and follow ethical principles proper, doing what is right, but we cannot infer this from self-interest without appeal to another principle. Self-interest is the principle of anarchy in ethics, of freeloaders attempting to get away with cheating; of the greedy grabbing more than their fair share, and the like. If I can get away with murder or rape, then it is justified on grounds of self-interest.[53] But it certainly cannot be justified ethically. Ethics should be a critique of self-interest not its glorification.

Hobbes is often misread to mean that we can establish a society on self-interest. His point is just the opposite: the society founded on self-interest alone is impossible and life in it would not be worth living. But more, the pursuit of self-interest has resulted in the environmental crisis we now find ourselves in: global warming, changes in the weather that affect the food supply, deforestation, and so on. Pursuit of self-interest may interfere with duties and virtues, especially duties to the environment. The consumer society is the culture of death, ironically, self-imposed. The assertion that self-interest can be the basis for any ethic much less an environmental ethic is wishful thinking.

Hobbes argued that entering into the social contract is in our self-interest, based on a natural law of self-preservation. Preserving our life is in our self-interest. However, it may equally be in our self-interest to break the contract when we think we can get away with it, or if we no longer perceive it to be in our self-interest. Hobbes argument that the contract is inviolable and binding on future generations is thus fallacious, for this is inconsistent with the self-interest premise. It is as much a "natural law" to pursue my self-interest, my desire, after the contract as it is to preserve my own life. If the two are in conflict, as Hobbes admits, I may try to preserve my own life. What he does not deal with are conflicts between my self-interest and the contract. Since the contract is ultimately based on my self-interest, I only will honor it so long as it remains in my self-interest to do so. If I judge it in my self-interest to cheat, especially if I think I can get away with it, I will do so, given the premises.

Worse, self-interest is relative, for it does not even take the interest of the greatest number into account. Thus it cannot be the source of any universal principle. Consider the principle that everyone should act on her own self-interest. But this may contradict my self-interest. The self-interest of a torturer may be at odds with mine: the two are in conflict. From this we can only derive contradictions, not universality. In this sense, the "ethics" of self-interest is deeply irrational. Ethics cannot be based on interests, only on principles, which are universal. Indeed, Kant argued for the role of reason in ethics as the faculty of principles.

Prof. Garrett Hardin has shown the irrational paradoxes of self-interest in detail in his concept of "the tragedy of the commons." Each herder uses the commons out of self-interest to the point where the commons is destroyed. Similar arguments apply to the fisheries, which have been overfished almost to the point of no return.[54] Using the commons to the point of destruction is not only irrational, but also anti-social. An anti-social policy cannot be universal and thus cannot be rational. The greatest tragedy is the use of the atmosphere and the oceans as dumping grounds out of self-interest, "externalization of costs," and resulting in global warming as we use the atmosphere as a commons exhibiting his pattern. Blind self-interest has meant that the nations of the world have avoided dealing with global warming, and have pursued a policy of laissez-faire on fossil fuels.

Despite the appropriation of the ethics of Utilitarianism by economics and the surreptitious substitution of "self-interest" for pleasure as its theory of value, self-interest as an ethic is distinct from classical Utilitarianism, especially that of Mill.[55] Indeed, Mill argued that only a "moral blank" could live their life based on self-interest alone. Self-interest is really the ethic of individual relativism, and thus has little or no theory of obligation. My only obligation is to myself, my own self-interest, not the greatest good of the greatest number. Jeremy Bentham and Mill were trying to reform classical hedonism, with its model of individual pleasure and tranquility, in the direction of altruism. The happiness of others was evaluated in moral action in the interest of justice. The insertion of "self-interest" by economic Utilitarians into the theory is reactionary, for it blocks moral progress and reverts to the amoral classical form of hedonism. I will return to this point below.

Acting out of self-interest can be evil; self-interest alone cannot make a bad action a good one or an immoral one a moral one. Moral principles may be the contrary of what some people perceive as in their self-interest, for example, the choice of embezzling money from a corporation despite obligations to stockholders. We may be obligated to reign in our self-interest in view of moral principles, rules, and values. Overcoming our tendency to act out of such self-interest is often the purpose of moral exhortation. We may have duties to others irrespective of whether they are in our self-interest.

Self-interest cannot give us a "decidability" principle, a principle by which to decide moral conflicts, since if my self-interest conflicts with your self-interest there is no neutral principle by which a moral interest can be decided based on self-interest alone.[56] If either party is favored, the decision is arbitrary from the point of view of self-interest, even assuming we could somehow derive "moral" interest from "self" interest. In other words, the "ethic" of self-interest is relative. Thompson's account of ethics uses a consequentialist model, in which ethics is based on value. However, alternatives to this model may be far more suitable for an environmental ethic, as Holmes Rolston, J. Baird Callicott, and others have argued.[57] Each of these authors has provided an explicit "decidability" principle, for example, the priority of the health of the land in Callicott.

In terms of an environmental ethic, self-interest has been a disaster. If animals are kept in a cage all their life, where they cannot move about, but this lowers the price of food, my self-interest is served. If they are experimented upon, and I benefit, why should I care, even if they are exposed to extreme pain? It is in my self-interest. Cosmetic and medical testing may be in our self-interest even if it results in torture of other species. If I want a large family, why should I restrain myself, even if I cannot afford them? I can freeload off society. The consequences for the environment of overpopulation are not my concern. Indeed, even the consequences for my own offspring are not a consideration for a purely self-interested theory. If global warming takes place after my death, why should I care, even if my descendants have to suffer? Self-interest from an ecological perspective gives away the store—an "ethic" with loaded dice. Self-interest is a rationalization of human domination, greed, and destruction. Rationalizations should not be confused with rationality.

Nor can self-interest provide any basis for dealing with threats to other species, including extinction, loss of habitats, biodiversity, and the stresses created by habitat loss. Preservation of species may not be in my self-interest; destruction of their habitat may be profitable. Thus the last remaining habitat of the Ivory billed woodpecker was destroyed for its logs during the 1940's. Despite offers to buy the last remaining habitat of this beautiful species, the owners elected to log the habitat and thereby destroy the species. Self-interest dictated the destruction of numerous other species, including the destruction of their habitats. If there is only one member of a species left and self-interest is the sole source of value, why should I preserve that species, if I decide I need its habitat for my own purposes? Since self-interest theories are anthropocentric, the question of the inherent value of other species cannot be raised within it. Other species only have value if they serve human self interests. But this shows the blindness of this theory. Humans are far more dependent upon other symbiotes than they are on us. The plants and animals we cultivate and tend would survive without humans, but we could not possibly survive without them. Moreover, other species are the condition of healthy air, water, and soil, necessities for all of life, including humans. Air is filtered and recycled by plants; soil is made and conserved by plant cycles. Water is often purified by other species and is filtered by unpolluted habitats.

Since the emphasis in this theory is on the self, it cannot possibly take account of the many threats to future generations posed by actions taken in pursuit of self-interest. The interests of future generations that conflict with my own must, if the theory is consistent, be sacrificed to present good. This holds even if I have children, for my interest as such does not include that of my children. Thus the theory taken in the strict sense cannot include any notion of obligations to future generations. The solution for the effects of global warming, which results in rising ocean levels that will make most areas of human settlement uninhabitable, would involve putting aside my self-interests, and thus are opposed to pursuit of my self-interests. Similarly, salinization of soils and erosion constitute threats to future generations. The Central Valley of California, some of the most productive farmland in the US, but which has poor drainage in many areas due to adobe soils, faces salt buildup. The Mississippi Valley, and the "outback" of Australia, breadbaskets to the world, face erosion problems of tremendous magnitude, due to a combination of wind, water, monocrop agriculture, and plowing practices. Yet farmers and regulatory agencies continue to ignore such problems generation after generation in pursuit of self-interest. Self-interest also works against environmental justice, since a disproportionate share of the costs of pollution is borne by the poor.

Similarly, the theory can say nothing about rising expectations and affluence as contributing factors in environmental destruction. If it is in the self-interest of consumer society to destroy habitats and extinguish species, then consistent with the self-interest theory that is to the good, for they achieved their interests. The theory cannot address the extent to which humans must be willing to make sacrifices to preserve rare species. For that is not in their self-interest. Nor can it call upon our duties and virtues, for these may contradict pursuit of our self-interest. In sum, self-interest is the worst basis for an environmental ethics I can think of and, far from helping with the solution, is the main cause of the problem.

Against such an "ethic," I would urge cooperation as the condition of the pursuit of any self-interest. Parenting cannot be based on self-interest and no self could survive without the benevolence of parents.[58] As Virginia Held has written, autonomy emerges from nurturing by parents and individuals are creatively actualized out of interaction with parents, especially the mother. Moreover, all the historical goods that we take for granted were discovered and built up over centuries through social cooperation and these are the condition of any present "self-interest." As Sam I. Hayakawa (1941) has noted, our ancestors have given all the legacy of the past to each self as a kind of gift. The self-interest of the individual is totally conditioned and dependent upon this legacy.

Cosmopolitan ethics extends moral considerability beyond the tribe, the in-group, the elite, and the "greatest number" to a generalized other. The ethics is cosmopolitan because it does not prioritize the in-group and treats other cultures under moral norms. In this sense it is consistent with moral progress, the extention of rights, and moral consideration to larger and more inclusive groups. Moral progress would extend this consideration to the biosphere, which is the condition of self and other: the world or cosmos.

The health of the environment is the condition of self-interest and thus in our self-interest. We cannot pursue our self-interest without a healthy environment. So even with regard to long-term self-interest, preservation of the environment is basic. To destroy our environment is to destroy ourselves: suicidal, as well as genocidal. At the least this involves a Kantian contradiction, since it means that we ought to pursue our self-interest, but by destroying ourselves we cannot pursue our self-interest. In other words, blind pursuit of self-interest is deeply irrational. But if the theory is "rescued" by bringing in "rational" or "long-term," this defeats the theory.

4. The Immorality of Utilitarianism[59]

In this section I will argue that Utilitarianism cannot provide ethical guidance since it can itself be used as a justification for immoral practices. The logic of its theory of value, the relation of value and obligation it utilizes, and its theory of obligation could be used to justify the most horrendous immorality. I will argue that these consequences morally compromise the theory irrevocably. Such a theory can hardly provide moral guidance. I will begin by examining the theory of value in Utilitarianism, hedonism, then the relation of value and obligation, and finally the formula of Utilitarianism, "the greatest happiness of the greatest number." I will concentrate on Mill's form of the theory, which is the most well known and even addresses several criticisms which less sophisticated forms of the theory fail to meet. I will also consider some more recent revisions or reforms of the theory. Finally, I will make some criticisms of the use of the theory by economists.

Our goal will be to show that the Utilitarian theory that lies in the background of "satisfaction of interests" is unsatisfactory and its reformulation as "satisfaction of interests" does not escape the criticisms of the former.

The theory of value that unites ancient Epicureanism to modern Utilitarianism is hedonism. Hedonism is a theory of value in which pleasure is the highest, if not the sole good and pleasure is equated with happiness. As a species of the affective theory, happiness is accounted a (positive) feeling by its advocates and opponents alike, a feeling of pleasure. As hedonism involves a feeling of pleasure it reduces all value to random episodes in a relation that refers back to the subject who feels the pleasure.[60] Thus hedonism is a *summum bonum* theory with subjective pleasure justifying all value. The moral issue is whether hedonism, as a *summum bonum*, can be used to justify immorality. If so, it cannot justify all value.

As with self-interest, hedonism is not completely wrong, since pleasure is indeed a part of life. What is implausible about the theory is the attempt to account for all human activity in terms of pleasure, to make a *summum bonum* out of what is only a part of life. At times in our life we don't pursue pleasure, whether from the duties imposed by employment, the necessities of health, hard physical labor, the development of our character, and so on.[61] Indeed, even prudence and considerations of long-term good may enjoin us to forego pleasure. On moral grounds, we may forego pleasure because it would have immoral consequences. We might obtain pleasures from pederasty or incest, but do not take advantage of children or relatives even if these are not minors. Other values, moral values, supercede the single-minded pursuit of pleasure.

Pleasure cannot justify a man taking advantage of a five year old, and thus cannot be a *summum bonum*, justifying all value.

I would like to suggest that the same arguments apply to hedonism in the context of environmental values. A person might obtain pleasure from shooting the last two hundred or so whooping cranes left on earth. It might bring them happiness to know they had successfully aimed, fired at, and hit all the remaining birds. Would anyone say that this justifies killing the last of this perilously endangered species? If not, can happiness or pleasure serve as the sole end? Or can other ends override pursuit of pleasure? The argument in Mill is that pleasure is the "only" end. If pleasure is a distinct kind of value, a distinct category of good, this cannot justify any arguments for it as a *summum bonum*. Pleasure is a distinct good and therefore not "the" good.

Hedonism as a theory of value is morally deficient, since it could be used to justify cruelty. If it makes someone happy to torture animals, then this is good, if the theory is consistent. If it gives us pleasure to kill rare species that are "in our way," then we are justified. Thus hedonism of itself reduces morals to values: an action is right if it has pleasurable consequences. The classical Utilitarians, to their credit, attempted to correct the moral deficiencies of the ancient theory with the greatest happiness principle. However, if the theory of value were to be qualified to rule out such immoral actions, it would no longer be a hedonistic theory. Pleasure would be an element of value but not the *summum bonum* of value. Moreover, the happiness principle is the foundation of the ethical principle, the greatest happiness of the greatest number. Because it is basic, no amount of tampering with the ethical principle can overcome the potential immorality of the theory of value.

Like other subjectively based theories, hedonism denies all inherent value except that in pleasure, that is, it denies the value inherent in "objects." The value of a tool is not inherent but, according to hedonism, only as an instrument of pleasure. Yet Mill argues that virtue, a means, may become part of the end, that is, have intrinsic value, once it becomes bound up with pleasure; and that feelings of justice are essential to Utility. Thus he implicitly recognizes that some actions that are not done out of pleasure are intrinsically valuable, just actions, contradicting the theory. Either pleasure alone has inherent value, or the theory is inadequate. But a tool must have inherent value as a tool, to be of value as a tool. A tool is valuable whether or not we take pleasure in it: how well the tool functions.

Some goods are good regardless of our individual relationship to them. Water remains valuable when our thirst is quenched; water in other parts of the world has value even if we never personally drink it. The theory is defective as a theory of value, then, since it cannot tell us what is valuable about things other than pleasure. It conflates pleasure with value, but many kinds of value cannot be explained in terms of pleasure. Because all value derives from pleasure, the theory is reductive, but fallaciously.

As a *summum bonum*, pleasure would have to tell us what about the world is good. But the world is good apart from whether it gives any one individual pleasure, if only as the condition of everyone's pleasure. There would be no value at all without the world, and thus the world is more than just an instrument of individual happiness. The world is not a means to my pleasure as the condition of my pleasure. I am in the world while experiencing pleasure and thus the world is partly constitutive of my pleasure. The value of the environment transcends any pleasure we may receive from it. For it is the condition of our life, which, in turn, is the condition of any pleasure we get from life. Even if we do not get pleasure from it, we must value it, for we are in it and our pleasures are in it.

Hedonism also devalues the world. Hedonism strictly speaking entails that all value lies in the random feelings of the subject. The world only has value as an instrument for enhancing my pleasure. Thus the world has no inherent value. Even feelings of pleasure that arise without any cause are more valuable than something in the world to which I am indifferent. But it is odd to argue that the world is only of value for the subject who experiences pleasure, in that the subject could not exist without the world. Since the subject is in the world the world is "objectively" necessary to the existence of subjects. The subject actually depends on the world for a life in which such feelings, pleasures, and so on are possible and without which no subject or valuings could occur. The grounding of value in the subject is ultimately circular in this view, for the world is only of value to subjects, but subjects require the value of the world as their precondition. What is required for an end has inherent moral value, since it is a condition of other inherent values.

Hedonism also confines value to humans who feel pleasure, as only human lives are considered worth living. All the many wonders of the world, its flora, fauna, landscapes, and resources are reduced to instruments of the whim of a subject who may or may not take pleasure in them. Indeed, if destroying or defacing natural beauty were to give more pleasure than retaining it, then it would be justified. This is the actual reasoning that was used to fill in scenic canyons along the Colorado River in order to dam the river for water. Grounding the inherent value of species or environments in a hedonistic value theory is impossible. Species do not have a value of their own, but only as instruments of pleasure.[62] If I have pleasure without ever experiencing art, the beauty of landscapes, intellectual achievements, the development of talents or other goods then all these are without value. Only my pleasures have value. Pleasure taken in being a "couch potato" would be equally valuable with pleasure taken in the Grand Canyon and all other external goods of the world for such a purely subjective theory. But if art, the life of the mind, natural landscapes, and other goods do have a value of their own, then hedonism must be erroneous.

Does pleasure alone have intrinsic value or the life which includes such pleasures? But is a person of less value when they are not feeling pleasure, such as the victims of illness or of the Nazis? Hedonists might argue that such a person is living for the day when the pain will have ceased: living on hope. But were their lives of less value just because they were not happy during that period of their lives? The accounts of the lives of the slaves from earlier centuries suggest otherwise. The slaves in the American South did not lead a life of pleasure, but still clung to life. These persons persisted and found reasons to live other than immediate pleasures. Indeed, life may have value apart from pleasure. As I argued elsewhere, a person who was bedridden but had their brain hooked up to electrodes that caused pleasurable stimuli would not have a happy life—few of us would trade our lives for that life.

A hedonistic theory of value would lead to relativism, since there is no way to compare pleasures between individuals. What gives me pleasure may not give you pleasure. Only the individual could judge whether anything actually gives him or her pleasure. Thus hedonism as such cannot be used to condemn those with deviant tastes, especially child molesters and adult sexual predators. Given this relativism, it would only be by chance that the majority could agree on what makes them happy. If one person gets pleasure from a landscape but the majority wants to turn it into a shopping mall, how can a justification for preserving the landscape be established?

Mill argued that the pleasures of the educated should be taken as the standard. However, this undermines the theory, introducing extra-hedonistic elements from intellectualism. Moreover, the educated do not always agree on the highest pleasures. Mathematicians prefer solving problems; poets the sublimity of verses. Which educated judgment is to be taken as definitive?

Apologists for Utilitarianism argue that the spirit of the theory is benevolent: how can wishing that people be happy be bad? Hedonism is only wishing that people be happy. I am not arguing that hedonism is always bad or that pleasure is never justified as an end. The issue in a moral context is not this benevolent wish but whether this can ever justify immoral consequences. Since obtaining my own happiness at the expense of another, in the form of doing harm—torture, rape—can never be morally right, pleasure can never be a general principle or norm of value. Happiness as a value can be seen as a part of value, not the whole, and not the *summum bonum* or justification of all values. It does not fail as one kind of value but fails as the only intrinsic value. Unqualified, hedonism justifies sadism, crime, and cruelty. Qualified, it cannot be a *summum bonum*.

A Utilitarian would point to other factors besides just pleasure, especially the greatest happiness of the greatest number. I will address this formula below, but here I hope to show that the greatest happiness theory is incompatible with the hedonistic theory of value as such. Bentham argued for the theory under the rubric of the theory of "utility," but utility implies utility for something. That is, utility is measure of means, not ends, of function and efficacy, not what they function for. But in speaking of a highest good, we do not mean a function but a standard. How then can means give us ends, much less a highest good? Use can never be a measure of value as such. Moreover, my happiness does not justify the happiness of others, thus cannot be a check upon it. The relation of value and obligation is problematic for Utilitarianism as well.

The ultimate question for such a consequentialist ethic is how social obligation can be derived from the theory of value. It is at least problematic whether obligation can be derived from, in the sense of justified by, individual feelings. There is a *prima facie* contradiction in Utilitarianism between its theory of value, which is egoistic and nominalistic (pleasure can only be my pleasure) and its theory of obligation, the greatest happiness of the greatest number—which as an obligation is toward others and may preclude my pleasure.[63] Why are we obliged to any duties based on happiness: how does a feeling create or warrant an obligation? No obligation can be drawn from pleasure as such, not even the obligation to pursue pleasure or happiness. Are we obliged if we do not feel happy? Or if such obligations end our happiness?

Utilitarianism's principle of obligation cannot be logically reconciled with its theory of value, for how does a person get from individual pleasure to "the greatest good/happiness/satisfaction of the greatest number?" Obligations derived from hedonism must be either universal or particular in form, to wit, all pleasures are equal, thus pleasure is universal in extent; or all pleasures are particular, contextual, or individual. A majoritarian/social thesis does not follow from this, but a universal pleasure principle, that is, universal self-interest, universal egoism with a hedonistic quality. Utilitarianism is not only consistent with a life of self-interest, it promotes it, since happiness as pleasure is individual. From individual pursuit of happiness no universal norm can be logically inferred, especially if happiness is individually different. Mill concedes these points and argues both that the problem of ignoring conscience is not confined to utilitarian ethics and that "control by reason" may be required.[64] In this case, happiness is not a sufficient basis for ethics and rationality is again required, as in his qualitative value theory.

Nor does a theory of obligation follow, for only if pleasure entails obligation can it justify obligation. Since obligation is to others, it does not at all follow from my pleasure that I have duties to others. On the contrary, responsibilities to others may interfere with my pleasure. But what of the greatest happiness, presuming we could somehow get from my pleasure to the equal pleasure of each. Does it follow that I have obligations to others? Utilitarians could argue that since I want pleasure, my obligation is only to be among the greatest number or to try to persuade them that my way will make the greatest number happy. But what if my happiness conflicts with that of the majority? Does it follow from the hedonistic premises that I have obligations to the greatest number? But how, given the premises could it? An additional minor needed to derive a universal (social) obligation from an (individual) feeling. Whether obligation to others can be derived from, in the sense of justified by, individual feelings is at least problematic. As happiness is a feeling within an individual, it does not of itself entail such an obligation, and may work against it. As the good of others is not individually based, the sanctions of society must be based on the interests of society, not those of individuals.

While obligation cannot be derived from the pleasure or happiness of individuals, the reverse does not follow. Pleasures are compatible with the ethics of duty and the ethics of character development, so long as the pleasures are within moral limits or in accord with a good character. Indeed, such pleasures may accompany the pursuit of duty or the practice of virtue, as Kant and Aristotle, respectively, argued. This argues both that pleasure is only a part of good, and is insufficient for an ethic. Right and wrong, moral values, cannot be based on pleasure and happiness, but can encompass happiness. Pleasure may deter us from excellence, but excellence may include happiness. Thus the ethics of duty and of character development are morally superior, for they include strictly moral values, rule out immoral pleasures, and encompass more values.

Mill argues both for instilling social feelings and for some basis for social feeling in human nature, that there is something in humans of a social instinct of some kind. Yet if his premise is that each person desires "his own happiness," then as individual, how can social feelings be instilled? Mill argues that social habits are instilled by education and rearing, thus they are not traceable to the natural desire for happiness. A further premise is required to show that these feelings will make a person happy or are based on happiness. Mill might argue that the greatest happiness principle argues for a more social criterion, but as based in individuals, it cannot be social as such. Happiness is individually located. Further, even if a social sentiment could be introduced, Mill has not shown how it could be extended beyond intimates (family and tribal loyalties). Extending "moral sentiments" to the environment and rare species would be even more problematic.

Mill argues that virtue, a means, may become part of the end, that is, have intrinsic value, once it becomes bound up with pleasure; and that feelings of justice are essential to Utility. Thus hedonism is not enough. But the issue is whether we can use happiness to ground obligation if happiness is insufficient? A vicious circle emerges in which happiness becomes the standard of decision for competing values, but in turn requires moral values to avoid immorality. Virtue is justified by happiness, but happiness requires virtue.

The question for us is how can immoral pleasures be excluded by this theory of value? A sadist may judge that he derives pleasure from torturing innocent people. He may get further pleasure from prolonging the pleasure. Similarly, children have been known to taunt, torture, and hurt animals and judge the activity "fun." If it makes them happy, then how can hedonism argue against it? If some man gets pleasure from raping women, and especially likes to frighten them by holding the decision over life and death and emphasizing this to the victim, what is wrong with such an activity if it makes the rapist happy? What if a pederast gets pleasure from sodomizing little boys? Or a thief from robbing helpless old ladies? Or others from defacing beautiful landscapes? Derivation of the greatest happiness principle does not follow from the desire for or feeling of happiness.

5. The Theory of Obligation in Utilitarianism: The Greatest Good of the Greatest Number

Utilitarianism addresses the question of the egoism of the earlier forms of hedonism, in Epicureanism, with the principle of the greatest happiness of the greatest number, bringing in social considerations in the form of the happiness of others. Mill distinguishes moral obligation explicitly; morality consists in the rules or duties that conduce to the good of pleasure or happiness.[65] Thus right and wrong are separated from good. Morality is a theory of ends and the means that best obtain such ends, generally realized through action. Morality does not consist in motives: "the motive has nothing to do with the morality of the action,"[66] since it is governed by the end. He also distinguished morality from expediency. Although duties are distinct from their end, duties also consist in feelings, the "feeling of duty" (Mill, 1973, p. 434). This feeling is the pang of conscience, which Mill claims is "natural" at least in the secondary sense of within human capabilities. The moral sentiment is based especially on social feelings, which Mill calls a "powerful principle in human nature." However, even if the greatest happiness principle did follow from the desire for or feeling of happiness, the problem is that the greatest number may prefer to pollute, to devastate the environment, to kill helpless creatures, and use them to test products. The majority may pass laws that allow motorcyclists to roar across the desert and kill everything in their path, including rare and endangered species. The principle does not protect the species and may justify such "pleasures."

Another problem with Mill's form of the theory is that Utilitarianism is ultimately anti-equalitarian, despite the qualification Mill makes that each person's happiness is to count as much as any other person's. Some, not all benefit. For Mill also argues that the Utility principle of "the greatest happiness of the greatest number," is supreme in cases of conflict. What this implies is that the happiness of the majority overrides that of minorities and individuals. The "greatest happiness of the greatest number" is majoritarian; the "greatest number" is a majority of a society. The happiness of some is thereby worth more than that of others: the happiness of the "greatest number." If the majority is made happy by actions or policies that make the minority unhappy, their happiness is of greater weight. Thus it cannot be used to uphold minority rights and may even supercede them.

Moreover, it could be argued that hedonism and Utility are immoral, since they can justify immoral acts. Aiming at the happiness of the majority could conceivably justify extreme immorality. If the Nazis made most people in the world happy by the Holocaust—anti-Semites live in many European and Middle Eastern countries; the Nazis sometimes used them in the death camps—would this make it right? Yet Utilitarianism could justify the Holocaust, since if it makes the majority happy, it would be "good." If the majority enjoys destroying a minority, then it is justified. If the Klan get pleasure from torturing minorities, and the majority does not lift a finger to save them, how could Utilitarians object? Slavery in the American South also made the greatest number happy.

Again, gross inequalities in compensation, in employment, and in benefits are justified if the majority likes them. Acquiescence in violence at the margins of society—lynchings, acquiescence in violent crime in poorer neighborhoods—would be utilitarian *de facto*, for the violence would be accepted in the interests of the majority—or at least the complicity of the majority. They could consider it too expensive to eliminate, or even as a check upon minority aspirations. Even "going along with" crime against poor African-Americans, gays, lesbians and others at the margin of society, instead of indifference, could be justified by "the greatest good of the greatest number," as it indeed was in the Jim Crow South. Only those with majority values, the normal life-style, and conventional norms are worth saving or protecting (compare minority support for anti-gay legislation).

Another practice consistent with Utilitarianism would be the use of humans or animals for grisly forms of entertainment, whether modern snuff films or ancient Roman spectacles. Feeding humans to the lions makes the majority happy: the "greatest good of the greatest number," as does the torture of animals. If this principle were universalized, randomly killing innocent persons would be justified if the majority were happy about it. The suffering of innocents could also be justified by this principle. Those who favor the death penalty could also use utilitarian arguments as a justification. Ernest Van den Haag argued that even if innocent persons are occasionally executed, that society still benefits from the practice (Ernest Van den Haag, 1998). However, such injustices would undermine the entire criminal justice system, which in theory attempts to separate the guilty from the innocent. Similarly, if eliminating a species for sport or as a "nuisance" to humans make the majority happy, it would be justified.

Utilitarians might argue that average or total happiness would be less in such cases. But what if the happiness of the majority was so great that it counterbalanced the unhappiness of the minority? What if the extreme affluence of the majority made them indifferent to the fate of minorities? Since the principle as it stands calls for the greatest happiness of the greatest number, it favors the majority as it stands. Bringing in average or total happiness is a *deus ex machina* to try to save an immoral theory. Reforming

Utilitarianism by reformulating it to incorporate minority rights or even attempting to get around the worst excesses of hedonism and majoritarianism by resort to "total" happiness or "average" happiness would change the theory, undermining it. There would no longer be a supreme principle of morality, the "greatest good of the greatest number," since the theory would now be reformulated as something else. Moreover, the "total happiness of the greatest number" would not prevent the excesses, since slavery, torture, mass entertainment, and other evils might well add to the total happiness or the average happiness.

Mill also accepts the self-characterization of utilitarianism as equalitarian, based on the above principle that the happiness of each is to count for one, but ignores how the "greatest good of the greatest number," as majoritarian is thereby an implicit undermining of rights, as Tom Regan, *inter alia*, has argued. The notion of rights as justified liberties and claims includes those minorities have, or should have regardless of the wishes of the majority.[67] Similarly, if each is to count for one, as each should according to Mill, then this principle is inconsistent with the majoritarian implications of the greatest happiness principle. Benefits accrue disproportionately to the greatest number. The minority does not benefit equally; the majority benefits disproportionately.[68] Hedonism, even qualified by the more altruistic treatment of Utilitarianism, can be used to justify the violation of minority and individual rights. Only deontological ethics are egalitarian, since duty is universal and rights are reciprocal with duties.

In our own time, Utilitarian arguments are useless against the "happiness" of consumer society, which would rather destroy the earth than give up gas-guzzling SUV's, ludicrously large houses, absurd electrical devices such as "electric carving knives," and other trinkets of the affluent society that conduce utilities to burn more coal and accelerate global warming. On the contrary, the happiness of the majority, the "greatest good of the greatest number," may justify pollution, global warming, destruction of the ozone layer, and many other ecocrimes. It would be "too expensive" to clean up and thus be a threat to "the American way of life." Again, overpopulation may be good for the greatest number of humans, and thus destruction of habitats to increase human food consumption or provide lumber (Amazonia, the rain forests of Southern Asia) are "justified" by this formula. The use of animals in research for humans is another example. Even though medical research on animals is useless, since their physiology differs from that of humans, and the medicines must be tested later on humans anyway, researchers in pursuit of fame and fortune persist in torturing animals for "human good."

We should also note an inconsistency in this rationale. Animals and plants are in the majority, so Utilitarianism cannot be used to justify research or human interests. If the greatest good of the greatest number is the supreme principle, then human happiness should be sacrificed for the majority, the non-humans, wherever there is a conflict. Nor will limiting pleasure to human pleasure work, since animals also experience pleasure. Since few would wish to pursue this consequence in practice, we can safely ignore Utilitarian arguments in any ecocentric Cosmopolitan Ethics. Utilitarianism cannot be an adequate basis for any environmental ethic, since the greater number by far are the simplest organisms, bacteria, protozoa, and the like. To make an ethic based on such creatures would be to ignore the higher forms of life entirely, such as the consumption of plankton by large whales and other fish, and undermine species preservation, to speak nothing of human good.

6. Satisfaction of Interests

Some contemporary Utilitarians have substituted "satisfaction of interest" for the classical hedonistic value theory, in a shift toward economic usage.[69] Interests here could mean curiosity, as in an interest in botany. In this form, it could be taken in the sense of the conative theory of values, that is, striving and desiring. The Perry-Prall interest theory of value connects interest to desiring, willing, and other forms of conation.[70] Indeed, the theory has much of the flavor of Perry and Prall's version of a conative theory.[71] The theory of "satisfaction of interests" could be interpreted as satisfaction of desire in the sense of fulfillment of desire. Insofar as the element of desire is stressed in this theory, and thereby the ties in to conativism, the theory is not affective and is a break with classical Utilitarianism. In this form the theory loses its primarily affective tone and may be treated distinctly from classical Utilitarianism, as a variety of the self-interest theory or of conativism.[72]

The relation of the theory of "satisfaction of interests" to other conative theories is somewhat muddy. Given the context, it likely means "interest" in the sense tied to self-interest. "Satisfaction of interests" as a theory of value and the self-interest theory share the term "interest." There is also the connection to demand, in that a desire could be interpreted as a demand in the economic sense. This makes the relation of the economic forms of utilitarianism to classical economics even more apparent.

However, "satisfaction of interests" may posit value in the satisfaction more than the interests. In this interpretation of the theory, neither the desire nor even the interest as such have value, but instead their satisfaction. But "satisfaction" is equally ambiguous. "Satisfaction" implies more than simple completion, but perhaps an element of feeling, that is, a feeling of being satisfied in the sense of being happy. "Satisfaction" could be taken as either a synonym for happiness or for a watered down version of happiness. Although we are not happy, we are satisfied. In either case, the theory retains an affective element. Thus the classical theory and the contemporary theory overlap, but are not identical. Thus issue is satisfaction of interests more than hedonism.

Satisfaction could also be tied more directly to interests, and their satisfaction in the sense of fulfillment or completion. The interest is fulfilled in the same sense as desire is fulfilled. Desire and interest come to an end with its satisfaction, which alone has value. However, as in Perry and Prall's theory, interest determines the value of what has value, since these are in relation to the interest and must satisfy it. In this form the theory cannot be rational, since what will interest each is relative, not universal; and may be infinite. Moreover, the theory requires a relation to a valuer, a subject who values. But the theory also argues that values outside the subject will satisfy their interests. Thus the locus of value is ambiguous.

The combination of "satisfaction" with "interest" in economic forms of utilitarianism does not indicate which is regulative. Is something valuable because it *satisfies* interest or because it satisfies *interest*? Is the feeling or sense of fulfillment of value, or the interest? Or does this mean the theory is a composite with both affective and conative elements?

We may first ask, is "satisfaction of interest" of value at all? Regardless of whether the emphasis is on satisfaction or on interest, the theory is subjective and relational. Thus it has all the problems of subjective or psychological theories. These include the problem of what is the value of the satisfaction of interests themselves? Are they truly valuable? Or are they subject to critical evaluation? (the critical question). Satisfying an interest in dangerous addictive drugs is hardly of value.

As both George E. Moore and Dewey have pointed out, some further reasoning is needed to derive the desirable from the desired.[73] Just because the achievement of some value is possible does not indicate whether it should be achieved or actualized; whether it is worth achieving. That some end is desired tells us nothing about the worth of achieving it. If "satisfaction of interests" is taken to mean satisfaction of desire, the same problem arises for the satisfaction of interest theory. The issue is not so much if self-interest or "satisfaction of interest" is ever worthwhile, but which "satisfaction of interest" is worthwhile? Which is of the greatest quality? For consistent self-interest theorists, if drinking alcohol to excess satisfies interests, the life of an alcoholic is as good as Mozart's. Quality of satisfactions may be at odds with quantity, just as quality is itself problematic.[74] Diversity of satisfactions is problematic and if satisfaction of interests in the sense of what is desired is the standard, then quality suffers. If quality is the standard, the rationale for satisfaction of interests is undermined, since most people do not pursue refined satisfactions and quality may require criteria and standards that go beyond simple satisfaction of interest or self-interest.

Again, who is to decide which interests, of contending ones, are correct? Could the satisfaction of interest alone be used as the criterion? What if a third party had to choose: would they judge according to their own interest, the satisfaction of interest of a third party? But then it the satisfaction of interest of another party and still another interest would be required to choose between the three, and so on *ad infinitum*. Human behavior cannot provide standards for the worth of such satisfactions, since humans are anarchic in their pursuit of pleasures. It won't do to argue that it is not in their "ultimate" interest to be an alcoholic, since this would be to go beyond simple satisfaction of interests and adopt another standard, whether longevity of life, quality of life, or the like. Satisfaction of interest alone is relativistic.

What is the location of the relation that is involved in the satisfaction of interests and that which the satisfaction of interests requires or intends? Does the value reside in the good that fulfills or satisfies the interest or in the person who is satisfied? But without the good, the interest would never be satisfied. Since the relation is essential, it cannot reside entirely in the person or the good. The result is difficulties in location and the resulting worldlessness of such theories.

The theory confuses satisfaction of such interests with other values, the category mistake: something may be beautiful but satisfy no interest of mine. What is implausible about the theory is the attempt to account for all human activity in terms of "satisfaction of interests," to make a *summum bonum* out of what is only a part of life. "Satisfaction of interests" is only partially true as a theory of value. The theory is not completely wrong, since "satisfaction of interests" may indeed be a part of life. "Satisfaction of interests" may hold in the economic sphere and there may be times in which we pursue "satisfaction of interests," for example in purchases. No one would propose that we be forced to buy worthless goods. Here we deal with one area of our lives, which may even include necessities of life like food, with "satisfaction of interests" in mind. But at times we don't pursue "satisfaction of interests." "Satisfaction of interests" is not the motive of most of our activities; we pursue other ends in life and "satisfaction of interests" is only one of these. Indeed, as the example of exchange shows, we sometimes deliberately pursue "satisfaction of interests" as a distinct activity apart from other activities of life, just as we pursue other kinds of goods apart from "satisfaction of interests." The argument in some forms of the theory, including Thompson's self-interest version, is that "satisfaction of interests" is the "only" end. If "satisfaction of interests" is a distinct kind of value, a distinct category of good, this cannot justify any arguments for it as a *summum bonum*.

The satisfaction of interests theory confuses evaluation and value, where we may rank some good in terms of its satisfaction, without consideration of its inherent good. Similarly there is a reduction of or confusion of the language of interest and that of value. I may not rank the Grand Canyon high in terms of beauty, but that does not detract one iota from the inherent beauty of the Canyon.

Actions may achieve their result and thus be of value, whether or not they satisfy any interest. "Satisfaction of interest" confuses such satisfaction and action. Actions are public, but interests are not. Even if satisfactions of interests have validity, only the creative actualization of goods that will satisfy interests has value. The creative actualization of new art works is also required in order for them to satisfy our interest or to choose them. No interests could ever be satisfied, then, without creative actualization of goods. At best some objects that satisfy our interest are a species of value, and at worst a devaluation of values. In other words, creative actualization is the mark of value, not satisfaction of interests.

It is difficult to see what value the notion of "interest" adds to value theory even if we could determine what the value of interest themselves are or the legitimacy of this norm. If steel is superior to paper for construction this has to do with the inherent character of steel, not interest, "good" or not. Steel does not lose value when our interests change. Again, beauty has little to do with interest. The Grand Canyon is beautiful independently of our interest.

Above all, there is the egoism of such theories involving the devaluation of world in terms of a *pros hen* to a small part of the world (reductionism).[75] It would be the height of folly to think that the value of air, water, and food depends upon my ego! Stop breathing for even a few minutes and the ego will disappear. An ego can no more dispense with such necessities of life than with society as a whole. If from a theoretical point of view "satisfaction of interests" is the height of folly, from an environmental point of view, "satisfaction of interests" has been a disaster.

Satisfaction of Interest could be bad or evil; interest alone, whether in the sense of taking an interest or satisfaction of interest, cannot make a bad action a good one or an immoral one a moral one. Moral principles may be the contrary of what some people take an interest in, for example, the choice of gas-guzzlers despite pollution, global warming, and obligations to future generations. We may be obligated to change our interests in view of moral principles, rules, and values. Changing such interest is often the purpose of moral exhortation: instilled as a result, not a basis. The social consequences of actions are more of an issue for morals regardless of personal choices or whim. We may have duties to others irrespective of whether they are what would have satisfied our interest. Such social obligations precede any specific interest.

Other bad consequences follow from the "satisfaction of interests" theory of values. The theory is allied to a deterministic view of human nature in which humans are "blank slates" that must be filled. Advertisers have embraced this model and view humans as consumers of economic goods, which are justified in terms of "satisfaction of interests." In this view, if some "good" satisfies an interest, it is justified. The destiny of humans is as passive consumers of whatever trivial product corporations with the capital to advertise can sell to them. This is quite a different destiny than that implied by the pursuit of moral excellence of character or as practicing duty for its own sake. A society based on satisfaction of interests is inferior to a society of character.

The theory as it stands is compatible with habitat destruction and species extinction. It just may not satisfy my interests to prevent habitat destruction or species extinction. My life will be over before the effects of global warming, resulting in rising ocean levels, swamp the lowlands of the planet. Why should I care? I get more satisfaction driving a gas-guzzler than preserving the planet for future generations, to speak nothing of preserving other species. Similarly, threats to all life due to ozone depletion, destruction of forests from logging (acid rain) may satisfy my own interests. The ecological costs of cheap electricity, including acid rain, global warming, nuclear waste, and so on, may be irrelevant to my satisfaction. Cheap electricity satisfies my interests. The loss of biodiversity is not on the radar screen of satisfying my own interests.

Proponents of this theory would no doubt argue that some epithet or qualification such as "good" or "moral" is needed to complete the theory. Satisfaction of interest alone is insufficient: only "good interests" have value. This qualification defeats the theory however. Interests or their satisfaction must be axiologically or normatively corrected. Thus a standard of value or of right other than and beyond interest or its satisfaction itself is required to give interest value. "Good interest" cannot come from satisfaction of interest itself since it needs correcting. But then satisfaction of interest is superfluous for determining value, since any psychological state, actuality or action can be corrected according to the non- interest standard. It can only be because of their superior quality, or the value of life as such. But these are both standards that appeal to other values beyond the theory.

Why should I pursue my "long term" self-interest or my "long term" satisfactions? What about my long-term happiness? Arguments to the effect that we should pursue the long term at the expense of the short-term have been used to save these theories from obvious problems, such as spending today and lacking tomorrow, eating today and dieting tomorrow, and so on. I would argue that these qualifications are fatal to these theories, since they all argue for a highest good and then argue that it is not the highest good. A child should pursue pleasure, but should not pursue pleasure, since it may not be in their long-term good. A person should murder if it is in their self-interest, but should not, since they may get caught in the long run. This "long-run" brings in notions of pragmatic consequences, that is, what will be the conceivable practical effects of the action. If the harm of some of these consequences outweighs immediate gratification or self-interest, then gratification and self-interest should be overruled. This is to argue that consideration of pragmatic consequences constitutes a higher good than pleasure, satisfaction of interests or self-interest, and may thereby outweigh them. Alternatively, pursuing my immediate self-interest may conflict with my long-term good, a contradiction, and thus would be irrational.

Proponents of selfishness—for my own pleasure, self-interest or its satisfaction are but species of that—might argue that long term happiness or self-interest outweigh short-term. But there is a hidden argument for meliorism in this approach that again appeals to pragmatic consequentialism. The argument for long term happiness or self-interest is that although going to a dentist will result in short-term pain, which will not satisfy my immediate interests, my situation will improve with time, amelioration. Ultimately, I will be out of pain. Thus amelioration trumps immediate satisfaction of interests, happiness or self-interest. But then pragmatic considerations, amelioration, and practical consequences, override the *summum bonum* of these theories and invalidate them.

Moreover, arguing for long term self-interest or happiness over short term self-interest defeats the theory. Again, this presumes rationality and thus goes beyond self-interest to another norm or principle: that we should act rationally. (I argued that these are not equivalent in section one). If we should act rationally, or if calculating our long-term interests or happiness requires reason, then reason becomes the supreme norm: we must use reason as a standard of what is in our interest or will make us happy. But then self-interest or happiness are not the *summum bonum*, reason is. Without reason, we can neither calculate our self-interest nor act on our long-term good. But isn't reason only a means to calculate our long term self-interest, albeit a necessary means? But if not all self-interest or pleasure are good, but only rational ones, then these theories are false.

7. Economic Utility

As I stated above with regard to satisfaction of interests, economic utility, or the capture of Utilitarianism by economists and their sympathizers, represents a subtle change from the classical theory that has gone largely unheralded in the literature. The change lies primarily in the theory of value: from hedonism, that value consists in happiness, to some form of the interest theory. Thus it has transformed an affective theory to a conative one and substituted demand for pleasure. This change was not in response to the moral failings of the classical theory, for instance that if rape makes the greatest number happy Utilitarianism could justify it. For the economic theory is subject to the same objections, as I will argue below. Instead, it was an attempt to combine the theory with classical economic theories that spoke in terms of self-interest. Thus happiness was revalued to satisfaction, and specified as satisfaction of interests. Economic Utilitarians who substitute self-interest or satisfaction of interests for happiness, are subject *a fortiori* to the above critique of classical utilitarianism and the self-interest theory.

The economic form of the theory mixes desire and feeling and thus is either ambiguous or confused. I noted in the previous section the connection of the "satisfaction of interests" theory to the Perry-Prall interest theory, and thus to conation in general, including the self-interest theory, and economic demand theory. These include different theories used by economists in modeling but have implications for theory of value, society, and philosophical anthropology. Economic models that posit "needs" are a form of this type, as are economic "demand," "utility" or any model of the life-style of or for consumers. The theory of interests does not fare any better if the qualification "satisfaction of" interests is added. Satisfaction of interests is a form of desire, and suffers from all the problems of the conative theory. One problem for the theory is exactly this ambiguity, for its proponents could defend the theory by equivocation, hopping from desire to satisfaction of desire, to satisfaction of interests, to interests, to will, to self-interest, to happiness, and Utility all the while claiming that they are defending the same theory.

Economists have frequently embraced utilitarianism as a model of humans for economics. The models make assumptions about humans that are self-fulfilling, since they affect public policy. "Utility" has two meanings in this context, the philosophical one of "the greatest good of the greatest number," and the more economic notion of "net aggregate utility." Insofar as utilitarianism is a theory of value, I am dubious if it can be encompassed by demand, since happiness or satisfaction is a feeling, not a desire. We may desire the feeling, but may as much lack it, which is why we desire it. Generally economists are referring to means, not ends, but conflate the two in their models. Nevertheless, demand, self-interest, satisfaction of interests, and the like are species of desire.[76]

The theory of "satisfaction of interests" is unsatisfactory as a theory of value and still less as the foundation for a moral theory. But what its adherents have not noticed is how little satisfaction of interest is compatible with the principle of utility: the greatest good of the greatest number. For example, the sale of illegal and dangerous drugs may satisfy the interests of both buyer and seller, but are hardly consistent with the greatest good of the greatest number. Similarly, "externalizing" the cost of pollution satisfies the interest of the polluter, and perhaps even the customer but is in conflict with the principle of utility. Paying taxes may not satisfy my interests, but may be good for the society as a whole, or, in Utilitarian terms, "the greatest good of the greatest number." The theories are often in conflict and such disparities may be irreconcilable.

There is also the critique of capitalism that comes from the work of the economist Thorstein Veblen. Veblen argued that what is in the interests of the owner of machines may not be in the interests of society as a whole. In particular, the factory owner may shut down production instead of sell goods below a specific price, thus, in Veblen's remarkable revaluation, "sabotaging" the full capacity of the machine to produce goods.[77] It may satisfy the interests of oil companies to conspire to raise prices, by deliberately withholding production, but not be for the greater good of the greater number of the society. Similarly, as one journalist put it, "private" enterprise is not always identical to "free enterprise." Collusion to raise prices and increase profits can take the form of reducing competition, by buying competitors, as the airline industry did with People Express. Collusion between energy providers in California, by the simple tactic of shutting down power plants for "maintenance," produced ruinous rates just a few years ago, which almost bankrupted "the greatest number." These are examples of the economics of "satisfaction of interests" that are not for the greater good of the greatest number but all in the self-interest or satisfaction of interests of the few and detrimental to society. This is to speak nothing of managers who loot companies or the public out of self-interest, notably Bernie Madoff.

Whether Utilitarians would abandon the Utility principle to accommodate the "satisfaction of interests" is moot. But if they did so, they would change their principle into an unrecognizable mutant. Mill argued that only a "moral blank" could live on self-interest alone, but this is what the contemporary proponents of the theory are, in effect, proposing. Utilitarianism was supposed to be a reform of the excessive egoism, selfishness, and the amoral implications of earlier forms of hedonism. The scrapping of the principle of Utility for the dubious benefits of the "satisfaction of interests," to speak nothing of self-interest, would be a regression, not an advance. However, Utilitarianism has so many problems as a moral theory that perhaps little would be lost.

Economics can no more make a moral theory out of Classical Utilitarian ethics than could the best efforts of philosophers. Is throwing the least skilled out of work justified in terms of the greatest good of the greatest number? Some economists, bolstered by the "Phillips Curve" advance this rationale, and the Federal Reserve Bank, which raises interest rates until enough unemployment is achieved, enforces it. Is deflation or reduced inflation justified by it? Is slavery, which after all, was an economic institution? Indeed, slavery shows just how little Utilitarianism is capable of becoming a moral theory. Slavery served the South well: it enabled a privileged class to live elegantly and prosperously.[78] Indeed, they were willing to lay down their lives to defend it. Since the whites were the greatest number, and believed it was their "greatest good" to preserve slavery, the economic institution of slavery is thoroughly consistent with the principle of "the greatest good of the greatest number."

The economic schema of values undermines moral value with *price*. The origin of this sabotage of morality lies in Utilitarianism. Cost-benefit analysis in Utilitarianism involved applying the theory to economics, putting a price upon human life, just as the slave owner put a price on kidnapped Africans. Economization of ethics began with Utilitarianism.

The amorality of this model is astounding. Unrestrained demand for drugs would end with a society of addicts. Demand for atomic or biological weapons by terrorists to blackmail the society would bankrupt it or result in McVeigh-style acts of destruction but at a much larger scale, the destruction of large cities and their markets. Satisfaction of interest is not desirable, as these two are not coextensive: a "demand" for ransom money by a group that threatens to blow up the world. The "satisfaction of interest" theory is actually self-defeating as it stands, since if individuals satisfied their interest alone, the market and the society utilizing such a market would collapse internally or be destroyed from without. Indeed, the events of 2007-8 provide evidence for the ruinous economic consequences of the "satisfaction of interests" taken alone.

The "satisfaction of interest" theory of value would be better described as the "amoral" theory since the argument is that anything that satisfies an interest has value. This theory has been tied to theories of price, since the relation of price and value is an ongoing issue for economics. In this version, anything to which a price can be attached has value. Since there is a price for addictive drugs like heroin, heroin has value. Pederasts form a "consumer market" for children, therefore there is a "market" for "kiddie porn," kidnapping, and sex with children. Organized crime satisfies some interests not otherwise fulfilled by the market, therefore they exemplify the model of "rational" economics. Whores satisfy an interest, ergo, whores have value. Examples like these, to which we could easily add many more, reveal what is wrong with this and all the theories surrounding conativism. They are indiscriminate and would justify the life of a moral midget, freeloader, moral hypocrite or outlaw. But consumer capitalism, based on greed—or unlimited "satisfaction of interests"—has cultivated just such creations, the corporate managers of Enron, Tyco, World Communications, and other companies whose management put greed ahead of morality. Economics has indeed spawned "value free" managers, or instead, those with purely economic values, Randian-Marxist values. The life style is one fit for a master with a concubine, that is, a model suitable for "managerial society."

Some economists argue that economics is a "science" and therefore value-neutral, but the model they use is a selection, reflecting evaluation of alternative models. Economists generally argue within the context of the "market," that is, they presume the superior value of one model of economic life over others. The utilitarian value of their work is also obvious in the use made of it by governmental, financial, and corporate agencies and institutions. Thus the work of economists is evidently not without value; more, they explicitly aim for valuable models. A "value-free" model cannot critically evaluate such "satisfaction of interest" since moral values have been deliberately excluded from any "rational" model for economics. However, economic models are normative in the sense of how "rational agents" (at least according to the theory) ought to behave. Therefore, they are not value free, insofar as they incorporate ideal norms.

Most economists, conceding that the market alone is insufficient in policing society, argue that economics must be supplemented by morals if the market is to function, whether to prevent economic fraud, theft, or a market for immoral or dangerous commodities (kiddie porn, addictive drugs). But if so, it cannot be value free, since the (dis)value of such commodities is required for the theory, or the practical functioning of the market.

"Satisfaction of interests," "self-interest," and happiness as pleasure spawn a consumer life-style, not a virtuous life-style. Mill read Aristotle and tried to incorporate virtue into Utilitarianism, in an implicit critique of Bentham. But why is Mill's version is superior to Bentham's on Utilitarian grounds? If sadism gives as much pleasure as self-improvement through study and diligence, or even more happiness, there is no reason to improve oneself. Nor can any but the individual measure such pleasure: hedonism is a relative standard. Thus it cannot actually provide the universality requisite to obligation, for there is no universal measure of happiness.

Economic materialism could also be accused of pure quantification of value: that human life has no value without pecuniary means. The implication is that poor people have no worth or intrinsic value by themselves (compare the phrase "human resources"). It may be in my self-interest or make me satisfied or happy to treat others as means, but it is hardly just. Social justice requires another norm of value than quantity or "willingness to pay." There are higher values than money alone; and quantification is a perversion of value. Indeed, utility in the sense of "useful for…x" implies good for something else and hence is incomplete as it stands, since that which it is "useful for" is not specified and cannot be determined by utility alone. Indeed, economic activity should be a means to a good life, and thus cannot provide any standard to evaluate the worth of ends, especially ends in themselves such as beauty. As Roderick Nash put it, "One does not consider the price if someone threatens to rape one's daughter."[79]

Thus approaches to environmental ethics based on self-interest, Utility and its offshoots, whether "satisfaction of interests" or economic forms of Utility, are deeply mistaken, as well as deeply irrational.[80] Both self-interest theories and Utilitarianism confuse having grounds for acts—self-interest, pleasure—with having reasons. Moreover, they mistake having reasons with having *good* reasons: they neglect critical evaluation of self-interest and pleasure from an ethical standpoint or standard.[81] Ultimately they confuse self-interest and self-absorption with ethics, ignoring the crucial role of the other required for ethical worth, to speak nothing of the environmental conditions of pleasure, self-interest, and other small parts of a much larger whole.

The theories are also anthropocentric. They cannot be the basis for preservation of species, or the environment as a whole. Such values only obtain value from satisfying human interests in the Utilitarian view. Attempts to compromise with such theories, for example Bryan Norton's, are a deal with the devil. An ecocentric cosmopolitan ethic must be independent of economics and form a sufficient check upon unrestrained desire, self-interest, and other manifestations of greed. Otherwise it is not an ethic at all, but a rationalization of such greed. Compromise with greed is appeasement. Cosmopolitans need virtues that help them resist desire, not more consumer trinkets. The whole world is at stake.

In this chapter we have revalued all theories based on interests and self-interests as economic, and not moral theories. The phrase "rational self-interest" has been exposed as an oxymoron; self-interest is an irrational, not a rational "foundation" for economics. An economy based on self-interest alone, without moral restraints, would end with the public looted, defrauded, deceived, and bankrupt, and with the environment necessary for both economic and other activities destroyed.

Seven

POPULATION: A CRITICAL EVALUATION OF THE "RIGHT" TO UNLIMITED PROCREATION[1]

> If the earth must lose that great portion of its pleasantness which it owes to things that the unlimited increase of wealth and population would extirpate from it for the mere purpose of enabling it to support a larger but not a better or a happier population, I sincerely hope, for the sake of posterity, that they will be content to be stationary, long before necessity compels them to it.
> John Stuart Mill (1848)

> Perhaps the main reason we find it so hard to shake off the traditional view that farming was unequivocally good for us is that there's no doubt that it meant more tons of food per acre. We forget that it also resulted in more mouths to feed and that health and quality of life depend on the amount of food per mouth…Forced to choose between limiting population growth and trying to increase food production, we opted for the latter and ended up with starvation, warfare and tyranny. The same choice faces us today, with the difference that we now can learn from the past.
> Jared Diamond[2]

> Humanity has already overshot Earth's carrying capacity by a simple measure: no nation is supporting its present population on *income*—that is, the sustainable flow of renewable resources.
> -Paul and Anne Ehrlich[3]

The United Nations consensus on reproduction and thereby population is that decisions on having children are private decisions that should be made by parents. Parents alone have the right to decide how many children they will have; the state should not interfere in this most "intimate" of all decisions. Most political ideologies, in a rare show of unanimity, concur. Our argument in this essay is that this view is radically mistaken and will hinder moral progress. It ignores the impact of population on the wild, on social resources, and other humans. I will argue that population control, and thereby reproductive decisions, are a public concern not a private decision. The negative impact of unlimited population growth on society, children, future generations, and on the environment raises issues of duties to control population. Since freeloaders might ignore such duties, public policy should include enforcement mechanisms.[4] I will argue against traditional forces that favor an increasing population, including culture, religion, and individual desires. My arguments will be pragmatic, and I will appeal to human good as well as the good of the whole biosphere and individual habitats. But I will not limit myself to holistic arguments, since overpopulation can be considered a problem for any ethical theory. This is not inconsistent, since I am not arguing only for or against a holistic or an ecocentric approach, as I stated in chapter one, nor am I utilizing it here. My pragmatic argument is that overpopulation can be seen as an ethical issue even in egoistic theories, to speak nothing of theories that take duty seriously, ethics proper.

Prof. Frank Derringh stated (personal communication) that the UN has also sponsored conferences addressing the problems of population and endorsing, or at least not objecting to governmental programs to control population, for example, the Cairo Population Conference. The Conference is a positive development, but the notion of a "right to reproduce" still must be critically evaluated.

1. Population Ethics

The condition of society is a healthy environment. All cultures require clean air to breathe, clean water, fertile soils, and the interrelation of species in symbiosis.[5] The survival of a culture requires the survival of the homeland of the culture. The homeland is the originating territory of the culture with a specific "carrying capacity" (Garrett Hardin) that is, a limited capacity to feed and sustain a limited population, even if technology can increase this capacity to some degree. Destruction of this habitat of a culture would entail destruction of the culture. In other words, all societies and cultures require an environment. Moreover, a particular culture and its practices may have an intimate relation with the homeland. Doubtless, the arctic environment of the Inuit impacts their practices as much as an alpine habitat of the Swiss and a tropical rain forest that of the Yanomamo. Some cultural practices might prove impossible outside a specific habitat. A homeland usually includes more than one habitat, or one climate zone. But in a larger sense, a homeland is an environment and no culture can be separated from its environment. Cultural values require environmental values; cultures have a duty to the environment.

The most obvious duty, qua duty, is to society in the sense of a generalized other. All choices that are good involve as one consideration the relation of oneself with society, including coordinating our actions with others'. Stealing and murder harm *others*. Duty is primarily social as duty. Further, the impact or consequences on society are paramount from an ethical standpoint. Society is the condition of individuals: the parental relationship.[6] But more, we decide how our potential is best fulfilled within society, within a particular social situation. Self-advancement requires a society with specific capabilities. Consideration of what in our personal potential will best benefit society, and hence oneself is morally requisite. Even virtue is within a social nexus. Thus all ethical considerations, duty, consequences, and character involve social considerations. Decisions on having children are not entirely autonomous or without consequences for society.

Economic optimists assume we can expand production *ad infinitum*. But we cannot if resources are finite: production of food, in particular, is proportional to the carrying capability of the land.[7] Humans already utilize over 50 percent of the land and almost all arable land.[8] There is little reason to believe that food production can be increased significantly, since almost all arable land is already being utilized for food production.[9] Expansion of agricultural land would be at the expense of wildlife, further endangering other species. Moreover, and despite its pretences, economics is not a science and does not make good, accurate predictions consistently. Economic trends are no guide to the future, with regard to economic growth, availability of resources, or other forecasts. The fact that we now can live at an affluent economic level says little about our long-term prospects. The depleted fisheries are instructive here: in the nineteen-fifties the seas were promoted as an unlimited source of cheap protein. As of 2007, ninety per-cent of the large fish are gone and many once-common species are so rare that they may be endangered.[10] Further, a trend is not destiny: we are no more required to follow the trends of the past than to reinstitute the "Holy" Inquisition.

Such optimism also presumes we can treat other species as either instruments of humans—despite our symbiotic relationship with them—or as expendable. But animals not only have intrinsic value, as self-actualizing, but are required for human life.[11] We live in symbiosis with them, and therefore humans are dependent on them, not special. Humans are one species among others and we live in a larger whole, the biosphere, on which we are dependent. Humans require other species: they are our food as well as the condition of healthy air, water, and soil. They are practically necessary, and we thereby have an obligation to them. Wild species also have ecosystemic value: the value of a part to a working whole. The interrelation of species and habitats involves symbiosis and holistic relations, including the role of predator species such as birds of prey and wolves in a healthy environment. Thus we cannot and should not destroy other species willy-nilly.[12] As parts of a larger whole, species are included in the whole to which humans, as part of the same whole, have duties.

Population pressures are capable of causing ecocide, both regional and worldwide. To feed more and more people, a society must eventually cultivate larger and larger lots of land. More and more resources must go to food. Production can barely keep up with demand. One small change in the weather pattern, as in that which occurred in the late twentieth century in the Sahel, can mean starvation.[13] Moreover, recovery of fragile desert, forest, and other marginal habitats can take centuries. If humans persist in attempting to grow crops, herd animals, and misuse such marginal lands, the habitats may never recover. Collectively, regional ecocide adds up to world-wide ecocide as the areas fit for life decrease overall. The historic impact of humans on several habitats should also be considered. The Great Plains were once covered by a tall-grass prairie, with herds of bison, and many other indigenous plants and animals. This habitat was almost totally altered by humans for agriculture—a situation crying out for redress. Restoring at least some prairie in order to preserve a unique habitat and its flora and fauna is a demand of environmental justice. I have also mentioned the desertification of the Fertile Crescent in ancient times, caused by the build-up of salt in soil that has not been properly drained.

Critics might argue that humans have increased yields exponentially during the last hundred years, and have been able to support much larger populations, more than double the population of one hundred years ago. The "Green Revolution" has resulted in both superior strains, which are more disease resistant, and increased yields per acre. Since the same acre produces more yields, increasing the area under cultivation may not be necessary.[14] But the issue is whether the success of such technologies will continue forever or whether there will be future limits to increased yields. There is no guarantee that we can continue to increase yields, and at least the possibility of potential limits to yields. Just as there may be genetic limits to how long humans can live, there may be genetic limits to how much we can increase yields. The rate of increase in yields has tapered off in recent years. Should we gamble our descendents' future on the slim possibility that technology will forever guarantee increased yields? The result might be mass starvation, as population increases faster than food supply. Moreover, many of the gains have resulted from the use of fertilizers that are manufactured from petroleum. As petroleum stocks decline, will the price of fertilizer go up to the degree that its use is prohibitively expensive?

We can make similar arguments against statistical claims that have been used by optimists to argue for non-interference. One is that with rising incomes and affluence, the birth rate goes down. Another is the percentage of poor in the world has declined with affluence.[15] But such numbers, as is often the case with statistics, are misleading. The decline in the birth rate has not meant a decline in births, and the population will continue to increase in the future, to an estimated nine billion humans by the middle of this century. Similarly, the decrease in the *percentage* of poor obscures the truth that their numbers overall have increased, to over 1.4 billion at present.[16] Some estimate that one in three go to bed hungry. The gains for the affluent are more than offset by the increases in population in poor countries. Whatever the actual number is, poverty in absolute terms has increased, we cannot feed present numbers, and there is no guarantee that affluence will result in a decline in births, as Garrett Hardin forcefully reminds us.

In the meantime, other countries are expected to make up for overpopulation by a culture on "humanitarian" grounds. But it is hardly humanitarian to have a one way relationship that could mean the death of a regional habitat that gave rise to the culture in the first place, without agreements to limit population. Further, areas that can support crops in other countries are under stress. The Great Plains of the United States face acute water shortages in the near future, as the Ogallala Aquifer dries up. The Sacramento-San Joaquin Valley of California suffers from periodic droughts, despite the largest system of dams and canals for irrigation in history (Marc Reisner, 1986/1993). Australian farmers and livestock must also deal with sustained droughts. In other words, these areas are marginal. As Al Gore (2006, p. 78) has written,

> The lesson from this experience is almost unbearably obvious. Our global civilization, which after the many thousands of generations up to the end of World War II had reached a population of fewer than 2.5 billion people, may, by quadrupling in the space of a single lifetime, dramatically increase our vulnerability to the extreme climate changes that we ourselves are now setting in motion. The signs of increased vulnerability are already evident, not only in the Sahel, the Amazon, and the Aral Sea but in California and Florida and the High Plains states, which are now using up the underwater aquifers….

Other countries are also facing acute water shortages, including those that can barely feed themselves or must import food. Farmland is also being gobbled up for development of the suburbs, which involve the most wasteful, energy-intensive life-style in history. New Jersey, once the "garden state," is now the suburban state, as hordes of "ex-urbanites" flee overcrowded, underfunded city schools for the affluent life in developed farmland.

From an environmental point of view, unlimited population growth has been a disaster, which, unfortunately, will probably accelerate in the future, as tropical forests are cut and global warming increases. Worse, there has been an unprecedented shrinking of habitats for wild species, many of great value to humans. Reduction in the numbers of birds of prey, for example, has meant an increase in their favorite foods, including species that are a nuisance to farmers, like mice, rats, and rabbits. Plants that absorb carbon dioxide quickly, and which might have helped to alleviate global warming, have been cut down to make way for human development and commercial profit. The seas have been overfished to the point that once common fish species are endangered. Tropical forests, home to the greatest diversity of species, have been cut down to provide cheap beef for fast food franchises, many of whose products are thrown away half-eaten, resulting in loss of biodiversity. The forests are also carbon sinks and their destruction increases global warming.

Nor have we factored in past devastation of habitats that were converted to agricultural use, notably the plowing under of the long-grass prairie in the Mississippi valley and of rare indigenous plants and animals in the

Sacramento-San-Joaquin Valley of California. Similarly, the conversion of land in the growing of wheat, rice, and other crops in dry areas of Australia has devastated native wildlife there. Inter-specific justice argues for reduction in the amount of land taken from once common species and some attempt to restore their habitats so they can increase their numbers to safe levels. Restoration of these would reduce the land available for farming even more. By the standard of inter-specific equity, we already have overpopulation, since humans have had to devastate other species and their habitats to make way for one species. But even by anthropocentric standards, this has not worked, since few nations can feed their own populations. A norm for overpopulation lies somewhere between these two standards, allowing enough space for humans to thrive but also for other species to thrive in their own way, and preserving rare habitats. A country is overpopulated if it cannot feed itself, especially if it cannot do so without destroying other species: genocide.

Another way in which decisions on having children affect others is reduction in future quality of life. Global warming will result in a net reduction of land, since sea levels will rise. There will be less land for more people, who require more food. This will create pressures for greater use of wilderness for food production. The reduction of wilderness will mean greater stress on wild populations of animals and plants, a situation that is already occurring in Africa and India with respect to elephant, tiger, and other great cat populations. Similarly, in North America bison and wolf populations, rare endemic plant species in Southern California and the Everglades in Florida are in danger, among others. Moreover, the areas inhabited by humans will be more crowded, with fewer opportunities for such amenities as gardens, parks, wilderness areas, and other contacts with wild species. Crowding creates its own forms of stress, as numerous experiments have indicated, lowering quality of life as well as fueling social conflicts.

Increasing the "quality of life" can be interpreted as a modern reformulation of Jefferson's "pursuit of happiness," which he regarded as a divinely given right. The pursuit of happiness will be potentially much curtailed by population increases. Natalist decisions could have an impact or will have one with respect to the reduction in the quality of life for others. The decision to have too many children thereby violates a basic right. Unfortunately, such decisions may ultimately impact others in a way that is irreversible, even though they may have no say in the matter. (Technically, "pursuit of happiness" is a right given by the "Creator" a divine right. Since divine and natural rights are problematic in a state of civilization, a point addressed in this essay, I have taken the liberty of translating such rights into the notion of "human rights," which is consistent with Jefferson's meaning, although he did not use this language.)

Finally, overpopulation will decrease the quality of life for future generations: our own children, grandchildren, and subsequent generations. The consensus is that even by optimistic projections, all petroleum production will peak by 2015. However, demand from major developing countries, especially those with large populations, such as India and China, will increase (Paul Roberts 2005). Energy costs "ricochet" through markets as producers pass on the increased cost of transportation, heating, and other hidden energy overhead. The net result will be that oil will increase in price to unprecedented levels, creating recession in oil importing countries and a reduction in quality of life. Even if the tar sands prove competitive with the rise in the price of oil, they also are finite and contribute to global warming. Deferring the problem of switching to wind and solar is foolish. Moreover, the number of products made from petroleum is astounding, including plastics, fertilizers, and many other chemicals. Whether substitutes can be found for such products is problematic.

The result will be that our children, and subsequent generations, will experience a reduced quality of life, defined in economic terms, as the cost of food, housing, and other goods rises with demand, and the amount of land shrinks with the need to increase food production to feed more mouths. Rising temperatures due to global warming will make summer a season to be endured not enjoyed. Traffic jams will become a way of life, not a temporary inconvenience. Wilderness will shrink and overcrowding to accommodate the increased numbers the norm. "Getting away from it all,"—from urban stress and crowds—will be a thing of the past for the densely packed inhabitants of choked cities. The simple pleasure of observing wildlife may become the luxury of a few, not the heritage of everyone. Unspoiled landscapes and scenic vistas will be eliminated to feed more and more mouths. The quality of life of our descendants will be reduced because of their parents' indifference to the obvious problems of overpopulation.

A country that cannot feed itself, without destroying the environment that sustains it, and the wild, is over-populated. Prof. Derringh (personal communication) has raised the point that in a globalized world, every country does not have to feed itself. This point is practically valid, but does not affect my standard for overpopulation. A country that must import has already failed. Moreover, the dependent country is at the mercy of the good will, and the weather, of exporting countries. In 2007, the rice crop was reduced in Vietnam and Australia, with the result that very poor people in importing nations like Haiti were eating "mud cakes," in which mud was mixed with other food to fill stomachs.

The alternative to genetically modifying crops to feed more mouths, which would threaten the environment with risky, untested science, is birth control. Population control is the ground of possibility of prosperity, that is, the ground of possibility of greater wealth. The choice is between quantity of people and quality of life.[17] Almost all of the individuals born prefer an affluent life-style to a poverty-stricken one. This is a demand for education, a job, meaningful work, an automobile, house in the suburbs, and so on.

Thus we have compelling social, cultural, environmental, and quality of life reasons for limiting population. However, we are up against strong competing traditions in favor of allowing free rein to couples to have as many children as they wish. Our argument is that since the environment is required for humans, and what is required is obligatory, that obligations to preserve the environment as a whole are obligatory.[18] The environment is not a luxury but a necessity for humans. Degradation of the environment considered as a whole, and such essential elements in it as air, water, and soil, are as immoral as murder by strangulation, for they affect other humans and their habitats. The destruction of the environment is self-destructive on the part of humans. This clearly constitutes a moral problem, since all human values, including "subjective values," require the maintenance of the environment as a whole in good health. Pollution of the environment as a whole is problematic for all life and in particular human life. Increases in population also affect the environment in problematic ways; raising enough food to feed increased populations has negative impacts on the wild in terms of destruction of and disappearance of habitats. Soils have been degraded and diminished, air made unbreathable, and water, polluted. Narrow interests, justified in terms of economic demand alone, aggravate the problems for the worse. Consideration of a situation in terms of narrow issues, for example, economic factors or the value of one single species, humans, ignores moral consideration of all the many factors involved in a healthy environment: the whole. Thus overpopulation is a problem from the perspective of ecocentric cosmopolitanism as well as that of anthropocentric environmentalism.

Although I am obviously not upholding any ethic of self-interest, utility or any other simple or foundational consequentialism, we could easily make the case that limiting population serves the greatest good of the greatest number and is in everyone's collective self-interest.[19] It can hardly be in anyone's interest to live in an overcrowded world, with fewer resources, lowered expectations, inhospitable climate, and a poorer quality of life. Nor is it in the interest of the greatest number that a few individuals freeload off the majority and produce too many children; or that some cultures do so and expect the others to provide for them. Unfortunately, freedom of reproduction will favor the freeloaders and force their decision on those who favor restraint. Thus from the perspective of either self-interest or that of utility, public steps to discourage births are required.

In sum, human expansion due to overpopulation is a threat to all life. Yet the environment is a requirement for all of life: we cannot survive without air, water, and soil. Overpopulation has been the main cause of destruction of the environment, including rare habitats and the species endemic to them. Prof. Derringh points out that consumerism and the pursuit of wealth are also contributing factors. I would agree, but the vast increase in population during the last two hundred years has increased the demand that make consumerism and limitless greed more dangerous. The quality of life of all humans would increase with a smaller population. Ironically, the technology is at hand to limit population, despite the advances in technology that have contributed to its increase. Several birth control options are available and both international agencies and private organizations distribute them and educate the public in their use. Why then does the population continue to increase?

The first problem with any attempt to advance a population ethic is that rights and duties are frequently assigned to individuals not collectivities. Reproduction belongs to individuals, not the "public." However, a family is already a social relation. For "it takes two to tango," that is, reproduction already involves more than just one individual. Marriage and intercourse involve a special social relation, recognized in all cultures as especial by marital ceremonies, taboos on adultery, and other cultural norms. Further, children "belong" to a family in terms of kinship ties. The children of our own family are distinct from all other children. If families have rights, they are not the rights of individuals. If parents have duties toward their children, they are collective and apply to both parents in their relationship to each other and to the children. Many other examples of collective rights and duties could be added, including "the people" of a state or municipality in criminal court cases, of business corporations, and others.[20] Rights and duties can be social, as in the duty of a society to provide an education to its children and their corresponding right to an education. From an ecocentric standpoint, the individual actor, the "autonomous moral agent" depends on the environment, including air, water, and soil and symbiosis with other species: autonomy in terms of other agents does not extend to autonomy from the environment. The whole model of the individual moral agent as the basis for morality is undermined by consideration of the social and non-human environment, by environmental holism. The issue here is whether there is an unlimited "right" to reproduce that supercedes any duties to society, children, and the environment.

Another issue this topic raises is the relation of private to public. But the contrast is misplaced as applied to population. Families are not wholly private, since they have social recognition and a social impact. Indeed, marriage is a social institution wherein a society recognizes a couple as married. If we recognize economic, education, and health "rights," then every society has a duty to provide food, jobs, schooling, and health care to all children, including children of large families. Those with large families have a disproportionate and thereby unequal impact in comparison with those with smaller families. Families with large numbers of children receive more social aid than smaller families, whether in health care, education or job opportunities. But the number of parents contributing to society is the same. The two parents of families with large numbers of children give only as much back to society in the form of contributions like labor as those with smaller families. Thus they make unequal demands on society: a form of freeloading. Critics of this view might argue that family income determines many how services will be distributed, and that larger families must support themselves. But this view is myopic, for it ignores the right to public services and the demands for social services made by all. Education is usually paid for out of public funds; health care, from a pool.

Prof. Derringh argues (personal communication) that families with more children may give more back in the form of child labor. But it can hardly be an increase in the quality of life of these children to work from a young age, foregoing an education. In many cases, children are bound to some employer at an early age, work for less and for longer hours, or are sold into prostitution, because their families cannot afford to care for them.

The other sense of privacy is the issue of intimacy, in the sense of private acts of copulation. However, the romantic view of male-female intimacy is inapplicable to the issue of population. Human intimacy is one thing; population another, although they are related. A couple can be intimate without producing children. Pregnancy, despite attempts by some cultures to hide it, is a public event. So are new children. The argument that decisions about whether to have children involve "intimacy" is only partly correct. The act of mating may be intimate, although its intimacy is a matter of collective self-deception in one sense, since everyone knows what is going on. But the result, pregnancy and birth are not private. Children are within a society and have an impact upon society. Private decisions affect the public, affect society. New births make claims on the community and resulting new people are a part of the community. The issue of births is public, thus subject to restraints. Thus population control, as opposed to intimate acts, is a public concern. Having too many children affects others. There is still another impact, which is the flow of immigrants from overcrowded countries to those with more jobs and affluence in our time. I will discuss this issue below.

Freedom and morality may appear in conflict here, or rights and morality. However, rights are based in morals, a point I will discuss below. Where morality and freedom conflict in a social setting, freedom must yield; only in a solitary setting should social morality yield to freedom. Thus intimate acts are private and should only be governed by public standards where they result in harm to one of the parties, as in rape. But childbearing does have a social impact that is now considerable, resulting in strained social services, overcrowding, reduced wild habitat, and contributing to global warming and other problems. Thus we are not arguing that there are no norms for a limit of ethics, but that consequences for others include the impact of population.

Having more children than a "decent minimum" not only has an economic impact, but a social one as well. Overcrowding and overdevelopment reduce quality of life for all. Decisions about having children impact society in the form of overcrowded cities, strained social relations, and reduced opportunities. Contrary to the cheery optimism of advocates of the "affluent society," overpopulation will not result in more opportunities, since there will be more candidates for every position, and some are bound to lose in the competition. Thus many in the society will end up on the losing end, with dreary, low-income, dead-end jobs and little hope.

Is saving the environment always of less value than saving a person? Some have argued that people come first and that in any conflict of creating new human persons and saving other animals, humans should always come first. But we may question the implicit speciesism first. What if that person is Hitler? Would anyone argue that Hitler led a worthwhile life, when he was responsible for the deaths of millions of Europeans? Or that preserving his life has more value than that of a rare species like a whooping crane, whose numbers are marginal? But then sometimes the environment is of greater value than saving an individual person. But what if the person is innocent? That person still has an impact on the environment as they require the environment. Moreover, this is a phony contrast, since we are weighing present goods—other species—against possible future persons, yet unborn. No value attaches to possible persons until they are born or at least have developed to the point of viability. But if decisions on birth control are contingent upon parental decisions, as the UN and others argue, then possible persons could as well not be born as produced by those decisions. Thus if everyone decided not to have children, those possible children would have no actual value: they would not have been creatively actualized. More decisively, more people are not more valuable than the environment, since they depend upon that environment. The relation is reciprocal.

Another objection is that technology will allow us to outgrow our problems. New sources for commodities will enable societies to provide jobs, food, health care, and other goods and services to an increasing population. Thus the dangers of environmental degradation have been overblown.[21] This theory was used to argue against Global Warming. President George W. Bush even appointed skeptics about the science to an advisory committee on Global Warming in the hopes that they would debunk the science. However, even they were forced by the evidence to conclude that human use of fossil fuels is responsible for increased levels of carbon dioxide, methane, and other culprits in the melting of the Arctic ice caps and the polar tundra.[22] The result will be felt for centuries, as the seas rise, inundating islands and coastal areas. Thus optimists cannot be taken at their word, since they ignore actual losses of habitats and other danger signs.

Economics: energy generation versus the ecology is the new value conflict here; the next phase of technological history will be the extent to which prosperity will supercede environmental beauty as a socio-cultural value—the degree to which energy will be an ultimate value (a ground of possibility of technological prosperity). Energy is only now emerging as a distinctive economic value in history—previously energy was more easily available, but in the future, energy will supercede other economic values. The problem is the ecological costs of cheap electricity, including acid rain, global warming, nuclear waste, and many others. Population control is the easiest overall solution for society. Energy costs will form a limit to society and its members—small families will form a social value.

However, we have more compelling reasons to think that economic solutions taken alone will not work. A Utopian increase in production, whether through state ownership and planning; or new technologies and managerial planning, would take the squeeze off the dispossessed. The increase in goods, if no other constraints were introduced, would result in an increase in population that would outstrip resources, such that poverty would again arise. Nor would utopian production levels ever produce enough top-line goods. These would either have to be rationed or proscribed altogether as "luxuries." Without controlling populations, increasing goods and services will effect an increase in consumption by increased numbers of people. Moreover, increased consumption, as James Speth has noted (2005), uses the environmental equivalent of "principal": forests, fisheries, and so on, not "interest" or sustainables. Thus growth comes at the expense of our children, since these assets will be used up for present consumption. We achieve growth using the assets of future generations and other species. It is unconscionable to have children and use up the goods that they will require for life.

The environment has been treated as property by many cultures. Thus "property rights" advocates have opposed so-called "takings" by governments to save species and habitats. This stance is at least ironic in the New World, in which the property of Native American tribes was "taken" and given to Europeans and Euro-Americans. From this history we may safely infer that property is based on the forceful eviction by, or fraudulent tricks of the European settlers, not any "rights." However, the idea that property rights are inviolable and absolute has no basis in law, as John Cronin and Robert F. Kennedy Jr. noted: courts have consistently ruled that if use of private property adversely affects the larger society, for example in the form of pollution, it can be regulated.[23] Similarly, ground or running water contamination by a single private property owner can be restricted. The state has the power of eminent domain, which preempts private property rights. They also restate Rachel Carson's arguments for a right not to be a victim of pollution.

Moreover, the same argument was made about slaves—originally kidnapped from Africa—as modern property owners make about their "private" property. Since they paid for humans, they had a "right" to their property, regardless of how the agents originally obtained the slaves. In truth, they have attempted the trick of reversing the actual relation of humans to the land. Humans occupy a habitat to which they belong, they do not own or control the habitat. The other species that occupy the habitat should no more be the "property" of humans than African American slaves, for the other species too require the habitat to live and unlike humans, are native to the habitats.

Conservatives only wish to protect private property "rights," not those of free speech or other rights (The "Patriot" Act). Conservatives are *de facto* in favor of increases in populations, since they realize in their economic hearts that increased population means increased demand, lower wages, and also pressure to exploit the few remaining wild lands for human profit. They are all in favor of coercion when it comes to protecting their own interests; opposed when it is not. They love government when in power; but attempt to undercut it when they are not. They oppose court-mandated solutions when they go against conservative ideology, for example on the abortion issue, but defend court decisions when they favor conservatives (2000 election). They oppose taxes and governmental spending on social programs, but favor spending that increases profits especially for defense industries and military supply concerns (Haliburton).

True conservatives would try to conserve, especially conservation of the land or nation, with its peculiar habitats and species, that is the condition of their nationalism. Raping the land is not "conservative."

On the contrary, the environment, like health is a public good, because its use affects everyone. Polluters who dump toxins in rivers and lakes poison it for all. Smokestacks that send pollution "downwind" poison the air for everyone. Mining concerns that dump tailings over the surrounding landscape destroy the land for generations. The many are made to suffer for the actions of a few, especially many other species. This is the contrary of the utilitarian ethic of the greatest good for the greatest number, to speak nothing of lacking either virtue or moral value. It adds up to the least good for the greatest number or the greatest evil for the greatest number. Killing is morally wrong and pollution kills. As Hardin pointed out, the air, water, and soil are "commons," that is, public goods, taking public in the wide sense to include the conditions of any society: not just air, water, and soil, but other species, habitats, and the biosphere as a whole.[24] However, even the despoliation of the private property itself can have deleterious consequences for others, whether from groundwater pollution (New York State's Love Canal) or for future generations, denied the use of the despoiled land.

Having too many children conflicts with others' rights—especially to a good quality of life, including the quality of life of other species. It also interferes with liberty and even life itself, and may do so even more in the future. Thus the issues in population are not a matter of an ostensive "right to reproduce" versus authority but instead a conflict of rights. Classical political thought argued that the liberty of each must not infringe on the liberty of others. But too many children will do precisely that. Moreover, the quality of life for all would be reduced to satisfy the few. Too many people detract from the quality of life, especially that of our descendants, who would pay the price for our moral cowardice in not confronting the problem.

Society has a right to collective self-defense, the basis for law. Society defends itself against rape and murder with laws. Similarly, society has the right to protect itself against threats to its well-being that derive from over-population.[25] The choice for our society is between quantity of people and quality of life. On the positive side, if all cultures were to limit births and effect an actual decline in population, as some have actually done, there would be more resources for all. Since the need for workers would exceed the number of positions in a declining population, real wages would go up. The supply of resources per capita would increase. Luxuries would be within reach of many, since their price would go down. Housing would decline in price as fewer people needed housing—indeed, no new housing would be needed, only repair. There would be more opportunities for solitude, wilderness viewing, and other eco-friendly recreation. Technology would serve, not regulate our lives. Pressures on rare habitats, rare and even common species would ease.

2. The Basis for Encouraging Population: Military and Strategic

One reason for encouraging population growth is military strategy. If a neighboring country has a larger population, it has a potentially larger army. Thus for strategic reasons, it would be "rational" to increase the population for potential military recruits or draftees. An opponent's large population may be offset by superior technology, strategies (the ancient Greeks versus the Persians), strategic alliances, and other purely military factors. But there is no guarantee that a perceived opponent will not adopt the same technology and bold military tactics to offset one's own, as the Soviets apparently did during their struggle against the Wehrmacht in "The Great Patriotic War."[26] Thus the nations of the world seem locked in a tragic conflict without any ultimate winners, as young men are sacrificed to the follies of generals who ignore the simple truth that both sides are evenly matched. This was the situation in World War I, as tens of millions of young men on both sides were slaughtered by the machine guns.

From a moral perspective this is insanity. The burden of caring for children, of feeding many more mouths than would be practical, is thrust upon the citizenry. Women become baby factories to feed the jaws of death. Moreover, the habitat is exploited to the point of no return as more and more land is brought under cultivation to feed the exploding population. All this for the immorality of killing thousands of innocent combatants—innocent because they did not start the conflict and generally will not benefit by it—to speak nothing of the "collateral damage" inflicted upon innocent civilians. War is morally evil. Moreover, war reveals the "contradictions" inherent in overbreeding young people only to have them destroyed in battle.

As John Dewey and James Tufts (1932, p. 508) said, a century ago:

> Today we are not so sure, as humans once were, that the world needs to increase its population. Rapid growth of population with a large birth rate

> has been characteristic of cruder civilizations which sought to gain a superior advantage in man-power or to compensate for the high death rate incidental to ignorance and lack of sanitation. Further the overpopulation of certain countries has been alleged as a justification for invasion of other lands, or for planting colonies where they are not wanted. Why then add to the evil of overpopulation? In answer it may be freely granted that there is no good reason for increasing the total population of the world, and that some countries are already overpopulated. The question is one of quality rather than of quantity....

There can be no rationale for the immorality of war as a rationale for increasing the population. On the contrary, this is an immoral reason to increase population, since its ultimate rationale is to kill off innocent persons. We create overpopulation only to kill it. But what of self-defense? We may not like the fact that we must arm to defend ourselves, but in the face of an Adolf Hitler, what else can we do?

Hitler is a hard case, since most wars are not defensive. Thus increases in population cannot generally be justified on military grounds, especially for offensive wars. However, the lesson to be learned is the folly of war, since most wars do not accomplish much, despite the deaths of thousands. The exceptions prove the rule: although World War One meant the end of the empires in Central Europe that had precipitated the war, and thereby the end of an era, it precipitated World War II. The purpose of statecraft should be to avoid the immorality of war, not prepare for one. This can be achieved best by economic and cultural ties, such as the European Community, which has slowly overcome nationalism in favor of cooperation. A federal system might reassure nationalists and regionalists, especially if cultural and educational affairs were left in local hands, while allowing the elimination of national armies and thus the military temptation. Hopefully, humans will eventually evolve to the point where national differences will no longer be grounds for political borders. Looking at history, there is much reason to be pessimistic.

One possible source of overpopulation is that one tribe attempts to increase its numbers relative to that of others. Like the arms race, this is a dangerous game, which no tribe can ultimately win. For it will only result in population spiraling out of control, exceeding the carrying capacity of the land and worse, dangerous dependence, as in Ireland, upon a few crops for feeding an excessive amount of offspring.

3. Relativism

One argument against any cosmopolitan ethic is relativism, whether cultural or individual relativism.[27] Some have argued that ethics is learned within a culture and individuals follow the ethics of their culture. However, cosmopolitan ethics is not a denial of cultural ethics. On the contrary, a cosmopolitan ethical model recognizes the distinctness of other cultures, their unique value. Contrary to this view in many ways is individual relativism, the "ethics" of self-interest, that individuals only act on their self-interest. With respect to population, the argument becomes that the decision to have children is either culturally mediated or based on self-interest.

There are many species of relativism and I will not examine all of them here, nor address all the issues raised by relativism of different kinds.[28] Nor will I go into individual relativism in detail, since I argued against the "ethics" of self-interest at length in chapter six.

One conclusion that can be drawn from the discovery of different cultural mores is nihilism. There can be no morality, as different individuals and cultures have different morals. No moral knowledge is possible due to multitude of value-systems in culture. Each cultural choice is as good as any other. As Sam Harris has put it:

> Many intellectuals tend to speak as though something in the last century of ratiocination in the West has placed all worldviews more or less on an equal footing. No one is ever really *right* about what he believes; he can only point to a community of peers who believe likewise.[29]

The consequence of this view would be that torture is as good as consensus; concentration camps as good as voting. After all, right and wrong are "all a matter of opinion." In view of this, they can all be viewed as nihilistic. As all values are relative, nothing is superior. As no hierarchy of goods can arise, there can be no true values.

I will not examine moral or value nihilism or scepticism in this chapter, since their claims are not at issue. Moral and value scepticism is little different than intellectual scepticism of ancient and modern times (deconstruction). Skeptics of all varieties deny that knowledge is possible—but this is an untenable view after successes of modern science, in which some form of knowledge has been creatively actualized. Similarly, moral and value skeptics deny that moral knowledge is possible; or that we can have any knowledge of values.[30] Value skepticism is untenable after Auschwitz, despite the refusal of some to admit the failures of moral relativism. Their stance is similar to those, including fundamentalists, who refuse to acknowledge the successes of science. This is to speak nothing of pragmatic contradictions—those who value some choices over others in their own behavior. They prefer steak to cow manure, lettuce to poison ivy, sex to abstinence, and so on. Actions speak louder than words.

However, ethical relativism could be distinguished as such: of competing values, whatever the source, whether cultural or individual, each is equally valid, or worthwhile. Yet each species of relativism creates difficulties for the other. If I can know only my own perceptions, I cannot know the values or cultures of others. Moreover, my own self-interest may be in conflict with my culture. Similarly with the other two species: individual self-interest may be in conflict with the historical "spirit of the times," just as the *Zeitgeist* may portend a change for a culture. Thus a general criticism of relativism is that its species are mutually contradictory.

Specific characteristics are common to the different forms of relativism, taken as an ethic. First, equality is fundamental to the notion of relativism, either equal difference or identity. Either each individual, each culture, or each token of the theory is equally an instance of that token, identical in some respect; or equally different from the others. The relation of each token is that of the equal value of each token: each culture or individual or perception is equal in value. Second, relativism is atomistic in structure in some sense: each unit is discrete. No hierarchy of cultures, individuals, or any more general identity can arise, as each unit is equal. Relativism is atomism without any resulting structure (unlike materialism, which argues that larger bodies are built up out of atoms, not against larger bodies). Where all are equal, there can be no overriding abstract structure.

Finally, the relation of units, instances or tokens (cultures, values, individuals) is extrinsic, as it usually involves differences, which for that token are alien. Since the units are equal, none can regulate another in a relation of ground and consequent; nor is there a hierarchy of cultures; nor an identity principle. This creates theoretical difficulties for relativism itself is a principle and an identity, but it tries to disestablish grounds for principles. Its egalitarianism is in conflict with its atomism—both theoretically and practically.

Cultural relativism in an ethical context argues that moral norms are "folkways" that each culture has evolved and that in these folkways lie the origin of ethics.[31] Moreover, a strong corollary of the cultural origin of ethics is that we cannot escape our own or another culture: each attempt at a universal ethic is a disguised cultural ethic masquerading as universal. John Ladd has called this the "dependency thesis," and distinguished it from the "diversity thesis," which only argues for different values and norms in different cultures.[32] Another premise of cultural relativism is that each culture is equal, or equal in worth, or equally different. Thus an ethical relativism is justified, for which, of competing cultural values, each is equally valid, or worthwhile.

The first question that we can raise with respect to the validity of the model of culture is, is culture a valid norm? *Hypothetically*, "a" culture is in a relation of autonomy from other cultures in mutual difference. But practically, the limits of one culture from another may be harder to draw. China has hundreds of dialects and regional customs. Is it one or more cultures? Similarly, Bantu has been called "the language with a thousand dialects." We may differentiate or identify subcultures as much as is practically requisite, given some end, such as knowledge. Thus Joseph Greenberg and his students have grouped all the Native American languages other than the Inuit and

Athapaskan languages as generically alike: the "Amerind" group (Merritt Ruhlen, 1987). Others have contested this grouping. But we can validly raise the question, are cultures as unique as cultural proponents assume? Contiguous and even distant cultures often share a religion, economic life, several institutions (marriage), and so on. The majority in Indonesia shares a religion with distant Morocco, as does the majority in the Phillipines with Italy. The animism of many cultures may not be original to each, but adapted and embellished by each. Why then should it be respected? Animism is not intrinsic to any one culture any more than the automobile, but adopted by them. Switzerland does not have a distinct national tongue but shares the languages of its neighbors. With modern trade, economies are integrated across national borders. With respect to nations, almost all of them have several cultures within their borders.

Are they as important in the life of the individual as assumed? What of autonomy, and individual decisions? What of eccentricity? Conflicts between individuals and society are often over cultural transgressions, as in the case of the "apostate" in Afghanistan who converted from Islam to Christianity and was arrested and put on trial. Cultural values are often treated as absolute by a society. Cultural values may thereby conflict with the notion of individual rights. The question we can raise is whether each person must follow his or her culture. Do Irish-Americans have to wear kilts, the traditional dress of Ireland? But are they Irish or American? And even if they have an Irish background, the present day Irish in Ireland only bring the kilts out of storage on festive days. Most of the time they wear more modern clothing. If the Irish themselves do not wear kilts for the most part, why should the Irish diaspora? The same points apply to others within a culture and those who have emigrated, whether Africans, Chinese or any other group.

The conclusion we may reach from such considerations is that an "individual" culture is ambiguous, and it might be better to speak of "webs." Moreover, culture is not a "given," but like race, a construct—creatively actualized by its members. However, unlike race, I mean "construct" in the sense of actually constructed and creatively actualized by its members. As actual, a culture is not "socially constructed" in the epistemological sense, although there can be elements of the social in actual construction.

Cultural identity is not evident from appearances, biology, or religion but is carried by the individual in the form of habits, perhaps a language and cuisine, a way of looking at the world and the like. Given how difficult it is to define "a" culture, we may at least ask why cultural values should always be respected where they impact others.

As to value: culture is pragmatically justified, as cultures have survived over time and developed and aided individual members in achieving goals. Cultural identity may constitute an important part of individual identity. Culture provides the framework for individual identity, the source of moral guidance, and norms and other benefits. Moreover, it is probably valid to state that everyone has some habits derived from their natal culture. We cannot escape the demands of culture entirely. I will not argue against all respect for culture, and I am aware of the cultural influence in the development of all human individuals. I am not denying the importance or potency of value criticism of cultures within moral limits, as I will make clear below; but a cosmopolitan ethic should also acknowledge the importance and value of culture and attempt to avoid excessively formal or ideological standards.

A culture is also a society's choice, as is its morality. The individual learns both from society. The cultural value should prevail here at the discretion of the cultural members. If cultural participants value a cultural mode, its value is manifest, for by pragmatic criteria, those who culturally value a cultural mode will participate—providing criteria for their genuineness. But where the case is not so clear-cut and the participants are themselves divided over cultural values, and where conflicts do arise, how can they be settled? What of conflicts between cultural and other categorical values: the standards of economics for example?[33] Should culture again prevail as the local source of more generalized categorical standards? The best way to resolve such disputes is by ethical reasoning. Ethical reasoning requires universals, a point we will address below.

What of value conflicts between cultural values and moral values and standards? What if a cultural institution is not up to a standard? In terms of a cosmopolitan environmental ethic, what of cultural practices that are in conflict with the health of environment—a cultural-categorial value conflict?

Ethically, cultural relativism is a disguised absolute: either one must accept fully the morality of our natal culture or become critical of it, wholly or partly in the name of a higher standard.[34] The uncritical acceptance of a cultural ethic is an absolute in itself. To invoke a higher standard is to go beyond relativism. Cultural relativism either asserts that we ought to live according to the norms of our (own) culture—an absolute—or that there is or can be a moral imperative (duty) higher than any one culture, or outside a culture, which is a denial of cultural relativism. Some might argue that this is a straw man—that few in contemporary times would defend this "ethic." But some have advanced new formulations of it in the form of arguments from cultures that there can be no universal ethic. Moreover, many students come to class persuaded of the truth of some form of relativism.

If we take different cultures seriously, then according to cultural relativism, we should live according to cultural norms. If we should not live according to cultural norms because they are contradictory in some cases, then cultural norms are not necessarily ethically or axiologically relevant. They must be evaluated further. (This is also a universal norm.) Thus no ground for relativism or moral skepticism is provided by the actuality of different cultures. Moreover, if we should treat cultures equally, as equally of value, equality becomes a universal norm. But then cultures that treat women as less than equals, such as the Taliban, or those that practice female genital circumcision, violate norms of equality. The premise violates the application. Universality is again asserted, contradicting the premise.

The practical issue is whether we can escape the norms of our culture. We have in the past in the form of cultural evolution. New norms are formulated that supercede those of the older culture. Moreover, individuals may violate cultural taboos, thereby escaping the norms of the culture. Values differ from actualities—at least before action has transformed the world, the actual—because humans have a choice: they are agents. A "moral science" must differ from a physical science in this fundamental way: a "moral science," or better, philosophical ethics cannot be solely descriptive, much less value independent. Descriptions of moral behavior, for example in the science of ethnology, describe cultural principles as they are practiced, for example puberty rites. Because a society or individuals may reject that culture in part or wholly (change), ethnology cannot predict moral behavior, though it may predict behavior. There is nothing in a value pattern (culture, hence esthetics) to provide grounds or evidence for that pattern's acceptance or rejection by individuals within a culture, especially one that places a high premium on individual rights and autonomy. There is a limit as to how much behavior can be predicted, and even this is difficult, since behavior may reflect a choice of values that is in conflict with that of the culture. And the choice may or may not be affected by values or values in conflict; more material considerations often underlie choice. Hence values can be independent of culture.

The ethical issue is whether we should always follow the norms of our culture: is it morally right to do so? But what if our culture practices patently immoral actions, like slavery? Relativists do not avoid the problems in facing a Hitler, for example—should we accept such a cultural morality as that of the Nazis? We have a duty to oppose our culture in some cases. The same arguments apply for historical relativism. Moreover, if animism is a false hypothesis, it can be no more deserving of respect than 2+2=5, or the literal truth of the Bible. Cultures can believe false hypotheses.

Cultural relativism is also tied to an absolute ethic of tolerance. Cultural relativists argue that we should be tolerant of members of other cultures.[35] But how does this follow from cultural relativism? If we ought to practice the norms of our own culture, it does not follow that we should tolerate others. Indeed, as I will argue below, tribalism is precisely intolerance of other cultures. Relativism in the elitist form is unjust. It holds others to standards of conduct not practiced by or binding upon the elite, the few enlightened, and so on. This is a moral not a logical objection. Thus the position is incoherent. Either we should be universally tolerant, in which case some universal norms are not relative; or we should not tolerate other cultures, in which case the "ethic" proposed by cultural relativists is false. Either imperative bypasses cultural relativism.

Tolerance passes for nihilism: democracy and tyranny are equal, given the predilection of some, usually those in power, to increase their power. The viewpoint of "the other side" or "the other person" has as much weight as our own, since in truth there is no value system we participate in or affirm by our self alone: one is "alienated." Despite the benefits of democracy and freedom for the majority (free speech, press, freedom of information) freedom is not deemed superior to alternatives, as relativism precludes such judgments. The result is similar for each species: a reactive nihilism too timid to assert the worth of even the one value left to it.

But what if cultural relativists argue that we should generally be tolerant of culture, but not of specific immoral practices followed by some cultures? For example, should we tolerate the old Sicilian practice by young men of kidnapping young women, and raping them, to force them into marriage? Or do some cultural practices, such as slavery in the Sudan, fall outside what is ethically permissible? But then we are asserting universal rights or norms that trump culture, and cultural relativism is again defeated.

If we are to uphold tolerance for different cultures, then relativism implies duty, for it requires *autonomy* of the different cultures. Such duties go beyond the particular culture, respecting the value of an alien culture: cosmopolitanism. Relativism also implies reciprocal justice as regulative norm (limit). If relativism ought to be our moral practice, then mutual forbearance is a duty. If it is not a duty, an "ought," then cultural relativism is not a moral stance. As a moral stance, cultural relativism presumes reciprocal justice. But then it goes beyond cultural relativism. Since we have universal duties, relativism implies duty, and perhaps a deontological ethic. For either we do have a duty to respect the autonomy of different values, in which case such a duty overrides relative values, and pure relativism is overcome; or we do not have such a duty, in which case relativism is bypassed as morally irrelevant. We ought to practice the ethics of our cultures, or we ought not to. If we ought to respect the values of another culture, this is a universal imperative. If the universal imperative is denied, then the values of the other culture are not morally relevant. Such a duty is cosmopolitan, not cultural, not relative to a culture. More, the argument for tolerance between cultures is cosmopolitan: that we should in some sense celebrate the differences. But then not only does it argue for an absolute, as others have pointed out,[36] but also in some sense for a special case of cosmopolitan ethics.

The rights of culture presume reciprocity, for each culture must tolerate the others. Thus cultural autonomy is not absolute, but dependent upon mutual recognition. The rights of Nazi "culture," then, does not include the autonomy to practice genocide, since Nazi "culture" is dependent upon mutual recognition. Cultures have the relation of parts to a whole, that is the same relation as ecocentric cosmopolitanism. An intolerant, tribal culture may still be the source of values, and be evil, because intolerance is wrong: some bad values, some good. But a good culture is tolerant of other cultures, that is, cosmopolitan and inclusive not exclusive. Again, if competing values are equal, cultural chauvinists of are of equal standing with equalitarians; if competing cultures equal, benevolent cultures are of equal value with malicious ones. But the latter are not tolerant, either within or with regard to other cultures. Why then should they be tolerated?

The ethical issue is whether we are limited to cultural ethics. The advocates of cultural relativism argue that all ethics derives from some culture and that we cannot escape our own. But both these theses may be challenged. Philosophic and religious ethics are attempts to escape culture and formulate an ethic for all humans.[37] Historical changes in culture argue that we can escape our own: historicism contradicts this proposition. For example, the end of slavery put an end to the culture of slavery.

An ethic must satisfy social and categorial values as well as cultural values. To be ethical a culture must be valuable in a social sense, that is, it must recognize equality of individuals as a guiding moral principle, in the sense that each individual is able to behave morally. This is to avoid inconsistency in values: hypocrisy. If the same rules do not apply to everyone equally, we are confronted with a power relationship, not an ethic. For why should we regard all cultures as equal, if they do not? If like the Nazis they regard some as superior to others? Social rules must apply equally to all to be moral. That is, all societies must fulfill the condition of choosing egalitarian rules to avoid the corruption of power relationships and hence being a corrupt society, one without a morality. A good example is the case of clitoral circumcision, in which only women are deprived of sexual pleasure, not men. This is plainly unequal: the fidelity of men is not put in question, including the men who cheated with the women. This nasty cultural practice is hardly "moral." Since it violates norms of equality, and has no valid function—for women could still cheat, based on affection for a man they are not married to—why should it be respected? Moreover, if cultural values fail to serve the members, then their mere status as the historically accepted cultural mode gives them no privileged or sacred status.

I have tried to show that cultural relativism presumes some universals, and logically this is contradictory. If we should follow our culture, this norm is both universal and absolute as it stands. If we should tolerate other cultures, this norm is universal and a duty, as is reciprocity. But the uncritical acceptance of cultural relativism is a surrender of all critical intellectual functions. Some cultural practices are not morally justified and relativists would never accept them for themselves.

If we take different cultures seriously, then according to cultural relativism, we should live according to cultural norms. But this would constitute a universal ethic, from which it follows that even intolerant cultural ethics contain universal elements. Similarly, cultural values presume value, worth in general and also the value of truth in the sense of the truth of the worth of the cultural values and the truth of their being a part of the culture or truly cultural. Again, for cultural relativism, the relation of each token is that of the equal value of each token. But the notion of equal value presumes a universal standard and therefore presumes universals.

Still another issue is whether an ethic of cultural relativism follows logically from the actuality of different cultures? Both contemporary and primitive humans can add 2+2=4, or note that today is sunny. Only some attitudes differ from culture to culture.[38]

Ethics has a common root with ethos, the way of life of a culture. This way of life is the differentiating element that distinguishes one culture from another. Thus each culture recognizes the requirement of an ethics within the tribe. Jim Rachels has stressed this point in his critique of any ethics based on cultural relativism.[39] All societies must follow moral principles to survive. As Clyde Kluckholn has put it,

> The first thing to note is the universality of moral standards in general. To the philosopher and social scientist this generalization may appear too commonplace to require comment. But the universality may not strike the general zoologist as so obvious. Not only is human social life inevitably a moral life in theory and to a large degree in practice, but ethical principles are the fundament of most of the rest of the culture.[40]

"Society" means the condition of all cultures. The survival of a culture is contingent upon the survival of the society and hence at least some of its individuals at any moment in history. Murder is wrong not only because it destroys another's life and integrity, but also if all members of a society murdered, the culture would not survive. In the West, the individual is "sacred" culturally, but in any culture, individuals are sacred because they bear the culture. Though outsiders may not be included in this formula, this denotes an intolerant culture, not an exception to life as the basis of value.

The supposed skeptical consequences of relativism don't follow. For those who draw such consequences attempt to build a structure when the thrust of relativism is anti-structural. If different cultures have some different values, it does not follow that no moral consequences can be drawn. For a hidden premise is that we should follow the morals of one or another culture. Without such a premise, the existence of different cultural values is morally irrelevant. But if we should follow the morals of one or another culture, then ethical relativism is bypassed in favor of cultural morals or cosmopolitanism. Similarly, should we follow our own perceptions, tastes, values, etc? The true consequences of relativism, except cultural, is atomistic individualism, not to say solipsism.

The Greek philosophers dealt with the problem of cultural relativism by seeking a higher level abstraction that united humans of all cultures as humans, human nature. The human capability that transcended culture was the ability to speak. Speaking did not just involve production of sounds, since all higher animals are capable of producing grunts and other noises. The words had to make sense, to be meaningful. Thus Aristotle thought humans were separated from the other animals as *zoon logon ekhon,* the animal that can put words into an order (*logos* as "word" and "order"). The sounds made by humans differ in different cultural language areas. However, in all languages the words are put into an order that makes sense of the world, which is rational. Rationality distinguished humans as humans from the other animals, since humans of all cultures could speak, despite differences in the noises they made to make sense of the world. Rationality was indicated in putting words into an order, in making sense of the world.

Cultural relativism also appeals to a higher level abstraction that unites all humans as humans, namely, participation in a culture, *homo cultus* if you will. Relativists then make the judgment for the particular, that people should follow the differentiated culture, not the universal premise. But if all humans are united by culture, then there is at least as much reason to opt for cosmopolitanism as for tribalism, since the difference is inseparable from the identity. The universality of culture also argues for ethical universality in one respect. Cultural norms are more or less binding on all members of the culture. Individual differences are put aside in order to uphold cultural universals. But if universals can be creatively actualized within a culture, and individual differences put aside for the sake of the whole, why can't this universality be extended outside the culture? By the universality of principles *within* a culture? Cultural universals qua universals? If a culture can have binding principles on all within, why not a principle for those outside the culture? Cultural universals argue against relativism in this respect.

Respect for culture as a norm follows from the imperative of non-violence: to limit violence to a minimum. But violent jihad and fascism of any type, particularly nationalist violence, violate the principle of non-violence and thus the norm: they are wrong. A cosmopolitan environmental ethic, which is the only universal ethics, advocates respect for cultural values. However, such values cannot constitute a violation of universal moral norms, either with respect to other humans, other species, or the biosphere as a whole. For toleration of cultures has moral limits, and culture is not the sole source of value.

A culture that exceeds the carrying capacity of the land will destroy itself. Southwest Asia is dotted with cities that declined as their soil wore out or was silted up. Prof. Derringh has argued that these ancient cultures had a bit more of an excuse, since they were far more ignorant of the consequences of their practices on the environment than we are. Will we learn the lesson they failed to learn?

Overpopulation will have the same result, whether sooner or later. Just as humans in general have been a self-destructive species, so have several cultures. The death of the land would mean the death of the culture, and thus the uniqueness and values of the culture would be lost, defeating the goals of preservation of cultures. Moreover, unique habitats often play a part in the evolution of the culture; the desert, jungle, or tundra are woven into the fabric of the culture in obvious and subtle ways. Thus cultural autonomy cannot justify overpopulation, for overpopulation would result in the death of the condition and an integral part of the culture, its homeland. In sum, cultural relativism, which is often disguised cultural chauvinism, cannot justify overpopulation.

4. Historical Relativism

Historical relativism is the thesis that each historical period is distinct and not comparable to others.[41] Thus ethics is historical, in a temporal form of cultural relativism. What is ethical for us may not be ethical to other ages.[42] "Historicism" may be taken as a species of cultural relativism, since it argues the exclusivity of a historical period as some sort of unified whole, that of an age, and that specific marks distinguish that age from any other. Thus a historical instance is different in time from any other. Some add that it can be only understood in its own terms, notably Wilhelm Dilthey and his followers, including Martin Heidegger. However, the problem may arise in a more acute form in the historicist version of relativism. Historicism can note the changes even in entrenched knowledge and argue that no knowledge is final. Similarly, the practical categories are based on modern Western society. They are thereby culturally relative at least in the historical sense. The relevance of critically evaluating historical relativism to cosmopolitan environmental ethics is that we are arguing meliorism, that is, that taking environment into account is moral progress and thus against historicism in that sense.

Historical practices can be patently immoral, for example the kidnapping of Africans and their "sale" as slaves. Understanding the motives of the historical actors in terms of their "unique history" may be helpful, but is morally irrelevant. Moreover, I do not think that we need to search very far for the motive in enslaving Africans, for that motive is common to their age and ours, namely, greed. The greed of the oil companies, which if left unchecked may destroy the earth with global warming, is of a piece with that of slaveowners.

Relativism is anti-cumulative. We can see this in the historicist version, for which each historical unit, however defined, must be evaluated on its own terms, without reference to any other. While commendably sensitive to historical difference, and the *per se* value of each culture, this denies the value of cross-cultural categories abstracted from like characteristics of each. The experience of the past cannot then guide the present: each culture's experience will be different.

However, we do not starting from nothing but have access to history in a way our predecessors did not. We have Aristotle but he did not have us. Historical and other forms of knowledge involve accumulations. The alternative is not absolutism and relativism, since there is a middle position possible. This is the melioristic position that both ethics and knowledge can improve: epistemological meliorism and moral progress.

Against historicism in political garb it can be argued that though justice may be seen differently by different ages, all ages still view justice as desirable, valuable, the goal or condition of association. Against its elitist form, associated with Friedrich Nietzsche, I would argue that reciprocal justice is the condition of the elite also—merit is post-associational.

5. Individual Relativism

I noted above that conflicts between individuals and society are often over cultural transgressions. Cultural values are often seen as absolute by a society: inviolable norms, which can only be broken for a higher value. Thus cultural values may conflict with the notion of individual rights, as in the case of female "circumcision."

Individual relativism is not identical with cultural or ethical relativism and may come in conflict with culture. In this view, ethics is up to the individual. Truth, beauty, and goodness are relative to individuals and cannot be reconciled. Self-interest can be a species of this; so is esthetic relativism, which argues that each person's taste is different and different tastes cannot be reconciled.[43] This kind of relativism, ironically, makes of relativism an absolute. Absolute relativism is implicit egoism. The subject legislates according to individual ends. Duty is left to individual discretion.

I criticized this type of relativism extensively in chapter six, in the form of the "ethic" based on self-interest. Individual relativism as an "ethic" is as oxymoronic as one based on self-interest, since if good is left up to individual taste or preference, anything goes. Rape, murder, and ecocrimes are "right" if they accord with individual preferences. This is a parody of an ethic. Logically, an individual "ethic" is an oxymoron. Ethics proper, apart from value theory, has to do with right and wrong and with obligations to others. We learn right and wrong from our society, in the light of established norms that allow society to function and continue.[44] Right and wrong generally involve obligations; obligations are primarily to others. "Duties to self" extend these universally, but are "weaker" than duties to others. In sum, ethics arises in society of which each individual is a member in some relation and consists in obligations to other members of the culture, at a bare minimum, although most cultures also include provisions for travelers, guests, and members of other cultures in the form of hospitality.

Social values and cultural values argue against individualistic relativism Some activities are common in view of shared values, thus not relative to individuals. Meliorist cooperation will result in superior conditions for all, at least in some spheres of activity. Moreover, humans require society. To merely survive—to live, not to say live well—we must depend on society.

Even our birth depends upon social relations; our upbringing on customs, cultural particulars, and institutional norms all given to us by society. Individualistic relativism ignores socialization and acculturation in the formation of individuals, contradicting the lessons of sociology and anthropology. All societies require conformity to moral norms and if individuals all made exceptions of themselves, society would collapse. We learn moral values from society. Social and cultural imposition argues against individual relativism. Individual tastes, beliefs, and other characteristics are socially determined, for the most part. Expressions of individuality arise especially in an individualistic culture, which encourages self-expression. Even feelings are determined as socially distinguished and approved or discouraged (softer feelings in men in some cultures). A society's values are expressed in its total culture, not just its morality, if it is to have coherent values. Moreover, nothing in a society is value-free, since society is expressed as cultural choices that are the consequence of social choices in the light of environmental and other circumstances.

Advocates of individualistic relativism might argue that making moral arguments amounts to individual opinion. If one person tells another what to do, they are trying to impose their own individual desires or norms on other individuals. However, we can argue for social norms that are impersonal and involve either the good of all or the meliorist functioning of society. Impersonal norms are those that do not involve oneself, but involve judgments about moral situations. Moral condemnation is not purely relative: I may condemn wife beaters and not be a wife. I can condemn rapists and never rape. Arguing that the consequences for society of a proposed act would be better or worse is impersonal.

From another quarter, sociobiologists could argue that individual males are genetically "programmed" to create as many offspring as possible, as an aid in preserving and spreading their genes. This view ignores cultural mediation of individual behavior, however, which is essential to the survival of the infant. But cultures are generally, if not universally, against polygamy and some other sex practices on the part of individuals. Cultural taboos restrict individual behavior. Moreover, sociobiology confuses the proclivities related to genetic codes with genetic teleology. It is not as if genes plan our lives such that we procreate as much as possible. Genes do not deliberate, nor plan for the future: they are not teleological, however much they may function in some ways. Functioning for is not equivalent to planning for.

Another way of confronting arguments for individual relativism is to universalize the action. What if everyone did what that one individual does? What if everyone raped, stole, murdered or polluted? Society would come to a screeching halt. Individual relativism is thus functionally detrimental and morally a form of the freeloader ethic. Having too many children on grounds of individual relativism can only result in disaster for all.

One form of individual relativism that is germane to the issues of population is economic. This is the view that what is in my economic self-interest is good, regardless of its effect on others. With respect to population, the argument could be made that increases in population will increase the demand for goods, and thus that it will increase sales and profits. Since increasing profits is in a person's self-interest, increases in population are good. Further, an increase in population, *ceteris paribus*, will also increase the supply of available workers, the labor pool. In the absence of other mitigating factors, an increased supply of labor will result in lower wages (compare exporting the factories to countries with cheaper labor—cheaper usually because of high population). Lower wages will increase profits, so they are in a person's economic self-interest. Such arguments will mainly benefit the wealthy few, not the majority. Thus they are ideological. Critics might argue that wages went up with rises in productivity and other factors. But there is no guarantee that the extra will be shared with laborers, as it has not been during the last thirty years: real wages in the United States have remained stagnant except for the top ten percent. Moreover, labor is one cost of production, even if it alone cannot explain value or price.

To this we can respond with the arguments we have already made. No human lives outside of society, least of all economic concerns, which depend on customers, suppliers, and many other social relationships. Economic life in particular requires moral, as opposed to violent relationships. The condition of society, economic life, and even a person's self interest is a healthy environment. Moreover, it can hardly be in anyone's self-interest to have an overcrowded planet, where a person cannot enjoy life. Properly viewed, economics is a means to an end, generally to preserve and enhance a life—the good life. We do not normally horde money for its own sake, but for the sake of other ends, whether a beautiful house, travel or other goods. But all these goods require an environment, and in many cases would suffer from an overcrowded world. Wilderness areas that provide respite from civilization would disappear on a crowded planet. Beauty would be sacrificed to wealth. The good in the world would diminish.

There is no actual economic advantage on the whole to an overpopulated world, since resources will be stretched farther, diminishing the average amount per person. Increasing "growth" has limits, as others have documented (Donella Meadows et al., 1972). Indeed, we have known this for decades, just as we have known about the Greenhouse Effect (global warming), soil erosion, and many other environmental problems. Still, nothing is done by our short-sighted "civilization." Worse, growth will be at the expense of an already stressed environment, of other species and their habitats, and of future generations that will have to do without scarce, irreplaceable resources like petroleum. "Growth" at the expense of the future is not genuine, but a case of stealing from our descendants.

6. Tribalism

Tribalism has always constituted a challenge to any cosmopolitan, expansive ethic and continues to do so today in the form of national rivalries and Identity Politics. Tribal ethics is exclusionary, whether in its elitist or identity form. Identity politics is tribal in the sense of creating a tribe delimited by the identity, not kin relations. Like tribes, defining identities separate out a group as constituting its individual members. Tribal mores tend to the glorification of the group identity at the expense of others. Tribal humor creates victims (other ethnic groups, sexual minorities, the insane, and so on) and such targets are thereby separated from the group. Tribal ethics are relative, like cultural relativism, and exclusionary. However, a tribe is not the same as a culture and "tribe" or "tribal" can be taken in a more generic sense of exclusionary identities of many types. Thus an exclusive club, class, use of gender, or any other limiting identity can be called "tribal" in the generic sense, by analogy with social and cultural tribes.

A cultural ethics that is exclusive is the ethic of a tribe. A tribal ethic is a kind of cultural ethics: the ethic of a chauvinist culture. Such an ethic includes members of the tribe and excludes others from moral considerability, to speak nothing of rights. Thus a tribal ethic is not cosmopolitan. If the tribal ethic only takes the form in which one tribe is judged different than the others, even if distinctive, the ethic contains a basically equalitarian social view. If, however, one tribe is adjudged higher than others are, it is a hierarchical view with an aristocracy on top (Friedrich Nietzsche). The glorification of the tribe over other tribes is the source of chauvinism and nationalism. The world owes the "superior" tribe a living: slaves, servants, seizure of "private" property, and so on.

Contemporary ideologies, including so-called identity politics are like revaluations of tribal ethics: an ethic of a wealthy elite, of men, of women, of heterosexuals. Ideologies are so many revaluations of tribal ethics: from an ethic of a wealthy elite, men, or heterosexuals to their counterparts: poor, the workers, and women, to gays and lesbians. The extreme right- and left-wing counterparts are equally tribal. Some of these ideologies exclude the other from equal moral considerability; others are capable of doing so. They are exclusive of the poor, women, gays or other identities. Thus cosmopolitanism should include critical evaluations of both elitism and identity politics as forms of tribalism.

Anthropocentric ethics is a larger form of tribal ethics, for it excludes individuals of other species. Only humans make value judgments, have inherent value, or are morally considerable. What is left out of such perspectives is the larger relation to the environment required of all tribes. Tribal values require an environment for their attainment. Thus the attempt to meld environmental philosophy with the politics of identity subordinates the greater concern to the lesser. "Ecofeminism" takes feminism as the model for the environment, not the environment as the model for the relations of women to a larger whole. Similarly, "queering the environment" takes a movement that concerns a small minority within human society and attempts to model environmental ethics upon it. Obviously, both women and gays, in the generic sense of lesbians and gay men are humans. Thus there is a disguised anthropocentric premise behind such ideological posturing, however much we may wish for equality for these groups within society. The environmental movement cannot be only a movement for minority rights within society since the environment is the condition of any association, male, female, gay or straight. Protecting the environment is a much more universal concern, for all of life is at risk, not just a minority of human life. Moreover, rights only apply within human association.

From an environmental viewpoint, the empowerment of women can be beneficial, since it can result in more control over the number of births. Nevertheless, there is no guarantee that women will use their newfound freedom for the good of the environment, since cultural factors may intervene to move them in the direction of reaction, as is likely in Iraq, where some women vote for theocracy with as much enthusiasm as the men. Similarly, women in the United States have been in the forefront of the pro-life movement as much as the pro-choice movement. Women can buy into the consumer society, and join the unsustainable way of life in the suburbs, and thus contribute to the problem of human destruction of their own planet. As for gays, if being gay means not having children, then so much the better. But it frequently means pressure toward normalization, for example, gay marriage, artificially inseminated children of gays, a guppie co-op, and all the consumer trimmings. Thus neither movement represents an unmixed blessing for the environment, although specific tendencies in each may ultimately help, whether feminist caring for the earth, or the rationale for the gay life-style that argues there are too many people anyway, so that sex should be divorced from procreation and gays may help the overpopulation problem.

These points should not be taken as a criticism of such ideologies in themselves, only of the attempt to subordinate environmental ethics to anthropocentric ethics, in the form of tribal ideologies. Such attempts mimic other forms of anthropocentric ethics, in that they are speciesist. The good of humans has been the only concern up to very recently; the good of other species, of distinctive habitats, or of the biosphere were treated as instrumental to human good, if they were considered at all. In other words, "ethics" was selfish, since only human good counted morally. A selfish "ethic" is hardly an ethic at all, even if someone could make a case for not fouling our own nest on selfish grounds. For we may murder, rape, and steal on selfish grounds, if we think we will get away with it, but would anyone argue these are ethical? But this is precisely what we are doing to the environment: murdering other species, raping the land, and stealing habitats from native species.

Ultimately, anthropocentric ethics is tribal ethics writ large: for the whole species of humans. The tribe of humans will lord it over other species because it needs *lebensraum*, a barely disguised Nazi ethic. The Nazis' ideology was tribal: the Nazi rationale for Germany attacking surrounding countries was that the German *volk* needed "*lebensraum*" and thus the borders of Germany had to be expanded. One tribe, the Germans, was judged the "superior race" and others were to be either murdered or subjugated. "Manifest destiny" was the rationale for the expropriation of the Native American tribes and the attempt at their cultural genocide. Similarly, humans are judged the superior species, and others are either murdered or subjugated. Only because humans are blinded by their own greed do they not see that the relevant moral principles are the same when we are considering human destruction of the environment. Fascism means the violent subjugation of other peoples in the name of a tribe. Similarly, destruction of wild for sake of cultural *lebensraum* is anthrofascism. J. Baird Callicott was taken to task for denying rights in the wild by his critics, who labeled him a fascist. But the wholesale destruction of habitats and species by humans is surely an example of anthrofascism. The ethics of the tribe is the ethics of violence.

A cosmopolitan exchange of ideas, not tribal xenophobia, has produced the greatest advances.[45] Ethics as such is cosmopolitan if we exclude tribal ethics. Perceiving our duty to those outside our tribe—universality—involves both overcoming tribal attachment and actualizing morality as such in its pure form, consideration of the other as a person, as in Immanuel Kant, not the utilitarian morality of a tribe, that regards other tribes as instruments. The origin of cosmopolitanism lies in Greek philosophy, especially the materialist tradition, not any elite.[46]

A tribal identity is odd in any logical sense, since it is both arbitrary and inconsistent. It is arbitrary because it stops at a specific level of association and in doing so provides a false identity in some sense. Identities with a group are arbitrary since one identity is chosen over another, generally larger identity. Thus "identity" is equivocal in the sense that the individual has several identities: one's family, tribe, locale, nation, race, and species, among others. A choice by the individual or that of a movement to prioritize one involves subordinating the others. A person becomes "German," (or any other nationality) not male/female, human, a sports/music fan, animal, and so on. Here a selection is made in which a part of a person's total identity becomes the whole. Other aspects of people's total situation, their biological identity, gender, region, and many other characteristics are arbitrarily ignored for the primacy of the tribal identity. But why should an individual identify primarily with a tribe, and not all humans, or better, all of life in its interdependence? With much larger identities of which the individual is in more vital relations? Humans are animals if only because they live, directly or indirectly, off of plants. Prioritizing such identities are indeed partial and thus cannot be universal. Moreover, they are false if taken as the complete whole, since they distort the larger web of social and biological relations that intersect in individuals: their other identities, especially as humans in a biosphere. Prioritizing race or species not genus as the fundamental identity element constitutes an error in evaluation, in which the lesser is confused with the greater. It is also a bad classification, since it favors the more restrictive identity over the more inclusive—as if mammals were selected as arbitrarily distinct from all other vertebrates, animals, and other levels of biological classification. I am not arguing against cultural or tribal identity, only prioritizing one at the expense of other relations.

Tribal identities are thereby inconsistent because they affirm the value of association while at the same time denying it; providing an identity that is also a difference, creating a tension in identification. False identities nevertheless are differences, for they bring an instance under one identity different than another. To identify with is to ally instances, and thereby to differentiate them in some respect from other instances. A racial identity provides an identity, but by doing so differentiates that identity from other "races." Identity as difference involves some sort of contradiction in a person's identity. The value of association with those of like identity is affirmed, namely, identity with the tribe. At the same time it is denied, since the value of like identity with larger associations is ignored. The value of a person's narrow identification is affirmed, and its differentiation; the value of other identities is denied. The "logic" of tribalism, racism, and other narrow identities is contradictory.

Choosing an exclusive tribal identity involves a decision to conform to tribal mores with their limiting identities, not cosmopolitan ethics with a more inclusive identity. Or better: an environmental ethics with recognition of value beyond the human realm. False identities are tribal, as ingrown not extensive.[47] Obligation is extensive: it extends beyond tribe through identification with all cultures (cosmopolitan), universal love (Christianity, Moism), and so on.[48] Ethics is universal, and built on the premise of universality in reasoning. Thus Aristotle takes the universal nature of humans as the rational animal as the basis on which to examine the perfections or virtues proper to that kind. If we could articulate any Aristotelian moral "imperative" it would be to act from species difference (be human, be rational), not any tribal or partisan perspective. Reason as the identity element of humans undermines social, class, national, and racial differences—the fundamental point of the Enlightenment. Normative identities should be grounded in ethics.

How do we reconcile a critical evaluation of relativism with a cosmopolitan outlook? Ethics must change to keep up with contemporary situations. One measure of this is moral progress, in which rights and moral recognition have been extended to larger and larger numbers of people. The contemporary situation that is different from the past is the reduction in the death rate among humans. Modern medicine and the "green revolution" have meant that more people survive into old age, and the many children that used to die off from childhood diseases now live to adulthood and reproduce. The secularization of Christian ethics has meant that for medical professionals to go from rich to developing countries and help save lives is considered virtuous. Together, this has meant an explosion in populations in the developing world. Given these changes in medicine, it is now immoral to overpopulate.

The new intellectual consensus on the good life is relativistic, rather than absolutistic. For the most part, intellectuals regard values as subjective instead of hierarchical and competitive.[49] They are therefore more likely to be tolerant of minority and deviant lifestyles, such as those of gays and lesbians, than in the days of more rigid social norms. This outlook is wider, and includes more types, and is not directed to only one tribal more. Philosophy is also less judgmental than in ancient times. Philosophers today argue less for a model of a single best way of life than in ancient times.[50] Other ways of life than philosophical are given their due. Philosophy is less morally critical, more morally ambiguous. Does the inability of philosophy to judge what is the good life reveal a loss of nerve? Or does it reflect a more egalitarian view of different lives?

Still, there is some consensus. Bigotry is ruled out by tolerance. Tolerance comes from the reflection upon the existence of different views within philosophy itself. Philosophy cannot set standards for others while having none for itself. "Community" is more of a factor in "schools" of philosophy. Philosophers tend to feed off each other's ideas and engage in sometimes hidden dialogue, written as well as oral. Journal articles often address previous articles (compare Plato's dialogues). This social aspect of modern philosophizing requires some degree of tolerance for other views. Also, it means relations to other peers or equals, instead of "higher" principles. This trend means philosophy is more "other directed," despite the subjectivity of life styles.[51]

One argument for relativism is that it reflects a perspective, that of an interest, notably in Nietzsche and Marx. Marx, however, acknowledges that members of a class may nevertheless identify with other classes, as he himself did. Thus class identity is not fixed, undermining perspectival relativism, that statements necessarily reflect the interests of our own class. Moreover, it is not purely relative to make moral condemnations: As noted above, I may condemn wife beaters and not be a wife. I can condemn slavery and not be a slave, condemn rape and not be a rapist, and so on. Such arguments undermine a major rationale for relativism. Proponents of this view may counter with some notion of a perspective not based on class or self-interest—that all truth and values are perspectival, but this notion contradicts itself. For if all truth is perspectival, then this truth is also perspectival. If so it cannot be universal. But if it is perspectival, it cannot be universal and at least some truths could possibly be non-perspectival. The same arguments apply to the assertion that arguments for population limitation represents the outlook of one culture.

The philosopher articulates universal ethics by such rules as non-contradiction. Contradictory principles, norms or rules would be incoherent; that which is incoherent cannot be valuable, nor provide lessons or guidance to individuals. What I think philosophy has learned from cultural studies is tolerance for deviance as part of a universal ethic—a more cosmopolitan outlook. What it needs to address is the conflict between cultures and preservation of the environment that is their condition. Tolerance for cultures cannot extend to immoral practices, such as slavery,[52] or to destruction of the commons, the biosphere. A cosmopolitan ethic is universal, but that does not mean tolerant of immorality. Moreover, the revaluation of cosmopolitan ethics requires consideration of the biosphere in all moral evaluations, since a cosmopolitan ethics is an ethics of the world and thereby includes the world of life as a whole. Value conflicts within the confines of a cosmopolitan ethic must therefore consider not only tolerance for cultural values and ways of life, but for the conditions of culture in a habitat and the symbiosis between humans and other species within that habitat. Cosmopolitanism ought to be revalued to include other species and their ways of life, as well as the whole of which they are parts.

Moral progress involves different sorts of improvement. These include changes for the better within a culture, in individual behavior, or in trans-cultural cooperation. "Better" is measured in terms of moral consideration of larger numbers of agents, of better quality of life, and of decreased violence. Moral progress is melioristic in aiming for an improved moral climate. Our argument has been that the expansion of moral considerability to all the elements of the environment would constitute meliorism in our present age. A corollary of this includes population control.

7. The Basis for Encouraging Population: Religion

Another objection to limiting population has come from fundamentalist religions, and also those that make selective use of the Old Testament.[53] In the case of fundamentalists, every word in the Bible (Koran, and so on), including the Old Testament, is considered literally true. In the other case, whether or not the Old Testament is considered literally true, several ethical or religious injunctions are considered binding on non-Jews while others are not. For example, the injunction to "be fruitful and multiply" is treated as Gospel, while the duty to keep a kosher household is ignored. In this selective "cherry-picking" of scripture, an arbitrary use is being made of the Old Testament. Those injunctions that some denomination likes or find convenient are accepted while others are not. But if the Old Testament is the source of norms, then it should be followed in all its strictures. Since the Old Testament allows for polygamy, for example, polygamy should be allowed. The response of fundamentalists is that the customs in the Old Testament reflect a different time and place and that polygamy is not suitable for contemporary times. I could not have said it better: be fruitful and multiply may have been suitable when there were fewer than one billion humans; it is definitely unsuitable when humans predominate over two-thirds of the earth and their impact affects even the oceans, which have been decimated by human overfishing. On the contrary, to save species, as God commands Noah, it is imperative to limit human populations.

However, there is a more compelling case to be made against the Old Testament, which is, since Genesis is manifestly wrong, it is a weak source of norms for contemporary humans. The world did not begin in 4004 BC, as the Biblical account would have it, but billions of years ago in a Big Bang. Life also began long ago. To save the doctrine of Biblical inerrancy, some who are influenced by fundamentalism argue that dinosaur bones, radioactive carbon, background radiation, and other evidence for a much older world were created in place by a God seeking to test our faith. What this says about their God is an issue I will not explore, but this sleight of hand will not save the doctrine. For if God can and would fool us on the evidence what is to say we are not also fools to believe the Old Testament?

Since the Old Testament is compromised, and is revealed as having human, not divine authorship, it cannot be authoritative for anyone but those willing to suspend common sense for a dubious faith.[54] One alternative is to rule out theology as a source of norms, since if God exists he is morally culpable in not having prevented evil: the Crusades, the extermination of Native Americans, slavery in the U.S., the Potato Famine, Auschwitz, and so on. Since he is infinitely powerful, he could have prevented such evils; therefore he is morally culpable or accessory to such crimes ("will/mandate of heaven"). To argue for divine voluntarism (they are not crimes if allowed by God) is to argue that there is no morality only divine arbitrariness, what Kant called the ethic of "divine whim," and therefore no one is blamable for anything, as all is willed by God. If God "infinite" or "inner light" our "assent" to his "will" is not "ours" but *his* as he is everywhere. Moreover, some religions have tried to justify the immoral: as the Southern Christian Churches did in defending slavery based on the Old Testament. Again, could jihad justify genocide? But then this interpretation of jihad is obviously immoral as genocidal. Is a religion intolerant (Theo Van Gogh)? But then why should it be tolerated? This is morally contradictory. Those who advocate "pure tolerance" argue that only tolerance separates civilization from such fanatics. But they ignore the right to self-defense, without which there are no rights of any kind, including the right to practice religion.

In any case, religion will not provide norms for those who take a more balanced view of the Old Testament, as a first attempt at a historical view of cosmology. But since Genesis is wrong on cosmology, and the history of the universe, it can be wrong on ethics as well. The premier example of where it is out of date is on issues of population, since "be fruitful and multiply" may be a strategic plan to increase the population for a besieged people. In other words, the so-called command to increase population was military in motivation, not religious. The commandment is no more binding on non-Jews than the command to keep meat and dairy apart or to circumcise children. The same arguments apply to any other religion that claims a sacred command to overpopulate. The sacred text is used selectively to increase the numbers and thereby the power of the faithful. But such strategies presume that others will pick up the bill. We must hope that those who substitute tolerance for critical thinking and are blind to overpopulation in other cultures will wake up. Humans have no more responsibility to aid those who, based on religious strictures, overpopulate, than to build their houses of worship. Religion is not ethics, as holy wars, which all "faiths" have engaged in, testify. Religion must also meet an ethical test and in our times, this means an ecocentric test of not exceeding the carrying capacity of the land, and overpopulating at the expense of quality of life.

We should not live according to norms of five thousand years ago, any more than we could live by their science or world-view. If fundamentalists were consistent, they would stop driving, using modern technology, and live like the Amish, since modern technology is not in the Bible. But by using the Bible or the Koran or Hindu tradition selectively, fundamentalist of all stripes try to have their cake and eat it too. If we believe in technical progress, then moral progress must accompany it. The alternative is omnicide: nuclear war or environmental catastrophe.

However, there is something to be said for devotion to a cause higher than consumer greed, which despoils the environment far more than the ascetic life encouraged by many religions. Some religions may therefore provide a model for how to live in an eco-friendly way. Moreover, the spirit of many religions is universalistic, since the message of the revered prophet is taken as being for all humans, not one tribe, class, nation or narrower identity.

8. The Problem of Coercion

Limited social engineering of the Popperian type, which changes part of a culture but not all, has unintended consequences, especially overpopulation in the Third World based on introduction of modern medicine, among others. Unfortunately, the social engineering has not included population control norms or methods. Consequently, starvation has resulted, when sudden changes in the weather patterns result in withered crops (the Sahel). Even where there is no starvation, the quality of life remains poor, in both senses of that word. In the rush to save lives, which both liberals and conservatives consider virtuous, the habitat is ignored. Because modern medicine and agriculture are being introduced, it makes sense from a holistic point of view to introduce modern birth control methods. Since cooperation will not persuade freeloaders, it also justifies the use of the law to limit births for the public good.

One objection to population control is that limitations on an unlimited right to procreate will involve some form of coercion, however much this is disguised with a velvet glove. One question this raises is, what is legitimate coercion? —is society protecting itself coercion? Are rules that protect the public, such as laws against speeding in school zones, coercive? Are quarantines, which protect the public against the spread of infectious diseases? But if these are enforced through coercion, then some forms of coercion are legitimate from a moral point of view, namely, those that protect the public, such as laws against murder and public health regulations. Surely, protecting the quality of life of the public justifies at least some regulation. Then protecting species and habitats also justifies regulations that preserve them. Holistically, overpopulation would threaten the entire biosphere, and thereby be practically self-defeating. Since the state may intervene in cases of murder, that is, to protect life, it may intervene to control population, which threatens the life of all. As Daniel Callahan has stated:

> "Governments have as much right to intervene in procreation-related behavior as in other areas of behavior affecting the general welfare. This right extends to the control of fertility in general and to the control of individual fertility in particular...Governments have the right to take those steps necessary to insure the preservation and promotion of the common good—the protection and advancement of the right[s] to life, liberty and property...Since excessive population growth can touch upon all these elements of national life, responses to population problems will encompass both the rights and obligations of governments."[55]

The issue is not whether there will be coercion, but which issues we will apply it to: protection of private property or protection of the biosphere. From an environmental perspective, "private property" is arbitrary, for property lines may cut across habitats in a haphazard way. Animals and plants that dwell in and are an essential part of a habitat do not and cannot recognize such property lines. But the larger issue is: will we defend private property to the point where the environment is so degraded that it can no longer support life, except perhaps for a few weeds,[56] or shall we pass legislation controlling the use of land? Controlling use of the land to assure its preservation is requisite from several ethical theories, again including that of Utility, the greatest good of the greatest number. But for a deontological ethic, that is, ethics proper, misuse of the land is inherently wrong and thereby should be limited by the state.

As law requires enforcement, we must frankly admit, like Hardin, that limited coercion will be involved. But coercion to assure quality of life is no different than coercion to prevent theft, murder, and rape, all of which would lower quality of life. This supports the argument for a state solution to the population problem. If social engineering through medicine is justified, so is social engineering of population. Voluntary controls will not work, any more than voluntary restrictions on rape, murder, theft or other crimes. Moreover, there can be no rights without enforcement, as Thomas Hobbes argued. The equal rights of all are protected through universal laws that apply to everyone. The enforcement role belongs to government. What then of abuses of power of the totalitarian type? Such abuses transcend the issues raised by population, since they arose before environmentalism was widespread. Abuse of power is possible by any regime or any ideologically motivated party. The checks and balances that curtail governmental abuses are only as good as an alert citizenry, as the pattern of revolving door military coups in some South American countries demonstrates. The issue is a red herring in the population debate, since it applies to all governmental authority, not just this one issue.

Another objection is that public birth control regulations will violate "universal human rights," among which are private family planning decisions.[57] However, it should be kept in mind that rights are moral.[58] Rights that transcend legal rights, which have a government foundation, can only be moral rights; otherwise, they would be grounded in coercion. But such rights involve corresponding obligations as moral. Duties and rights are generally reciprocal, since my right involves your duty not to interfere with its exercise, and your right creates a duty for me, if only the negative one of non-interference. The principle of reciprocity applies between cultures as well as between individuals. The basis of rights is moral and thereby within the framework and bounds of duty.[59] In discussing any so-called right to reproduce, we should remember that rights are basically moral rights; freedom is moral freedom. Another way of putting it is that rights have a moral basis, hence involve others or arise in a social context of mutual rights and responsibilities. Since rights require responsibilities, the so-called right to reproduce also requires responsibilities, as a species of rights. Thus rights in the sense of freedoms[60] require moral evaluation including consideration of environmental impact. Also, as I argued in chapter four, rights are inter-human and do not affect our relations with wild species. As population is an area where species with rights can have an impact on species without rights in the same respect, moral evaluation is complicated. However, we have duties to preserve other species as well as the environment as a whole that limit our rights.

The same principle applies to freedom: the "freedom" to reproduce. Freedom is also moral freedom: it requires moral limits, or freedom would come to an end. Robert Kirkman (2002) argues that environmentalists have not considered human freedom enough. My response is that freedom is moral. I do not think Kirkman would argue for immoral freedom: the freedom of rapists to rape, of drug dealers to hook teenagers, and so on. But if freedom is moral, it is subject to moral limits and would include moral consideration of actions with respect to their environmental consequences. Humans are free within moral limits. The alternative is political freedom as the only freedom: which would justify immoral laws; or absolute freedom: anarchy. Anarchy is a species of immoral freedom, as Hobbes argued. My liberty is limited by the equal freedom of others. Thus neither rights nor freedoms are unlimited, but are limited by reciprocity and equality. All members of society should respect the rights of others on both moral and practical grounds. Thus limiting the right to reproduce is not inconsistent with the notion of rights, since rights are limited.[61]

Unlimited population increases are not feasible ecologically and unwarranted ethically. An unlimited right to procreation would require unlimited exploitation of other species, and is therefore wrong. From an ecocentric perspective, overpopulation is a disaster, with species extinctions, destruction of rare habitats, and the potential for alteration of the entire planet from Global Warming.[62] Humans have gone from establishing a place within alien habitats to the destruction of huge ecosystems, such as the prairies of the North American Great Plains. Our argument here is holistic. We are not surreptitiously appealing to the greatest good for the greatest number, but the good of the whole and the parts that constitute it. But even from the point of view of Utility or of self-interest, an overcrowded world, with fewer resources available to each, Global Warming, loss of species, and other environmental problems, is a turn for the worse and morally problematic in *any* ethic. In a crowded world, reproductive decisions are no longer an absolute individual "right" or autonomous decision. There is no "right" to overpopulate.

I have noted that the model of moral progress as the extension of rights has been challenged by the ecocentric model of humans as members of an ecological community. An ecocentric cosmopolitan ethic locates humans as part of a larger world that includes other species, many of which are essential to human survival. The ecocentric model describes our relation to the larger world, cosmopolitanism, more accurately; the model of moral progress is only valid within civilization. As I argued in an earlier chapter, "Rights are part of a larger whole." Within this whole we can distinguish between a right to procreate and an unlimited right. A limited right to procreate involves respect for other species and recognition of their intrinsic value, as well as their functional value: their part in the larger whole of their habitat.

Humanists—whether radical, liberal[63] or conservative—are in denial that population is the cause of poverty and that global warming is a crisis that dwarfs other "issues." Both liberals and conservatives are anthropocentric and only differ on how to divide the environmental pie: whether exploiting it to exhaustion for the sake of the rich or the poor. From an environmental perspective, there is little difference between these ideologies, inasmuch as both see the environment as a field for exploitation. Neither, then, can provide any basis for moral progress that advances to include the wider world of all the elements of the environment.

As for radicals, they are in denial, for they see poverty as the cause of social injustice to the masses. But the masses themselves can cause their own poverty. A family with many mouths to feed must divide its income among more members, while that same income divided among fewer mouths would mean more food per person. The same principle applies to other goods: the fewer people, the greater the relative supply. If "social justice" theorists like Murray Bookchin truly want a better life for the masses, they will favor birth control. They will not promote policies whose long-term effect will be desertification and starvation. A lower population will improve the quality of life of everyone, while increased quantity will lead to ruin, in which the poor will pay a disproportionate price. The vision of universal affluence would not mean social justice but its impossibility. For not only are we using up finite resources, but creating horrendous new problems, including global warming. The vision of a chicken in every pot and a car in every garage of everyone in the third world is impossible without control of population. The attempt to have your freedom to reproduce and eat your affluent cake would result in such an environmental catastrophe that current poverty would look like affluence and future generations would be worse off than their parents.

(Note for radicals: China limited its population to one birth per family and its economy took off. Since the Chinese government can hardly be accused of lack of radical credentials, a case can be made for population control within radical ideology. The principle of supply and demand supports the view that fewer people would mean a greater number of goods per person. Fewer people would reduce demand, *ceteris paribus*, and thus lower prices for the supply. Moreover, fewer people would reduce the labor supply and thereby increase the relative demand for labor, and wages would rise in response.)

Global warming would also create a massive refugee problem, as lowlanders are forced to flee the coasts, and less land is available to feed an increasing population.[64] Indeed, parts of three members of the United Nations, Tuvalu, Vanuatu, and Kiribati, have already had to be evacuated, while a fourth, the Maldives, had a temporary evacuation. Moreover, low-lying areas of such countries as Bangladesh and Indonesia have also been abandoned, while surges from a higher ocean recently destroyed long stretches along the coast of the Northeast United States.

To deny this is myopic. We cannot have sustainability, much less affluence, without getting control of population. There is also the effect of farming methods that increase yields, to feed more mouths in the short term, but whose long-term consequences will make it harder and harder to prevent widespread starvation. Erosion and salinization of soils are threats to future generations. Feeding more mouths now but destroying the soil is as shortsighted as can be. It holds out the false hope that we can just go on exploiting a finite resource forever. It attempts to put off the day of reckoning when we have to face such problems. It is in everyone's long-term interest to limit population in the short term, if we are to avoid Malthusian consequences in the long term.

Garrett Hardin (1993, pp. 169, 177) contrasts short term and long term interests, in a Malthusian context. Some have argued that the Malthusian results have not occurred, and did not in his lifetime, but they are ignoring the sad fact that about one-sixth of all humans go to bed hungry. They are also ignoring self-defeating prophecies, which prevent negative consequences through publicity, consciousness raising, and resulting changes in behavior. Several countries in Europe have negative population growth, in part due to publicity about over-population.

If we have a duty of beneficence, as Peter Singer (1976, p. 85, 92) argues, we should work to prevent too many births. An increase in population is an increase in the demands on non-human resources, to speak nothing of human resources. It is a demand for the conversion of wild space into agricultural land, to accommodate human desire. Human desire can never be the basis for an ethic, however, despite the efforts of economists to turn it into one. For "desire" requires ethical *restraints*. Indeed, excessive desire is called greed, which has had a recent heyday in the popular literature. Greed is a desire for more than a person's fair share—the desire for inequality at any price. The life style of greed is self-destructive, and thus refutes itself, for if everyone were greedy, the environment in which those greedy people lived would soon be destroyed along with the greedy people themselves. As for desire: the desire of a rapist cannot be given license. The desire of a spouse to get rid of a now unwanted partner—murder—is condemned by all cultures and punished with the supreme penalty by many. Legitimate desires are those that pass a moral test, as Kant, and to a lesser extent Dewey, realized. Desire must be evaluated in *terms* of duty. The function of ethics is to *limit* desires, not increase or rationalize them. I agree with Buddhism that one problem with contemporary civilization is that it is based on desires and even stokes desires through the use of advertising. Ethics means control of desires. Environmental ethics will include the moral evaluation of all desires. "Economics has dictated to ethics long enough."

In this chapter we have revalued economics, religion, and other institutions from controlling ends to means to serve society. Society is best served with a good quality of life, instead of masses of underfed, starving people with no chance to live a good life. Quality of life is of greater value than quantity of people. This can include, however, some traditional practices that lend themselves to increasing the quality of life or decreasing the population or both. For example, one such practice is voluntary celibacy, which may be revalued in an overpopulated world. Culture, while important in the formation of human identity, cannot supercede our obligations to the larger whole in which culture is possible, the environment.

Eight

IN DEFENSE OF SPECIES

> The greater variability of specific characters … those which distinguish species from species… Charles Darwin[1]

> Locking up burglars and armed robbers takes away their chosen livelihood. However some things are more important than jobs; Earth's remaining biological diversity is among them. Dave Foreman[2]

At the Dec. 2003 Eastern APA Conference, Daniel Elstein read a paper against speciesism, "The Social Construction of Species and the Moral Indefensibility of Speciesism."[3] In it, he argued against the actuality of species, in an attempt to attack speciesism. I will argue in this chapter that although speciesism itself is morally suspect, the paper went too far, and that species are biological actualities.[4] I will argue against extreme forms of modern social constructivism that species are not human constructs, even if humans name them. Nor would social constructivism make a practical difference in our interaction with animals: we would still be wary of bears. I will also argue against the use of arguments that attempt to build on cultural relativism, that species differentiation is part of most cultures. Further, their differentiation is essential in determining the part they play in the environment. I will argue that environmental studies, especially the holistic relations in the environment, have something to teach philosophy, especially that the isolated individual pictured by nominalism is highly artificial and abstract. Finally, I will argue that the moral consequences of abandoning the species concept would be a disaster for the environment, since the basis for saving rare, threatened, and endangered species would be undermined in theory. I will argue that one form of evil is genocide, the deliberate destruction of a kind.

1. Social Constructivism

One of the main arguments in Elstein's paper is that species are a social construct. In order to challenge this point, it will help to examine what social constructivism means. What is social constructivism and what does it have to do with species? Social constructivism is the thesis that some or all concepts are constructed by society, or, in its extreme forms, that all of reality is constructed by society. Since the emphasis is on society as a whole in some respect, that is social relations, this theory is distinct from and a counterweight to individual constructivism, to the idea that individuals construct their own reality in some way. As "constructivist" it uses the metaphor of a construction to argue for the hypothesis that concepts or even reality itself are not independent of what society claims they are, that society "constructs" reality, and that this social construct is fabricated over time, until a social consensus is reached.[5]

Many analyses of social constructivism have distinguished between "strong" and "weak" constructivists.[6] In its weaker form constructivists hold only that human representations of reality are socially constructed. Strong constructivists argue that reality itself is socially constructed and not independent of such constructions.[7] It is difficult to analyze exactly what the strong thesis means. Surely it does not mean that humans constructed the world, all the flora, fauna, and minerals in it, and the stars and the galaxies, in the sense of making them? For these all preceded human evolution and are the condition of it. But then the thesis must revert to an epistemological one, namely, that "reality" is a matter of human consensus (Manuel Arias-Maldonado, 2011). "Reality," in this view, is the end result of human reflection upon and social agreement about what will be considered real, as opposed to illusory.[8] "Real" is an ideal construct, perhaps in the minds of some or all of the members of society. In this form, the thesis is a type of idealism. However, this thesis does not involve the stronger claim that humans actually built the world, only that human judgment mediates assertions of what is real and what is illusion, and more, that "reality" is the result of such human consensus, not the "given." There is no pre-given "reality" but only a socially constructed idea of reality. In this form social constructivism is a communitarian form of idealism.

With respect to species, the strong constructivists' argument is that species are also social constructs. Social consensus and even a social construction decide what is or is not a species. There is no "reality" to any species, say elephants, but only a social consensus that there is a species, elephants. (The weaker form of social constructivism would only argue that our representations of species are human constructs). In this section I will challenge this claim primarily on biological grounds.[9]

Let us first define the meaning of "species."[10] According to *Webster's Third New International Dictionary*, a species is:

> "1 (a) A class of individuals having common attributes and designated by a common name: a logical division of a genus or more comprehensive class: a subclass designated by adding to the name or connotation of the genus some specific difference that limits its application to a specific group ("the triangle is a *species* of plane figure").
>
> "(b) a limited kind or group having a distinguishing characteristic; esp. : one capable of including variant individuals and of being subsumed in a more inclusive category ("mineral *species* are made up of varieties

> having common basic properties"). ("One *species* of tramp who wanders from workhouse to workhouse"—Osbert Sitwell)"
> (c) humans as a species
> "(d) (1) a category of biological classification ranking immediately below a genus or subgenus and being denominated in taxonomic usage by a binomial that consists of the name of its genus followed by a Latin or latinized noun or adjective which is usually not capitalized and agrees grammatically with the genus name: a group of intimately related and physically similar organisms that actually or potentially interbreed and are less commonly capable of fertile interbreeding with members of other groups, that ordinarily comprise differentiated populations limited geographically (as subspecies) or ecologically (as ecotypes) which tend to intergrade at points of contact, and that as a group represent the stage of evolution at which variations become fixed through loss of ability to exchange genes with members of other groups although formerly conceived to be the total progeny of a single distinctive specially created pair...
> (2): an individual plant or animal or a kind of plant or animal belonging to a particular species—not used technically.

The first two definitions are logical, but provide evidence that the notion of "species" is not arbitrary. On the contrary, species are basic to logic and reasoning, and therefore to the reasoning of social constructivists. Apart from the specifically biological meanings, "species" represent different logical categories, including "social." Unless the claim is made that logic is arbitrary,[11] or also a social construct, the logical meaning of "species" provides evidence against the social constructivist thesis. In the former case, social constructivism is itself a social construct, since it uses logical reasoning. But then it has no more claim to truth than any other arbitrary social construct.

The last three definitions are more relevant to biology; importantly, humans are themselves a species. Indeed, the thesis that human society constructs reality presumes this distinction, namely, that humans are a species that constructs reality. The theory thus presumes what it is trying to disprove, the actuality of species. For how can reality be a "social construction" of humans unless humans are actually a distinct group? A distinct species? Surely, the advocates of this thesis are not arguing that bees, lilies, fungi or amoebas have a part in this social construction? But then if constructivism is confined to humans, then humans are a distinct actuality, not a social construct. Otherwise the thesis fails, for humans would not be distinct and could not construct reality. "Human" is a species difference.

The argument in the paper is that biology "uses" species and may find them "useful." But taxonomy and botany do not only find the notion of species "useful": taxonomists and botanists regard them as biological actualities. Let us take elephants as an example. An elephant has tusks, a trunk, and big floppy ears. Other grazers, such as horses do not. This is not culturally relative, since such features are independent of cultural perspectives or "social construction." Anyone can confirm this for himself or herself, either at the zoo or in the wild. No one who is not either disingenuous or badly deceived would confuse an elephant with a horse or a gazelle. Elephant ears, tusks, and trunks are not a social construct, but part of the elephant physique. Social constructivists who do not believe in the independent actuality of these features could confirm it by trying to pull the tusks off of a living bull elephant. Moreover, these characteristics are universal: all elephants have these three features; no horses do. This separates horses from elephants with actual universal characteristics. These features are not arbitrary, since they are universal. Some might argue that there may be exceptions, such as genetically abnormal individuals, "monsters." But these abnormal individuals are still part of the genetic continuum, as I will argue later. They are almost always identifiable as the member of one or another species.

Color would be arbitrary as a differentiating factor ("race" in humans) and is thus a social construct.[12] Species differences are not. Many species exhibit variations in color within the species, but this range of color is not coextensive with that of other species. For example, *Calochortus albus*, a California native bulbous perennial, varies from white, the most widespread color, to, rarely, pinkish and red. *Calochortus amoenus* is pink to lavender. While the two species overlap in color, the geographic range of the pinkish shades of these distinct species do not overlap. That is, the color ranges of the species do not overlap, since the pinkish forms of *C. albus* grow in a narrow band in the Coast ranges of West-central California, while *C. amoenus* grows in the Southern Sierra Nevada. (The species also differ in morphology, geographical range, and other characteristics.) In any case, color may not be essential to species identification in the botanical sense, but only for practical identification. Yet occasionally color can be a sign of a deeper difference, a genetic one, as in the discovery of tetraploidy in *C. aureus*, a species originally thought to be a color form of *C. nuttallii*, the state flower of Utah.

More decisively, elephants mate with elephants, horses with horses. As the fourth definition above states, a species "interbreeds." Evidently, male elephants can distinguish female elephants for mating purposes, as can males and females of other species. If they could not, the species would die out. Elephants do not mate with gnats and vice versa. Mating is within a species: thus it helps define a species as a species. But this argues for species as a historically continuous genetic complex. Once they evolved, (African) elephants have endured over time as a distinctive species and will continue to do so in the future, if human predation does not put an end to them. There is a genetic as well as a morphological continuity of the species from its evolution to the present, despite normal species variations. Of course species that reproduce by division do not mate. However, they are genetically linked, perhaps having a single common ancestor, which subdivided successfully and has thereby become widespread.

Reproduction ties species to the science of genetics. The genetic evidence is distinct from morphology, but also reinforces it. If geneticists find a gene that is distinctive they will look for morphological features to match it and vice versa, as in the case of *Calochortus aureus*. What is evident is genetic blocks against mating between unrelated species, as the definition confirms. Elephants cannot mate with horses and produce offspring. Mating between different species that are of a closely related but outside species usually results in sterile offspring. If all individuals of a species mated outside their species, that species would die out, and there would be no elephants. The genetic and historical continuity of the distinct species would be broken. But the necessity of mating within a species argues for species. Reproduction is of a species, in which offspring resemble the parents in several decisive respects, and the mating pair recognize each other as identical in species.

A species can also go extinct for environmental reasons, including predation. Humans are considered responsible for the extinction of dodos. Species extinctions, however, provide further evidence of the actuality of species. Dodos are extinct, but not elephants. Is "dodo," then, a social construct? Or Passenger pigeons? But while all dodos are extinct, elephants are not. The species has gone extinct as a whole, not just as individuals. I will examine this point, and the related one that species are confined to a geographic habitat in more detail below.

To argue that these elements of species identity are only social constructs also ignores the relations individuals of species have to each other, including mating, nurturing, herd behavior, and other signs of species recognition among animals themselves. Without exception, elephant mothers nurture elephant babies not the babies of mice. Elephants move in herds of elephants and do not attempt to attract lions to their herd. Indeed, herd creatures of different species sometimes avoid each other, even if the other species is not viewed as a threat. Lions kill hyenas and cheetahs, perhaps as competitors. These examples could be multiplied by the millions. Whatever the basis for this behavior, it clearly involves species recognition that is not mediated by human society.[13] Indeed, animals that do not recognize members of predator species will soon be killed. Since animals recognize other species as species, apart from human mediation, species are not merely a human social construction.

Some species recognize other, different species, especially symbiotes. Just as human farmers tend favored species of plants and get rid of "weeds," other species may have a cozy relationship with one particular species. Ants herd aphids and will defend them from predators. Similarly, some species are parasites of other species. Some orchids are so specialized that they have only one pollinator. Such symbiosis involves the recognition on some level of one species by another. Bees do not attempt to pollinate rabbits, but flowers. Are flowers then a "social construction" of bee society? Are the birds that peck insects off of the back of rhinos "socially constructing" rhinos? The birds recognize rhinos, and do not attempt the same maneuver with lions. Since such species recognition behavior is independent of human mediation, it cannot be an arbitrary social construction of humans.

Most animals have a preference for specific foods and not others. Is it an "arbitrary social construction" that humans do not find wood appealing but termites do? Or dung to dung Beatles? Or that grazers prefer grass to meat while carnivorous species like lions prefer the grazers? Is it arbitrary that we separate humans from termites in this respect, or lions from wildebeests? If anyone still claims that this is a social construct, and that species differences are arbitrary, I would ask them: which cage would they prefer to be in at feeding time, that of a lion or a wildebeest?

"Social construct" implies an element of arbitrariness, that it is simply a matter of choice how species are separated from one another. However, species differences are not based on human will, since they preceded any human will. Humans evolved from other species. It might be argued that all scientific concepts are social constructs, and thus that species are social constructs. But then every concept would be a social construct and the term would have no real use. Moreover, to argue that all scientific theories and concepts are social constructs in this sense is to argue that all science is arbitrary, questioning the entire spectrum of sciences. As the late Prof. Stephen Gould pointed out, to argue against the Darwinian view is also to question geology, cosmology, and other historical sciences, since the evidence on which they are based is interrelated.

Still more evidence is provided by habitats. The definition of biological species I quoted above includes ecotypes and in some cases, geographic limits to species occurrence. Niche, habitat, and ecotype are distinct terms relevant to this context although there is no agreed upon definition of these concepts. Ecotypes are species that are confined to a specific habitat, whether xerophytic, hydrophytic, or some other distinct kind. Habitats include a combination of factors that collectively make up a distinctive living zone. Botanists and other scientists have separated these zones by type, for example, the "Sonoran" zone in the Southwestern United States, a dry, generally hot area with alkaline soils. In this zone, low rainfall has resulted in retention of much of the sodium and calcium ions creating relatively more alkaline conditions. In more rainy areas these ions would be dissolved and thereby move deeper into the subsoil, or washed away. In other words, the soil type is often related to the rainfall totals in any habitat. The species that can survive in such conditions will generally not thrive in zones with very different characteristics.

Social constructivists might argue that the differentiation of habitats is also a social construction. However, the evidence, once again, is overwhelmingly against them. Many species live in a particular habitat and will not grow or do well away from it. Particular water conditions may constitute a distinctive habitat. Some plants are "hydrophytes" growing only in water, for example, water lilies (*Nymphaea* sp.). Xerophytes like the Saguaro cactus will not do well in a rain forest, if they survive at all, nor do water lilies thrive in the desert. Other species recognize habitats as such, thus they are not a social construction. Ducks prefer wetlands to deserts; humans wear clothing in the winter. Temperature is another factor: bananas do not like temperate or Polar Regions. Most tulips, especially the *T. fosteriana* hybrids that constitute the great majority of common garden tulips, will not thrive or bloom without chilling in winter, since their parent species are originally from montane areas in Asia with frigid winters. Neither quantity of rainfall nor of heat is "arbitrary": it can be measured and its effect on species out of habitats in the broad sense is evident apart from human intervention. A sudden freeze will kill some plants that are not hardy; without sufficient rainfall (watering), all plants will die. These elements are independent from control by human society and long predated our species. Soil types are another factor, as any horticulturist will attest. Most plants have a decided preference for specific soils. Heavy clay soils are heaven for some plants but death to others. Some plants, epiphytes, do not even grow in the ground, but only in the duff that collects in the crooks of jungle trees, including many orchid and bromeliad species. pH, mineral content, and many other factors make up distinct habitats, in which distinct species are found. Some irises will not bloom in acid soils; others will not grow there at all (the entire Oncocyclus group). The combination of these factors produce soils with distinct characteristics (aridisols, oxysols, and so on) studied by soil science.

Are all these factors of temperature, water, soil, and other growing conditions a mere social construct? If so, I challenge social constructivists to grow water lilies out of water, cacti immersed in water, oranges outside in areas with a pronounced winter or epiphytes in heavy clay. It won't work. But these elements of a habitat provide evidence for species, since species are frequently found in particular habitats, an ecological niche.

The same considerations apply to animals: polar bears are differentiated precisely by their habitat. Polar bears do not live in the tropics. If social constructivism were true, this attachment to habitat would be arbitrary, since the theory claims no actual species apart from social construction. But few animal species do well out of particular conditions of temperature, water, and other elements of the habitat. Even humans bring their habitat along with them, reinforcing the point. We do not venture out in winter without our nice warm coat, gloves, scarf, and other clothes. We bring our original tropical habitat with us. Otherwise we would freeze to death. Cold is not a social construct.

Are there scientific problems with the notion of species? Elstein has argued that Charles Darwin thought that species were basically widespread or well-marked varieties, and in his article, he accurately quotes Darwin to this effect. However, Elstein has oversimplified Darwin's more nuanced view and ignored other statements of Darwin.[14] First, from species as widespread or clearly marked varieties, it does not follow that there are no species. Darwin did indeed note that "naturalists" often disagree over both the definition of "species", and which animals and plants should have specific, and not varietal rank.[15] The origin of species then lies in favorable characteristics of varieties that survive natural selection and become widespread. However, this constitutes an argument for a redefinition of "species," not its social construction. Different species may have a common origin and evolve apart. Darwin's target is "independent creation," not species as such, as others have noted. Darwin also defined varieties as "species in the process of formation," the opposite of the definition coined in terms of varieties.[16] From this perspective, varieties are stages on the way to speciation.

Even if Darwin thought varieties more important than species, however, this would provide scant support to Elstein's social constructivist views. For Darwin did not state that he believed varieties were social constructs. Different varieties would constitute a standard distinct from social construction.

The title of his famous book, *The Origin of Species*, indicates that he thought the notion of species should be retained. Darwin also noted that a species can go extinct, and so also for the validity of species. He argued that one species can evolve into another. If species can change into another, then he must have thought they could be differentiated. Indeed, the whole thesis of the book would collapse unless just this point were valid. For how could there be new species if there are no species, and thus no origin of new species?[17] However, Darwin mentioned but did not address the issue of interspecific mating—perhaps since he was not familiar with Gregor Mendel's work and modern genetics. Naturalists do not disagree that a species mates with other individuals of the same species.

One problem here is crosses whose genetic ties are unclear from an examination of visible characteristics. In general, trial breeding can clear up genetic ties. If the individual or variety can be crossed and produce fertile offspring with either suspected parent, botanists rank it as a variety, not a distinct species. Darwin argued that the distinction of species and varieties was somewhat arbitrary, and a "species" was only the more widespread form. However, botanists and zoologists have dealt with this problem by designating each variety within a species as a variety. The original species is then given the same epithet for both its specific and varietal forms, for example, *Tigridia mexicana* var. *mexicana*. Other varieties have another varietal epithet, for example, *T. mexicana* var. *passiflora* and *T. mexicana* var. *lilacina*. In this way, the varieties are not ranked but the specific tie is recognized, even if some varieties outnumber the species: *Tigridia mexicana* var. *passiflora* as far as is known is probably more common than var. *mexicana*.

Are varieties, races, and subspecies arbitrary, unlike species? This is not the view of many taxonomists. In the botanical family I am most familiar with, *Calochortaceae*, varieties are distinguished by major morphological or genetic characters. Marion Ownbey and J.M. Beal (1943) argue for (genetic) evidence for subgroups of *Calochortus* as triploid, for example, *C. longebarbatus*, var. *peckii*. This taxon is sterile but survives by cloning. Color varieties are geographically consistent, that is, individuals of a variety in a geographic area tend to have the non-normal color, such as *C. macrocarpus* var. *maculosus* and *C. gunnisoni* var. *perpulcher*. However, these are much weaker varieties than those with morphological or genetic differences. Thus variation may be a matter of degree, but is not arbitrary. Moreover, the question arises of how we identify varieties. Elstein argued that species are nothing but strongly marked varieties, but these still have a general identity as a variety. If we can generalize about varieties, we can also do so about species.

Sometimes species are distinguished by range, habitat (dry/wet), even bloom time or in rare cases color, *not* shape or other characters of the plant or animal. In this practice, modern classification differs from that of Plato and Aristotle, who tied species differences to "essences," which in turn denoted the unchanging form or substance of the species. However, species distinctions of this type have been challenged by experts. For example, polar bears are distinguished from other *Ursus* sp. mainly by color and habitat, not other characteristics. For this reason, some classify this taxon as a variety. Again, cross breeding and other genetic tests could provide definitive evidence of whether the polar bear is a distinct variety or its own species, if they could be mated successfully—which in the case of bears might be done more easily by artificial insemination.

One of the most difficult plant genera I know of to identify correctly is *Hymenocallis*.[18] To a novice, the differences in *Hymenocallis* can appear arbitrary. Classification is not aided color differences, as almost all of the fifty-some species are white. In some cases the differences are based on habitat: some species grow in wet situations, others in dry. For example, *Hymenocallis howardii* grows in *llanos*, seasonal wetlands in Western Mexico; and *H. azteciana* on dry slopes in the same area.[19] Morphologically they are extremely difficult to distinguish, but in terms of habitat it is simple. It could be that the two species are sympatric and took different turns at some point, specializing in different habitats. Be this as it may, crossing the two would provide good evidence for their relation.

An apparently even more arbitrary species differentiation is bloom time. *Hymenocallis howardii* and *H. jaliscensis* have overlapping ranges and habitats (*llanos*), and are very close morphologically, but *H. jaliscensis* consistently blooms before the other, similar species. Is this arbitrary? Crossing the two would provide good evidence of their relation.[20]

I would take the morphological definition as primary, not the genetic, "biological," niche or common descent criteria (for examinations of these other approaches, see Stamos). One reason is that this is how taxonomists and botanists separate species. Another is that morphological criteria allow for evidence from other cultures and other species, who may know nothing of genetics, but know the danger that the presence of a lion token represents. Nor is morphology separable from species tokens, despite their abstract separation. The attempt to separate morphological characters as the units of evolution from the species they characterize is an excellent example of Whitehead's fallacy of misplaced concreteness. Despite considerable field work, I have never seen a morphological character in the wild. Morphological similarly and species similarity are actually identical.

Genetics alone cannot delineate species without what appears to me to be arbitrary judgments. How many genes or chromosomes must be different to make a new species, four? Five? Six? Picking some number of genes as a standard would be arbitrary, to speak nothing of the fact that sometimes the genetic differences between species are few and other times many. In other words it is dubious if such a standard number could be established. If we separated specific genes to compare them we would need extra-genetic criteria to do so, or we would beg the question.

Moreover, one species may include specimens with different chromosome or genetic numbers, for example, *Calochortus luteus* (Ownbey 1940; Ownbey and Beal 1943). The enterprise of reducing living things or species to their genes is a category mistake, mixing up different perspectives: the micro and macro (Kauffman and Clayton, 2006). This point applies to other microscopic criteria as well, past and future. Moreover, even individuals have genetic differences, a further argument that we cannot differentiate species genetically except arbitrarily.

Similarly, Prof. Randall Auxier's argument for the "death of Darwinism" (Auxier, 2006), and his distinction of "economic," and "genealogical" hierarchies (involving demes vs. organisms) is reductive, since his claim that the economic speaks of "energy transfers" ignores the individuals of a species involved. "Energies" are reductive. My approach is phenotypic and therefore not genotypic: genetic hierarchy is not decisive in delimiting species, which Auxier concedes (Auxier, p. 11).

Other criteria are generally not universal, and do not apply to all species, notably, Ernst Mayr's "biological" or interbreeding criterion. (Martin Gorke agrees with Mayr's criterion of species, but David Stamos does not.) Some species do not interbreed, notably, practically the entire microscopic world of bacteria, protozoa, and the like. Nor do plant clones, such as bulbs; and sterile individuals or varieties, for example *Calochortus longebarbatus, var. peckii.* These cannot be tested by crossing to look at their offspring /fertility but are still of a kind. Moreover, many stands of plants are "reproductively isolated," but not distinct species. Such plants do not interbreed, but are the same species. Some species occupy one habitat or niche, but not all do. Some, like dandelions, occupy almost every niche, unfortunately. Clades and lineages are useful, but cannot tell us what is a distinct species without first looking at morphology, a point that applies to Darwin's criterion of common descent as well.

Relations, including the principles implicit in causal relation of genes/chemicals to biological forms are not enough, for example, the causal relation of genes to morphology. We cannot at present predict what form such genetic material will take at the macroscopic level. Although the principles of these relations are regulative as causal, the relation is contingent. Species change can come about due to genetic variation, but also due to environmental pressure (Wilkins, 2007). Genes are not decisive in driving species origins.

Morphs are characteristics of token types, that is, individuals of a kind or species. These morphs are normative for a species and can thus be variable. They are general but actual norms, and may therefore fall within a range. They are not absolutely identical but close enough in the relevant respects to fall within the norm for the morphology of that species.

Critics might argue that such species differentiation depends too much on the genetic test. In his review of Stamos's book on species, Joseph LaPorte argued against any monistic species concept by noting that for species that reproduce sexually, "interbreeding seems to be doing all of the important work delimiting species, and similarity seems irrelevant. In other cases, such as with asexual species, one is tempted to suppose that similarity is doing all of the important work" (LaPorte, 2006, 389-90). LaPorte suggests a form of pluralism in our species concepts. I would respond that LaPorte has confused a test for a species with its definition, concept or characteristics. Using sexual tests to determine if outliers are the same species is one test for a species. But as he admits, it cannot be used for all species. We can often test for specific barriers by such tests, but they may stand in need of further criteria. Horses and asses can cross but the offspring are sterile. Does this mean that there are no species differences between horses and asses? "Contrary to [the] popular view, different species can interbreed if they are related closely enough. But the offspring of such unnatural unions are often infertile…" (Gleitman, et al. *Basic Psychology*, 2000, p. 348.) Moreover, interbreeding is not a decisive test due to non-sterile hybrids. In some plant species, these form local hybrid "swarms" often with flowers of great variety within a single stand, for example *Calochortus superbus* x *C. luteus*, and also *C. nudus* x *C. minimus*, both in the Sierra Nevada mountains of Northern California. One stand of *Calochortus superbus* x *C. luteus* I witnessed has a marvelous riot of patterns and color combinations. But these species are normally easy to distinguish and only interbreed in a few locales. Since hybrids can be fertile, interbreeding is not a decisive test, only one tool providing some evidence.

However, this test is only brought in for quite difficult cases. Normally, habitat and form are obvious considerations: the water lily or the elephant. We could try to grow *Hymenocallis howardii* in drier conditions than normal, to see if it would thrive, and if it did, this would tie it more closely to the similar *H. azteciana* (or grow the latter in wetter conditions). This would provide additional evidence apart from intra-specific crosses, for or against the species distinction. Extinction of a whole species is another indication.

Without morphology, we cannot identify species at all. Chimps have about 99% of human DNA. Would anyone argue that chimps are human? Again, the genetic similarities are not enough. Nor is bimorphism as serious a problem as LaPorte (2006) makes it out to be, since there is not an interbreeding species on earth whose males and females do not differ, if only in their reproductive organs. In some species this is more pronounced, but it is present in all sexually reproducing animals and not a few plants. Bimorphism is a difference of degree, but does not make a difference in kind. We can still see that male and female elephants are the same species from morphology despite differing genitalia. Similarly, infants are of the same species as their parents, despite differences in size.

What bimorphism does reveal is that species concepts are general, and thereby admit of exceptions. *Calochortus albus* is generally white but rarely pink or red. Some bird species have different color patterns for males and females (cardinals, peacocks/hens). Generality does not provide an exceptionless principle. Generality means that no two individuals of a species are exactly alike, in every respect, but are morphologically similar in the relevant respects. Male and female elephants differ in their reproductive organs, but both have the distinctive characteristics of elephants, namely, a large body, big ears, tusks, trunk, and other characteristics that distinguish them from hippos, rhinos, and other large, local herbivores. Where bimorphism is pronounced, there is generally another principle at work, for example, the peacock display involves sexual selection by females over the centuries.

However, as a good pragmatist, I do not exclude the other sorts of evidence, but with a "both-and" approach, welcome supplemental evidence from genetics, population studies, lineage studies, habitat studies, and so on. Such studies may help to delineate or even prove species in problematic or "messy" (Stamos) situations, for example, the discovery of tetraploidy may force us to go back and look for morphological differences at a macro level. From a pragmatic point of view, species criteria are not necessarily mutually exclusive and are actualized in the same tokens. I am not a pluralist when it comes to defining species, but can accept plural modes of evidence in helping to delineate them. However, I am agnostic when it comes to habitat differences without morphological ones, as I indicate in the text, since mere difference in habitat or niche is not enough to validate specific rank.

I am not denying the problems: that is what makes classification such a difficult judgment. But tests can back up such judgments and these tests are independent of human society. Science is the attempt to go beyond convention to establish non-arbitrary standards. However, I am not arguing for a species "nature," as Plato and Aristotle did, for as I indicated, morphology is not always decisive in differentiating species in practice. Instead, as I will argue below, conventions regarding species characteristics do not involve a "nature," but an actual norm, that is, a general principle identical for each individual of a kind. Elstein may imply that the only alternative to social constructivism is essential realism. But we have a host of different alternatives, including, as I argued above, Darwin's.

2. Cultural Relativism

"Social construct" implies an element of arbitrariness, as if different societies would construct concepts of species in different ways. One argument used to support the notion that species are social constructions is that different cultures classify in different ways. Thus each culture constructs its own view of what constitutes a species.

Frequently, arguments from cultural relativism are taken as self-evident. No actual evidence from anthropological studies is presented. But there is evidence against this interpretation of cultural relativism in taxonomy. Cultures frequently may agree on genera or even species separation, invalidating the cultural relativism argument.[21] Other languages have words for "elephant," "horse," and other common generic distinctions. These generally coincide, point for point, and unequivocally, with their English equivalents, as a glance at any bilingual dictionary will confirm. The Inuit may have no word for the vanilla orchid, as it does not occur in their habitat, just as the Yanomamos of the Amazon do not have a word for polar bears in their language. But the Inuit may well agree that polar bears are distinct from their prey and are far more likely to be wary in the presence of the bears than rabbits. The taboo in almost all cultures against bestiality is the recognition of the difference between humans as a species and other species, to speak nothing of the norms of reproduction involving distinct species. The separation in practice for sexual reasons is a universal of culture.

However, even if cultures did not agree on the exact same separation of all species, this does not invalidate the actuality of species. Cultures frequently do not distinguish species at all but only genera. For example, different species of *Calochortus* were indiscriminately referred to as "sagaw" by Native American tribes of the Southwest.[22] The Spanish explorers indiscriminately referred to several different species as "Mariposa." "Species" is a scientific concept, and in many cases, only scientists separate different species, not cultures. However, in other cases cultures are well aware of species differences.

Similarly, most members of society will look at two different botanical species and call them both "flowers" or "lilies," without being in the least aware of species differences between the two lilies. How then can species differences be a "social" construct, when few members of society could tell *Lilium tigrinum* apart from *Lilium superbum*; or even tell different species of cats apart, a familiar genus? The average person may not care about the differences between *Calochortus gunnisoni* and *C. nuttallii* and may no more know the difference between them than that between *Homo sapiens* and Mencken's *Boobus americanus*. The point is that cultures and individuals do not divide species differently, since *they may not divide species at all.* "Species" is a system of classifying living things tied to precise methods and expertise. Indeed, even scientists from other fields like physics may not be able to recognize different species.

George Lakoff and Mark Johnson have documented this point.[23]

> Berlin, Breedlove and Raven (1974) and Hunn (1977) in detailed studies of Tzeltal plant and animal categories, note that at the basic level (they call it the folk-generic" level) Tzeltal speakers are extremely accurate (in the 90-95 percent range) at identifying plants and animals relative to scientific biological classification. At lower levels—the species and variety—their accuracy drops off precipitously to around 50 percent and below. In short, we are better equipped to recognize plants and animals at the level of the genus, that is at the basic level, than at lower biological levels.

Nevertheless, we would regard any culture that classified elephants and gnats as one species odd, to say the least. Is the social constructivist arguing then that there are no differences between elephants and gnats? That all classification is arbitrary? But this is absurd. Social constructivists make the same mistake as other cultural relativists, that is, they presume that no culture can be wrong, or make an error. But it is not at all obvious that they cannot. "Nazi" culture argued that there is a pure race of "Aryans," but there is no scientific evidence whatsoever for such a fictional "race."

It is not necessary to make the value judgment that science is superior to culture in the area of knowledge, since science is trans-cultural. Many cultures have contributed knowledge to science and science has drawn its practitioners from many cultures. The argument that species are a "social construction" presumes that species are a construct of "society." But which society is meant, one particular society, or all of them collectively? If one society is meant, namely modern Western society, the charge is erroneous, since other cultures have contributed their knowledge to trans-cultural science.[24] If all societies collectively are meant, then the argument cannot appeal to cultural relativism, since all cultures have contributed to the consensus.

3. Would Social Construction of Species Make a Practical Difference?

Even if social constructivism were true, would this thesis affect our practical interactions? Let us say for sake of argument that "bear" is a social construct. Would anyone be less wary of how dangerous bears can be armed with this insight? Would we not back off if confronted with a bear in the woods, especially a mother with her cub? Wouldn't our behavior reflect our belief, however socially or culturally mediated, that bears are far more dangerous to most species, including our own, than earthworms are, and that we need run only from bears, not earthworms? Don't many other species exhibit the same behavior? But if this is true, what is the practical difference between social constructivists and non–social constructivists? Pragmatically, isn't this making a mountain out of a molehill? We socially construct bears to be dangerous and other species "socially construct" bears to be dangerous. Pragmatically speaking, then, on both accounts bears are specifically dangerous. But if we can recognize an individual in the woods as a member of a dangerous species, is this not practical evidence for species? Again, pragmatism would argue that since we would prefer to eat nutritious not poisonous species, there is a practical difference between the two.

This practical point ties in with Charles S. Peirce's discussion, following Alexander Bain, about pragmatic qualities on which we are willing to act. If we are willing to act on a principle, then we believe in its actuality in some sense, for example, crossing a bridge. We are willing to act on believed characteristics that are yet unrefuted: including those of people (trust). Since we act on the principle of species differentiation, as the distinction of bears from earthworms indicates, we believe in species differences in a practical sense. Most prefer roses to dandelions in gardens and dogs to slugs as pets. Moreover, various characters of each kind are categorially distinct: the diamond has aesthetic and economic as well as scientific properties. Women generally prefer diamonds to glass engagement rings in our culture; industry uses the former and not the latter for cutting tools. Thus scientific classifications are not the only relevant ones from a practical point of view.

Mating provides more evidence. Normally a species mates with its own kind and not other kinds. There is evidence that sexual attraction may exceed this norm, at least among humans, but this is not the norm. Bestiality is a sexual practice, but is rare among both humans[25] and animals. For this reason most cultures adjudge it a perversion. But be this as it may, mating is for all practical purposes inter-specific. Crosses between closely related but distinct species are generally sterile; if they are not they provide evidence of varietal status.

Medicine also provides evidence. Disease germs are treated as a kind, as are their biotic cures. Strep is caused by a kind of bacillus, where regardless of the individual organisms, any such individuals of the kind can cause strep throat. Similarly, antibiotics are kinds: penicillin and related molds. Doctors do not prescribe individuals as cures, but tokens of a kind or class of organisms. It does not matter which individual penicillin mold is used as an antibiotic, only the kind.

4. The Challenge from Nominalism

Some biologists have argued against species on the ground that only individuals exist and thus species do not.[26] They thereby raise two distinct issues, (1) the validity of metaphysical nominalism; and (2) its applicability to biology. I will first examine the validity of metaphysical nominalism. I will then reexamine nominalism from a new angle, by raising the question: what do environmental science and philosophy have to tell us about metaphysics? Finally, I will examine the last question, namely, the relation of metaphysical nominalism to biology.

The thesis of the late Middle Ages, that individuals alone exist, has been called nominalism. This term can cover several distinct approaches to the issue raised by the controversy over universals, for example, conceptualism, termism, individualism, and so on. However, I believe that nominalism may be a disguised or modified type of essentialism: that individuality is a universal characteristic of "reals," or of existents. Everything that exists is necessarily an individual; the "nature" of things is to be individual. While this may not constitute a concrete universal, it is a nevertheless a universal characteristic of individuals; and "individual" is itself posited as universal. Thus the theory contradicts itself, for if nominalism defines itself in contradistinction to essentialism, but being an individual is essential to existing, then it posits and denies an essence.

The primary issue is, does the world consist, as nominalists maintain, only of individuals? We must first be clear about what is an individual. But this is by no means clear. We might mean that an individual is numerically one. But the world contains conceptual and logical distinctions that are not one, at least in the same sense. For example, Mt. Everest is the highest "mountain" peak in the world. But is a mountain peak attached to a range an "individual?" It is certainly not an individual in the same sense as either a mountain that stands alone or an animal that can move about. For it does not have a distinct base, but shares its base with other peaks in a range. An animal, however, does not share its body with another animal, so is more nearly one, as more nearly detached from any other individual. The "peak" is not physically distinct from a shared range, thus not physically "one" but a part of one range.

Similarly, a mountain is in all cases attached to the ground: isn't it arbitrary to separate it as numerically one, when it is physically attached? How can it be "one," when it is not separate? The mountain is not one in the same sense as an animal, for it cannot be separated from its specific locale. "Individual" is based on the paradigm of the substance in the ontology in Aristotle: that we can separate substance and attribute (via Ockham). We cannot "ontically" (as "*per se* beings") separate substance from attributes, however. These are not actualities but abstractions. We do not encounter individual substances apart from attributes but "individuals" of a kind.

If we encounter a rock that is loose but part of a mountain, is it "one" rock or a part of a larger formation? Where does the "rock" end and the mountain begin? The same arguments apply to plants, which cannot be separated from their place in the ground, their soil, without killing them. An animal, however, can act as a unity, as a "one," since it is separate, and free to move about. But even animals have microscopic organisms living in their gut with which they live in symbiosis, and without which they could not survive.

What we think of as one may actually be a collection of many. Is an individual human "one" or, as Friedrich Nietzsche put it, a collection of many organs with a "common digestive system?" Is our liver "one" liver? But isn't it arbitrary—as much of a social construction as race—to separate out an organ by its distinct function, when it is not physically separate? Similarly, "one" organ consists of many cells and many more molecules, atoms, quarks, strings, and other sub-atomic particles. Is it then "one" organ or instead many cells, molecules, atoms, and so on as reductive materialists might argue? At what level, then, is it one?

Is the atmosphere one or many? It consists in a huge number of individual atoms and molecules, so how is it one? Each of these may change its position relative to the others in a chaotic mix: thus it has no overall shape. It has layers, distinct components (gases), and even "currents" like the Jet Stream moving through it. In what sense, then, is it an individual? The same arguments apply to bodies of water, like Oceans. Is the Atlantic distinct from the Pacific? But they meet and commingle in the Southern Hemisphere. Again, fluids have no "units"—nominalism only can arise by ignoring fluids, and thus is not universal. "One" liter, gallon or other standard measures are all conventions—normative standards by which to measure fluids. The same arguments apply for energy and forces. This tells us that "individual" is totally normative and contextual.

"Individuals," are above all complicated. For (1) we are composed of multiple atoms, and thus plural in one sense not individual. (2) Again, living things are composed of individual parts (organs) in a larger whole. (3) "Individual" is a term of both identity and difference. It identifies oneself as an individual while differentiating oneself from other individuals. (4) The application of general terms or concepts to singulars is problematic. (5) Individuals act in the singular sense only to a degree: compare the division of labor, which is a social or cultural determination of individual acts. (6) The designation of individuality applies most neatly to separate individuals, that is, animals or detached rocks, less to plants and geological formations, and least to fluids, if it applies to the last at all. Thus nominalism leaves most of the world out of account, in its abstraction of one model as the model.

If nominalism were true, we would find a world consisting either entirely of individuals that were absolutely identical or absolutely different. In the first case, there would be no differences and everything would be a carbon copy of everything else, but still singular in quantity. In the second, there would be no elements of similarity or any relations between individuals, whether genetic or any other kind. The world we live in has neither of these characteristics. Insofar as we find "individuals"—an arbitrary classification of itself—we find tokens of a kind: elephants, lilies, and granite where their characters are inseparable from their actuality.

These different degrees of separation argue for degrees of individuality. Degrees of individuality argue against nominalism. We are "one" in some respects (action, intake) yet not "one" in others (cooperative action, dependence upon environment). A peak may be one peak, but it is not one in the same sense as a peak on a mountain that is completely separate from any range. A plant may be a single one, but is inseparable from its soil ball. Indeed, as Gorke (2003, p. 27) noted, "individual" does not apply to plants that divide, which includes most perennials. An organ will cease to be one without relations to the other organs. Liquids and gasses are not "one" in the same sense as solids at all. Quantification of our lives or of other individuals is a pragmatic contextual affair.[27]

Without endorsing nominalism, however, we can argue for many actualities as "individuals" of a kind. This is clear from the above definitions that argue for species as consisting in individuals that are members of classes: "a class of individuals having common attributes." It is not a contradiction to say that only individuals of a kind exist. Kinds are used in all sciences: kinds of forces, kinds of elements, of rock strata, of stars, and so on. All sciences early in their history go through a descriptive phase, which consists in the discovery of these kinds and their relationships. Individuals "exist" as kinds: namely, individual humans, elephants, lilies, rocks, and so on. A chemical element is of one kind, whether oxygen, chlorine, sodium, or another. There is no "individual" that is not of one kind or another. Let us take an individual rock. If the rock is ultramafic, then it is not feldspar or basalt, and vice versa: an individual rock of a kind. Individual rocks are still individuals. The outcome of science taken as a whole is that an individual can only be actual as a specific kind, for no individual actuality can be pointed to that is not of a kind, whether a kind of physical force, a kind of chemical, of mineral or of organism.

Individuals of a kind are individuals contextually. Plant specimens in a herbarium are detached from their soil and locale but still of a kind. A liter of water is not a liter of mercury. Individuals of a kind may still be individuals in some sense, but composed of atoms, molecules, and, if living, organs. We never encounter individuals as such but only token-kinds, that is, individuals of a kind. "Individual" is an abstraction from a composite actuality—an individual instance.

"Species" is used as a logical differentia, as definition 1 indicates, and is not limited to biology. With regard to biological species, an individual elephant may be physically distinct from another elephant and yet share specific characteristics that separate it from other individuals with their own specific differences. It is indeed an individual but only as a kind: an individual elephant, an individual organism characterized by its immense size, grayish skin, a trunk, tusks, and large floppy ears. Horses do not have these individual characteristics. These characteristics are also individual, that is, an individual trunk, two individual tusks, and ears, in the qualified sense that I discussed above, that is, as involving degrees and relations.

The metaphysics of "tropes" and "instances" is also compatible with nominalism of a modified sort.[28] Species differences could be tropes, not actual universals. The species differences mark individuals as of a kind, but they are individual tusks, trunks, and floppy ears. These individual characters are attached to individual organisms. They nevertheless distinguish the group of individual elephants from the group of individual horses. I would argue that tropes are abstractions, not actual.

We should ask whether the universe described by nominalism is too austere. What is lacking from this universe that we know very well is "real" or actual? I will argue that nominalism cannot account for either relations or scientific laws, and is therefore inadequate.[29]

The most obvious relation that nominalism leaves out of account is the causal relation. In biology the sexual act is the cause of pregnancy, and pregnancy precedes the birth of offspring. Reproduction is causal, involving an actual relation between individual things in addition to those things. Again, this relation cannot be based on a social construction, since it occurs in all animal species and preceded human society. Otherwise, human society could not have evolved. Living individuals are made by other individuals in a relation. Darwin, who treated the concept of "species" critically, nevertheless argued that no species is formed autonomously, but differentiates itself by adaptation to a niche (1991, p. 87 et al.). Such a view requires actual relations of a species or variety to a surrounding habitat, including competition with other species and adaptation to a specific habitat, whether tropical, temperate or polar; dry or wet.

The argument of nominalists against the actuality of relations is that they are not individuals *per se*. But individuals of specific types would not be individuals of that type without relations. Relations then become the condition or cause of individuals and are required in any world inventory. Such relations may emerge through a complex interaction of the players, as in the symbiotic relations of pollinators and flowering plants. But to deny such relations to save an oversimplified theory like that of nominalism flies in the face of established knowledge.

Stamos (2003, ch. 5) argues that species are relational, but this is implausible. While species are bound up with relations, they cannot be reduced to relations. An elephant has some actual characteristics regardless of other elephants—they belong to that "individual." Yet the norm "species" does point to relations as well, since "elephant" connects different individuals in a relation.

Spatial and temporal relations are also left out of the nominalist universe. Space is place, that is, the distance relations between any two things or a multitude of things, all of which are in space. In biological terms, space is vital in the form of space to grow, and feed. Competition among species includes competition for space, the "territorial imperative" so vital for both feeding and reproduction. As for time, nominalism leaves out events or changes. Being subject to gravity is an event, as is a cause-effect relation, but both of these are an external relation.[30] Events are not properties of individuals; properties involve an internal relation. Properties are not separate from the individual that they characterize, but one "billiard ball" or individual atom hitting another each involve distinct individuals in a relation. Changes happen at a particular time, while the duration of atoms or properties does not.

Nominalism also leaves symbiotic relations within an "individual" organism out of its account. Humans and termites both host intestinal organisms without which they could not digest their food. Are these organisms, which depend upon their hosts as the hosts depend upon them, separate individuals? But neither could survive without the other. As Lynn Margulis noted, "We people tend to see and name only the most superficial or largest members of the consortium; we then act upon the self-deceptive construction that the consortium is an independent 'individual'."[31] Indeed, she has pointed out that some lichens may have become so symbiotic that their species cannot be separated.

Another actuality in the world that is left out of the nominalist account is principles. This is strange, since nominalism can be taken as a principle, the principle that the world consists of individual things. Strategically it is less odd, considering that admission of principles into the nominalist world would upset the principle of nominalism itself. For principles are not individual things. The law of gravity is not an individual thing, but actual, governing all individual things. Those who argue against the actuality of principles do not actually jump off high buildings or bridges. They are prudent enough to recognize the principle in practice.

Echoing this point, Stamos (2003, p. 10) argues for the reality of laws. He points out that measurements of the size of a giraffe's neck are also part of what is "real." Both of these are normative actualities, since measurements, especially, are gauged in terms of normative conventions, namely, the metric,

British or other systems established to provide a standard. Normative standards apply to measurements of liquids, gases, temperatures, and many other actualities as well. These conventions are not arbitrary in the sense that they are actual and provide a standard of comparison of actual distances, temperatures, and so on.

One could argue of principles like gravity that such a principle is a characteristic of all individual things. But gravity is not itself a "thing," but a force. Forces are not things. On the contrary, forces are the condition of things, if the Big Bang theory of the universe is correct. In any case, the principle can be separated from the individual, since the principle of gravity will hold of individuals that are not now actual, future living things not yet born. Scientific laws argue for principles apart from individuals (Peirce). We do not have to endorse realism, as Peirce did later in his life, to realize why he thought that "realism" was valid. For without principles, scientific predictions would be impossible. Scientific predictions involve the actuality of laws, that the same principles operative now will also operate in the future. Also, that the same laws that govern our part of the universe are operative in other parts, universality. Nominalism would also be impossible without principles, since the actualities at the other side of the universe could not be individuals unless the principle that the world consists of individual things was valid. But this refutes the attempt at denial of the principle. Nominalism is thus faced with a dilemma: if it denies principles it is false; if it admits them it is also false.

The denial of form was historically connected with the denial of any teleology in the "objective" world.[32] Form is connected with an immanent *entelecheia*. If genes function properly they achieve a slightly variable but general morphological similarity between parents and offspring. The denial that the offspring resemble the parents and that this can be predicted using norms of genetic science is simply wrong. How genes achieve such an outcome is a more complicated form of a chemical equation: $NaOH + HCl \rightarrow NaCl + H_2O$ or any other chemical equation. I am not arguing here for the thesis of teleological design by genes, like sociobiology at its worst, but for normative laws in science that we can use to predict regularities. Prediction is the ghost of *telos* in that a specific outcome can be regularly expected. Species reproduction is one such regularity. Elephants do not give birth to gnats, and lily seeds do not produce roses.

The reconstruction of the past by the historical sciences, which include cosmology, geology, and biology, would be impossible without principles that hold over time. For example, genetics provides evidence for radiation patterns in the evolution of plant species, outward from a common source and locale. The principles of genetics must hold in the past for these patterns to occur. Similarly, the expansion of the universe after the big bang presumes that the laws of physics held from just after the big bang to the present.

The denial of the reality of relations and principles that is a corollary of nominalism cannot be seriously maintained. Spatial, temporal, and causal relations characterize all "things"; environmental and genetic relations all living things. William James argued that these are as much a part of experience as individuals. Moreover, pragmatism includes the recognition of "contexts," since what is true in one context may be dubious in another. An organism is an individual viewed at the macroscopic level. But at the microscopic level the organism is a collection of many individual cells of a kind such as liver cells, brain cells, and so on. At the submicroscopic level it is a collection of different atoms and molecules, whether organic or inorganic; of sodium, carbon, oxygen, and so on. Even these are composites at the subatomic level, consisting in still other individual forces and particles (protons and electrons, quarks, superstrings). Nominalism is relative to contexts, since what we call an individual at one level is a collection at another.

Nominalism also leaves out different functions of words other than naming. Nominalism originally meant that a word applied to a thing is only a "name" we give it. After Wittgenstein's work, the claim that words are mere names is untenable, for words play many different roles in different functions ("language games").[33] Language is not naming. Names apply to individuals, for example, Peter or Alice. "Concepts" apply to kinds and thereby reveal relations between individuals. From a linguistic point of view, nominalism is an over-simplification. Apart from pointing to an instance, can we ever speak of instances except as of a kind? Once we have identified the kind, and if we know it, we refer to it by its kind. Instances are thus normatively mediated in language; and instances, not "individuals." Despite Wilfred Sellars (1956) arguments against the myth of the given, he insists on arguing for nominalism. But if all "givens" including individuals, are normatively, that is conceptually mediated, how can we speak of individuals at all except in terms of quantity? Georg Hegel distinguished abstract concepts from notions and argued that the notion grasps the instance in all its concreteness. Indeed, there is little left of the individuality of the individual once all its characters are elucidated.[34]

Genera and species denote relations in the world, not individual actualities. There is no separate individual actuality in the world consisting of a genus. Instead, individuals have the requisite shape or other universal characteristic to be classified under a kind or genus. But these characteristics of individuals that bring them into relations with kinds are actual.

Nominalism views concepts as arbitrary conventions because it starts from the model of difference—different languages have different symbols for the same meaning, and thus which convention we adopt to conceptualize is arbitrary.[35] I agree that it is a convention but it is not arbitrary.[36] There may be good grounds for our verbal conventions. The conventions reflect agreement among members of a culture, or among experts, that a norm is valid.

"Individual" is such a norm. Principles are actual norms, that is, general principles in the world, independent of human will. Scientific principles are an example, but there are many others. Again, we are willing to act on such principles, and therefore on norms. Coming under, for example, the law of gravity is "participating" (Plato) in one or another principle or law.[37] While we reject essentialism of the Aristotelian sort, in the notion of general kinds he was on the right track. Normativity is required to understand and identify species, given individual variation. A "norm" or "type" is indicated for identity: identification. A species differentiation is a normative identity.[38] But this does not mean it is arbitrary, since as we argued above, such normative identities, in whatever language they are expressed, have a relation to actualities in the world. Nominalists should keep in mind that "matter," "individual," "thing," and other ostensive individualities are all normative identities. "Individual" identifies a universal general property, but is not arbitrary, given the nominalist premise.

Since the advent of indeterminacy in physics, scientists have revalued physical laws. The revaluation consists in the rejection of absolute certainty, whether in prediction or characterization. Predictions and regularities lie within a range—outside that range they are invalid. Many other examples of such ranges fall within a norm in general but not tidily. For one, color lies in a normative range—"green" is a continuum. "Green" may validly characterize anything from pale to dark green and from a green verging on yellow, to one verging on blue. "Green" in this sense is a normative identity with a wide range. Even species that show little color variation show slight variations within a single color. Similarly, the size of normal elephant ears, tusks, and many other features are variable without bringing their species identity into question. Indeed, the male and female of a species may be quite distinctive but still the same species: what is normal for that species.[39]

Norman Hansen noted another aspect of individuals, namely that "individual" is theory relative. Individuals are prescribed by a theory and therefore not autonomous from it: including individuals for nominalism. What a theory defines as an "instance" is itself normative. Thus "atoms" were posited or prescribed by atomic theory, but were later proven to be divisible. "Atoms" were normatively laden, not "real." Thus they were not an independent variable to which attributes "attach." The theory and the instances stand and fall together. If there were no theory specified individuals, then the theory would be "empty." As the late Prof. Reuben Abel asked (lecture), "Were there gravitational phenomena before Newton?" (Alternatively, if no new individuals are specified by a theory, then other "individuals" of competing theories are adopted.)

If "individual" is a determination, then theories cannot arise by abstraction from individuals. Theories themselves define the individuals they explain: they create new classes of individuals. "Individuals" are specified by a theory with which they stand in a mutual relation. Concepts at this level are in turn covered by "higher" order concepts such as genera, *ad infinitum*, but all are hypothetical. Competing conceptual hypotheses still must be evaluated, but do not represent competing abstractions.

The appeal to nominalism also conflates and thereby confuses metaphysics and biology. Nominalism argues against the "existence" of universals, not against the actuality of species of animals and plants. Species differences separate groups, not individuals. We could at least argue that the importation of metaphysical disputes into biology is a category mistake. On one level, biology is not concerned to deny individuality, however defined, or to define "reality." However, biology does appeal to actual principles and relations in the world, regardless of its relation to metaphysics. "Species" includes a genetic link, involving relations.[40] An individual organism is in one respect an individual as an agent. In another, this individual is a part of a genetic continuum, stretching back to the origin of that particular species. Species differences reflect relations between individuals, insofar as individuals are separate. In still another sense, the concrete individual is a part of a habitat, out of which it could not survive. The relation is required for the survival of individuals in both cases: an argument for relations in biology. The dichotomy of individual/species is another false dualism, inapplicable to biology.

Moreover, I do not think one can separate "horizontal" and "vertical" species as easily as Stamos thinks (2003, ch. 5). A species at a slice in time is an abstraction from actual relations, including temporal and sexual relations. Reproduction is an ongoing affair at all times for all species and this historical element cannot be dropped from any species notion. Stamos's (*ibid*, ch. 5) argument that the "horizontal" notion includes a large slice of time defeats the proposed distinction, for it includes the "vertical" within the definition of the "horizontal." This incorporation of a huge block of time is the incorporation of a vertical slice, as are breeding and habitat relations. The "pattern" is inherited from parents and must be for any species concept to have validity.

Nominalists in biology raise an interesting point: do "species" evolve? According to Darwin, species are replaced through natural selection, so do not evolve (Stamos, 2003, contests this interpretation). But compare breeds in Darwin as "artificial" selection. Alternatively, "morphs" evolve, according to some biologists (Stamos). But these are not *per se* actualities but characters of individuals of a species. A favorable variation may indeed be confined to one individual of a species, but this individual must back-mate with other individuals and its favorable variation persist as a dominant trait, for it to become predominant. If this trait becomes predominant, the daughter species may replace the parent, or a new variety may arise in which they compete. But if the new trait is sufficiently different one could easily say that it implies a new species. In this case, species do not arise, but are replaced. Yet it is rare that one trait makes a new species, as opposed to a variety. Instead, the species may be on the way to evolution in the form of a new variety that may further evolve into an entirely new species. But since the trait marks a new variety, it will not do to say that species evolve and yet remain the same species.

There is also the relation of chemical events in an organism to biological results: biochemistry. Changes in brain chemistry may cause differences in behavior. The science of genetics is at the border of biology and chemistry; its principles also involve chemicals such as DNA that cause biological changes at the molecular, organ, and organism levels.

There is a new argument from Ghiselin and Hull (Stamos, 2003, ch. 4) that species themselves are individuals. This *ad hoc* definition is at best an equivocation—for species do not act like single organisms, or single anything. The law of gravity means that if one organism falls off a cliff it will die, but species do not die in the same way. Individual elephants die but the species does not go extinct from the death of such individuals. Species are not individuals in any of the normal senses of "individuals," but only by expanding the definition to mean something other than what the term originally designated and its connotation. This redefinition can only create more confusion than solution.

We can see that environmental philosophy and science can also teach us about metaphysics[41] because environmental relations are not between discrete individuals, but instead consist in the relation of parts to a whole, a habitat. Saguaro cacti are not only a distinct type of plant in a distinct desert habitat, but provide a home to several birds, food for pollinators and also for species that eat its fruit. In turn, these other species serve the plant, by checking predators, fertilizing its flowers, and distributing its seed. Without the birds that live in it, predator insects might devastate it. Without the pollinators, it could not produce seed and thereby reproduce. Without the fruit eaters, its seed could not find a new place to grow. Each of the species depends on the others in their forbidding habitat, and could not survive in the long run without the others.

If relations are distinct from substances with their attributes, or from material things, a pluralistic universe is required, with more than one principle. From an environmental perspective, the isolated individual of nominalism is at best an abstraction: the individual removed from an environment could not long survive. Even animals, with their freedom of movement, are in a habitat that conditions their individuality with relations. These include relations to other species, especially symbiotes and prey. They also include relations to non-living elements of the biosphere, such as air, water, and soil. The animal is only "one" in the abstract, since in practice, if the animal were removed from its habitat, it would die. The animal cannot live except with the relations it requires in a habitat. An "individual thing" is an abstraction from the environment, not just living things but even an "individual" rock. Nominalism is an abstraction from the actual world of relations.

The problem of individuals, like that of universals, is quantitative. It concerns the identity of a singular—its unity. But conflating quantity with actuality may be a category mistake. "Individual" refers to quantity—but this mathematical property is one abstract, contextual property among others.[42] Why should it be judged *primus inter pares*? Only as an evaluation: just as in substantialism, stability is valued over dynamic properties. Nominalism contains a hidden evaluation.

We can conclude that environmental philosophy is pluralistic.[43] Environmental philosophy includes individuals of a kind, habitats, their relation and elements, the relation of organs to the individual of which they are a part and other distinct aspects of the whole, as well as principles. Individuals apart from relations are abstractions from actual conditions. No such individuals are actual, only normative instances, that is, elephants, lilies, granite, and so on. A normative identity for a species includes plural characters: genetic, morphological, reproductive, habitat/niche, and others.

5. The Moral Issue

The attempt to argue against speciesism by arguing that species is an arbitrary social construction is a well-meaning project, in which the goal is to eliminate mistreatment of other species by our species, based on species differences. The problem is that the result of this approach might be just the opposite of the one intended by its proponents. It might well undercut attempts to preserve animals of a kind, that is, species preservation. Without such distinctions, made by science and established scientifically, the ground could be cut from under attempts to preserve rare, threatened or endangered species and along with it biological diversity.[44]

Are bison distinct from cattle? Whooping cranes from eagles? *Calochortus persistens*, a species that grows only on one ridge in North California, from *Calochortus nuttallii*, a common species in the Rockies? If so, then the notion of species is more than just a "social construct." As environmental groups have stressed, "extinction is forever," meaning that a specific kind goes extinct, not only a random collection of individuals.

The overriding moral issue is species preservation, not the preservation of individuals. Individuals die in any case, but species can endure through reproduction. However, a species can also die out if its distinctive group characteristics are not recognized and its habitat preserved. Undermining all actual distinctions of species could be used to justify the extermination of individuals of rare species. Why preserve what is not unique? Some species are rare while others are common, which is a practical argument for the actuality of species. Preservation should concentrate on rarer species, especially endangered ones (J. Baird Callicott, 1989). The rarity of some species provides greater weight for recognition of the inherent or intrinsic value of a distinct species.

Why then preserve species? Species constitute a distinct kind of good in the world, may be irreplaceable, and also contribute to habitats in some unique way.[45] Each species is a unique genetic continuum that has actualized itself over historical time. It has its own unique good, since the species preserves itself by itself, self-actualization of a whole kind, the source of "good of its kind." "Natural variation" in Darwin has good and bad results. Once gone it may be impossible to bring it back. This uniqueness is entirely separate from human constructivism for it preceded human constructivism and is the practical condition of human constructivism.

Above all, other species have their own life, look after themselves, and have their own kind of life. They have their own value to themselves and try to pursue their own good. They creatively actualize their own lives just as humans do: they are self-actualizing and thereby have intrinsic value. While we might not prefer or like to exchange our human life for that of a pig, as John Stuart Mill argued, that judgment may represent the perspective of our own species. Polar bears would not want to exchange their comfortable habitats for our nice warm hearth, nor would fish. Moreover, diversity of goods is itself good, since it expands the inventory of good to include more kinds and a greater quantity of different goods.

Humans require other species for their own survival: every time we eat, we eat another species, unless we are cannibals. If humans are good, then using other species as instruments of our survival gives them instrumental value. If our life is good then its conditions must be good, at least instrumentally. However, if we require them, they are practically necessary and a practical condition of our own survival. They thereby acquire moral significance or considerability. Even from a utilitarian perspective, the "greatest good of the greatest number" requires preservation of species, since non-humans are greater in number. Similarly, if we require other species, preserving them is in our self-interest.

Often the relation of species in a habitat is as parts to a whole.[46] In general, predators require prey; the prey might overpopulate without the predator and starve, for lack of its prey. We may not know the function that a particular species plays in a habitat, and it would be perilous to endanger the habitat as a whole by eliminating a species in it, if only on precautionary grounds (Ian Smith, 2010). The role of predators in keeping a habitat healthy was not well known until recently. The reintroduction of wolves into Yellowstone National Park reinvigorated plant species that had been decimated by elk grazing, particularly along the Yellowstone River.[47] The habitat was in decay due to the absence of such predators and this affected many other species. Species and their habitats interact, and thus could be described as symbiotic. Each species may play a vital, if minor, part in a larger whole. Thus preservation of species also requires preservation of related species, and their habitats, including such non-organic elements of the habitat as water and soil.

There is also the reciprocal relation of symbiotes, as in the food chain. I discussed the role the Saguaro cactus plays as a home to some species, food provider for others, and symbiote for still others. Similarly, pollinating insects and angiosperms arose together and require each other. Carnivores require grazers and other food; grazers require plants that in turn require pollinators. Humans require domesticated animals and plants. Human scientists have to pollinate a species of flower in Hawaii that grows on cliffs or it will go extinct. Apparently, its insect or avian pollinator has become extinct. These symbiotic relations are required for these groups of animals and plants, for they would die without them. Thus biodiversity is also a good, since symbiosis requires it, and humans are symbiotes. More, biodiversity is good because it means more species, that is, more kinds of distinct goods: more goods. Biodiversity means a larger whole, with more parts in more distinctive habitats and niches.[48]

The necessity for biodiversity was recognized by a world congress of scientists convened by the United Nations. They stated:

> Biodiversity represents the very foundation of human existence. Yet by our heedless actions we are eroding this biological capital at an alarming rate. Even today, despite the destruction that we have inflicted on the environment and its natural bounty, its resilience is taken for granted. But the more we learn of the workings of the natural world, the clearer it becomes that there is a limit to the disruption that the environment can endure.
>
> Besides the profound ethical and aesthetic implications, it is clear that the loss of biodiversity has serious economic and social costs. The genes, species, ecosystems and human knowledge which are being lost represent a living library of options available for adapting to local and global change. Biodiversity is part of our daily lives and livelihood and constitutes the resources upon which families, communities, nations and future generations depend.[49]

Human destruction of the environment is also a threat to other species. I argued above for the inherent value of other species. Are humans one species among others or special? As symbiotes, we are not independent, but have other species and other inorganic elements of the biosphere as our own condition. Since these are practically necessary to human survival and well-being, they are required or practically necessary. What is practically necessary is obligatory: we have duties toward other species and other inorganic elements of the biosphere in order to survive ourselves.[50] We cannot dismiss such obligations if we are to survive. Nor are we as different from other species as we would like to believe.[51] We evolved from other species and are thereby related to them. We have many of the same biological functions, including nutrition, reproduction, elimination, illnesses, and so on.

Our rampant destruction of the wild and of other species is also a threat to future generations. As I argued in chapter four, if we have children, we have obligations to them. We must also assume that they will have children who in turn will have children, and so on. If we continue on our destructive rampage, our children will have a worse life, if they have a life worth living at all. Their children may have it worse yet, having to live with the effects of global warming; namely, rising ocean levels and temperatures. Our overuse of pesticides, fertilizers, and poor engineering of irrigation may result in poisoning and salinization of soils, erosion, disruption of ground water and riverine habitats: more threats to future generations, who will have to live with the consequences of our misuse of the earth. In other words, they will be the victims of environmental injustice as much as the poor are now the disproportionate victims of pollution. Rising expectations and affluence are also contributing factors in destruction of the earth. The spread of the affluent suburban life-style to the entire planet would result in environmental catastrophe, as global warming is accelerated just in proportion to sales of gas-guzzling autos, electrical gadgets, and other status symbols of affluence.

Under this head we can include destruction of landscapes, for example by flooding Hetch-Hetchy Valley and Lake Mead to allow energy generation, as well as commercial exploitation of beautiful landscapes. Flooding such landscapes means that future generations are permanently denied the experience of observing these beautiful sights just so contemporary humans can play with trinkets powered by electricity: "blow-dryers," "electric can-openers," and "electric carving knives." The point is that none of these devices are necessary and all of them are wasteful, since we could easily dry our hair, open cans, and carve by hand.[52] There are also the hidden ecological costs of cheap electricity generation, including acid rain, global warming, nuclear waste, and so on. Our descendants may value preserved landscapes, and be more grateful for preservation of these far more than the piles of junk formed by our discarded electrical appliances.

There is no need to throw out the baby with the bathwater. We can recognize the distinct actuality and intrinsic value of species without being speciesist, just as we can acknowledge the value of a "rainbow" of distinct cultures and identities without being racist, sexist, or homophobic. We no more want a world with only one species than a world with one "race." Diversity, not uniformity, should regulate our relations to other species: cosmopolitan pluralism. The fight against speciesism should not end the fight to preserve species.

To what extent should humans be willing to make sacrifices to preserve the environment? To what extent should humans be willing to make sacrifices to respect the wild such as preserving rare species? Apart from the duty to preserve rare species and all habitats, humans must undergo an ethical revolution of the sort to which this book is, I hope, a contribution. A new cosmopolitanism will mean a new appreciation of the actual place of humans in a much larger world, and as symbiotes along with other species. This will mean a complete revaluation of traditional duties and virtues.[53] If the human spirit can expand outward to embrace life as a whole, our species may prosper. I fear, however, that greed will overcome generosity of spirit and we will not only be the cause of our own destruction, but bring thousands of other species down with us.

6. Evil

How can we approach the issue of evil, in an ecocentric context? The Nazi Holocaust provides the new norm of evil in the modern period. Despite other genocides in the modern period—the Armenian Turks, Joseph Stalin's treatment of the Ukrainians, the Khmer Rouge, and others—there is nothing that quite approaches the Nazi record of deliberate, systematic, and mechanical slaughter.[54] Since their evil was not in the least abstract, but entirely concrete, evil need no longer be considered transcendent.[55]

Nevertheless, we can still raise a legitimate question about why the Nazis are the new norm: what is especially evil about them? After all many other cases of relentless slaughter could be used: why not Stalin or Mao tse-dung? Stalin killed as many or more than Adolf Hitler; the number of Mao's victims is unknown, but may be comparable to Stalin's. There are still other cases in recent memory. Why not the British in Ireland? Why not the Native Americans in the United States? Or the African slaves? Or the Tutsis in Rwanda? It can only be because these powers won and the Nazis lost. Is the oblivion of such cases itself evil? If evil is the norm created by the Nazis, then the Gulag, the British Empire, and the U.S. treatment of Native Americans and Africans must be comparable in many respects. Even the call in the Old Testament to kill all of an enemy's women and children—and even to extirpate "every blade of grass"—is evil.

So what is specifically evil about these cases? I think that we can say that evil in one sense is genocide: the deliberate elimination of a kind.[56] It is evil to indiscriminately lump together a vast number of individuals, to posit some arbitrary kind of bad quality to them all, and to kill the innocent along with the guilty. At Auschwitz babies, who could not possibly have committed any wrong, were executed along with their mothers. Thus the Nazi period provides pragmatic contextual parameters for ethics: the Nazis represented evil for our time (Auschwitz) a judgment which is not seriously disputed. This provides a parameter or negative standard: a fixed standard of evil *in history*, a basis for evaluation, and a contextual parameter for moral judgments as a standard. The lack of a will to act to save Hitler's victims while they could be saved is a corollary evil. More local cases of evil are still the elimination of a kind, for example, killing a whole family, or threatening to do so, in order to force a parent to do something evil against their inclination. The destruction of a family eliminates a local kind.

Some might argue that Hitler was "crazy," not evil. Perhaps they do not want to face the truth that anyone could deliberately do what Hitler did. Treating Hitler and his henchmen with psychiatric categories may have its value, but we cannot medicalize evil if we are to deal with it adequately.[57] Avoiding the issue of evil would provide a blank check for moral failings and an excuse for moral indifference: allowing even rapists to claim that they were victims of their environment. Indeed, science cannot judge Auschwitz wrong; thus it is limited in value: inadequate as a basis for moral philosophy.

By projection, species extinction is a genocide: the elimination of an entire *kind*. Just as the Holocaust involved the attempted destruction of an entire kind, and fundamentalists of different religions would attempt to eliminate lesbians, gays, and other sexual minorities as a kind, so is species extinction, and it is just as evil. For if it is evil to kill an arbitrary kind, including the totally innocent, it is even more evil to kill an innocent species. Indeed, species are not an arbitrary kind, but an actual kind. And most species are innocent of any harm to humans. Thus we can adjudge as progress in moral evaluation the extension of moral consideration from the victims of Auschwitz and other intra-human genocides to species.

Another question we may raise is: is saving the environment always of less value than saving a person? What if that person is Hitler? If Hitler is evil, and killing him in the thirties before he was able to commit his evil acts would at least in retrospect have been right, then killing humans is not always wrong. Advocates of non-violence would argue that we could develop other ways of dealing with potential tyrants and evil regimes than assassination. If enough people had opposed Hitler early on, he could not have committed his crimes against humanity. I put this issue before my students and they almost always respond that non-violence would not have worked against the Nazis. Indeed, socialist opponents of the Nazis within Germany did rally against the Nazis for as long as they could but were then suppressed. The limits of the efficacy of non-violence are at least debatable.

More importantly, sometimes the environment can be of greater value than saving an individual person such as Hitler. But what if the person is innocent? That person still has an impact on the environment as they require the environment. But if killing an innocent person is wrong, how can it be right to kill an innocent species?

What then of species that are harmful to humans? Consider smallpox or the Aids virus. (If some kinds harm humans this provides another argument for the actuality of kinds.) Why should we preserve such obviously harmful kinds? This is a difficult and hard borderline case to decide. But the first point is that these exceptions cannot possibly justify extinction of species that are not harmful to humans. Yet the extinction of harmless and even beneficial species is proceeding apace, in the clear-cutting and burning of the tropical forests, desertification and other evils practiced by humans of which most humans are oblivious. Secondly, we can distinguish degrees of harm. Some species cause only minor harm to humans; others only indirect harm, for example the corn borer. As Rachel Carson pointed out, human agricultural practices contribute as much to the increase of such pests as their own appetites, for after all, they are just feeding. Other harmful species can be controlled by biological means far more efficiently without resort to genocide. Lady beetles are an extremely effective check on aphids, and do not harm beneficials, unlike indiscriminate use of chemicals. Since they can be checked and the harm they do minimized by their own predators, genocide is not justified. This reduces the number of species that is harmful in any significant way to just a few.

A corollary to evil as the deliberate destruction of a kind is such obliviousness: the refusal of society to deal with the ongoing genocide of other species. In our time, the continued use of poisons and the refusal to face the problems of overpopulation and global warming exemplify this obliviousness. People in pursuit of affluence refuse to deal with looming environmental catastrophe. They continue to have children as if there were no problem, knowing that there is a problem—or they encourage it, like the Roman Catholic Church. Overpopulation contributes to species extinction, for example, in the Amazon and other tropical forests, where most species live. The consumption of resources by more and more people has as a result the penetration of such areas and their destruction. Willful obliviousness is itself, then, a corollary of the evil of the genocide of entire kinds, comparable to the refusal to allow refugees asylum during the Nazi period. The excuse that Hitler's general staff were "following orders" is a variation of this evil.

Another corollary is deliberate exclusion of any kind from moral consideration: creating artificial arbitrary barriers like skin color: willful and superficial.[58] As examples we have the exclusion of Jews, Gypsies, and others from citizenship and human rights by Nazis, even those who had fought on the German side in World War I, or for whom Judaism, whether as an ethnic tie or a religion, was of little significance. The common humanity of Jews and Germans was ignored in favor of imaginary differences.[59] Similarly, the Native Americans were deemed "savages" and excluded from property rights: their land was stolen from them. The slaves in the United States were treated similarly. Last but not least, gays are excluded by the Old Testament, and by Paul in the New and Mohammed in the Koran.

Speciesism would include exclusion of other species from moral consideration. By extension, this is as evil as the other cases of exclusion. Excluding some "other" (Levinas) denies (axiological) sameness. Defending human chauvinism based on some arbitrary and morally irrelevant difference between humans and other species is as immoral as positing an "Aryan" race, using it to exclude other humans, and then deliberately slaughtering them. Exclusion is evil as the prelude to genocide. The work of Darwin should not yield the lesson of the Social Darwinists but that of the mystics: that all life is interconnected as One, since we are all descended from the first life.

A person might argue that some cases of exclusion are not judged evil, and thus defeat the thesis. One counter-example is that Native Americans often retain a tribal affiliation voluntarily. A tribal relation or organization in some sense excludes those not of the tribe. Many bands and tribes do allow non-members to become one of the band under some circumstances. But tribes of themselves can be exclusionary. For this reason, we require the qualification of voluntary and involuntary. Jews who considered themselves good Germans, who spoke the language at home, who had little contact with the religion, and who even had fought in the German army were involuntarily excluded. Moreover, they were not allowed to leave or escape, but were slaughtered. Nevertheless, some cases of involuntary exclusion do not involve violence and may not be evil. Some Native Americans will not let outsiders into their bands, but do not go around exterminating others, either. Thus involuntary exclusion is not sufficient.

Another case we might examine is greed. Is greed exclusionary? One might say that it excludes others from equal consideration for goods. A disproportionate share of goods is attained at others' expense. Perhaps greed is considered a vice for this reason. But the analogy does not quite hold. Greedy individuals may put goods before other people, but they do not exclude people involuntarily from consideration. Nor, generally, are they violent toward others.

Evil also consists in violence. This can consist in generalized violence, whether in a civil war in which innocent civilians are targeted (Lebanon), or a society based on terror, as Hannah Arendt (1951) argued Nazi Germany was. The terror and violence does not have to be political; organized crime sometimes creates an atmosphere of terror, as it does in the drug wars. The corruption of officials begins with threats against their families. Similarly extortions of shopkeepers by ethnic gangs in New York City and elsewhere involve threats of violence, as do kidnappings for ransom when parents are coerced to save their children from threats. But evil can consist in individual acts of terror and violence, or the threat of violence, as when women are raped at the point of a gun, notably in the eastern Congo, where rape has been used as a tactic of intimidation, brutality, and degradation of the victims.

One problem here is that it may require violence to counter some forms of violence, despite Mahatma Gandhi's and Martin L. King's lessons to the contrary. Hitler's power reached the point that only Marshall Zhukov's and General Patton's tanks, and allied bombs could stop him. If violence is needed to counter some forms of violence, we can still judge it evil, but less evil than the violence that began it. And we can at least hope that Gandhi and King were right—that if enough strategy goes into non-violent tactics, they can stop evil.

One problem in defining or delimiting evil is to show how evil depends upon the good. Murderers, rapists, extortionists, and others who commit evil acts depend upon the morals of others, for they themselves are not the victims of murderers, rapists, and extortionists. On the contrary, they depend upon the morality and non-violence of the great majority to perpetrate their crimes. Like freeloaders, they take advantage of the virtue of others to exploit them. When this gets out of hand, in the form of generalized violence, and the harmonic whole is the ultimate ground containing strife, when strife destroys it, it destroys itself. Civilizations then stand in ruin, as Beirut did after the eighties, Germany after World War II, Mayan civilization, and many other cities and countries.

We might wonder, does this go deep enough? But the argument is that this is an example of evil, not its whole. Evil is difficult to delimit, for it is always lurking. But we can say some things and make certain judgments, based on what we know.

Humans are now engaged in generalized violence against the wild, indiscriminately threatening thousands of species. The rationale is exclusionary, dividing humans from other species on which humans directly or indirectly depend; and claiming that the earth exists for human exploitation. Facing the judgment that this is evil is difficult for humans, but it matches our criteria in every respect. And like civilizations that destroy themselves by generalized violence, that arbitrarily excludes the innocent, we will destroy our entire civilization across the entire world, if we persist in destroying our own home, the earth. We do not need to know everything to know that destroying our own habitat, overpopulation, and global warming cannot but have disastrous results. Hiding our heads in the sand, as German civilians did during the Nazi period, is irresponsible. We who live after the Holocaust can no longer make such excuses. The responsibility of all humans is to wake up and face our responsibility to the environment before it is too late.

Nine

THE VIRTUES OF THRIFT

> Infantilism is a state of mind dear to McWorld, for it is defined by "I want, I want, I want" and "Gimme, gimme, gimme," favorites from the Consumer's Book of Nursery Rhymes. And that is not just "a little sad." It is a lot sad.
> Benjamin Barber[1]

> The efforts of the entrepreneur to create a wonderworld are, in fact, creating a wasteworld, a nonviable environment for the human species.
> Thomas Berry[2]

In this chapter I will argue for the priority of ethics over economics, including demand and "needs." I will contrast the affluent consumer society and the thrift economy, and argue that the former is unsustainable. I will revalue economic value by the standard of Ecocentric Cosmopolitan ethics: a virtuous life-style, not a consumer life-style. As an alternative, I will propose the virtue of thrift. I will argue that thrift involves conservation of wealth and thereby a reduction in impact on the environment, so there is less waste of resources. Thrift provides a traditional alternative that could also be a model for sustainability: thrift, not "growth."

1. Economics and Ethics

How can we separate "economic" activities from other activities of a culture? Economic institutions engage in trade in our culture but not necessarily in all. And the government, a political institution, can also engage in trade. Economics is not a fixed category, but a possibility within specific cultural wholes, as a possible combination of values.[3] "Economic" activities are united by the ends and values they ultimately serve and by the categorial value of such activity. There is no "essence" of economic activities, but a historical continuum uniting some cultural activities as a contingent possibility actualized at specific historical junctures. Categories like "economics" are united by (1) the social or human end that is served and the (2) mode of evaluation, that is, means, ends, and so on.

Philosophy of economics reduces to the question of the *good* of prosperity. Are economic goods worthwhile? What is their value for life? Aristotle argued that economic goods are means, but part of a good life. He thought that some level of prosperity is conducive to virtue and happiness. Nevertheless, economic life should never be judged an end. "Leisure is better than occupation and is its end" (*Pol* $_{1337}$b33). By contrast the Stoics did not consider economic goods part of a good life; the material realm is neutral in value—neither good nor bad. Christianity, with its otherworldly orientation, for the most part accepted economics reluctantly, but merely as a means to salvation. Modern materialist philosophers noted the importance of economic factors in historical change and as historical causes, culminating in the work of Karl Marx, whom some have interpreted as an economic determinist.[4]

Economic activity is rarely worthwhile in itself, but rather as a means to fulfillment. Humans do not descend into dark coal mines, breathe in coal dust, get themselves filthy with soot, and endure scrapes and scratches for its own sake. They do so to make a living and to contribute to the overall good of society by mining a useful commodity. The evaluation of economic life and products as an end in itself is a bad revaluation.

To demonstrate and argue for this point, we may contrast economics with beauty as an end in itself. Beauty is the ultimate motive of economic life, which is why economism (Karl Marx, Ayn Rand) is such a perversion. Why does anyone pursue wealth? Do they wish to contemplate piles of gold or currency, like Midas? There may be a few who do, but generally money is judged as a means to some extrinsic end, often aesthetic. Money is used to purchase beautiful houses to dwell in, surrounded by beautiful flowers and landscapes. Wealth is also used to buy beautiful clothes, autos, furniture, art works, and so on. No one with wealth purchases what he or she considers an expensive but ugly home or other goods. The end is esthetic—so pursuing wealth for its own sake is an axiological perversion: treating a means as an end. Some might argue that money also enhances status and power, and thereby other ends than just beauty. But this only reinforces the point: money and economic life aim at ends outside themselves. They are extrinsic in value. Status and power are social ends. Indeed, money may be a means of attracting a beautiful wife or husband, whether because of high status or wealth. But then its use is again as a means. What then of crafts—do these provide an example of economic life that is done for its own sake? Is the creation of beautiful pottery, art glass, and other art works carried out for its own sake? Here beauty is the goal, not the craft as such, although the work may be enjoyable on its own.

An alternative view is implicit in economic determinism. In this view economics is central to life, since economics deals with the basics of existing—subsistence—and is the true motor of change. In one form of this view, morality is thereby determined by economics: economic realism. Economic ends determine morals and economic causes are the motor of history. Vulgar economism is a species of this view, where economic causes directly influence moral choices[5]; more sophisticated economism only argues that economics determines morals in the long run, or "ultimately."[6]

Not everything is life is determined by economic ends and motives. Indeed, it is not even necessary that those who desire such material goods should also produce them directly. They can perform other needed tasks, for example, entertaining or providing services for the producers. But are the goods of economic life determining of the shape of morals in the long run? Does moral progress in the long run represent economic progress? Against this form of economism or basic economic determinism I would argue that "economic" here has two distinct elements. One of these has to do with subsistence, or basic biological needs. The other involves goods above basic needs and the standard of living. Economics in the second sense is not concerned with basic biological needs ("materialism" in one form), but with a rising standard of living over time, the spread of prosperity.[7]

The problems of economics no longer reduce to scarcity or a lack of economic goods in advanced societies. They involve instead a conflict of values: economic goods and their concomitants, that is, machines, organization, and so on as opposed to quality of life, finding a solution for alienation, human values over technology and bureaucracy, spiritual values, and so on. Diversity is essential for (1) alternative values and (2) changes in values: from economies to other values. And how can we decide which goods should depend for their production on machines? What of primitive methods of production? Glassblowing is not competitive with machine production but produces beautiful art works. The end is different and this defines the worth of the activity.

Economic materialism could also be accused of pure quantification of value: that human life has no value without pecuniary means: that poor people have no dignity or worth without money. However, the economy is supposed to serve life, not life the economy.

Invention is not a copy of production: imagination is a condition of progress (a form of "consciousness"—or at least non-material), since it includes the power or ability to recognize and articulate new "needs" as well as to invent.[8] Imagination cannot be determined by the past and still be inventive, for invention involves the genesis of ideas about what is not present or past, but an altered future. Moreover, imagination is truly creative, inventing whole new categories of useful devices—to speak nothing of works of fine art—which had no analogue in the past. Since economic progress is dependent upon technology, and new technology depends upon imagining it, economics is more dependent upon imagination than economic determinists recognize.

Despite theories to the contrary, economics is not the central value, but depends totally upon morals. If economics were not limited by morals, embezzlement would be permissible; exploitation of the workers, just. The division of labor also requires morals in the form of cooperation and trust: it is not grounded only in economics, much less biology.[9] The ascent from nature requires morals in the same respect: trust in others and cooperation. Productive humans as such would still be within "nature" (bees, ants, beavers, and so on.)

Surely, no one would seriously commend an economically efficient but morally corrupt set-up as superior, whether the use of slave labor in Nazi Germany, or the plantation-slavery system in the US South. On the contrary, such corrupt and immoral societies demonstrate just why economics should never be the end. The lies perpetrated by the oil companies provide a more recent example. Their advertising and their funding of interest groups has the motive of undermining public efforts to limit global warming, by erroneously stating that there is no consensus of scientists about global warming. Truth is a norm for accounting as well, as well as non-theft. Thus economics *requires* ethics, since trade and the market require trust, cooperation, honesty, non-violence and truth, among other moral norms.

The economic model of humans is deeply flawed, even as it helps create the society of homo oeconomicus that it models, a self-fulfilling norm.[10] First, the value it posits as the *summum bonum* is axiologically false.[11] It would be deeply irrational for us to make economic goods the *summum bonum*. Not all problems are economic, contrary to economism. Economics is a distinct category of value, with distinct norms and principles, but should never be supreme. The methods of economics are inappropriate for other goods. Few hold economic goods supreme, that is, as the highest good or end (misers); for people value happiness or other extrinsic goals like health and friendship, not the wealth itself, and these are not economic. Health and happiness cannot be purchased, although means that may aid these goals can. To make wealth the end would be irrational, for it confuses ends and means. It treats means as ends and thus is incomplete, for what is the wealth for? Surely not itself? Can one eat gold? Economic values are properly only a means to an end, not an end in itself, and should be valued only as means.

As for pursuit of wealth as a life-style, a critical assessment, using a pragmatic criterion of values, would reveal that people are rarely motivated by earning great wealth, which is not to say that they would reject wealth that fell into their lap. Few people spend all their time pursuing wealth or take the steps necessary to acquire it.[12] They spend disposable income on family, entertainment, and life style choices, whether political, religious, or some other kind, instead of accumulating it. They are rarely motivated to change their economic status unless they are pressed by necessity, or the cost of other values, whether social prestige, fine art, or other goods. Few even pursue costly hobbies like collecting fine art or taking up flying; thus these do not provide an economic counter-example or "sub-structural" motive for most humans. The great mass of humans are content with a more modest life-style and incorporate other values.[13]

The economic life as a paradigm would reduce life to its barest level, the values of survival, and not its finest values in the life of the mind, the spirit and artistic creation. Our higher capacities remain undeveloped in a life fit for pigs who consume. The source of value lies outside the self in instrumental goods that have been purchased, a quantitative measure of quality and a perversion of intrinsic value. The consumer becomes a means of increasing wealth and measures self-worth in terms of a means, despite the obvious truth that economic worth is extrinsic to the person. The value of economic goods is as a means of sustaining life and a means of pursuing other values. A society that does not recognize this axiological perversion and encourage the creative and virtuous life is doomed. It will spawn a population of soulless consumers. Confucius called this type the "small man."

Economic life, far from providing any model for morals, provides a model of what morals should limit. Realistic commentators have noted that two emotions, greed and fear, dominate Wall Street investment capitalism. Against those who argue that economic life, especially investment promotes virtue, we can argue that it promotes greed and fear, not virtue. Greed is precisely not a virtue but a vice: egoistic not virtuous. On the contrary, for the ancient philosophers, moderation and generosity are virtues. The ancient philosophers did not talk about greed, except in the negative sense of a vice, but discussed moderation at great length as well as generosity. Moderation means that we control our greed; generosity, that we share what we have. Societies run solely on monetary incentives lose virtue—the fabric of society.

I hesitate to address or mention someone like Ayn Rand, even in a negative sense, since her influence has been so pernicious. But Rand's contention that greed is a virtue and her claim to be following in the footsteps of Aristotle shows how ignorant she is. Aristotle would have immediately dismissed the idea that the life of greed should be emulated. Instead, the greedy are slavish, ruled by desire, as he thought the inhabitants of tyrannies were. Paradigmatically, virtuous humans are ruled by reason, not desire. Aristotle gave an extensive analysis of moderation, the key to the virtues, in his ethics, as well as liberality and other forms of generosity. Greed is a vice, a form of excess.

Fear is an emotion that can be rational or irrational but is neither virtuous nor a vice as such. Instead, a virtue is needed to control it, namely, wisdom, temperance, and courage. These motives provide another reason why economics should not be an end, for as the chief purpose in life it does not bring out the best in humans but the worst. The sole remaining habitat of the ivory-billed woodpecker was cut down for lumber despite an offer from conservationists to buy it. The rationale was that its owners were self-described "money-grubbers," who would sacrifice species for wealth—for an increased standard of living. Status and power are enhanced by wealth, expanding its appeal. I would not deny that there are ethical capitalists, but the issue is whether capitalism as such either involves virtue or promotes virtue. I would argue that virtue is an extra-economic social value developed in the individual by their culture, and not by economic practice. They are virtuous in spite of the incentive to cheat customers, only because they learned virtue and duty from their culture and were impressed by it, apart from fear of the law.

Recently some have argued that greed "works" and thus that greed is good. In this view, greed has given us economic prosperity. But this is to judge only in the short term and without consideration of social factors. What are the results of greed as such? In 2009 the market went into meltdown, lost investments, while banks foreclosed homes, and others depreciated. Long before the recent "great recession," we witnessed overgrazed ranges, loss of fisheries—since they are overfished out of greed—and forests clear-cut and not replanted. Other results include polluted air, acid (dead) lakes, rivers whose fish cannot be eaten, and other negative, life-destroying effects. Judging greed by its consequences, the pragmatic test, reveals it is pernicious as the engine of self-destruction, not increased wealth, since it destroys the means of wealth.

Greed, or economic freeloading, can only operate (surreptitiously) within moral limits. For large concerns require the division of labor, cooperation, trust, and other properly moral values and virtues. Moreover, managerialism is the cause of increasing consumption: using advertising to increase sales. Managerialism uses sex to sell products that are often not needed, cultivating desire at the cost of virtue. Also, ecocentric virtues may be a threat to profits, for example, aluminum recycling, which is thrifty, and reduces electricity and thereby global warming gas use, also reduces demand for new aluminum. Again, technology, which reflects imagination, has brought prosperity, not greed. Greed as such can only bring vice.

Demand, the term economists use to measure the sum total of purchases, is a form of desire. Talk by economists about "utility" comes under this type, as does any model of the life-style of or for consumers. Generally economists are referring to means, not ends, but conflate the two in their models. Nevertheless, demand, self-interest, and the like are species of desire.

The model as such is amoral, since any demand is as good as any other. A demand for drugs is a market for drugs and thereby economically warranted. A demand for sex with children is as legitimate as any other. A "value-free" model cannot critically evaluate such "demand values" since moral values have been deliberately excluded from any so-called "rational" model for economics.

"Demand" as a substitute for value, like greed, would destroy itself if it were actually practiced, since demand alone would have the result that the society utilizing it would collapse internally. Unrestrained demand for drugs would end with a society of addicts. Demand for atomic or biological weapons by terrorists to blackmail the society would bankrupt it or result in McVeigh-style acts of destruction but at a much larger scale. Unlimited demand is not desirable, as these two are not coextensive: consider a "demand" for ransom money by a group that threatens to blow up the world. Examples like these, which could be easily multiplied, reveal what is wrong with this and all the theories surrounding conativism. They are indiscriminate and describe the life of a moral cipher, freeloader, hypocrite or outlaw. As I argued in chapter six, consumer economies, based as they are on greed, have cultivated just such creations, the corporate managers of companies like Enron and Tyco, whose management put greed ahead of morality.

Proponents of this theory would no doubt argue that the epithet or qualification "good" is needed to complete the theory. Desire alone is insufficient: only "good desires" in the sense of moral values have worth. This qualification defeats the theory however. Desire must be axiologically or normatively corrected. Thus a standard of value or of right other than and beyond desire or will itself is required to give them value. "Good desire" cannot come from desire itself since it needs correcting. But then will is superfluous for determining value, since any psychological state, action or even actuality can be corrected according to the non-conativist standard.[14]

Since desire is infinite, it cannot create virtue. The soul of virtue is wisdom and moderation in choices. Desire is unlimited and thereby cannot be moderate: it can never of itself limit appetite with what is only moderate. Moderate greed is an oxymoron. Wisdom is essential to choosing moderately, that is, virtuously. Wisdom is needed to determine what is appropriate under the circumstances. Wisdom ruled by desire is mere prudence and a subversion of wisdom, since wisdom must at times oppose desire, and should regulate it. Those circumstances are especially prevalent in the present era, when greed has denuded the environment at unprecedented levels. Wisdom, moderation, and learning how to limit desire as traits of character are relevant as never before.

Philosophy has too often taken the "natural" model as the most real—including the model of primitive desires. Civilization, society, and culture instill some "desires," regulate others, and prohibit still others. Raising children is both a socialization and an acculturation process, and includes filtering out some desires from others—of informing and regulating such desires. Desires are not basic, but informed by social values. I am not denying there are desires, but only that desires are "given."

An ostensive alternative describes economic life in terms of "needs," instead of demand.[15] However, "needs" talk substitutes norms for what are in truth values. "Need" is a normative standard of what is needed or required for human survival. This standard is based either on some model of "natural" humans, or a historically evolving model positing new "needs" that ostensibly arise in response to new conditions (Georg Hegel and his followers).

What few have noticed is that these models contradict one another. The "needs" model also involves a category mistake, for it uses some model of a "natural" human to define economic life. But this conflates biology and economics. Our biological needs are few: food, sufficient warmth (in non-tropical habitats), and perhaps medicine. Economics may build on biology but what ends we pursue above what is required to keep us alive are purely contingent. All other activities to improve the quality of life are values, not needs, the attempt to make life better. Most economic commodities are not in the least required for survival, whether personal automobiles, electric razors, or virtually all of the goods available in modern economies. So-called human "needs" are values, as humans "need" much less in the strict sense than they pursue. Economics should not be confused with biology. Similarly, other social categories do not serve needs but values. Religion, art, intimacy, and other institutions serve valued ends, but are not required for biological survival.

Most human pursuits, and especially economic goods, are not biological *per se*: instead they are the products of evaluation. The emergence of new values is creative, not based on survival. Wealth, a better standard of living, involves evaluation, not a utopian model of human nature based on life in the jungle. A need is such only in relation to an end: contingent, not necessary. Some might argue that an automobile is a luxury unless a person lives fifty miles from work and there is no public transportation. Then she "needs" a car. Or, an occupation requires a car: sales work, cab drivers, and so on. But even in this equivocal sense, "need" is not a basic category but derivative from other values: ends, such as the kind of work a person has chosen, the kind of life, and the place one lives. There can be no "system of needs," only a system of values. The phrase, "new needs arising" disguises relative improvement in living conditions behind a façade of "changing material conditions." "New needs" implies a rising standard of living, a relative improvement. Special things, before thought rare or beyond reach, come to be expected as a concomitant of a way of life.

The talk of "artificially stimulated" needs, in terms of advertising, ignores the artificiality of all "needs" except very basic, biological ones. The consumer economy is based on the artificial inflation of "needs" by selling images—exploiting sex, for example—using social sciences.[16] But the latest fashions, machines, and other trinkets are hardly needs—our ancestors survived for centuries without them and our life reflects that survival.

Economic values are contingent and culturally determined. Thus they are not necessities and economic life is not a "necessary" category, but contingent upon values. The diverse goods people make, purchase, or exchange reflect evaluations of what will improve life in some respect. The actions people take in pursuit of usable goods—in other words the motive for economic activity—must be values, not desires, wants or needs. History and culture point to this: different epochs, culture, and persons prize different things which are not needs in any biological sense and whose differences point to contingency and cultural determination. Novel things go far beyond the "given" connoted by "wants." Most desires are for artificial goods. The point is that such goods have been *created*; they have been of sufficient worth to supercede primitive "needs" and to generate the activity required to make such goods on a large scale.

Economic "needs" are far from the only human "necessities." Anyone who believes that economic production and consumption alone are the only necessities would have to be willing to do without other categories' benefits—to work at an assembly line, eat, and have no recreation, association, spiritual relief, satisfaction of curiosity, and other intellectual goods, pleasure, beauty, and so on. In other words they would have to do without ends other than economic ones, which economic ends normally serve. We do not work for its own sake, but for the sake of its benefits. That everyone doesn't need all these "needs" shows that they are not actually "needs." They are distinct, categorial goals made through association, pursued by people in recognition of the value of the non-economic categories.

The paradigm of needs accompanies ideologies that advocate a planned economy in one form or another. (That this is not "socialism" in the classic sense, but managerialism, I need hardly remind readers of George Orwell and James Burnham.)[17] A totally planned economy, run ultimately by a bureaucratic minority without alternative means, can prescribe "needs" without the problems of competing with alternative supplies for demand. Oligarchic technocracy (John Kenneth Galbraith) establishes needs by planning, just as in bureaucracy—that is, prescribing "choices" through market domination.[18] The paradigm of "needs" is used by bureaucratic managers, whether governmental or corporate, to control markets.

The "needs" paradigm, like the demand paradigm is technocratic: man has needs, technology alone can fulfill them, therefore prosperity and progress equals technology. "Needs" means the needs of a society organized like a machine: total interrelation of parts. Needs thereby reinforce technocracy. Imposed needs become a public standard of technocrats: technocracy. Needs are rationalized desires and therefore the other side of the modern economic coin. Thus the critique of needs can be a part of the virtue of thrift.

Both capitalists and socialists overvalue economics due to the force of interest politics. Economics is but one category of social life, and more on the side of a means to an end than valuable for its own sake, as some find friendship, companionship, good conversations, beauty, pleasure, justice, and other ends. Economics is a means to these other ends, not an end in itself. All else is a prescription: a positing of normative standards of what is "needed." In practice, this means what is in the interest of the bureaucracy ("planning," "marketing," "welfare"). The basic economic choices are not capitalism versus socialism in any case, if it ever was, since both of these systems have tended toward oligarchic managerialism. Moreover, both are anthropocentric in their view of values and regard the non-human world as a field for exploitation. They are both growth oriented, not thrift oriented; consumer, not virtue oriented. The issue is whether economic pragmatics will be within moral limits or ruled solely by greed and other vices—whether the goal of general prosperity will be a norm for evaluating any actual or proposed economy.

The main opposition to environmentalism has come from economic interests. They have proven a fanatic foe, to the point where they have financed misinformation about global warming, pollution, and other ecocrimes. They are fanatic in the sense that they would rather let global warming proceed and ruin the planet then take the necessary steps to prevent it. Any threat to their profits is interpreted as a view that must be discredited, instead of a challenge to overcome. Moreover, modern economies have sold the idea of the affluent consumer society through advertising, and thus have enlisted much of the public as allies in the war on the environment. This faction views wilderness as land to be exploited, rivers as irrigation water or a dumping ground, other species as potential food, forests as resources, and the environment in general as a means to obtain wealth. The value of the environment as such to a symbiotic species such as humans is ignored—to speak nothing of the intrinsic value of other species.

2. Thrift

Anyone who mentions the word thrift in modern conditions might appear to be dealing with trivialities. Some might regard it as naïve and presumptuous to try and resurrect such an archaic virtue. The idea of thriftiness echoes with faded images of bygone times, with tattered clothes made to last, and the saving of odds and ends, which may never be used. Thrift is superfluous in the "throw-away society." "Saving for a rainy day" and "waste not, want not" are outmoded principles given easy credit, overproduction of food, and welfare state handouts. In sum: grandmotherly virtues. Our attics, closets, and basements may be full to overflowing with goods that are no longer used nor worn out. The government pays farmers not to produce. Economists have noted the dated value that attaches to thrift. Prof. Galbraith implied that this virtue was showing signs of age in the face of affluence and the abundance of machine production.[19]

Thrift might suggest a virtue confined to a time and a place, that is, a virtue whose time has passed. It might indeed constitute a virtue but only under much earlier economic conditions, in which there was less wealth. Thrift was a virtue at the time of the rise of the "Protestant Ethic" and the resulting "Spirit of Capitalism" (Max Weber, 1977) with its high regard for thrift as a means of capital accumulation. It may appear like a historical curiosity today, an era of multi-national corporations with vast financial resources.

Yet thrift is an underrated virtue. In the Third World, thrift is still widely practiced. The poor, including students who first get by on limited means, may practice thrift as a necessity, even a way of life. In view of contemporary ecological concerns, thrift may prove to be the central economic virtue of the future. There may have been more than a bit of virtue in Grandma's attic. This is not to deny the economic side of thrift, however. In the modern context, thrift is not only an economic principle but also environmentally friendly. Given that humans live in an environment and conduct their economic lives in an environment, thrift has moral significance. Thrift may be a required virtue: a duty, which we are obliged to practice. In turn this may serve as a model for evaluation of choices in the face of ecological concerns. I will address these points below.

Some have argued that the motive of affluence, including spending, is the motor of modern economies. However, such spending is based on an initial thrift, which allowed accumulation of a surplus. Capitalism began as ethic of thrift, not only consumption. By this standard, managerialism is a corruption of the primal virtue that made capitalism possible. But if this virtue was required for the development of early capitalism, then independent economic life requires specific virtues and thereby virtue in general. I am not arguing that capitalism is equivalent to economic life, the mistake of the bourgeois economist. The point is the virtue required for any economy independent of government. Spendthrifts, who are profligate with their income, will soon come to ruin and their enterprises will fail.

The first question is what thrift means and whether it is a virtue. Thrift involves saving and the wise use of resources. To take saving first, saving may take the form of preserving or caring for goods that can be reused, whether monetary or other goods. Its opposite or contrary is wastefulness, the deliberate or reckless misuse of goods or resources. This is an economic virtue, since wealth and prosperity are preserved. I would argue that thrift is also an environmental virtue, since by saving and preserving we are using fewer resources. The saving of resources results in a reduction of predation on new "resources," including animals, plants, and habitats. This is wise, since both wealth and wild are conserved. Wise use of resources involves deliberation, evaluation, and choice of the best alternative for preservation and saving. Just as paying less for the same item is economically thrifty, so is reduction in the cost to the environment ecologically thrifty. Saving old clothes may save us the cost of new ones and the cost to the environment of producing new ones, whether overgrazing by sheep or use of pesticides and fertilizers for cotton.

Is thrift a virtue? No mention of it can be found in the originary document of character ethics, Aristotle's *Nicomachean Ethics*, one of the most comprehensive treatments of virtue in Western Philosophy. Aristotle thought that a virtuous mean of generosity or liberality lay between profligacy and stinginess.[20] If thrift is identified with stinginess or meanness, it might even be a vice. However, the context of Aristotle's treatment of this virtue argues against this view. Liberality with money involves social relations, in which generosity may indeed be a virtue. The virtue is concerned with the "spending and giving" of wealth, not the "taking and keeping" or "possession" of it (*N.E.*, $_{1120}$a8-9). More telling is Aristotle's treatment of profligacy, which is treated at greater length than stinginess. Aristotle implies that the profligate does not husband resources wisely, and thus soon falls short of money. The profligate is wasteful, by not preserving enough resources, and unwise in the care of them. The implication is that generosity requires some thrift and the wise use of available resources.

The modern period regards thrift as a virtue.[21] Thrift traditionally had to do primarily with excellence in economic affairs. This older economic form of thrift includes the idea of saving or conservation and shares with "ecological thrift" the principle or obligation to reduce or eliminate waste. Thus it may be a basis for both recycling and achieving a sustainable economic life, or be consistent with sustainability. A thrifty person may be the kind of person who is suited to a deep green or eco-friendly life-style. Thrift may be a virtue or even the virtue appropriate to conservation, sustainable economics, and being a good neighbor to the wilderness. Thrift may be another word for conservation or at least a major part of it. In another tradition, eco-feminists have stressed caring for the environment as a suitable approach. Thrift has the result of caring for the environment, since less use is made of new goods.

Waste denudes the environment; if not the only source of misuse, it hastens the process. Such waste becomes ramified through the interlocked economy, contributing to environmental stress in several ways. The more meat which is thrown out, uneaten, the more animals which must be slaughtered to fill this demand and the more grain used to feed them. In turn, this involves greater use of fertilizers and pesticides, greater demand for water, less land available for wilderness, possible erosion and contamination of water supplies through animal wastes, and many other bad effects on the environment as a whole. This is to speak nothing of the misuse of animals as "resources," to borrow a phrase from Tom Regan. To slaughter animals needlessly, because the food is thrown out anyway, is beyond callous.

Waste is not only an economic vice. Wasted paper means destruction of trees, devastation of their soils and habitats, threats to wildlife dependent upon such habitats, and the like. Sharing with economic thrift the avoidance of waste, ecological thrift would exhibit a greater concern with the effect of waste on the environment. It takes as much as eight pounds of grain to make one pound of beef. Clearly this is not a thrifty way to produce food, since a great deal of land must be used to produce a small amount of product. Thrift demands less cost for more goods, feeding more people for less. Beef has questionable health benefits, utilizes animals as a resource, and adversely impacts numerous habitats, particularly in fragile desert lands and Amazonia. Moreover, waste is also esthetically wrong in its misuse and scarring of landscapes, and devastation of animal life. Pictures from the nineteenth century showing slaughtered bison rotting on the plains are morally outrageous, not just wasteful.

Ecological thrift would result in less waste but this waste would be evaluated in larger terms than economic costs alone. The cost of waste to the environment would be factored in. If goods are misused, they are not being used wisely and this is not the best use of them. Would a cheaper process of manufacture mean greater production of toxic, hazardous or otherwise dangerous chemicals? Then it would not be ecologically thrifty, as its cost to the environment would be prohibitive, especially if detoxifying such chemicals is impossible, hazardous or excessively costly: if it exceeds that of their manufacture. The use of plastics made from petroleum wastes a finite resource, when recyclable alternatives are readily available.

Waste is economically wrong because it does not make the best use of resources. The best use of goods involves evaluation of what would constitute their optimum use. Their optimum use includes the least impact upon the environment as well as the least costly production and disposal of the good. However, recycling materials is generally cheaper than manufacturing goods from scratch, both in economic and ecological terms. Further, if plastic, glass, and other materials are not recycled, they are wasted. Recyclables are themselves resources or potential goods and overlooking them is ultimately wasteful. "Landfills" are wasteful uses of habitat, since almost everything can be recycled. Recycling is economically thrifty but even where it is not, it may be ultimately thrifty because of reduced impact upon the environment.

Evaluation of what constitutes waste must include considerations of the energy costs involved in recycling. Not the least of these is increased demand for electrical power, which has usually been met by construction of coal or nuclear generating plants. If the danger of these forms of power is granted, then we have an ecological "imperative" to reduce demand for electricity. This should override attempts to increase supply which utilize environmentally unfriendly means. One way to accomplish this is more thrifty utilization of it: both saving and reducing demand. Again, hydropower, while cleaner, has devastated the salmon runs in Western rivers and eels elsewhere. This is not ecologically thrifty, since the cost to the environment is unacceptably high.

Not all eco-friendly authors regard nuclear with horror, since it is renewable and does not cause a buildup of greenhouse gases. However, it does cause a buildup of some of the most toxic and dangerous substances known, and these cannot be detoxified. Nuclear power is not an optimum alternative, even if it does have its points. Construction of nuclear plants involves the use of other toxic materials, such as lead for shielding. Moreover, the longer-term economic costs of waste would be weighed against immediate costs. This would include considerations based on obligations to future generations. Our affluence should not be at the price of that of our descendants. Here the economic meaning of thrift still has significance in terms of generations: not our generation having enough, but all generations having enough through conservation.

A further consideration in this context for ecological thrift involves the distinction of recyclable from irreplaceable goods, finite from sustainable resources. Larger cars not only cause greater pollution but also waste irreplaceable gasoline and oil. And it is well known that the world's supply of crude oil is being depleted more or less rapidly. An economy based on irreplaceable goods is not sustainable.[22] The use of some resources will doom them in the end: their replacement cost is so high that it would be prohibitive, even given the requisite technology. While coal might someday be used to manufacture petroleum, the long term cost of this would be impossible to meet, since this too would eventually give out. In the long run both petroleum and coal are finite and irreplaceable, even if functional substitutes are found. But their environmental costs are even higher: the cost to the environment in terms of global warming from the use of fossil fuels. Use of irreplaceable resources when alternatives are available which can be sustained or recycled is ultimately wasteful, its ecological cost unacceptably high. The alternatives may also pose environmental risks. Further, their use may be denied to future generations.

In a sense of course, the law of conservation of matter and energy implies that nothing is truly lost through use. However, its value may be lost. This is an argument for recycling, since the ingredients of goods can be recycled into new goods.

Is there an obligation to be thrifty, or is thrift meritorious but not required by moral agents? Is it commendable but not dutiful?[23] Economic thrift is usually considered commendable but not required.[24] However, in some cases professional duties require it, for example, investment trusts, or any situation in which the funds of others are entrusted to professionals. If money is wasted there may be grounds for lawsuits. Although proprietors are not required to be thrifty, in conditions of severe competition thrift may become a requirement. I would argue that thrift is a duty in an environmentally conscious age, since we are required to be thrifty in order to minimize our impact on the environment, conserve for future generations, sustain and make the optimum use of resources. Thus thrift is a worthwhile goal, not only a means.

To what degree should we be thrifty? Can too much thrift be a vice, as Aristotle might argue? Thrift is a universal virtue, which covers all of economic life. More, it concerns our entire relationship with the world around us. It breeds respect for the environment since environmental impact is proportionally minimized to the extent that we increase our thrifty ways. Although wearing rags to save the cost of new garments might seem excessive, it is difficult to see how thrift could be a vice, unless it were to become harmful, for example, if insufficient clothes were worn which caused illness. Yet sewing an old garment or fixing an old car to preserve it is virtuous from an environmental point of view

Is economic growth that includes maximum thrift a legitimate goal? What if it conflicts with our obligations to future generations? Should social utility outweigh our obligations to future generations? Does "best use" refer to immediate or long-term use if these are in conflict? Sustainable growth is the optimum goal.[25] However, utilizing the standard of ecological thrift that we have outlined, the question becomes whether growth at the expense of living things is virtuous? The use of trees for construction may be sustainable, but also involves costs to the environment and use of living things as resources. More, there is considerable waste in present construction practices as a glance at any trash receptacle at a construction site would reveal. Since alternatives to lumber are available for use in construction, including brick, stone, stucco, sheetrock, glass, and metal for framing, the use of trees may not be acceptable from the perspective of ecological thrift. Growth has had the result of the rape of the environment; only if it were reformulated to include environmental costs, the use of eco-friendly alternatives, caring for the environment, and sharing with our descendants would it be a legitimate end.

Is thrift equivocal in meaning? Does thrift in the sense of saving conflict with thrift as wise use in the present? Is the best use conservation or "maximizing utility?" Hopefully, both are feasible, and wise use includes recycling for future use for example, in the supply of glass. In the case of irreplaceables as little use should be made as possible, so that such resources are shared with future generations. However, if saving and maximizing utility conflict, saving must take precedence. The story is told of a plant that used to grow in the Mediterranean region, which induced abortions. The plant was harvested into extinction and is no longer available to us. A similar fate may await the *Taxus* trees of the northwest, if their anti-carcinogenic properties are confirmed. This maximization of utility is hardly a wise use.

The counter-argument against any eco-friendly sense of thrift is that this may decrease sales, so that prosperity will suffer. In the long term, however, ecological thrift would be economically beneficial since it would be both sustainable and result in reduction of waste. This could be phased in over a period of time such that negative economic impact would be reduced. It should be kept in mind that thrift is originally an economic virtue. Thrift is an alternative way of creating more wealth.

Both thrift and demand should be weighed. However, saving may involve sacrifice and giving up desires for the sake of virtue. But desire raises the issue of whether all of the desires in the modern market are worthwhile. The evidence is they are not: consider the desire or demand for drugs. In any conflict, the standard should be the requirements of life as a whole, rather than desires. Only such a standard is compatible with an eco-friendly way of life. This principle covers the desire for an excessive number of offspring, population policy. The desire for too many children may be economically feasible for some individuals but is ecologically harmful. Thus its ecological costs are too high and impact upon the environment too great.

We can adopt an expansive definition of needs which gets larger and larger, for example the "need" for an automobile in the suburban life-style. But this is hardly a need in any genuine sense, anymore than a "need" for a yacht. The question arises of who is to determine such needs? The standard for "needs" must remain biological, since need implies necessity, and all necessities are biological, needs of life. However, this does not rule out all fulfillment of desires. But many of the ostensive “desires” people have are created and implanted by advertising and other media.

3. Consumer Society

How does our age measure up, then, by any standard or on any scale of virtue? With some qualification, the consumer way of life promoted by the media, and paid for by economic interests promotes the pursuit of self-indulgence at the expense of virtue. Many have chosen a life of consumption, including the consumption of drugs and alcohol, not a life of virtue, or even of self-development. The use of mind-altering drugs was initially for the purpose of consciousness expansion, especially in opening up the spiritual dimension. The spiritual message of the sixties was misinterpreted as hedonism, with drugs replacing alcohol as the substance abuse of choice. Indeed, in some cases both are used together! This substance abuse is the life-style that the counter-culture of the sixties began as a rebellion against! Consumers have in some cases gone into debt to pay for their consumer goodies, even to the point of bankruptcy.

Above all, the consumer way of life as practiced means the sacrifice of the environment, necessary to all life, in order to lead a life of conspicuous consumption. The nihilism of consumerism almost defines decadence, wherein the result of consumption at unnecessary and unsustainable levels will be death. In other words, the consumer way of life is decadent, since it involves preference for a slow death to moral improvement. With notable exceptions, consumers as such are indifferent to any but economic values, especially intellectual and aesthetic values. Despite an earlier exploration of spirituality, the spiritual message of the sixties, liberation from dead rituals and traditions, was sublimated into consumer materialism; liberation became self-indulgence. Choosing among market trinkets, cheered on by Wall St. and encouraged by Madison Avenue, is somehow supposed to make a person free, to promote self-expression. Free will is reduced to choosing between marginally different automobile models or hairdryers.

Despite well-known and well-publicized warnings about the effect of greenhouse gases in accelerating global warming, contemporary humans keep right on consuming, rationalizing greed. They pursue status, with huge gas-guzzling cars, knowing about the greenhouse effects and other deleterious impacts on the environment. They live in ridiculously large houses, despite the negative impacts on the environment of suburban sprawl, and housing shortages among the less fortunate. They use immigrant labor in their homes and businesses, knowing it is illegal, and knowing that it will result in too many people, to avoid paying a decent wage. Greed has won out over concern with either the environment or fellow citizens.

The premise of the Enlightenment and of economic materialism of all types is that economic lack is the cause of vice and economic prosperity would of itself increase virtue. There were visions in the eighteenth century of as many Shakespeares walking the streets of Paris as of paupers. But this vision did not come to pass. (Only Jean-Jacques Rousseau saw through this fallacy.) Economics became a sovereign category, just as politics had been, and shaped humans into consumers, not moral agents. Economic life did not lead to excellence but to a "life fit for a pig."

The desire for a lot of money traditionally meant wanting to control necessity.[26] A life ruled by necessity, not freedom is an inferior life. It does not include the pursuit of future possibilities and novel creations, but only a reaction to want, determined by the past. With the advent of the "affluent society," the desire for money has meant controlling the means required to pursue a life of vice. The middle and upper classes complain about high taxes and then many use their disposable income to gamble, buy intoxicants, and other forms of vice. In other words, they want more money to pursue their vices. This sea change is hardly in pursuit of virtue. "Freedom" has meant the freedom to pursue a person's self-interest at the expense of others, in a consumer life-style.

Or in so-called identity politics. Women rightly rebelled against a second place role in society. Some have, like many other groups in our history that have achieved success, left reform behind to live in the suburbs and affluence. African Americans rightly put an end to discrimination but some fail to see that their social conservatism on issues of gay rights is of a piece with discrimination against any other minority. In turn, gays themselves have often pursued self-indulgent affluence, ignoring social justice to speak nothing of the environment. As I noted in the Introduction, "Eco-feminism" subordinates the whole, the environment, to one part, human females. Similarly, "queering" environmentalism puts a small part before a much larger and more valuable whole, because the environment is the condition of all life-styles, including gay ones. Both of these kinds of identity politics are anthropocentric, not environmental so much as an attempt to make some form of identity politics relevant to other issues of far greater purport.

No doubt I will be criticized as anti-feminine or anti-gay. But such a response would miss the point. As tactics such criticisms only disguise the fact that the movements for liberation have, with some notable exceptions, failed to come to grips with the problem that affluence and self-indulgence have deleterious effects on the environment; that "liberation" has been interpreted to mean raping the environment, and that their own calls for liberation must be within an ecocentric ethic, not its source. Believing in liberation for one group, without respect to others is hypocrisy. The "others" in this case are other species. Nevertheless, these movements have hardly had the impact that corporations have had in making war on and destroying the environment in the name of prosperity. Asking for justice for others is not anti-liberation.

In terms of freedom and justice, the non-consumers must be allowed their values if the consumers are to be allowed theirs. If their relative life-style suffers, that is appropriate, so long as non-consumers are allowed their freedom; anything else is unjust. Thus preservation of habitats of rare species should mean that they are off limits to economic development as the price paid by advocates for the affluent society.

Even if such minor virtues as thrift were the product of economic life—itself a dubious conclusion, since they are its condition—such virtues only stand in the way of consumer society and an economy dominated by Wal-Mart and other giants, which have shut down small businesses in every town they enter. Consumers do not save, they spend. They do not defer gratification they demand instant gratification. As for small businesses: the business either comes up against Wal-Mart and fails or, if it succeeds, becomes either an acquisition target for a corporation or becomes a corporation itself (Apple computers). The waste of capital in small business failures, even in economic terms, is manifest.

Economists also use a "consumer" model without virtue, that is, people without character. The "consumer" model goes hand in hand with economism since the ideal is a one-dimensional, passive human that spends his/her time consuming. Economism as demand complements consumption.[27] The model is of "other-directed" humans without a self, who define themselves by the number of economic commodities they have purchased. Their "self" is the number of external goods they possess, in an implicit denial of inherent value. Since their value depends upon externals, it cannot be inherent. The virtue of their characters, to speak nothing of the virtue of their purchases is extraneous to this model. Becoming excellent and developing a good character is irrelevant to economism.

Is the distinction of consumer culture and virtue as I have drawn it a straw man, even a satire? But even if I have exaggerated some features, they do describe trends. The contrast highlights the difference between the way of life promoted by Franklin and that of Madison Ave. The virtue of thrift stands opposed to unlimited growth. Thrift includes preservation of goods for as long as feasible and maximum use, not consumption and waste, the "throwaway society." Only the ethics of economic thrift is sustainable, not growth and consumerism. Thus only thrift and virtue make economic sense. Using up every drop of oil and destroying the earth for our children or grandchildren is contradictory, to speak nothing of irrational. We cannot care for our children and despoil the habitat at the same time.

Consumption at present levels is not sustainable, especially given the simultaneous increase in population in Third World Countries. At some point, population will exceed the carrying capacity of a finite planet, and worldwide disaster for our own and other species will result. Along with population control, a reorientation of the entire economy is requisite to avoid long-term disaster. We should replace models of "growth," as measured in increases in "gross national product," that is, massive destruction of the environment, with models of long-term sustainability. The measure should be "gross national sustainability," including recycling of resources. Automobiles should be built to last, not to fall apart, since the cost of manufacturing their components is thereby earned over a longer period, and the cost of recycling their parts is delayed. In other words, thrift is a part of sustainability. Such recycling could include use of such energy sources as wind and solar, which while not infinite, are free, non polluting and abundant.

"Humanists" will argue that this is "heartless," as well as an interference with "human rights." I can only wonder how "humane" it is to condemn our descendants to grinding poverty, all to preserve a dubious right to have as many children as a person wishes and use up resources at an unsustainable rate at the same time. I can only look on in despair as more and more forests are cut down to house more and more children. Be this as it may, the day of reckoning can only be postponed so long. As Garrett Hardin has pointed out, we can prevent births now or watch as starvation becomes the norm in a not-distant future. Moreover, anthrofacism masquerading as humanism tries to avoid dealing with the problem of human destruction of other species. Up to one-third of other species may be imperiled by global warming.[28] There must be a balance between conserving other species and human affluence, if there is to be justice. One-sided "justice," which only takes humans into account, is no justice at all. Humans have already altered over fifty per-cent of the land, depleted the seas and forests, and use much of the remainder for other purposes, such as grazing. True justice would restore the balance. What is heartless is extinction for the sake of affluence.

Nevertheless, making the poor pay the price for affluence is also unjust, as they will according to scientific projections. This is especially true since they have not been the cause of global warming and will be its worst victims. I will address this point below.

The economy of an ecocentric cosmopolitan ethic will require thrift and other virtues, not consumption and growth, so that human impact on the environment will be minimized, not destructive.[29] Human expansion constitutes a threat to all life, not just human life, through its most conspicuous effects, global warming, resulting in rising ocean levels, the loss of habitats and biodiversity, and the poisoning of the environment. But the environment is a requirement for all of life: we cannot survive without air, water, and soil. A consumer life-style takes without giving back to the world. Moreover, it does not bring out the best in humans and can bring out the worst. Economic interests may accelerate the cultivation of appetite through clever advertising that appeals to fanciful sexual gratification. Cars are sold as potential "chick magnets," instead of on their merits. All of this is the opposite of virtue, in the sense of excellence. But becoming good environmental citizens of the world will require good ecocentric virtues. Moral character allows us to resist consumer lures.

4. Social Thrift

Is thrift a personal virtue? Thrift should indeed be practiced at the level of individuals. However, an imperative of personal thrift is not enough, since this would only be meritorious. Social thrift is required, in which the society organizes to conserve, recycle, use non-living resources, and reduces the use of environmentally costly foods like beef. Regulations should govern the use of irreplaceables so that future generations can have a share in them. The least possible use of resources is the most thrifty, since they are saved for other uses, for a "rainy day." Wholesale destruction of habitats to increase the standard of living is unjust to other species and future generations. Thrift is not the only eco-friendly virtue, but is one of the most important if we are to be good citizens of the planet.

A good society is a moral one, that does not allow freeloading by either a criminal or an exploitative class, but rewards everyone. However, social justice must be within the limits of environmental conditions. We cannot divide the pie more evenly if we destroy the pie. Nor will there be much pie left if we continue our wholesale slaughter of other species. Ninety per-cent of the world's large fish have been caught and sold; in some cases the devastation has been so great that the stocks of some once-common species may never rebound.[30] Destroying the oceans to feed more mouths will not work in the long run and is therefore foolish. Environmental thrift is the condition of social justice and therefore must take precedence. Again, if "social justice" means families with eight, ten, and even more children in the Third World, then the long-term consequences will be famine, especially in agriculturally marginal areas like the Sahel. Social justice theorists who do not recognize this are at best shortsighted.

Why use government? We need government intervention and strict regulation of pollution, carbon emissions, and the like in order to offset the advantage freeloaders might obtain. Otherwise one company would do its duty and install anti-pollution devices or carbon scrubbers, while another would not. The one that did not would have a cost advantage over the conscientious one. Only legal regulations can get around such a slow death to virtue and conscience.[31] Proposals from ideologues like "carbon-trading" and "free-market" incentives are smoke and mirrors, since they will not accomplish the objective of a significant reduction of greenhouse gas emission. They are fraudulent attempts to hide the issues and divert attention from real reform.

In sum, an ecological ethic should include thrift, the least wasteful use of resources, with an eye on the environment. "Waste not, want not," might be replaced by "waste not, destroy less and sustain more" in the ecological view of thrift. Being thrifty is saving and this provides a basis for conservation, both for the sake of non-human life and of future generations. Thrift is doubly excellent as the condition of genuine economic growth and also for living in an eco-friendly way. Thrift is a "win-win" virtue since it increases wealth while reducing negative impacts on the environment. Long-term growth requires a better use of resources, and the replacement of finite resources with sustainables. Indeed, it might be argued that the condition of possible economic growth is precisely the condition of possible sustainable economic life. Sacrificing long-term prosperity for short-term costs is not truly thrifty, since less is saved. However, ecological thrift has priority in any conflict with economic thrift. Ecological thrift involves minimizing environmental impact, by reducing demand or desires for goods, especially the use of irreplaceables. The effect will be that they can be shared over time with our own kin. Thrift is both a personal and a social virtue that is morally required: everyone should practice it. Indeed, thrift for the sake of the environment can even be justified on egoistic or self-interested grounds, since it increases personal wealth and may contribute toward the conservation of the environment in which we all live.

Far from constituting an archaic virtue, thrift is the excellence of an environmentally sensitive age. It is the heart of wise use of resources. We should revalue it radically by adopting it as a premier virtue for the coming age of sustainability. Thrift is both a good trait of character and an ecocentric virtue.

Ten

THE CULTURE OF DEATH[1]

> Nor is there much satisfaction in contemplating the world with nothing left to the spontaneous activity of nature; with every rood of land brought into cultivation, which is capable of growing food for human beings; every flowery waste or natural pasture ploughed up, all quadrupeds or birds which are not domesticated for man's use exterminated as his rivals for food, every hedgerow or superfluous tree rooted out, and scarcely a place left where wild shrub or flower could grow without being eradicated as a weed in the name of improved agriculture.
>
> John Stuart Mill[2]

> Ecologists recognize that reducing the planet to a resource base for consumer use in an industrial society is already a spiritual and psychic degradation. Our main experience of the divine, the world of the sacred, has been diminished as money and utility values have taken precedence over spiritual, aesthetic, emotional and religious values in our attitude toward the natural world.
>
> Thomas Berry[3]

1. Human Expansion

I believe that there is a profound sickness in human culture. When humans spread to a new area they destroy or alter the environment with an easy conscience. European culture is not alone in its rush to destroy. Humans in general have hunted species to extinction, for example the death of the mammoth in North America before the European conquest. It could be that this was the result of a meteor shower, as some recent evidence suggests. But the coincidence of the arrival of humans from across the Bering Straight and the extinction of the mammoth is hard to ignore. Agriculture also altered habitats in many different cultures. Thus the European conquest of the New World is only a prominent example of a much larger tendency. As Jared Diamond has stated:

> Undoubtedly, two simple reasons go a long way toward explaining our worsening (environmental) mess: modern technology has far more power to cause havoc than did the stone axes of the past, and far more people are alive now than ever before. But a third factor may also have contributed: a change in attitudes. Unlike modern city dwellers, at least some preindustrial peoples…depend on and revere their local environment.

> [However] it's now clear that preindustrial societies have been exterminating species, destroying habitats, and undermining their own existence for thousands of years. Some of the best-documented examples involve Polynesians and American Indians, the very peoples most often cited as exemplars of environmentalism.
> While courses in the history of civilization often dwell on kings and barbarian invasions, deforestation and erosion may in the long run have been more important shapers of human history.[4]

Diamond notes that human predation includes the destruction of habitats around the world, from the destruction of the southwest Asian forests—along with the desert lion—and desertification of its lands to that of the North American Southwest. Other examples of decimated species decimated include the moa in New Zealand and the mammoths, native camel, and others in North America. (Diamond, 1992, ch. 17). There is also the rise and fall of the culture of Easter Island, which chopped down all the trees it needed for fishing boats, and then was unable to feed itself. The destruction of tropical rain forests is a current example of the same tendency.

When the British and Spanish colonists came to the "New World," they brought their social, economic, and religious institutions with them, including especially the European animals and plants with which they were familiar. These included the typical farm animals that characterize Euro-American livestock: cattle, horses, pigs, goats, sheep, and chickens. Only the turkey, among native species, found a permanent place at the colonists' table. Similarly, wheat, barley, rye, oats, and other European staples were imported to replace native plants; only maize and potatoes were adopted into the European cuisines (the tomato and pepper were adopted by the colonists ruled by Spain, but only spread north later). The legacy of this mass importation of non-native plants and animals is the widespread occurrence of such non-natives as wild carrots, (*Daucus carota*, Queen Anne's lace) along North American roadsides, an escapee of colonial gardens; and of wild mustard (*Brassica* spp.) in most of the milder regions of North America, which was first spread by the Spanish missionaries.

Similarly, non-native fauna sometimes escaped their pens and went wild. Among others, the pigs and horses became "naturalized" in North America, which allowed modification of the economies of the Native Americans of the Plains. The Lakota and other cultures used the horse as a more effective means of hunting and raiding. The wild boars wrought havoc in the native landscapes, consuming bushels of acorns, especially in California, among other consequences. Predators that might have kept the wild hogs in check, notably the wolf, were killed off.

A more direct consequence was the massive uprooting and destruction of the native flora and habitats in favor of European agriculture over much of North America. Europeans found a landscape covered in "the forest primeval." In what was to become the United States they proceeded to cut down almost every tree in a massive area from Maine to Florida, and west to the Prairies, leaving only tiny remnants of Old Growth forest. Forests were replaced by farms, even in areas not well suited for farming, like the glacially formed, rocky hillsides of New England, where miles of stone walls in the woods attest to early attempts to wrest crops from thin soils in areas with short growing seasons. As the forests were felled, the land they grew on was subordinated to human purposes. Aliens replaced native species in a herbivorous genocide. Attractive native species like the lilies of the northeast were ploughed under. Wetlands, with their diversity of species, were filled in. The Euro-Americans even attempted the massive project of destroying coastal estuaries, the probable source of all life, such as the New Jersey Meadowlands, previously teeming with birds and other species.

After the invention of a better plough, the prairies were also brought into agricultural production, as were the drylands of the Far West and Pacific Coast. Agriculture in California and other dry states required massive water projects. Euro-Americans were equal to the task: they drained entire lakes and rivers to provide water for the farms of the Central Valley of California and the Orange orchards of the South. Tule Lake in the lower San Joaquin Valley and Owens Lake in the Owen's Valley were completely drained, killing off species that used to thrive in and around them (Tule deer). The forests of the West, home to even larger trees than those of the east, like the massive Giant Sequoia, the Coast Redwood, and the Douglas Fir, were seen, as Tom Regan put it, as resources to be harvested, not as prizes to be treasured. Only ten percent (10%) of the Old Growth forests of the West, home to rare species like the Spotted Owl, remain, but they are not yet sacrosanct. Lumber companies see the old trees and drool, and have attempted to harvest the small remaining forests by paying off the Bush Administration. Despite their propaganda concerning replanting and the like, the lumber companies efforts have created tree farms, not forests, as anyone can see who visits many of the areas "replanted" by the companies. An acre containing only one species of tree, selected for its commercial value and planted for maximum yield, does not a forest make.

The subordination of a Continent was totalistic, including concentration of Native Peoples on reservations, elimination of problematic animals like the wolf and bison, uprooting and killing of native plants in the clear cutting of the forests, and total alteration of native habitats throughout the Continent for agriculture. Since the Native Americans of the Plains used bison for food, the U.S. government, with pressure from agricultural interests among others, adopted a deliberate policy of destroying the species that once covered the plains. General William Sherman, who stated that "the only good Indian is a dead Indian," argued for this form of economic warfare against the original inhabitants of the land. He hoped that by eliminating the Natives' "larder" to bring them to their knees and force their military capitulation.

Subordination meant total destruction of the original North America. Whereas Native Americans for the most part adopted their culture to their habitat, Euro-Americans, perceiving the wildness of America as a threat to be tamed and subordinated, were not content until almost every tree was chopped down.[5] They were not satisfied until almost every wolf, whooping crane, eagle, grizzly, and bison was shot. Christopher Manes, in the Epilogue to *Green Rage*, also notes the destruction of the passenger pigeon, the reduction of the bison herds from millions to a few thousand, the elimination of the grizzly from most of the U.S., and other ecocrimes. He notes that "it had taken ten thousand years for the inhabitants of Europe to replace the post-Pleistocene splendor of their ecology with the squalor of industry, domestication, and overpopulation. Their American offspring were much more diligent: we accomplished the same task in a mere three centuries" (Manes, 1990, p. 244). Euro-Americans would not stop their total subordination until almost every acre of the prairies was under the plough. They would not end their genocidal rush until almost every vestige of wild America was contained, its "wildness" restricted to tiny "preserves" (David Foreman, 1991, p. 38). Lands too marginal for agriculture were not left alone but used to graze livestock, resulting in a degradation of massive proportions in the fragile desert landscapes of the Great Basin, among others.

I chuckle when I hear claims of how much Americans love their land. In truth they hate it, for they do not want to leave even a small fraction of America unaltered by methods and practices motivated by ideologies that originated in European countries. What they actually love is to change America to what it is not: home to European species, methods, and other foreign introductions. American patriotism is a salute to the massive destruction of North America effected by Europeans and their descendents. It was not enough to deforest the British Isles and Spain. The expansionist and elitist ideologies of Europe and the pressure of overpopulation meant that the life-style of these empires had to be exported to the colonies. The colonies were to be subordinated economically, socially, and spiritually (Mercantilism). The New World would become a clone of the old.[6]

Both the English Puritans and the Spanish Catholic Church were hierarchical in their worldview and in organization. Both believed in a hierarchical universe with God at the apex. Both excluded women from the ranks of the clergy and believed in theocracy. This theocracy consisted more in a theocracy of influence and guidance, than direct rule by clerics, but was no less severely enforced than a more direct theocracy: the Inquisition and the Salem witch trials. The different ranks in society, from the monarchy and nobility down to the peasants, were viewed as divinely sanctioned. Both English and Spanish societies were hierarchical, with a land-owning nobility of great privilege and with laws designed to retain such privileges. Finally, both the Puritan and Spanish male clerics believed that they were possessors of an exclusive truth, available only to an elect, and that perpetuating this truth in society, and even among the "savages" was a divinely ordained duty. This hierarchical mode of organization facilitated the subordination of the environment to economic and political purposes.

Patriotism is a salute to the genocide of the Native Americans and appropriation of their land. Despite all the fuss made about the "right" to "private property" by the ideologies of European origin, religious and secular, the "private property" of the natives was considered fair game for theft. The remnant were concentrated on "reservations," the origin of concentration camps in other, more recent events, from the Boer War to World War II. Worse, American patriotism represents the triumph of cutting down almost every tree in the Eastern United States over the course of 300 years and many of those in other parts of North America, Canada, and Mexico. It is a salute to the extinction of the passenger pigeon, the Carolina parakeet, and several other species; and the near elimination of many others. Even the bald eagle, adoptive symbol of the new nation, recently came close to extinction due to agricultural practices. The "beautiful for spacious skies" of the patriotic song are willingly sacrificed for cheaper power to run air conditioners to keep Americans happy in the climate that their own practices indicate that they despise—for why alter a climate you love? The "fruited plain" is not America, but the triumph of materialism and consumerism over the bison, the flowers of the field and the "savage" Native Americans of the Plains. "Manifest destiny" is the ideology of murder and death, but is far more totalistic than mere conquest of Native bands. Euro-Americans have attempted to destroy native flora and fauna, and their indigenous habitats: continental genocide

When I visit the Catskills or any other range in the East I am struck by two thoughts. One is the evidence of the "resilience of nature," since despite the near clear cutting of these mountains in the past, the forests have endured. They have returned in areas that are not kept constantly clear. The other thought is the mentality of those who would destroy a forest for greed. What sort of persons would cut down every tree in a huge forest for monetary gain?

Human culture suffers from a profound illness. This cultural sickness pervades all the areas of culture. In philosophy it appears as egoism in different forms, as if the subject were autonomous from society and the environment. Making the subject "happy" by subordinating the value of the entire world to this ego is seriously defended.

In religion the sickness is exhibited in the ideology of "be fruitful and multiply," of subordination of all of creation to the creature in the "divine image," and in the idea that the earth does not matter, since humans are here only "temporarily": their destiny lies in another world. Collectively, this means that religion is used to justify the transformation of the earth into a giant factory for human consumption. Overpopulation is encouraged, since in theory, other species live only to serve humans; and in practice, each sect wishes to outbreed the others, for power and influence. In our age, we should not be held captive to the deluded imagination of some ancient scribe in which we are forced to live by the outmoded religious standards of a bygone age. If we can ignore their polygamy, their practice of slavery, stoning adulterers, and many other practices, I do not see why we have to overpopulate.

It may also be indelicate but truthful to mention that the view of creation in the Old Testament is wrong, throwing the entire Judaeo-Christian-Islamic tradition into doubt. Humans are not in the divine image, but lucky apes. (One would hope that God does not have the face of George W. Bush). Humans are a symbiotic species, albeit a successful one, that evolved along with other species and is not especial. As Diamond put it, "Humans used to have a typically mammalian restricted distribution, in warm, nonforested areas of Africa. As recently as fifty thousand years ago, we were still confined to tropical and mild temperature areas of Africa and Eurasia."[7]

We are African in origin and thus our epistemological capacities are universal among humans—they evolved in Africa—but our metaphysics is diverse, based on local habitat. The Inuit are Africans facing a "frigid" environment or reality. Culture is a response of Africans to habitats. As Diamond put it:

> This geographic fidelity of animals is reflected in their geographic variability: populations of the same species in different geographic areas tend to evolve into different looking sub-species, because most breeding remains within the same population…
> In these respects, we humans have been typical animals throughout most of our evolutionary history. Like other animals, each human population is genetically molded to its area's climate and diseases. But human populations are also impeded from freely mixing by linguistic and cultural barriers far stronger than in other animals.[8]

In economics, the destruction of the rainforests, the world's fisheries, and the use of fossil fuels continues unabated. Expansion of grazing and agriculture to feed excess populations has resulted in desertification of large areas of Asia, Africa, North America, and Australia. Everyone knows that global warming is upon us but act as if business as usual is not only possible but desirable. Aquifers are being depleted, despite excess food consumption in North America and elsewhere. Overproduction of food with agricultural practices that degrade the soil is considered economically beneficial. Science has become a tool of economics, creating poisons to control other species, and bombs capable of destroying the entire biosphere in minutes. Science is equated with technology, and the drive to knowledge subordinated to the drive for profit, luxury, and prosperity. In medicine, keeping humans alive at all costs has meant research into genetic enhancement, cloning, in-vitro fertilization, and other expensive techniques for prolonging human life.

Quality of life is again sacrificed to quantity. Death, a normal part of life, is considered so heinous, that vast sums are spent keeping even very elderly and comatose patients alive.

The latest tendency in Western living is the atomic family living in its suburban box. By contrast, the typical rural tradition involved multiple generations living together, and in urban areas, the sharing of walls between buildings.[9] The atomic pattern in the suburbs has meant an astronomical increase in average energy consumption, since each family must consume separately from every other. The suburban home is inefficient in its use of energy.

To build the typical modern suburban house, a contractor or developer will typically "knock down," that is, kill all the trees on the property. They do not build a path through the trees and clear the area where the house will sit, but kill all or most of the trees on the lot. Houses are generally built of wood, that is, involve the use of lumber derived from still other dead trees, despite evidence that such trees absorb carbon dioxide and, if left to grow, would help alleviate global warming. Moreover, only a minority of houses is made of stone, a much better material, since it insulates better, lasts longer, is not subject to rot, resists fire, and is more widely available. Housing in the older cities of the Old World is often of stone, since stone is more durable and a far better protection against both winter cold and summer heat, to speak nothing of fire danger. Indeed, air conditioning is a rarity in Europe. Similarly, the Native Americans of the Southwest, an area of intense heat in summer, used adobe to construct their pueblos, reserving wood for, at most, the roof beams. Experience proved the advantages of stone to both Mediterranean and Native American peoples.

Suburban houses are also designed to maximize energy consumption. Older homes, built before World War II, often had wings that could be shut off from the rest of the house in winter to reduce heating bills. The kitchen was in a semi-detached shed at the back, so the heat from the cooking would not enter the rest of the house in summer. By contrast, contemporary homes do not have wings and the kitchen is often at the center. This means that central heating and air conditioning are standard, since such homes require that the entire building be heated or cooled in season. Since all the trees that previously surrounded the house were cut down, there is no shade to block the summer sun, increasing energy bills and thereby global warming.

Mimicking, if unknowingly, the castles of the English aristocracy, whose lawns were developed to accommodate grazing animals, suburbanites plant lawns, and use energy consuming devices to keep them in trim. The lawns are the symbol of status, despite their actual status as the sign of the conquest and the destruction of native species that they represent. Alien flowers often adorn the edges of the lawns, replacing more carefree native plants. The drone of mowers is a frequently heard noise in suburban neighborhoods.

"The affluent society" has meant that in many cases, each child has his or her own bedroom, increasing the size of houses and the corresponding use of energy. Similarly, each child of some affluent parents has his or her own TV, computer, auto, and other possessions. Such parents do not encourage the sharing of such appliances. Frequently, more food is prepared than consumed, and the excess is tossed out with the other garbage. The environmental costs of producing the meat that is wasted is incalculable. Millions of acres are grazed, including those that were burned in the Amazon to create grazing land, to produce food that will be tossed away. The killing is *useless.*

The modern symbols of affluence include the "throw-away" society, despite waste and destruction of resources. Recycling or repairing are only tolerated; it often costs more to repair than to replace consumer goods. We create mountains of trash instead of recycling it.[10] Consumer goods are designed with "dynamic obsolescence," not to last but to decay after a fixed period. Again, new consumer goods are created to replace them that strain the environment, increase energy consumption in their manufacture, and create acres of garbage. In truth there should be no trash, for everything could be recycled. Even dangerous chemicals could be chemically converted to harmless ones, if economics did not dictate morality.

Advertising "works," for it has proven effective in creating desires and training the society in wasteful consumptive habits. People buy into the life-style of status, becoming what they can purchase. Frequently they go into debt to buy the status symbols, and sometimes go in over their head, unable to meet their financial obligations.[11] The life-style promotes desire over character, for gratifying every desire, no matter how trivial, is considered more "realistic" than control over desires. In other words, a minority has grown wealthy promoting bad habits over good ones with relentless and unceasing advertising for trashy consumer goods.

Despite the words of Jesus that followers should "give all your money to the poor," religion is not perceived as a barrier to excessive consumption. Protestants, Catholics, and Orthodox alike live in the affluent society with an easy conscience. Indeed, if Max Weber's thesis is correct, modern Christianity may be a factor in support of this affluent life-style, since some take wealth as a sign of divine grace or favor. Weekly religious services are an opportunity for lavish displays of wealth, "conspicuous consumption" with high-priced automobiles, fashionable clothing, and other signs of status.

The environment has paid the price for the affluent society. Increased population has meant the exhaustion of soils, aquifers, and the necessity of fertilizer use. Tropical forests are being clear-cut. Native species are being exterminated to make way for food and other commercial crops. Predator species, that check overgrazing, are eliminated. The atmosphere is heating up, creating the need for even more energy consumption to keep humans cool in summer. Petroleum is being used faster than it can be pumped. Deserts are spreading; water disappearing. Yet few consumers are aware of the price paid by the biosphere to keep them "happy," that is, to keep them consuming. Ignorance is bliss in consumer society.

Human culture is the culture of death: killing billions of plants and animals is considered not only acceptable, but also justified. Not content with North America, humans are now aiming at cutting down all the trees n South America, especially Amazonia, Africa, and South Asia. The consumer life-style is the latest installment in the culture of death. What is novel is that consumption is increasingly removed from production. Prior to the twentieth century, people killed their own food. Now it comes packaged. Unfortunately, other cultures have tended to adopt some of the worst trends in the West. Especially in Asia, they hope to mimic this lifestyle, creating an even greater strain on the biosphere and more killing. In the minds of corporate executives, the spread of the American suburban life-style to other parts of the world is an obvious good. Millions of more automobiles polluting the air of China, India, and other developing regions is acceptable, if it means an increase in the bottom line. Millions of detached houses drawing upon decreasing sources of energy are for the good. The unsustainable should be increased and multiplied, since it will increase profits.

The bourgeois, sophiacratic, technocratic civilization that originated in the West is one culture among many and is not superior qua culture. Its mastery of technology does not make it more advanced. Western civilization is a result of choices among alternatives: evaluations, not historical determination. No civilization is either inherently superior to alien cultures or inevitable: manifest destiny on a worldwide scale. Nor is its evolution to consumer culture inevitable, but the result of complex historical causes and factors.

Human culture certainly has its merits as well: DaVinci, Mozart, Gandhi, and other cultural icons. But this does not alleviate or justify the culture of death, the murderous violence against the biosphere—which is required to create art—as well as other species. Even the mass murder of humans themselves at Auschwitz, the Gulag, and Africa (Rwanda, Darfur) is but the latest episode in this genocidal tendency, which included women, Jews, Native Americans, and Africans. Millions of Africans died on route to slavery, but little was done to save them: this was considered an acceptable price for importation of slaves, despite the founding document of the United States, the Declaration of Independence, which stated that "all men are created equal and endowed by their Creator with certain inalienable rights…life, liberty, and the pursuit of happiness." In the West, hunting animals is considered "sport." What is tragic is that none of this is needed. Humans do not need meat in their diet and thus do not need the use of millions of acres for grazing and the production of food for livestock.

Prof. Nannery has argued that humans do not exploit other species any more than other predators. His legitimate point is that if other species can feed off their prey, why should humans be held to a higher standard? One point here is that no other species has achieved the domination that human culture has—spreading to every part of the earth except the poles. Moreover, the other species have not altered their habitats to either the degree or the extent that human culture has. Beavers create ponds out of streams and drown some flora in the process. But beavers are limited to riverine habitats and have not altered fifty per-cent of the habitable biosphere in the process. The massive extent of human destruction distinguishes human alterations of the environment, as the example of the United States indicates. I argued above that much of this is unnecessary, since humans do not require meat in their diet. The argument from animal predation is irrelevant. Moreover, humans claim enlightenment and humane morals. But this is contradicted by massive extermination and human-caused extinctions.

The problem for environmental ethics is the environmental impact of the affluent society. The main symptom of the culture of death in contemporary society is the refusal to deal with the obvious problems all of life faces as the result of pollution, namely, global warming, habitat destruction, loss of biodiversity, soil loss, and other ecological problems.[12] What then are the legitimate claims of human culture with respect to its place within a larger whole of the biosphere, on which it is dependent?

2. Threats to All Life

The main causes of our contemporary ecological crisis are overpopulation, the consumer economy, and destructive technologies. Our overpopulation problem is unprecedented. In the twentieth century human population increased from two billion to over six billion, a threefold increase that happened in only one hundred years. It took millennia for humans to achieve the two billion mark, but less than one hundred years to triple it. Moreover, one third of those six billion are in two countries. Overpopulation has caused rapid habitat destruction and with it loss of biodiversity—destruction of species. Technology has made this destruction feasible, whether the use of huge nets to capture all the fish in an area of the oceans or the clear-cutting of entire forests. The pursuit of affluence has been the cause of the shrinkage of the wild to an ever-diminishing sphere: humans, one species, directly or indirectly use about seventy per-cent of all the available land and have fished the oceans to the point where ninety per-cent of the large fish are gone.

The issue here is human impact in general and the use of technology in destructive patterns.[13]

1. *Energy generation: the ecological costs of cheap electricity, including acid rain, global warming, nuclear waste, and so on.* Energy and the environment will constitute the new value conflict within the economic sphere; the next phase of technological history will be the extent to which prosperity will supercede all other values, including environmental beauty as a socio-cultural value and the degree to which energy will be a socio-cultural

force.[14] The Bush administration created the phony "war on terror" in order to grab energy supplies. Energy availability has only emerged as a distinctive force in Western history in the last hundred or so years. Previously, energy was more easily available, at least in North America, but in the future energy may supercede all else. However, this value conflict may have precedents in the exhaustion of forests in many civilizations. The forests provided the main source of energy. Chinese cuisine, someone has remarked, evolved as it did to enable quick cooking in response to a lack of fuel. Energy costs will be a cause and limit on prosperity, general wealth. Yet population control will be also be a force, even if hidden, a norm for society. It will constitute a limit to society and its members—small families will be required as a social goal. Garrett Hardin has remarked that this factor is hidden, since it is like the 800-pound gorilla in the room—a problem no one wants to face.

Yet this conflict could easily be avoided if society turns towards what are in the long run the cheapest and most plentiful sources of energy: the wind and the sun. According to Prof. James Speth, 120,000 acres of wind farms could generate all the electricity needed in the United States (Speth, 2004). This may sound like a huge amount of land, but is actually quite small, given the area of the US as a whole. New York State, which has fewer square miles than most states, has about fifty-four thousand square miles, far more than are needed for the wind farms.

The big lie is that wind power is not practical. But Denmark already generates thirty per-cent of its electrical needs through wind. Our ancestors already used wind—a proven technology. Substitute technologies have already been creatively actualized. To handle the problem of variability in wind speed—for there is always some wind—storage batteries could be used for when wind dies down. When the wind is strong, excess electricity could be stored for use when gusts are weaker. Also, engineers could install propellers of different sizes on the same windmill to take advantage of light breezes. Windmills could be concentrated in especially windy areas, like the Great Plains, mountaintops, and coasts. If enough extra windmills were built, there would be electricity to spare. Finally, solar could supplement wind and could be collected from panels on each building. Electric vehicles could use home-generated electricity and could recharge at night. This would eliminate about twenty to sixty per-cent of pollution caused by fossil fuels, based on the statistic that transportation contributes twenty per-cent and electrical generation forty per-cent.

The truth is that wind is a threat to fossil fuel considered as an asset of energy corporations, who have an investment to protect and foolishly do not care if the environment is ruined as a result of fossil fuel use. If they were wise, they would invest in windmill farms and make money soundly. Moreover, if windmills were installed on each home, they could at least generate some of the household electricity. However, wind and solar can be decentralized, and this may also be perceived as a threat by the shortsighted and greedy. As for costs: the cost of fossil fuel use is higher if hidden, externalized costs added, including threats to health from pollution, forests (acid rain), and others.

2. *Rising expectations and affluence as contributing factors in destruction.* Some argue that the rich cannot expect countries in the developing world to sacrifice their affluence when the rich countries have shown no sign of doing so. There is considerable justice in this view. China and India, poor for centuries, are just now beginning to taste the fruits of affluence. Why should they have any less prosperity than the citizens of Western Europe or North America? Of course these two countries have the additional problem that they have contributed an undue share to population growth, just as the U.S. has contributed the lion's share of global warming gases. China, at least, has tackled the population problem to some degree. India has not been as effective. But if a country chooses quantity of people, it has only itself to blame if the quality of life declines. The root cause is not unjust distribution, but excessive population. Too many people on too little land. And if everyone lived the life of the average U.S. citizen, the result would be environmental catastrophe. Nor would it be sustainable, since fossil fuels would be used up all the more quickly if everyone in developing countries drove a large car and had a big suburban house. Thus affluence would again be out of reach for the masses of Asia.

A better approach to rising expectations would be to pay the long-term price for affluence with population control, sustainability, and environmental preservation. Installation of wind and solar power might be slightly more expensive in the short term, but its long-term value is obvious and is the only sustainable approach.

3. *The effect: global warming, resulting in rising ocean levels.*[15] There is a "contradiction" in our society between long-term and short-term good. The short-term good is increasing prosperity, but such prosperity has come at a price. Rivers and bays that once provided food are now so polluted that the fish and other seafood cannot be consumed safely. Air that is necessary for all of life is so bad in some cities that it directly causes deaths, notably, Los Angeles, Mexico City, New Delhi, Beijing, Shanghai, and many other large urban areas. Soil is laced with pesticides and often runs off into rivers, endangering a vital resource. The whole is being ground up to increase a very unimportant part, namely, the wealth of a few. We can judge such excess wealth unimportant because luxury is not a vital necessity and the sacrifice of all of life for the trinkets of the affluent society is the soul of foolishness. Destroying the world for the affluence of a few is not only foolish; where done despite knowledge of its environmental consequences it is evil.

Wisdom includes good judgment about future courses of action and therefore requires calculation of long-term consequences. Global warming poses the greatest long-term threat to life, including human life, of any long-term threat. Global warming will not be local but world wide, altering entire continents in ways that can only partially be foreseen. One of these is that up to one-third of present species may go extinct.[16] Many areas in Africa will be dryer; tropical forests may disappear as increasing heat stresses the ability of plants, insects, and other living things to cope. Glaciers will melt, putting water supplies all over the world at risk.[17] Agriculture will become increasingly difficult as climate patterns alter while water becomes scarce. Meantime, there will be more mouths to feed with less arable land. Shorelines will shrink as oceans rise and lowlands will be inundated, forcing millions to flee to higher ground.[18] All this so that the affluent can own such trinkets as a second home and a yacht.

Two theoretical difficulties accompany consideration of long term consequences, namely, ignorance of the future and unintended consequences. I examined the first topic in chapter five. As for unintended consequences, no amount of wisdom will enable us to foresee all consequences. On humane grounds, medicine was used to defeat diseases like polio all over the world. The consequence was the unprecedented rise in population during the twentieth century, from two to over six billion. The value of our solutions is finite: we do not have a God's-eye view. Nevertheless, if we would be wise, we must collectively try to consider as many consequences as we can imagine and deal with the negative ones in our decisions. We must try and deal with new circumstances, anticipated or not, as best as we can. Our wisdom is finite but not futile.

Global warming will itself result in mass extinctions, as species, already stressed by habitat destruction, are unable to adopt to higher average temperatures.[19] Human refugees from flooded coasts may be joined by animals fleeing the lowlands in a desperate attempt to survive. Even more wild lands will be destroyed to make up for the loss of the lowland crops. Who will take in all of these victims of the affluent society? Who will be responsible for feeding them?

The IPCC report, if anything, understated the threat.

> The scientific consensus most people are familiar with is the IPCC's "Summary for Policymaker's" reports. But those are not a majority [scientific] opinion. Government representatives participate in a line-by-line review and revision of these summaries. So China, Saudi Arabia and…the Bush Administration—get to veto anything they didn't like. The deniers call this "politicized science," suggesting the process turns the IPCC summaries into some sort of unscientific exaggeration. In fact the reverse is true. The net result is unanimous agreement on a conservative or watered-down document…. *Scientific American* noted that objections by Saudi Arabia and China led the IPCC to remove a sentence stating that the impact of human greenhouse gas emissions on the Earth's recent warming is five times greater that that of the sun.

> "How else does the IPCC lowball future impacts? The 2007 report projects a sea level rise this century of 7 to 23 inches. Yet the IPCC itself stated the 'models [of sea level rise at the end of the twenty-first century] used to date do not include uncertainties in climate-carbon cycle feedbacks nor do they include the full effects of changes in ice sheet flow....The result is that compared to the 'consensus' of the IPCC, the ice sheets appear to be shrinking '100 years ahead of schedule', as Penn State climatologist Richard Alley put it in March 2006."(Joseph Ramm, quoted in Thomas Friedman, 2008, p. 121-2).

Destruction of vast areas of the biosphere is genocidal and thereby evil. Since global warming constitutes a threat to all of life, including human life, it is wrong even from ideological perspectives, that is, anthropocentric "ethics." Even in terms of the greatest good of the greatest number of people, or our own long-term self-interest, or that of the human species, poisoning our air, reducing our water supply, destroying our soil, and threatening all the other species that we depend on symbiotically is sheer lunacy. But these would be the consequence of global warming. Avoiding global warming is a "no-brainer," to borrow a popular phrase.

Wisdom dictates that we cut out the excesses of the affluent society, especially since global warming is not a victimless ecocrime. On the contrary, those who caused the problem will pay the least: namely U.S. citizens, who account for thirty per-cent of the energy use, which contributes to greenhouse gas generation, but make up only five per-cent of the world's population. Similarly, today's affluent will not be alive when the worst of the consequences of global warming are felt in the future, even though they have children, and thereby a responsibility to them. Thus from the perspective of social justice, ecocrimes are an immense injustice to all the people.

4. *The effect: habitat destruction and species extinction.* Human expansion into wild habitats is a potential threat to all of life; an actual threat to most of life.[20] The unlimited expansion of human population and the invention of habitat altering technologies together have meant that the earth has been transformed to the point where over half of all habitats have been disturbed or destroyed to feed more and more human mouths.[21] One species has transformed the earth for its own benefit, threatening thousands of other species. The impact of global warming, with its resulting rising ocean levels may have catastrophic effects upon the earth and its species. Our species is indeed "playing dice with the universe."

The environment is a requirement for all of life, not just human life: neither we nor any other species can survive without air, water, and soil. Habitat destruction is wrong, as habitats are the condition of species, just as soil, water, and air are required for all life. Indeed, all species, including humans, also require other species as the condition of healthy air, water, and soil. Air is regulated in a symbiosis between animals and plants, with animals consuming oxygen and creating carbon dioxide and plants reversing this process. The interaction of numerous organisms, from earwigs, dung beetles, and earthworms to microscopic bacteria, creates soil by breaking down organic matter into soil humus. Water vapor is in part exhaled by plants, which also anchor soil from scouring winds and erosive floods. Other organisms clean water.

There is an unconditional interrelation of species and habitats involving symbiosis and holistic relations. Some species require other species, while some require generalized others. For example, there is the role of such predators as birds of prey and wolves in healthy environment, which keep the herbivores in check but are thereby dependent or symbiotic on at least some other herbivores.[22] By contrast, koalas and pandas eat highly specific foods. Destruction of other species for money is a form of genocide as heinous as any in the twentieth century.

5. There is also the *threat to all life from the destruction of forests due to acid rain.*[23] The forests are the earth's "lungs" and their destruction might wreak havoc with circulation of air and absorption of carbon dioxide. Forests serve many functions. They provide a living area for many species, create and anchor soil, absorb and store carbon dioxide, release oxygen, convert sunlight to energy and thereby sustain the food chain, and many other functions. The forests face multiple threats including acid rain, clear-cutting, and human expansion for housing and agriculture. Acid rain is caused by the use of dirty fuels whether for transportation or energy generation—hence the importance of replacing the use of fossil fuels with non-polluting sustainables, wind, and solar.

Their exploitation for human use is neither necessary nor environmentally friendly. Wood could be replaced as a building material with more efficient inert materials, especially stone. Stone, ceramic, and other contemporary building materials, such as foam or glass insulation are far more eco-friendly, since they are far better insulators than wood. Moreover, the utilization of inorganics does not involve destroying any forests, and thereby preserves the functions that forests serve, particularly as a habitat for wild species. The use of inorganics is a win-win situation, reducing the need for heating and cooling and preserving carbon soaks.

U.S. national forests are now used for grazing, which is destructive of understory, and may destroy forests in the long term. Cattle also contribute to the problem of global warming with the release of methane, a major global warming gas. A vegetarian diet is healthier and would require less space. While ranchers pay a nominal "grazing fee," this charge comes nowhere near covering the costs of administration, to speak nothing of the hidden costs to the environment, and is another form of externalizing costs. The actual cost of raising cattle in the West is not paid by the ranchers but the environment, and thereby the public.

6. *Pollution, ozone depletion, salinization of soils, and erosion: threats to future generations.* Humans are awash in their own filth. Pollution of our own environment for the sake of affluence is a symptom of decadence. A society that is willing to destroy the conditions of its own life, and worse, impose the results of its foolishness on its descendants, is blinded by greed. Poisoning the air and water that we and all life require for our continuation is self-destructive.[24] A species that would rather destroy itself than give up its ways is suicidal. What are we to make of such a self-destructive species—one that would pollute itself to death instead of give up its trinkets?

Another shortsighted practice is the destruction of soils by erosion. The thick soils of the North American prairies took centuries to build up but are being eroded at a cataclysmic rate. Dave Foreman notes that "five hundred million tons of soil erode annually from Western public range lands—many times faster than the soil is replaced" (1992, p. 95).

The long-term effect may be that our descendants have only thin soils to feed ever-growing populations. Another form of this habitat destruction—of habitats that we require for our own food—is salinization of soils in areas of irrigation with low drainage or high amounts of chemical salts in the water supply. Whole civilizations came and went in Southwest Asia due to soil loss from salinization of their soils. The same could happen in the United States in the Southwest, and already has happened in some areas, notably the Kesterson area of Central California and some areas of the desert in Arizona—to speak nothing of the "Fertile Crescent." Such practices are a threat mainly to our own species, which depends upon healthy soils for food production. As I noted, to have children is to take on a responsibility to them. To have children and take away the means for them to feed themselves is the height of irresponsibility and foolishness.

Finally, there is the threat to all life of ozone depletion, caused by the release of artificial chemicals. In one respect, ozone depletion is a success story, since the chemicals that cause it were banned by international treaty. The treaty constitutes proof that humans can change their ways and deal with an obvious threat to all of life while maintaining a healthy economy. In another respect, it is the story of human foolishness, since some industrial countries, led by the United States, attempted to block the needed changes for as long as possible. Chloro-fluorocarbons were only phased out gradually even when alternative technologies were already at hand for the solution.

3. Technology

A great issue becomes so by re-forming our lives. All other values become revalued or regrouped around and in relation to the new issues formed by the new value. Our lives are now technologically enhanced—we communicate, travel, learn, work, and so on in a technological setting. This differentiates us from less technologically advanced ages.[25] Technology is a common epochal factor in all the categories of our age. Karl Popper has argued that we "can't go back" to a pre-technological culture. This historical irreversibility, while not absolute (in principle we *could* go back) gives technology greater significance than it might otherwise have.[26]

Technology has a history and its development in the West to the point where it is central and had a decisive impact on Western culture involved a series of decisions. One is the liberation of nature and knowledge from myth in Hellenic civilization.[27] While the rise of Judeo-Christian spirituality during the Middle Ages eclipsed this development, it was never lost. Greek philosophy was revived during the Renaissance and has lived in tension with religion ever since. The rise of science—the revival by early modern philosophers of ancient intellectual standards independent of religion—meant that the biosphere would be increasingly controlled and made subservient.

The Enlightenment use of Hellenic "reason" thrust Western civilization away from the spiritual toward power. Technological domination included both strategy and economics as affluence also fed technological domination. Technology liberated humans from nature to some degree, and some visionaries hoped that humanist values would similarly liberate human institutions socially.

We now know that technology has been a mixed blessing.[28] The overuse and misuse of technology has been a threat to all life. In our age the main threat is energy generation and its consequences: the ecological costs of cheap electricity, including acid rain, global warming, nuclear waste, and so on. Due to global warming, ocean levels are rising, which has had the effect of inundating low-lying countries and threatening coastal areas. Another effect has been the destruction of forests due to acid rain and the spread of pests into areas where they were once rare. As a result of global warming, the pests can invade virgin forests and cause devastation, as pine beetles have done in the Western United States. As I noted above, other effects include salinization of soils and poisoning of ground water, lakes, rivers and other fresh water sources; and air pollution and ozone depletion.

Some might argue that pollution is part of the logic of capitalism. It makes economic sense to externalize costs, even if the effect is to pollute the local air and water. Costs of clean up are thereby reduced, decreasing operating costs or overhead and thereby increasing profits. Thus Ralf Brand argues that free markets "create incentives to maximize the use of free resources and free sinks, usually common goods such as water and air—a strategy known as 'externalization of costs'" (Brand, (2005) p. 99). The problem with this argument is that socialist countries are as avid polluters as capitalist. The former people's republics of Eastern Europe had fewer pollution controls than their Western counterparts, and the low grade coal of East Germany in particular at least contributed to dieback of foliage in the forests of Germany. Pollution does not respect political boundaries. Moreover, the government of the People's Republic of China has exploited the coal deposits of that country as avidly as any capitalist state, and refuses to sign on to the Kyoto accord. In truth, externalization of costs is part of the logic of technology, in which efficiency is the standard. Efficiency does not include absorbing costs that can be externalized, even if they are life-threatening. Pollution is part of the culture of death, which puts profits and economics before lives, the biosphere, and even human well being. Technology has also made habitat destruction and species extinction easier; and has been a contributing factor in erosion of soils. The misuse of technology is a threat to the entire biosphere.

Technology is less dangerous than technocracy. Technocracy is the worship of and domination by technology. Technocracy involves the appropriation and transformation of large portions of the earth by and for technology, and the reign of the one best, most efficient, rational method over all cultural variegation.[29] Technocracy means that nations, businesses, economic organizations and individuals create conditions for technology to transform the earth. Thus like totalitarianism, technocracy is a metaphysical theory, in which reality is transformed.[30] Despite the use of technology by modern economies, indeed their dependence upon it, technocracy is neither capitalist nor economic per se, even if some economic motives are factors. Its origin lies primarily in intellectualist values. The origins of technocracy lie in: (1) Cartesian mathematization of nature (including analytic geometry) and his mechanical view of nature. (2) The rise of modern experimental science. (3) The Enlightenment's equation of progress with material and economic quantitative increases. (4) Karl Marx's identification of historical change with change in means of production; (5) Auguste Comte's and Edward Bellamy's Utopias (compare Ayn Rand and Howard Scott in Akin, 1977). (6) Thorstein Veblen's distinction of business and industry. Technocracy requires planning by an elite of educated specialists; thus it is a form of sophiacracy.[31]

Technocracy may not be an alternative to bureaucracy, but only coordinate with it. Karl Wittfogel's (1957) "hydraulic society" is both bureaucratic (at least in China) and technocratic (massive irrigation projects, the Great Wall, etc). These examples coincidentally refute the idea that technical transformation of the earth is exclusively modern or Western. Most Asian agriculture depends on irrigation, which is an artificial transformation of the environment. The technocratic fantasies in Scott and Rand are anarchistic, despite their Platonic type elitism; yet this may not be a viable alternative.

The rule of technology is often judged as a conservative ideology in the United States, as technologists are often allied with industry, that is, private capital (John McDermott, 1969). But strictly speaking, technology is not conservative, however much self-described conservatives have embraced it as their own or identified with it. In defense of this point we need only point to the common roots of conservation and "conservative." Moreover, some equate progress with technical progress. "Technical progress" is progress as technological and progressivism as technocracy. In other words, technocracy is a progressive ideology that was challenged by many romantics.[32] Finally, if Communism is "radical" then technology is radical, for communism has meant rapid industrialization. In sum, technology is not conservative, liberal or radical, although technocracy may be.

But I would argue that technocracy is more subtle, and only indirectly political.[33] Technocracy blurs boundaries of the political, economic, and social. Instead, it is a way of life, as Winner has argued, with repercussions for the economy, government, and society.[34] The issue is whether it subject to democratic recourse or not—especially as whole nations tie themselves into technology and thereby become dependent upon it.

John Kenneth Galbraith (1967) advanced the "convergence thesis" in which he predicted that communism and capitalism would increasingly look alike, despite ideological differences. In a sense, this has happened in the People's Republic of China, which while ostensibly communist, leads the "capitalist" world in factory production. However, our thesis is not the convergence of capitalism and communism but the melding of technocracy and bureaucracy. Corporations that utilize or produce technologies are organized bureaucratically; bureaucracies have "techniques" of procedure, operation, production, and so on, in Jacques Ellul's wide sense of technology. Oligarchic technocracy establishes "needs" by planning, just as in bureaucracy—prescribing "choices" through market domination and advertising.[35] If technology by itself dominated the way Jacques Ellul (1964) thinks, there would be less of a dark side: pollution control would be technologized. But this has only happened if at all, due to pressures extrinsic to technology. Economic, political, and social forces also affect the organization of technology, as the social constructivists have argued.

In another sense, technocracy is planned abundance. Notably, planning is not confined to "socialism." Technocracy tends to oligarchy, in the literal sense of the rule of a few. Galbraith has christened this the "technostructure." Technicians are needed to run modern society, whether as engineers, managers or other specific forms. More subtly, technocracy involves a public standard, ideologically derived, which has direct bearing on public policy, that prescribes activity, resource allocation (defense: guns not butter), and so on. But the public standard, though a call for political action using political institutions, is not strictly political. For although a defense buildup, as an example, ensures the future of the Galbraith's "technostructure," they do not rule direct, nor desire to.[36] Their views are ideological, but this ideology is not imposed *by them* through government agency. They transform the world with technology, but the government is in independent interaction with them. Instead, their ideology establishes political priorities, planning policy, and the shape of the future of society as technological. Experts decide government spending priorities, not the public.

However, technocracy is not only an oligarchy of managers. Positivist technocracy is the ideology for which advances in science and technology "require" changes in society, as technical progress outruns social. We label it "positivist" since knowledge is the determining factor in social change.[37] Positivism also includes the idea that universal education will help allay the problems of human culture. The idea that social and political institutions have not "caught up" with technical "progress" is a technocratic demand: a "need" created by technocrats to account for the failures and debacles that involved technology: gas chambers, the threat of nuclear war, and widespread environmental devastation. Justifying massive destruction of wildlife in terms of the need to bring technical progress is ideological—as is the so-called "end of ideology" as some sort of consensus on technical progress.

We can become so locked into and dependent upon technology that we cannot give up even a harmful technology. The addiction to oil, coal, other fossil fuels, and nuclear power provide an object lesson on the danger of technocracy. The economies of many nations are so tied in to fossil fuel use that the obvious dangers to the entire world from their use, notably global warming and pollution, are virtually ignored. Nations and corporations plan their strategies around such technologies and thus are ruled by them.[38] Technocracy reigns by becoming technological domination.

Modern war is completely technocratic. Historian Babara Tuchman examined the closed inner logic of technology in the *Guns of August*. She described how the railway timetables in Germany helped bring on World War I. Technology becomes technocracy, regulating human affairs with disastrous results. Technocrats both planned and were created by technology as the military-industrial complex in the US and USSR: wealth was wasted on non-consumable weapons. That they may well end civilization as we know it would be the denouement of the culture of death.[39] Strategic "necessity" is also based on technical means. Thus both the United States and the Soviet Union rushed to corner the German rocket scientists and Japanese biological warfare researchers and their results for themselves after World War II. They did so despite the contribution of these scientists to war crimes as bad as any at Nuremberg, including shelling British civilians, and experimenting with lethal medicines upon Soviet, Chinese, and other civilians.[40]

The "balance of terror" should provide a warning to us regarding the dangers of both technocracy and strategies based on technical progress. The United States and the Soviet Union came close to nuclear armageddon during the Cuban missile crisis. More recently, Pakistan and India also came close to a nuclear exchange. The frightening logic of "use 'em or lose 'em," "launch on warning," fail-safe, and other precipitous nuclear strategies reveal their dangerous edge. James Burnham and others argued that the "balance of terror" has, paradoxically, made war out of date.[41] The fallacy of Burnham's "Pax Atomica" is the fallacy of induction. Just because there is a large statistical probability that something will, or will not, happen, such as atomic warfare, based on past experience, says nothing about it happening in the future. Chernobyl and Three Mile Island should have settled that point long ago. The "balance of terror" means we can no longer settle human disputes by violence. A new ethic of non-violence is thus called for, ironically, by circumstances in the development of warfare.

The dreams of eliminating technology are utopian given mass consumer demand. For the *sine qua non* of the demand is technology. And the technocratic managers will cater to the largest value, demand, even where they "create" demand through advertising. Moreover, the two feed off one another, as many advances in technology can be applied to consumer goods and vice versa. The pressure to increase technology will thus increase in the future not decrease; while third world poverty and nationalist conflicts will make technological advance—not other values—receive priority. The "question" of technology is whether it shall be servant or master: technocracy, including the "logic" of nuclear strategy—notably the "logic" of Mutual Assured Destruction—but also economic life: the dehumanizing work on the assembly line, and other failures of technocracy.

Technocracy especially means consideration only of technical solutions to problems—and sticking to technocracy instead of environmental concerns. But what "works" is not merely technical: technology may be problematic (pollution). Moreover, assuming *a priori* that progress as technology will be universally better is ridiculous: consider the pollution left behind in the "Love Canal."[42] The good or bad effects of technology cannot be known with any certainty; we cannot compare humans of different times and cultures with a uniform standard; and we cannot bring a human from the past to compare. A better solution may be needed to deal with and curb technocracy than *techne* itself.

Human culture is totally dependent upon the environment. Jared Diamond emphasizes this point in his comparison of different culture areas and the ascendance of some over others:

> I'll argue that continental differences in level of civilization arose from geography's effect on the development of our cultural hallmarks, not from human genetics. Continents differed in the resources on which civilization depended—especially in the wild animal and plant species that proved useful for domestication. Continents also differed in the ease with which domesticated species could spread from one area to another.
>
> Storable food surpluses grown by some individuals permitted other individuals to devote themselves to metallurgy, manufacturing, writing—and to serving in full-time professional armies. Domestic animals provided not only meat and milk to feed people, but also wool and hides to clothe people, and power to transport people and goods. Animals also provided power to pull plows and carts, and thus to increase agricultural

> productivity greatly over that previously attainable by human muscle power alone (1992, p. 236).

Diamond notes that Europeans had sheep, goats, pigs cows, dogs, and horses by 4000 B.C. Humans on other continents could not domesticate as many animals, with as many functions, either because there were not as many available (New Guinea, South America); or they were harder or impossible to domesticate (Sub-Saharan Africa, North America). The resources available in a habitat were decisive in allowing some cultures to conquer others.

Culture has allowed humans to transcend their original habitat and adapt to other habitats. Moreover, the ability to learn has allowed humans to copy inventions from other cultures useful in new habitats: corn and potatoes from the New World were adopted in Eurasia; wheat and rice from Eurasia are now grown in the New World. In the past this has been the culture of death, as humans exterminate weaker natives. Since human culture depends upon the environment, wisdom demands the revaluation of the relation of human culture and the environment. Culture is within a larger whole upon which it depends. Ethics should include the environment as the other without which all life, including human life, would be imperiled.

Eleven

PRACTICAL PROGRAM FOR IMPLEMENTING THE ENVIRONMENTAL ERA

> Highly important is the question of the *waste of natural resources*...Under the influence of the profit motive, forests were cut off with no provision for replanting. It would not pay a private owner to make such provision for the needs of future generations, and so nothing was done to replace forests, even on land unsuitable for any other purpose...Attempts have been made to guard against waste in oil, but the desire for immediate profits has been so strong as to encourage a large amount of wasteful exploitation. Even more disastrous in its ultimate possibilities is the exhaustion of the soil in certain parts of the country by the washing away of exposed bare soils into the rivers, and so into the ocean or the Gulf of Mexico.
>
> John Dewey and James Tufts[1]

I have outlined cosmopolitan duties and virtues in previous chapters. These include obligations to other species, to future generations, to keep the air, water, and soil clean, to keep habitats intact, life diverse and population down. The main issue for the economy and society of the future will be: Is it sustainable? In light of this question, I have included a few practical suggestions below. I make no claim to originality in these proposals, since other authors may have previously proposed them.

First, a note on wisdom. Humans can continue in their foolish ways. But if philosophy is the love of wisdom, then philosophers have the duty of warning their societies about short-sighted foolishness and of attempting to publicize alternatives, namely sustainable, eco-friendly methods that do not involve destruction of habitats and other species or endanger future generations of our children. I have argued throughout this book that the environment is a requirement for all of life: we cannot survive without air, water, and soil. We also require other species: our food as well as the condition of healthy air, water, and soil. Species and habitats are interrelated in symbiosis and holistic relations, including the role of predators in a healthy environment, such as birds of prey, wolves, and others. The most vital issue for our time is whether we will wake up and realize the danger to ourselves, to other species, and to all of life.

As for defining healthy environments: a healthy environment in general is one that allows living things, both humans and other species, to thrive. We can add that it fulfills the environmental necessities that we mentioned in the introduction, based on the need for air, food, and water. A healthy environment has clean air and water; its soil is unpolluted and we should take steps to preserve it in that state. It will hold a variety of native species and wild habitats in good condition.

A practical program will have five main components:

1. Relations to Other Species

We have a duty to preserve habitats and species, which I argued for in previous chapters.[2] The argument is that cosmopolitan revaluation of reciprocity includes treating other species ethically and thereby avoiding genocide.[3] Species extinction by human hands is genocide: the elimination of a whole kind. (Thus the attempt by fundamentalists to eliminate homosexuals is genocidal, although genocide is a much more widespread problem, extending to many species.) Since species have inherent value, as self-actualizing kinds, they are morally considerable. Habitats are the condition of any species, including humans. Critics might argue that habitats continually change and that therefore there is nothing to preserve.[4] I would respond that the "stability" of a habitat means it is still viable or suitable for life over historical time. Thus it still contains available water, clean air, and soil that is suitable for typical plants. By contrast, human grazing and agriculture have expanded barren areas: desertification. Habitat dynamics can be short-term or long term. An example of the first is plant succession, from pioneer to climax species—or at least from earlier to later species.[5] Long term dynamics include any more radical change from one kind to another kind of habitat, such as desertification or, even more long-term, from temperate to polar (Antarctica). But the stability of the land over the middle term remains: as suitable for life. Moreover, short-term habitat dynamics do not affect the overall character of the habitat. Woodlands remain woodlands; prairies, prairies; and wetlands, wetlands through short-term plant successions.

Conservation includes conservation of other species for their own sake, for their ecosystemic value, for the sake of future generations, and on grounds of prudence. We do not know the function of every species in the environment, and their conservation will ensure that function is fulfilled.[6] But even if a habitat would recover without a species, conservation of other species is an imperative of environmental justice: treating other species as we would like to be treated. It is also imperative for a symbiotic species such as humans, dependent upon other species for well being. Humans should sacrifice excessive luxuries to conserve and preserve rare and endangered species and threatened habitats, especially rare wetlands. This would above all demonstrate and actualize the virtue of thrift. (Self-interest has little place in an essay on the virtue of thrift. But to give the devil his due, even on grounds of self-interest, preservation of species that could possibly benefit humans in unknown ways is vital.)

Mistreatment of animals must be addressed. A vegetarian diet would be preferable for all, as it is healthier and would allow more animals and more humans to live. Less land would be required for human food. By some estimates, it takes eight ounces of grain to make an ounce of beef. If humans gave up beef, many more humans could be fed on half the land now used to graze cattle. Moreover, non-meat sources of protein are superior and more easily digested, whether dairy or plant (beans and legumes). Factory farming should be banned and consumption of beef discouraged. A side benefit would be a reduction in methane from ruminant digestive processes. Fishing should be strictly regulated and genetically modified fish outlawed. Individual animals of rare species should be protected with heavy penalties for violations.

Other measures that can be enacted include:

A. Set aside more wilderness. One effect of the expansion of human settlement and land use has been habitat destruction. The result has been species extinction, whether actual or potential, and thus a loss in biodiversity. All species, including humans are interrelated through symbiosis and holistic relations.[7] Humans also require other species as our food as well as condition of healthy air, water, and soil. Other species are generally adapted to some niche and in all cases require suitable habitat.

Thrift includes conservation, almost by definition. Alternatively, conservation has a corollary of thrift. Conservation as a virtue would mean conserving for its own sake, not for the sake of profits, desire, or human exploitation. The wilderness should be conserved for its own sake, not only for human use—just so that there will be some unspoiled wilderness. Some land should be set aside on esthetic grounds (parks), and also ecological grounds (wilderness, watershed). Not all land has economic use. Some land is simply not worth cultivating. Humans now use or have transformed at least fifty per-cent of all the land on the earth. Since they are only one species, this figure should gradually be reduced to create more space for other species. One happy benefit would be more room for plants to grow and absorb carbon, sunlight, and create more oxygen.

Wilderness is almost by definition land that is not and preferably has not been commercially exploited. Loggers and developers salivate at the thought of exploiting such land. A virtuous society will conserve it despite such temptations; giving into greed is a mark of corruption and vice. It will also resist commercial exploitation of beautiful landscapes. Nothing is more depressing than the presence of hotels, restaurants, gas stations, billboards, and the other paphernalia of "civilization" at scenic areas like the Grand Canyon.

Such landscapes are valuable for their own sake, apart from their commercial use. Subordination of beauty to economics involves a category mistake. The destruction of landscapes by flooding Hetch Hetchy Valley and Lake Mead is a tragedy. The use of such water to alter arid habitats reveals how little these lands are loved by their occupiers. As I noted elsewhere, evil is the attempt to eliminate a kind. Insofar as we transform the desert, using piped in water, we are genocidal. Wholesale destruction of habitat to increase the standard of living as genocidal, that is, evil. People who do not like the desert as it is should not live there.

Some advocates from third world nations argue that it is wrong for those in the rich countries to lecture poor countries about preserving wild areas when the rich countries have created most of the problems.[8] Thus United States citizens consume up to 35% of the energy worldwide, with just 5% of the population. Such imbalances in consumption indicate that global warming has more to do with the consumer life-style in rich countries than the development of poorer ones.

I sympathize with the poor, but it does not follow that going the way of the rich countries is the solution. Taking the road of development in emulation of the consumer life style will exacerbate already acute problems. Training local fishermen in the methods used in factory fishing will not create more fish—on the contrary, it will hasten the decline of already overfished oceans and rivers. Cutting down more forests will not delay global warming, it will increase it. Nor will any increase in fossil fuel use. That rich countries have created the problem of global warming, or at least disproportionately contributed to it, does not justify adding to the problem. Arguing that because one nation has done something wrong, others can do the same is a fallacy.[9] This is like saying that because one person raped a woman that it is unfair to other men—they too should be able to rape women. The argument ignores the original point, the wrong done to the planet. The argument is like saying we should become part of the problem, instead of looking for solutions. On the contrary, all peoples must unite to stop the rape of the planet. Diverting attention from problems by blaming them on others or pointing to problems elsewhere will not solve the problem.

Such attitudes also ignore problems of mass extinctions in such countries not caused by wealthy, as if only the rich countries practice animal and plant genocide. But on the contrary, many indigenous species are imperiled due to illegal poaching, loss of habitat due to agricultural expansion, and indifference to the fate of species that are not profitable. In particular, overpopulation drives the expansion of agriculture at the expense of both wild habitat and forests. If there were no rich countries, the poor ones would still face these problems. Poor countries should, on the contrary, learn from the past mistakes of others. The extermination of the wolf and bison in North America should motivate governments in other parts of the world to prevent the loss of species in their own domains by setting aside large and viable preserves. Many have done a heroic job of trying to stop poaching; a world-wide effort is needed to expand such efforts and boycott nations that continue to buy products made from banned animals, notably China's purchases of tiger parts and elephant tusks.

B. Prevent any more exploitation of present wilderness and remnant communities, and encourage their expansion. Humans can help endangered species by removing exotics from the community and thereby aiding rare species and helping remnant communities regrow. In the Midwest of North America, prairie communities could be expanded with little effort; in the forests, laissez-faire will eventually bring back remnant communities.

At least twenty per-cent (20%) of each state or nation should be set aside as wilderness, that is inviolable and cannot be used for commercial purposes of any kind. Off-road vehicle use and the like should be banned entirely, since it wastes fuel and destroys habitats. Outstanding landscapes should be preserved in as nearly a pristine state as possible.

In this regard, Dave Foreman has made a number of suggestions that we can repeat:

> Constructing a meaningful but politically possible National Wilderness Preservation System requires us to outline our goal carefully, and the steps to achieve it. I offer the following blueprint:
> 1. Draw the line on what is now wild. Not one more acre of old-growth or substantially natural forest should be cut. Not one more mile of new road bladed into a roadless area. Not one more Grizzly Bear murdered in Yellowstone. Not one more free-flowing river dammed. Too much has already been lost.
> 2. Recover native ecosystems. In many cases, to recover native ecosystems, to reintroduce extirpated wildlife, and to repair damaged landscapes, all that is necessary is to close roads, cease damaging activities, and leave the land alone. In others, minor hands on restoration may be required; physically reintroducing extirpate species, removing a few developments, and performing minor watershed rehabilitation. Some areas will require more expensive, long term, and active management to be returned to a state of natural wildness. These areas should be designated Wilderness Recovery Areas, with more intensive rehabilitation work allowed, until wildness is restored.
> 3. Restore large ecological wilderness preserves east of the Rockies (see his list below). These core areas and smaller Wilderness Areas and Wild Recovery Areas should be linked to one another by undeveloped corridors. Such corridors are vital for the transmission of genetic diversity between core preserves. Without such corridors, preserves become ecological islands, and populations of low-density species, such as large predators, may become inbred. When it is determined that suitable habitat exists, extirpated species should be reintroduced if it appears unlikely that they will return to the area on their own. The near extinction of mature American Chestnut trees (due to an exotic disease) leaves a gaping hole in the Eastern forest. The Forest Service and National Park Service should fund a research project to develop a

blight-resistant American Chestnut that could be reintroduced to its former habitat in these protected areas.

(Foreman 1991) includes as projected preserves:

• A 10- to 20-million-acre Great Plains National Park with free roaming Bison, Elk, Pronghorn, Grizzly, and Gray Wolf.

• A 5-million-acre North Woods International Preserve around the Boundary Waters Wilderness in Minnesota and Quetico Provincial Park in Ontario.

• A large deciduous forest Wilderness Recovery Area in the Ohio Valley with Elk, Bison, Gray Wolf, and Eastern Panther.

• A 10-million-acre National Park in northern Maine with Gray Wolf, Lynx, Wolverine, and Woodland Caribou.

• A 1.5-million-acre Bob Marshall Greater Wilderness in the Adirondacks of New York, with Gray Wolf and Eastern Panther.

• A 4-million-acre Wilderness Recovery Area in the South Appalachians, centered around Great Smoky Mountains National Park, with Eastern Panther and Elk.

• A 5-million-acre Everglades/Big Cypress National Park in Florida.)

4. Restore major roadless areas in the West. There are currently thirty-eight areas where minor road closures would create core roadless areas of more than a million acres:[10] These million-acre or larger wilderness units should then become the cores for even larger wilderness complexes linked to each other and to smaller Wilderness Areas by wild corridors….

7. Rehabilitate free-flowing rivers. Perhaps more than any other ecosystem type in the United States, rivers and riparian habitats have been abused, altered, and destroyed. High priority should be given to rehabilitating free-flowing rivers, eliminating disruptive exotic fish species, and restoring native fish and other riverine species where feasible. Not only should no new dams be built, but a program should be launched to remove dams and recover free-flowing rivers.

8. Discard the notion of static landscape preservation. What is being preserved in Wilderness Areas is the process of evolution, of speciation, of serial changes in ecosystems. Natural landscapes should be large and diverse in order to absorb catastrophic events such as huge forest fires, insect and disease outbreaks, temporary regional extinctions, and cyclical population fluctuations. (In a large enough preserve or complex, a specific habitat may be wiped out by a stochastic[11] event such as the 1988 Yellowstone fires, but similar habitats will continue to exist elsewhere in other parts of the area, or in other areas connected by corridors.) Wilderness proponents need to learn from conservation biologists, who in turn need to see grassroots conservation activists as their natural allies and the management of public lands as a vital opportunity.

> 9. Preserve wilderness for its own sake. Conservationists must develop a new (old) reason for wilderness, a new understanding of the place of humans in the natural world, a new appreciation for the other nations inhabiting this beautiful blue-green living planet. We should recognize that the true reason we favor wilderness preservation is Wilderness for its own sake. Because it's right. Because it's the real world, the arena of evolution; because it's our home. The Gray Wolf has a claim to live for her own sake, not for any real or imagined value she may have for human beings. The Spotted Owl, the Wolverine, Brewer's Spruce, the fungal web in the forest floor have the following of their own intertwined evolutionary paths as their due. Not only should Conservationists recognize that it is the inherent value of natural diversity that argues for its preservation in our hearts, but it is also the most effective argument for preservation; we should state that rationale forthrightly to the public. Unless we challenge our fellow humans to practice self-restraint, to share Earth voluntarily with our wild fellows, the wilderness crusade is pissing in the wind.

In justification of these proposals, David Foreman asks,

> Why does a man with a lifespan of seventy years think it proper to destroy a two-thousand-year-old redwood to make picnic tables? To kill one of thirty breeding female Grizzlies in the Yellowstone region because she ate one of his sheep? To rip through a five-thousand-year-old Creosote Bush on a motorized tricycle for some kind of macho thrill? To dam Glen Canyon and Hetch Hetchy for electricity and water to irrigate lawns? Until we learn to respect these others as our equals, we will be strangers and barbarians on Earth. Wilderness—real wilderness—is the path home. The articulation of that truth is the vital duty of the preservation movement. We cannot achieve it by hiding behind the anthropocentric arguments of monumentalism, worthless lands, utilitarianism, or primitive recreation. We can do it only by stating what we truly believe, and challenging humankind with that ethical ideal.[12]

C. Make wilderness areas off-limits to food production, development or any other human use. Foreman argues that we should,

> terminate commercial livestock grazing on the Western public lands. Only 3 percent of our nation's red-meat supply comes from public land, and the government spends more on managing this private grazing than it receives in fees from the grazing permittees. Grazing has been the single most important factor in the devastation of intermountain ecosystems: the widespread decimation of bear, wolf, Mountain Lion,

> Elk, Pronghorn, Bighorn, and Bison; destruction of native vegetation; and severe damage to watersheds and riparian systems.[13]

In other words present grazing practices are inefficient, harmful to wildlands, exploitative of public lands, and destructive of the habitats of wild species.

Humans already occupy, control, and use a disproportionate amount of land for one species. Human expansion into wilderness areas constitutes a threat to other species. At a minimum our obligations would include leaving the habitats of such species as little disturbed as possible; preferably it would involve attempts at restoration—which often will happen anyway by leaving the habitat alone for long enough. The wilderness should be preserved for its own sake. Like scenic landscapes, wilderness has a value of its own, that cannot be measured in economic terms.

If we are moral, we will attempt to restore habitats, not only preserve the few that are left.[14] I argued in chapter three that such restoration is a duty of rectificatory justice. Restoration will include "bottomlands" that are most suited to agriculture, not only "wastelands," since the former have peculiar species and these are the most endangered. For example, large sections of the tall grass prairies of the American Midwest can be restored from remnants of native flora and fauna. Such restoration would not significantly impact agriculture, since family farms are already failing in record numbers from overproduction, and could be bought at public expense. Similarly, the government could buy up habitats for rare species so landowners will not be impoverished.[15] Clearing of "swamps" and wetlands for Coastal development should be outlawed; the beach should be public, not a private preserve. The government should no longer provide insurance to such coastal housing at public expense—welfare for the well-to-do. Stiffer enforcement and penalties should attach to the importation, legal or not, of rare species, including mandatory jail sentences.

D. Favor native plants, and extirpate invasive exotics.[16] The preservation of natives will preserve the worldwide gene pool. While removing exotics is controversial, even among ecocentrists, they often outcompete natives, because they have no local predator. The natives have to compete at a disadvantage, since they have to fight off predators as well as struggle against exotics. Moreover, some exotics, such as the Eucalyptus in California, kill all other species around them, reducing competition. Such consequences may be acceptable in their home range, in Australia, but are unfortunate in alien habitats like that of California, which is the home to many endemics and contains one of the most treasured and unusual flora in the world, including the two tallest species, the two longest-lived, and many more unusual plants. California also includes some of the most beautiful wildflowers in the world, such as the *Calochortus* family. In sum, non-natives upset the "balance." (I have already argued for preserving communities, above). Since preserving native species requires preserving their habitats, such habitats should also be preserved and off-limits to development, especially wetlands.

E. Insure the survival of predators. Many studies have shown that predator species keep an ecosystem healthy.[17] In particular, studies of the Yellowstone River in the National Park indicate that the presence of wolves keep the herds of elk and other grazers on the move.[18] If grazers do not have any predators to fear, they will linger at a site and overgraze. In many cases this has resulted in reduced plant populations. Since plants keep the soil in place, loss of plant cover has meant erosion, such as the loss of soils into the Yellowstone River. As a result, the river is no longer clear and it contains fewer fish. Fewer fish means less food for fish predators, such as the bald eagle. Loss of plants also means fewer shrubs, and birds that use the shrubs for nesting decline. In sum, wolves keep the ecosystem as a whole healthy. Coyotes serve a similar role in California and other habitats where wolves have been exterminated. Lions, tigers and other large felines serve a similar role in Africa, Asia, and Central America.

Similarly, sharks and other marine predators keep coral reefs healthy. Coral reefs are one of the most diverse habitats in the ocean. A recent study of healthy and unhealthy coral reefs linked the healthy ones to the presence of predator fish.[19] Overfishing of tasty predators: sharks, tuna, snappers, and others—has led to a decline in the reefs.

2. Conservation for Future Generations

A. Increase the use of sustainable energy supplies and decrease the use of non-sustainables. Our society has to bite the bullet and finance sustainables, especially non-polluting sustainables like wind and solar. All the other alternatives are dangerous, polluting, or both. As a general rule we should never use non-sustainables if sustainables are available. Similarly we should never use living creatures, or material taken from living materials by killing them, if non-living is available. For example, we should use stone rather than wood; plant sources of protein instead of meat. Humans take too much already. Moreover, the virtue of thrift argues for preservation of sustainables for essential uses. I will expand on this point in the next section.

B. Outlaw factory fishing. A two hundred mile limit around each nation should be established and enforced to discourage using the ocean as a "farm." Drift nets should be banned. In areas where the fish have been depleted all fishing should be banned by international agreement and the ban enforced. The seas are not resources, but habitats. One side benefit will be that local fishermen will be able to earn a sustainable living. While it would be better yet if humans adopt a vegetarian diet, including the elimination of seafood, selective fishing is better than overfishing, if some people must eat fish.

C. Tax all non-sustainable technologies on a geometric scale with an environmental impact tax, including energy generation. This will help the transition to sustainables, as well as encouraging inter-generational equity. More non-sustainables will be conserved for future generations. Also, as Thomas Friedman has argued, a carbon tax should be instituted that reflects the full cost of production of electricity from carbon, including pollution and the impact on global warming.[20] Some have suggested cost accounting that includes the costs of using up sustainables and "natural capital," of impact on future generations, and on wildlife.[21]

D. Decentralize energy generation with highway and rooftop windmills, and solar panels to power homes. One hidden source of opposition to renewables is that they can be decentralized: that corporations will lose control of energy. But for a cosmopolitan ethics, this is a move in the right direction. Decentralization means that use and abuse will be localized. There will be no more Chernobyls, Fukushimas, or Three Mile Islands. "Brownouts" and "blackouts" will similarly be localized as will price gouging by energy companies, which occurred in California at the beginning of the century. People will have more control over their own energy use and more choices on how to spend for resources. They will know the actual price of energy and can make their own decisions on how to apply their income. So-called "free market conservatives" should be applauding such moves, but are actually no more in favor of the free market than they are of freedom in general. Nor are they conservatives, since they show no interest in conserving the land, other species, habitats, clean air, or other treasures.

Moreover, sustainables are cheaper, require less massive capital outlay, and thus can be utilized in poorer areas of the world with limited energy needs. These areas could then skip over the pollution stage of the past as they largely did with cell phones instead of landlines. *The New York Times* recently contained articles about how building a solar array takes only two years, while building a nuclear power plant takes up to twenty. Windmills can be erected in even less time.

Third World nations will not have to wait for "investment" by multinationals, but can construct windmills and solar panels village by village for far less cost and requiring far less maintenance and technical expertise. Solar might be especially useful in Northern Africa, where large stretches of the Sahara are almost devoid of any life. Huge solar arrays could be built and the countries in the region could fill their own energy needs without resort to fossil fuels. They could conceivably produce enough energy to export some to neighbors or even to Europe, if a way was found to decrease transmission losses to economically acceptable levels. Similarly, the deserts of Central Asia, the Southwestern United States, Western South America, and Australia could be put to good use, with a minimal impact on area wildlife.

E. Replant forests as carbon soaks. Forests are not tree farms; trees planted for paper or other wood products should be outside of and in addition to public forests. As public, wild forests should be considered wild and inviolable. Forests are plant communities, and therefore do not consist of a single species, a tree farm. Species that are adapted to the habitat are suitable members of the community, particularly natives, which evolved in them. Killing forests for profits is the equivalent of omnicide.

F. Use stone, cement, ceramics or other inert, insulating components as the normal building material, not wood. We need the forests as carbon soaks or sinks, as filters, as habitats and the home to many species, and just for their own sake. Forests are not resources. Stone, widely defined to include cement, stucco and other stone-like materials, is a better insulator, is widely available, less costly, and more durable. Stone structures are also less subject to rot, insect infestation, and other maladies of wood construction. In the long-term cutting down carbon sinks is foolish in a period of global warming, with rising amounts of greenhouse gases, when a cheaper, more durable, and better insulating material is available. Stone insulates both in summer and winter, decreasing demand for electricity. As for the aesthetic element: woodwork is indeed beautiful, but so is stonework. The cathedrals of Europe, the Taj Mahal, the Blue Mosque, much of the Forbidden Palace, modern skyscrapers, and many other architectural wonders are built of stone.[22] However, tree farms outside of forests might be beneficial as carbon soaks, so long as they did not protrude onto wild lands.

G. Non-organic farming should yield to organic, preserving the soil in a relatively unpolluted state. Organic farming includes as much use of organic fertilizers as possible. Organic fertilizer is cheaper in the long run, since it does not derive from petroleum, unlike many artificial fertilizers. The material costs nothing—it is recycled. Organics are easier on the soil, since they add humus and cation exchange capacity, while increasing aeration and drainage. It enriches thin and heavy soils alike. Organic farming also includes recycling organic materials, providing a good place for them instead of "land fills." Organics also feed several soil organisms that are involved in soil food chains, so it indirectly helps soil organisms that are vital to good soils.

Organic farming also includes soil enrichment and preservation. Monocrop agriculture depletes soils of nutrients that one crop uses disproportionately. Recycling organic matter restores much of these essential plant nutrients without poisoning the soil. Soil mass is slowly increased by such additions, offsetting losses due to runoff and other forms of erosion.[23] In other words, organic farmers build the soil up, they do not poison it, deplete it or lose it.

Another aspect of organic farming is the use of biological, not chemical controls for "pests." An old Arab saying has it that "the enemy of your enemy is your friend." This saying is especially appropriate in dealing with crop predators. Just as "one man's garbage is another man's treasure" so is a pest of one species the food of another. In the economy of any habitat, every species has a predator. Biological controls are based on this principle. Birds of prey and cats capture rodents, which might be a threat to grains, while smaller birds like starlings eat insects. Lady beetles, lacewings, and spiders prey on aphids and other pest insects. Having a flock of insectivorous birds in the area of farms is far more effective than to repeatedly apply insecticides. For one thing, the chemicals kill indiscriminately, as Rachel Carson (1958) pointed out over fifty years ago. Thus they kill essential beneficials, especially bees (Freya Mathews, 2010), which are required for pollination, along with targeted pests, and may end up in species higher in the food chain such as insect-consuming birds. It is foolish to kill friendly predators, who do valued tasks for you at no expense. Moreover, chemical residues may poison humans, especially developing children. They also poison soils, especially synthetics that do not break down easily and may poison soil organisms connected with decomposition. Finally, biological controls are cheaper in the long run, since they cost nothing: they derive from local habitats. This fact reinforces the argument for wilderness preservation, where wild species can live and from which they can attack pests. Well-regulated farms will contain woods, or other undisturbed areas where predators of pests can live and raise their young. Working with other species instead of against them helps humans and helps the other species: a "win-win" situation.

Advocates of pesticides will argue that biological controls are slow, unpredictable, and thus that crops could be lost by the time they kick in. But new companies now sell predator insects such as lady beetles. The lady beetles can be released just as aphids begin their cycle in the spring. Similar arrangements could be made with other predators. The Chinese are reputed to have built nests for over-wintering spiders and thereby decreased crop loss from pests that the spiders consume. Creating friendly conditions for insectivores of various genera, including many species of birds, would go a long way to reducing the menace from pests.

3. Ecology for Society

Consumer society should be discouraged and replaced by a sustainable economy.[24] The efficiency of cars and the energy system should be increased each year by law.[25] New homes should be heavily taxed unless they are built of sustainables and energy efficient. Taxis should be banned: public transportation increased. Immigration should be ended, or greatly reduced, especially from countries that have made no attempt to limit their own population and thus are exporting and externalizing their problems. Birth control devices should be subsidized at public expense, so they are as cheap as possible, and citizens educated in their use and encouraged to use them.

Population is a public issue as I argued in chapter seven. The presence of economic refugees, in the form of illegal and legal immigrants, shows the global scale of the population problem. Just as illness has no borders, justifying public health institutions, and coercive measures such as quarantine, so population pressures have no borders. I will make some practical proposals in this section to deal with population problems.

A. Defend laws that control population, and provide public financing for groups like Planned Parenthood. Laws that control population are like any other laws, whose purpose is to protect people from evils like murder, rape, and theft of livelihood. All laws limit freedom of choice in some respect, as Hardin has noted. Laws against drugs restrict the freedom of Hells Angels to peddle methamphetamine to rural teenagers. Laws against speeding through school zones limit the freedom of leadfooted drivers. The issue in population is whether an irresponsible freedom to breed outweighs other rights, not whether it is a violation of rights as such. The other rights are the right to life and the right to pursuit of happiness, in the form of future quality of life. Too many people would threaten the biosphere with global warming, deforestation, and many other evils, and thus all of life. The right to self-defense justifies laws against too many children. Two rights, then, that of life itself and quality of that life, justify coercive laws to control births.

It would be better, of course, if this end were accomplished voluntarily, though birth control. The public good here, the value to everyone of an improved quality of life, requires us to promote birth control.

Prof. Derringh, in a conversation, has raised the issue of whether in population policies there should be programs to prevent children above and beyond the negative disincentive of taxes on too many? At this point, more coercive measures, like those adopted in the People's Republic of China, are generally not needed, if birth control, female empowerment and education are available. However, individual countries could use stricter policies if these fail. They could best accomplish this goal through education of parents in the problems of having too many children, the reduction in quality of life and happiness for all of the members of the family, and thus the advantages of controlling births. International agencies, such as the United Nations, international development agencies or private organizations like Planned Parenthood could assist. Birth control costs should be subsidized for the poor and could be shared internationally, since reduced births would benefit everyone.[26]

Governments that refuse to limit births or that limit availability to birth control, but demand international aid to help feed their starving citizens should be treated as the international pariahs that they are: they are putting future life and quality of life at risk for everyone, freeloading off other nations.

In sum, we have a direct duty to try to limit population in all societies. This can take the form of political and social activism to change laws in such societies so that population control becomes public policy. We can also support organizations that work to control births, such as Planned Parenthood. Once population control becomes a part of the law, we have indirect duties to keep such laws in place, and see that they are enforced as well as to raise awareness among the public about the problems of overpopulation

B. No new immigrants without efforts at population control. Immigration, if allowed at all, should be limited to countries that allow and encourage birth control.

It may be politically incorrect to mention a delicate subject, namely that immigrants from the "developing world," where religions have contributed to the forces in favor of overpopulation, have poured into "first-world" countries out of desperation or for economic advancement.[27] That is, they want a life of affluence, an improved quality of life. As I stated above, however, quality of life is inconsistent with increased quantity of people, since to feed all the additional mouths, both quality of life and affluence must be sacrificed. "Be fruitful and multiply" is a recipe for disaster, as too many mouths result in famine, overcrowding, the exhaustion of energy resources, and the like. Immigration proves this point, since it reflects overpopulation in "developing" countries and thereby creates a lack of opportunity, as the price of labor decreases in reverse proportion to population increases.

Affluent countries that allow laborers from the third world in effect encourage their over-population. The third world governments can export their excess children, externalizing the cost and freeloading off people willing to limit births. Thus efforts at population control in a country will come to nothing if it allows immigration from over-populated countries.[28] Immigrants from countries where religions have contributed to the forces in favor of overpopulation have shown no sign of controlling births once they arrive in the host country. On the contrary, they have continued to produce too many children, straining social resources and creating social tensions. They import their religion along with their children and the strictures of the religions. We need not dwell on the fact that many in affluent countries profit from third world immigration, with cheap labor, maid service, and other benefits for the few.

Critics of this view play the race card and argue that it is racist to argue against immigration from the third world. In response, I would argue that the issue is not "race," whatever that is, but population and culture. "Race" is a social construction without any scientific merit, as I argued in previous chapters. However, culture is not, including the strategic argument for overpopulation by a culture mentioned in chapter seven, namely, the advantages of numbers in influencing the political process. By hiding behind "race" and the guilt of the first world, which often was racist, immigrant groups can get away with ignoring the problems of overpopulation. But a world overpopulated by previous "minorities" is still overpopulated. Their children will also be victims of overcrowding and other problems created by too many humans. The affluence that drew immigrants will disappear with infinite demand for economic goods, and finite resources like petroleum, with an expanded supply of labor. Moreover, privileging such social constructions as "Hispanic," which combines descendants of whites from Europe, Native Americans, African slaves, mestizos, and other "mixed race" groups shows the absurdity of appealing to "race." Hispanics are not a race by any criteria, and the attempt to play the race card by arguing that they would be discriminated against if their "race" were not recognized is absurd. What is racist is creating a race to justify immigration from already overpopulated countries. Immigration from every country should cease; the rich world already has too many people, whether native or not.

The intrinsic value of humans does not imply or require universal obligation to birth potential children. But what of starving children in poor countries? The case for not saving all humans was made by Hardin and Holmes Rolston: lifeboat ethics.[29] However, the duty to restrain population growth could be treated as a reciprocal obligation: an obligation to save present children *if and only if* all countries that are affected agree to adopt population targets and limits. Moreover, they would agree to take steps to restore damaged habitats, prevent excessive human impact on the wild, preserve rare species and habitats, keep a large portion of land as "wild," and other measures to increase conservation. The principle should be that population should not exceed the carrying capacity of any political formation ("nation") within these limits. Thus rich countries will give excess food to poorer countries that adopt a more cosmopolitan ethic. Countries that do not practice birth control should be boycotted.

To feed an overpopulated country or redistribute food from a country that can feed itself to one that cannot is to appease. Appeasement does not work with respect to population any more than with respect to the Nazis. Like families that hide an alcoholic relative and do not take steps to get help, modern societies are "enablers." They are also enablers of rulers in other countries that refuse to take their responsibilities to deal with population issues.[30] Again, countries that practice birth control, limit their population to within the carrying capacity of the land, and conserve wild areas are under no obligation to poor countries in which religious or cultural practices lead to starvation and disaster. To tolerate their religion or culture does not create any obligation to join it or support it. If having too many children creates problems for a culture, it is up to that culture to solve such problems. Indeed, any other approach would be patronizing: implying that such cultures cannot cope. If such cultures or religions prefer starvation of children to population planning, isn't that their "right?"

C. Eliminate tax incentives for excessive numbers of children. There should be no tax incentives for three or more children—on the contrary, a progressive tax on any child after the second should be instituted, to pay for the excessive load on public services. Tax laws should penalize any couple that has more than two children. Nor should any welfare benefits be extended to such couples. Encouraging births is another form of enabling.

The present tax system is supportive of more children since it provides deductions for each child, regardless of number. This system encourages freeloaders, who have more children than others, and thereby take a disproportionate share of resources and public services like education. The cost of educating ten children is far greater than educating two, but the tax burden is not proportional, whether income or property taxes. The principles of a cosmopolitan environmental ethic argue for the opposite: that after a two-child level is reached, no additional deductions should be allowed. On the contrary, each additional child should be taxed at a stiff progressive rate, to discourage additional births. Public services like welfare should also be limited to small families. Paying for the preferences of individuals who have large families is not a public good, and it is not a public responsibility to pay for them. The norm here is replacement numbers alone, that is, two children, which would mean a gradual decline if practiced. Thus such a norm is a non-violent way to reduce population. Our outlook should be cosmopolitan, not provincial.

Critics will immediately charge that this is "anti-children," but I would ask them: is sharing too few resources with other siblings more pro-children than discouraging births and providing the means, through publicly supported birth control education and devices, to prevent them? Is allowing ten children to starve worse than preventing their birth in the first place? Are the periodic famines that have swept many parts of the world, from the Biblical account of the Egyptian famine through potato famine Ireland through modern examples like Bangladesh and the Sahel, better than efforts to control population?[31] Is a poor quality of life for many better than a better quality of life for fewer? Is eliminating the wild to feed too many people better than preserving it for its own sake and for the sake of future generations to see and marvel? I know that some are indifferent to these concerns and all for too many people. But I think that more reflective and wise consideration and evaluation of the alternatives will result in affirming the public good of reduced birth rates for humans worldwide.

The "solution" proposed by the Bush regime for overpopulation is more development. But many third world countries are overpopulated and already using up what little wilderness is left. Development at the expense of the environment violates the conservation that actualizes the virtue of thrift. The Bush regime opposes birth control as a sop to the religious right. But for all their "pro-life" yammering, they have not and cannot provide the resources to feed all those extra mouths.

D. The affluent society combined with overpopulation is a declaration of war on our own descendants. We have an obligation to future generations if we produce children.[32]

Limiting births is not anti-children. On the contrary, I would charge that those who are against limiting population are anti-children. For they would decide the fate of others—of future generations—without the latters' input or consent. In other words, they deny the rights of future generations to choose their own quality of life by overpopulating the earth now. The crowding that future generations will experience will be the result of present decisions, but foisted upon our descendants.

E. More environmental threats to future generations include the salinization of soils and erosion. We have a duty to preserve the air, water, and soil in good, workable condition for our descendants. One such duty is prevention of the salinization and erosion of soils, since these twin evils threaten the ability of future generations to feed themselves. Deny our descendants soil and we deny them the means of life. Soils should be monitored by governments to prevent salt build-up. Soils with poor drainage should either be irrigated more carefully or restricted to specific crops. Farm practices that encourage erosion should be banned and some cover left to preserve soils. Organic matter can be recycled from cities to farms at public expense to renew and enrich soils.

Since ecological relations are holistic and require the interactions of multiple elements, we have a duty to preserve species for the role, known or unknown, that they play in preserving the health of the environment. The species themselves are required as the condition of future generations of such species. We cannot privilege human future generations; we also have a duty to preserve future whooping cranes and bighorn sheep as such, apart from any pleasure or "transformative experience" such species may have for our descendants. But this requires that population be limited, which is another restriction on the rights of future generations. There is no individual right to be born, only rights once the births have occurred.

The effect on future generations also requires that we end global warming, by eliminating the use of fossil fuels and reductions in energy waste. Global Warming will result in rising ocean levels that will reduce farmland and biodiversity, make farming more difficult, and create an unprecedented refugee problem, as millions abandon inundated lowlands for higher elevations. Al Gore has noted that "since one-third of humanity lives within sixty miles of the coastline, the number of refugees likely to be created will be unprecedented."[33] This is especially true if the oceans, in the worst case scenario, rise twenty feet or more, as they did during the last inter-glacial period and earlier, in the age of the Dinosaurs.[34] Estimates of how high the oceans will rise vary with different scenarios, depending on a number of factors. These include how much of the land-based ice on Greenland and Antarctica melts into the oceans. Predicting this quantity can only be a rough estimate. What is not in dispute is that if the ice melts there will be catastrophe, including coastal flooding, surges, seepage into present farmland, and many other consequences.

We cannot leave such matters to our descendants, as our own age is the direct cause. Since forests are carbon sinks, they should be preserved and indeed increased in size, to help reduce Global Warming, to speak nothing of their ecosystemic value and their inherent and unique values. Thus humans will have to place limits on technologies that cause global warming, as a direct threat to our descendants, to thousands of species, and to life as a whole. Rising ocean levels and higher temperatures may wreak havoc on both ecosystems and agriculture in ways we cannot foresee. What we can foresee is that the effects will be negative, interrupting established patterns for the worse. On this ground alone, its effects on future generations of humans, the technology should be changed. The ecological costs of cheap electricity, including global warming, air pollution, acid rain, nuclear waste, and other calamities are not worth it. Since we already have alternatives, it would be foolish to dawdle.

I argued for the virtues of thrift in the previous chapter. The practical measures that we can adopt to favor the environment are all thrifty.

F. Recycling: cans, bottles, plastic, and other resources should be recycled. Leftover food should be saved for another meal, fed to animals or composted. Machines should be fixed for as long as possible before they are replaced, unless replacements use considerably less energy. Above all, we should do without where we can, since buying things we don't need is wasteful. This limitation on our desires is the heart of virtue, in both the Western and Eastern traditions. Plato argued that desire must be restrained by reason; the Stoics that economic goods had little connection with virtue.

Since this work aims at a cosmopolitan ethic, the positive attributes of other cultures and perspectives must be given their due. Religions have been used as an excuse for torture, murder, war, and other crimes. They aim at aggrandizement by increasing the numbers of their followers through births, in order to increase their power. Nevertheless, some of these religions have made positive contributions as well, providing hope for the desperate, love between neighbors, and charity. The ascetic life-style that devolves upon the stricter religious may provide a model for or contribute to an ecologically healthy world. Christianity, Islam, Hinduism, and Buddhism all have ascetic traditions. The ascetic life-style is a thrifty life-style. Non-attachment to things of the world may provide a spiritual basis for ecological thrift, the virtue for an ecologically conscious age.

Hazardous wastes and sewage can be recycled chemically, by using catalysts, oxidation, acids, and other chemical means to turn them into fertilizer or other useful chemicals.[35] Since it might be hazardous to use human wastes on food crops intended for humans, they could be used for the flower trade, for food grown for other species, and to help regenerate eroded land. Other chemicals can be filtered or refined by some other means and reused. Similarly, carbon dioxide and methane could be trapped by hydrogenation into liquid or solid forms and stored or applied to other uses. Indeed, it has recently been proposed that carbon dioxide could be recycled into fuel, although this may absorb more energy than it produces. However, the process could be powered by wind and solar. The chemical plants could be financed by a tax on the sources of pollution: a disposal tax. Since the wealthy have benefited disproportionately from the profits of externalizing pollution, a luxury tax could be applied to pay to set up such operations.

"Garbage" should become an archaism since everything will be recycled, as it is in the economy of the biosphere. Food now thrown away can be either fed to animals or composted. Inorganics like glass and metals can be collected and recycled. Organic chemicals can be recycled or broken down into more useful ones. Machines can be taken apart and the components reused. There is ideally no such thing as "garbage," since everything can be reused or recycled. All that is needed is enough imagination.

G. Tax houses with more than two bedrooms and more than two additional rooms more heavily, with a geometric scale, unless they are "carbon neutral" (Thomas Friedman). Single family homes with more than three bedrooms should be outlawed.

In poorer countries, the whole family often sleeps in one room, or perhaps the parents alone have a separate room. Why does each child need his or her own bedroom, the suburban American ideal? By the average world standard, such a notion would be ludicrous. The rationalization of parents that they do it for their children would be comic, if the results were not so tragic; if they were actually worried about their children, they would work to prevent global warming, destruction of wild habitats, and loss of biodiversity. On the contrary, their children and grandchildren will suffer the consequences of their self-indulgence and status seeking. Many in our generation have let the environment go to hell, using air-conditioning, which has resulted in holes in the protective ozone layer; V-8 engines, which wastes fuel to a ridiculous degree and thereby denies it to succeeding generations; and suburban sprawl, eliminating habitats and living space for other species. They have chosen careers over children: that is, self-fulfillment over their moral responsibility to instill character in the next generation. Generally, such careers have the goal of more affluence and thereby more status: bigger houses, bigger cars, and a bigger negative impact on the environment.

The waste of space intrudes into the wild and impacts global warming. The extra building materials, heating costs, and other add-ons of large houses also impact the environment. If the affluent insist upon such a waste of space, if they are more concerned with status than survival, they should be taxed heavily for the privilege of wasting their income on such trivialities. Private structures that exceed a specific standard should be taxed proportionately by a geometric scale, such that houses with twice the average space should pay four times the taxes; three time the average should be taxed at nine times the average and so on. Status symbols are a hidden environmental killer. Similarly, large automobiles or those that run with poor gas mileage should be taxed progressively.[36] In some European countries, such taxes are already in place. The result is that small cars predominate as people try to lower their costs. Such a tax is not an infringement on freedom if it taxes hidden costs to the environment of fossil fuel use. There is no right to destroy the biosphere. People can still drive wasteful, large cars, but only if they pay the true price.

Our descendents should not pay the price of our excessive affluence and greed—our vices. If we have children, we have a responsibility to them. We must also presume that if we have children, that our children are likely to also have children. We thereby incur a responsibility to all future generations. The threats to future generations include habitat destruction and species extinction, loss of biodiversity, destruction of forests through logging and acid rain, and salinization of soils and erosion. Global warming is an even more severe threat to future generations. If we truly love our children we will not shower them with trinkets but teach them how to live sustainably. We will cut back on status symbols and learn to live with less. We will re-learn the lessons of our ancestors regarding thrift and pass them on to our descendants.

H. Encourage settlement in cities and discourage suburban sprawl.[37] Although this point is contingent upon reigning in population, given equal populations, cities are much more friendly to the environment than tract housing.[38] Energy use is only half as much in New York City compared with a typical suburb. Urbanites share walls, reducing energy costs. Public transportation is more common and more feasible, reducing energy use, and also the impact of fossil fuel used for transportation on global warming. Distribution costs are also reduced. By reducing suburban sprawl, land is freed for preservation. The quality of life in cities can be enhanced by greenways, tree planting along city streets and rivers, public parks and gardens, and other environmentally friendly urban policies. Meanwhile, cities have much more to offer in the way of museums, performing arts, varied cuisines, and other amenities than sterile suburbs.

Public transportation should replace private in any urban center, to the point of banning private cars. The spread of the automobile culture to the Third World has meant an environmental catastrophe in urban centers, whose air is so polluted that it has directly resulted in deaths, violating the right to life. Not only would banning urban autos reduce pollution, it would allow citizens to meet in a public setting and mix.

I. Tax other forms of waste. Nothing is more notorious than the imbalance in which the United States has five per-cent of the world's population, but uses thirty-five per-cent of its energy. In other words, U.S. consumers are using far more than their fair share of a resource that is in decline and also contributes to global warming. Moreover, electrical power production accounts for about forty per-cent of greenhouse gas emissions in the U.S, twice as much as private automobile use.[39] Electrical devices in every sphere of life that have replaced hand labor have accelerated this trend. I include electrical hair dryers, clothes dryers, bread makers, carving knives, can openers, garage door openers, and many others, all of which are wasteful of resources and environmentally pernicious. The use of these devices is superfluous, since the same functions could be performed at least as well by human labor, and was for centuries by our ancestors. Their use is dangerous, since it contributes to global warming, through electrical power generation, and creates problems of disposal.

From an environmental point of view, such devices should be banned, since they are a waste of money. The wind is a far more efficient dryer than any machine, as our ancestors well knew. Moreover, wind is free and does not produce pollution as a hidden cost. The hand can replace most of the other functions, except perhaps in the generally exceptional cases of a disability such as arthritis. Proponents of the so-called "affluent society" might argue that banning such devices outright would be too great an infringement upon freedom. Nevertheless, their use depletes a scarce resource and impacts others in the form of global warming and environmental degradation. There will be no such freedom in the future world that will be overheated, reduced in size, full of refugees, and with aberrant weather, if such practices continue. "Freedom" to destroy the earth is not anyone's right. As usual with such assertions, short-term gain is put before long-term consequences, the mark of foolishness. Thus at the least, there should be a strict consumption tax, and energy tax, a gas tax, and so on, to discourage wasteful use.

The hidden costs of energy generation are externalized. If they were added to the cost of the energy, alternatives would look far more attractive. Thus oil and gas exploration are given tax write-offs, externalizing their cost to the taxpayer.[40] Nuclear waste is buried at public expense. Other ecological costs of cheap electricity include acid rain, global warming, nuclear waste, and many others. Since much of the power is generated for use in devices whose value is dubious, electrical generation above a basic level should be heavily taxed. The costs of creating such energy should be borne by those who profit by them, not the public. Energy generation should never be allowed at the expense of the environment.

J. Require the use of renewable and non-polluting sources of energy, and replace coal, petroleum, natural gas, and nuclear with wind and solar. "Trading" carbon credits is only postponing the inevitable: that we will have to move away from a fossil fuel economy. Continued use of fossil fuels will destroy many species by global warming, violating the obligation to preserve other species and biodiversity. Moreover, a carbon economy is unsustainable, since fossil fuels will not last forever. Nor will "sequestering" the carbon dioxide work. One leak and all that money is wasted. Miriam Horn and Fred Krupp have documented the research on the use of algae to absorb CO_2 from the burning of fossil fuels.[41] But a more far-sighted approach would be to use the algae to absorb the excess carbon dioxide already in the atmosphere. The algae could be used as a food source for livestock, or buried in coal mines as a carbon sink. But why waste money on a polluting, finite source in the first place when a non-polluting, sustainable source is already available? Going with the fuels of the past will only delay the inevitable, when the supply is exhausted. Worse, global warming will increase. The continued use of fossil fuels might also encourage complacency. The move to sustainables should be our priority.

Some will argue that the French have made a deal with the devil and "successfully" implemented the use of nuclear. But this ignores the disposal problem: what is to be done with the waste from such plants? And what of the loss of fish and other wildlife from the vast intake pipes used to extract water from rivers and other sources to cool the towers? Moreover, it costs a great deal more to build a nuclear plant than a wind farm or solar array.[42] Why not go with a sustainable, cheap source instead of an accident waiting to happen? Indeed, private investors have avoided funding for nukes on the legitimate ground that one small mistake and "a billion-dollar asset becomes a billion-dollar liability," as one journal put it.

If nuclear power is to be used, as in France, the plants should have independent cooling systems, that do not use local water sources, whether rivers, lakes or the ocean. They should be large enough to have "overcapacity" for cooling and never draw on local water, which externalizes costs. The water can instead be recycled once it cools. The waste from such plants could possibly be used for city heating, as steam is already used in NYC; or to turn turbines. It could be placed in lead lined tanks that are used to heat water.

If each building were equipped with solar panels as a standard feature, particularly in sunny habitats, just as roofs are now standard, then energy consumption would decrease. These could be placed on both the roof and the sunny sides of buildings. Moreover, the energy use would be decentralized, saving the waste of energy involved in transmission. Batteries could be used to store the energy for night use, especially lighting, and for rainy days. Solar power could be supplemented by wind power—each building having a windmill hooked into a generator along with solar panels. The two would supplement one another, since the loss of power to the solar panels on stormy days would generally be made up for by the extra wind power. The wind turbines could be designed with multiple rotors, with a variety of sizes, to take advantage of minor winds. Moreover, the stored electricity could be used to recharge the batteries of electric vehicles. Since most automobile travel is local, the objection that electric autos cannot make long trips is irrelevant. It is also false: simply recharge after the batteries are low and away you go. Transmission lines from wind farms could be run along highways, reducing the cost of placing the lines, and allowing access from a major market. Also, windmills could line the highways, providing a source of power for autos. Power lines could be strung along major highways, street car style, and vehicles could draw power direct from the wind or stored batteries charged when the wind is blowing.

If enough windmills and solar panels were installed, the demand for electricity for home and office consumption might decrease by a sufficient enough amount that large-scale electrical generation could decline. Tax credits could be used as an incentive to install such technologies. While the short-term costs of going solar and building windmills may be a little higher, the long-term costs will be significantly cheaper, and ultimately consistent with the virtue of thrift. Better, the long-term impact on the environment will be less, not more. Electric companies could train their customers on how to conserve, including when to shut off appliances. The companies should be public, not "privatized," and should aim at reducing consumption, not increasing it.

Another objection to wind and solar is esthetic: some object that all those windmills in a residential neighborhood would be unsightly. But no one objected to TV antennas when those were widespread. Small windmills on each home would be no more unsightly than TV antennas used to be and would take up less space. Similarly, power lines decorate even the most affluent neighborhoods: why are these acceptable but solar panels and windmills not? Nothing is more hypocritical than people demanding electricity but objecting to sustainable power: the so-called "NIMBY" complex. If people want power, they should not externalize the pollution generated by energy production to others. As with epidemics, pollution has no borders: it affects everyone. If some will not accept energy generation in their backyard, a case of out of sight out of mind, they can pay higher prices for buying power from windfarms. Once consumers started saving money by having their own power source, objections of this sort might ease. Further, the health benefits of wind over fossil fuels could be a selling point.

Thomas Friedman[43] has pointed out,

> Telling every individual on the planet who wants of can afford a car that they cannot have one would be changing our lifestyle. But banning cars over a specific weight or engine size, or bringing maximum speed limits back down to 55 miles per hour, or banning taxis that are not hybrids—such efforts do not strike me as fundamentally cramping anyone's lifestyle. Telling people that henceforth we are going to ration electricity (you can only have so much per month) would certainly involve changing our lifestyle. But making it illegal for office buildings in America to leave their lights on after hours, as tens of thousands of companies mindlessly do, as you can see by driving through any major city after midnight, does not strike me as fundamentally crimping anyone's lifestyle. Telling people that they cannot have an iPod or laptop would certainly involve changing our lifestyle. But requiring all iPods and laptops to be made with easily recyclable materials doesn't strike me as fundamentally crimping anyone's lifestyle. Telling people they cannot live in anything more than a five-thousand-square-foot space would certainly involve changing our lifestyle (in the developed world, at least). But telling anyone who wants to live in a more than five-thousand-square-foot home

> that they can do so only if the home is energy net zero—only if it generates through solar, wind, or geo-thermal power as much clean energy as it uses—doesn't strike me as fundamentally crimping anyone's lifestyle. Forcing everyone to ride a bike to work would involve changing our lifestyle. But requiring municipalities to set aside bike lanes running from suburbs to inner cities doesn't strike me as fundamentally crimping anyone's lifestyle (and might make our whole society healthier). Implementing congestion pricing in every major downtown in America, as London and Singapore have done, could involve some lifestyle changes, but if it were accompanied by big new investments in mass transit, not only might we not be worse off, we might actually be better off…"

Biofuels can be carbon neutral in theory, but the cultivation of plants for use in biofuels requires clearing of wild habitats. Moreover, the goal should be to reduce carbon dioxide and other greenhouse gases, not keep them at the same level. Only by reducing greenhouse gases can global warming be reduced, and perhaps reversed. This would be best accomplished by planting more trees, allowing wilderness areas to recover, grow back, and expand; and the use of wind and solar as much as possible. Biofuels are not needed, since an environmentally superior alternative is available. Critics, perhaps financed by the energy industry, argue that the wind does not always blow, nor the sun shine. But storage batteries, plus a national grid of windmills and solar panels could easily overcome this problem. The wind blows more or less continually in some areas, including the Coasts and the Great Plains. Deserts have reliable sunshine. Massive arrays of solar panels and windmills in such areas could provide backup for areas with less favorable conditions. Moreover, on most days, in most areas, either the sun is shining or it is stormy. If it is stormy, the wind is blowing. Thus the lack of solar will be compensated by wind and vice versa. On some days, the wind is blowing and the sun is shining: a double effect that could recharge household batteries for still, gray days.

As Thomas Friedman has emphasized, we need to rethink energy holistically.

> We need a Clean Energy System that is always trying to optimize three things at once—innovation and generation of the cleanest and cheapest electrons, the most efficient and productive use of those electrons and other natural resources, and constant attention to protecting and conserving our natural systems and educating people about their material, spiritual, and esthetic value.
>
> Also, technological systems have to be rethought: e.g. how heat produced from one system as "waste" can be used in another system requiring heat. Again, the braking action of vehicles can be used to generate power.[44]

In sum, we need holistic interaction within energy systems.

Instead of storage batteries, excess electricity could be added to a large regional or even trans-national grid, so that areas generating electricity by wind could feed those without and later get some back from other areas if needed. The power lines could run along existing rights of way on the interstate and other major highways, allowing easy recharging of electrically run transportation, whether cars, busses or even trains. Indeed, windmills could be built all along the tracks, providing plenty of electricity for electrified trains. Braking could actually generate more electricity. On days with excess power generation, the extra power could be used for generating other forms of energy. For example, excess wind/solar could be used in electrolysis of water to produce oxygen (O_2) and hydrogen (H_2) for fuel. Similarly, wind especially could be used to make hydrogen at night when demand for electricity is less. Hydrogen could possibly be used for jet fuel, reducing the impact of air travel on global warming.

K. Collect water from rain runoff. "Waste not, want not," elementary wisdom.

L. Environmental education. As in the many classes on environmental ethics and philosophy of technology, education can help in unlearning bad environmental habits. Among others, children should be educated in the evils of excessive consumption, and the vital importance of conservation. It is hard to educate people on the dangers of continuing as we have, given the consumer society we live in. One way to do so is through appeals to the heart: what about your own children? Should they inherit a poisoned world? One with ongoing global warming? One effective approach I have used is to point out the intimate connection every human already has with their environment. Every time we take a breath or drink water we interact with the environment--indeed, are part of it. We cannot survive for more than a couple of minutes without air; of days without water. This brings it close to home.

Educators should stress non-violence in ethics and discourage violent games, sports, computer programs, and other activities. As Gandhi noted, progress is progress in non-violence, including progress in preventing extirpation of other species. The young should be acculturated to living symbiotically within their habitat, and therefore destructive habits should be discouraged. Since killing animals for gratification of human taste buds is a form of violence, we should encourage vegetarianism. Since humans do not require meat, the latter should be taxed. Some animals consume much more feed to produce meat than others do, for example, it takes eight pounds of feed to get one pound of beef, but only three pounds of feed for one of chicken. Beef should thereby be taxed so its cost is twice that of chicken and chicken three times the price of vegetable proteins like soy and beans. People who eat meat should pay the full price of the environmental impact of such habits. A side benefit will be that less land will be required to provide food for humans.

4. Reform of Institutions

A. Demand corporate responsibility.[45] The impact of global warming will be mostly on nations that had little to do with its causes. The long-term consequences of global warming will fall disproportionately on poorer countries, that produced little of the greenhouse gases that are its cause.[46] This situation is obviously unjust: it makes other pay the price for our affluence: national externalization of costs. Since the affluent cause the problem they have an obligation to help such countries cope with the affects of global warming, with the caveat that poorer countries must adopt some form of birth control.

Corporations, as part of society and profiting off of society have duties to others. Corporations are not removed from society but within and dependent upon society: symbiotic with society. Thus they have duties to others. If companies threaten to outsource their labor to get around regulatory legislation, their licenses to do business can be revoked; their goods kept out of the market. As public corporations, they can be pressured through their corporate charters.[47] Foreign companies and nations can be treated similarly, Countries that pollute at the cost of other nations could be boycotted; nations that ignore the boycott could also be boycotted. Those who argue that boycotts do not work ignore the lessons of South Africa and apartheid. The United States in particular has the clout to end global warming tomorrow, if it had the will. Unfortunately, it is in the forefront of polluters.

The nefarious influence of corporate lobbying on legislation is notorious. The insidious influence of money on politics should also be dealt with: elections should be publicly financed and no outside money—bribes—should be allowed any influence on the deliberations of democratic legislatures. Such bribes should be dealt with severely, with stiff penalties. While corporate spokespersons should be allowed to present their point of view, they should have no greater influence than the most modest citizen. The revolving door between governmental and corporate service should be outlawed, for example, the head of the EPA in the Bush administration getting hired by Exxon-Mobil. A ten-year ban should be in place on being hired by any company regulated or affected by decisions of any governmental agency that any employee worked for. Civil service should normally be for life. A government cannot regulate if its agencies are in bed with those it regulates. Political appointees should be carefully scrutinized to see that they do not pervert the law.

B. Prevent unlimited resource extraction and externalization of costs. Obviously, emitting pollutants, piling up waste, clear-cutting, and other anti-environmental practices affect others; they are therefore within ethical purview.

Garrett Hardin has noted that one strategy pursued by corporations is to keep profits "private," while externalizing costs. Thus the cost of cleaning up pollution is often borne by the taxpayers, who do not share in the profits. Even emission of pollution is externalization: businesses should never have been allowed to release toxins in the first place (General Electric polluting the Hudson River).[48] All businesses should bear the cost of dealing with their own pollution. If they claim that if they have to absorb such costs that they cannot earn a profit, then such a business should never have been started in the first place. It is not economically viable.

Clear-cutting affects wildlife, destroys forests, which are vital for soaking up greenhouse gases, and produces siltification in streams, lakes, and rivers. The soil that has taken millennia to build up is lost in a few years, victim of greed. Forests should be conserved, in accordance with the virtue of thrift, not destroyed. Logging on public land should cease altogether. Indeed, new indigenous trees should be planted in areas that have been clear-cut. The practice of turning forests into tree farms should be discouraged or limited on private land; forbidden on public land. Despite some advantages from clear-cutting, such as the removal of dodder, the disadvantages to the forest as a whole outweigh any advantages. Mining concerns should not be allowed to let leaching operations affect groundwater or streams and rivers. If the mine cannot operate profitably without poisoning others, it should not be operated at all.

C. Polluters should be responsible for and pay to clean up pollution.[49] In particular, energy companies should pay for the pollution and greenhouse gases they emit (as should the cattle industry: methane). This would reveal the true cost of energy and beef to the consumer and make sustainables more attractive. They should not be able to charge it back to the public by raising rates.

D. Set up a world court that deals specifically with ecocrimes, including those who kill endangered species for profit. Those who pollute should be imprisoned with stiff sentences, which may act as a deterrent to the greedy. Those who lie or deceive about global warming should also be prosecuted for ecocrimes.[50]

5. End "Management" of Wildlands

I have been thinking about managerialism for a long time, in part due to the stimulation of different "dystopias,"[51] but also because I was struck by the phrase "forest management." I was both startled and outraged by this phrase, which sums up human stupidity, arrogance, and greed in one phrase. Stupid because it implies that forests need someone to manage them. But this view is historically and biologically myopic. Forests long predated the evolution of humans and managed quite nicely without humans for eons.

It is arrogant because it implies that humans are knowledgeable enough to manage entire ecosystems (Aldo Leopold; Ian Smith, 2010). The opposite is closer to the truth: in numerous instances clear cutting, despite assurances to the contrary from “managers” and “experts,” did not result in rejuvenation but in devastation, for example in Northeastern California. The original trees did not come back. As Dave Foreman stated,

> Great concern is expressed about anthropogenic desertification in the “backward” nations of Africa, Asia and Latin America, but some of the worst desertification in the world is occurring on land managed by the United States Forest Service and Bureau of Land Management; 225 million acres of land in the lower forty-eight are undergoing severe or very severe desertification. In a 1985 inventory of 118 million acres, the BLM found 71 percent to be in an unsatisfactory range condition.[52]

The height of arrogance is thinking that humans can manage the forests better than leaving them alone, better than the forces of evolution. Aldo Leopold and Martin Gorke, among others, notes that ecosystems are so complex that even environmental scientists do not fully understand them.[53] The idea that managers who are not trained in ecology could do so is laughable. The “management ethic” is neither ethical nor scientific but economic—an oxymoron.

Unfortunately, even some well-intentioned environmentalists have bought into the idea that the wild has to be “managed” or that wilderness needs “management.” Joseph Des Jardins (2001, p 157) stated,

> Most wilderness areas are human constructs in that constant human activity is necessary to preserve and manage them. Thus even the decision to set aside and preserve a wilderness area involves the active management of the wilderness. A wilderness may be less something that we discover than something that we create. The decision to create, preserve, and manage a wilderness area, therefore, involves ethical questions of how we *should* manage wilderness ecosystems. Should forest fires be suppressed or allowed to burn? Should species be reintroduced to areas in which they once lived?

But this point is a rationalization of human chauvinism and greed. Wildlife long predated humans and is hardly dependent upon our species—on the contrary, humans are far more dependent upon other species than other species are on us. The wild does not need “management.” It needs to be left alone, with clear zones that are not managed. To argue that “setting aside” something that is already there is somehow “creation” of wilderness defies all logic. The entire universe was already here before the evolution of humans; did we somehow create it? Moreover, “management” often results in mismanagement. Managers think they know what is best for the forest, but they rarely do. Ecosystems are too complex, as Aldo Leopold pointed out long ago. “Management” of the national forests often results rape and devastation that cannot be repaired, at least in the short term, for example, in Modoc County California, where acres have been destroyed under Forest Service "management." Several areas of Nevada under BLM management are even worse off. They look like a moonscape.

Even Dave Foreman believes that restoration requires long-term "management (see quote above). But due to forest and habitat dynamics, this supposition is questionable. The mixed deciduous forests of the northeastern United States reestablished themselves after the invading European colonists initially cut them down. The Catskill mountains, although in second growth, have gone through the cycle of pioneering to climax species and are now primarily mixed deciduous forests again.

Prof. Derringh (personal communication) wonders whether "leaving alone" is even possible as long as people exist. Bill McKibben (1989) has raised the same concern, noting that due to wind and water circulation, there is no area of earth on which humans have not had some impact, especially in the form of pollution. I share their misgivings, but the point here is the narrower one of not managing any undeveloped areas. I would also suggest expanding such areas, and thus the "unmanaged" portion of the globe. While we cannot restore a habitat entirely, the regrowth of woods in the Northeastern United States, in areas that were farmed, shows that other species are resilient, given a chance.

The phrase "forest management" is in any case disingenuous, since the goal is not managing the forest with an eye to its preservation and flourishing, but exploiting it along with the rest of the environment that humans depend upon. In other words, "forest management" is a euphemism for killing thousands of trees for profit and replacing them with tree farms—or not at all. "Forest management" actually stands for destruction of forests. Forests filter the atmosphere, provide some new oxygen, absorb carbon dioxide, and keep the earth cooler. Their destruction would result in human self-destruction.

The mentality that treats people as "human resources" is the same as that which regards the entire earth as a resource. The view is totalitarian since nothing is outside the system of resources and the reach of the managerial "ethic."[54] Even the wilderness, which survived on its own for centuries, requires "management," in this ideology. In other words, there is no wilderness under managerial ideology, since for managers wilderness is simply one type of resource for exploitation. No one seems to question the basic premise: that a biosphere that survived for eons somehow now needs human "management." This is the greatest propaganda victory in history, especially combined with the notion that managers require more benefits and absurdly higher salaries than others do.[55]

Many of the problems of the environment can be traced directly to managerial decisions. One is their control of technology. Sales of technological goods receive priority without any attempt at long-term planning, particularly control of human population. Overpopulation and technology combined form a threat to all life due to human expansion. The effect has been increased global warming, resulting in rising ocean levels. Another effect is habitat destruction and species extinction. Managers evaluate an increasing population as a benefit, because a larger population is simultaneously a larger market and an increased supply of labor, and thereby a decrease in labor costs.

The environment is a requirement for all of life: we cannot survive without air, water, and soil. We also require other species: our food as well as condition of healthy air, water, and soil. Species and habitats are interrelated in symbiosis and holistic relations. Destruction of forests and of biodiversity due to acid rain and global warming are further threats to all life. There are also the ecological costs of cheap electricity, from energy generation, including acid rain, global warming, nuclear waste, and so on.

Another problem that can be traced to management is the attempt to externalize costs by pollution and other decisions that have resulted in a degraded environment. Modern management tries to externalize costs and thus avoid the environmental price of technology. Historical critics might argue that this practice began with industrialization, before the managerial revolution. However, the process has accelerated under managerialism, and is not confined to so-called capitalist countries: the "communist" regimes of Eastern Europe were notoriously responsible for at least as much pollution of the environment as their Western counterparts. "Externalization" of carbon dioxide, the result of burning fossil fuels, has resulted in global warming, the greatest threat to the biosphere, as well as to the well being of future generations. Christopher Manes (1990, ch. 10) argues that corporations complain of ecotage, but break the laws themselves. This is especially the case when they dump wastes into the commons: rivers, soil, or the atmosphere. Decisions made by managers now are thereby affecting the distant future of the whole planet. A true accounting would include the costs of cleaning up technology.

Residual chemicals in the soils, salinization of soils, and erosion are threats to future generations. As I argued in a previous chapter, if we have children we have a responsibility to them, obligations to future generations. Externalizing the costs of our use of the soils to the next generation is as bad as externalizing it to the poor. As one Third World sage put it, we should not "rob from the future."

Still another is the commercial exploitation of beautiful landscapes, which may be profitable, but destroys the beauty that makes it profitable, the unspoiled quality of the area. Some wilderness should be preserved for its own sake—just so that there will be some unspoiled nature—and for future generations to enjoy as a secondary benefit. This is to speak nothing of the beauty of such landscapes as valuable in themselves, and indeed, the source of the value of most wealth, since the purpose of wealth is to buy goods of quality.

Finally, the expansion of managerial control is a threat to other species, as the case of the forests indicate. Humans are one species among others, and depend upon other species for their own survival and well being. Cynics might argue that managers did not create human greed, and that rising expectations and affluence are contributing factors in environmental destruction. But the use of advertising to spread the message of affluence without responsibility is at least partly to blame. Moreover, managerial decisions to pursue affluence have been made in countries that have little public input, notably in Asia (China, Singapore).

To what extent should humans be willing to make sacrifices to preserve the environment? The environment is not a resource and limitations should be placed upon its use that do not reflect economic priorities. This means less pie, not a redivision of the pie.[56] People do not need the vast array of goods that characterize the affluent society. From an environmental point of view, the affluent society, at least the affluence of one portion of it, has been a disaster. The solution is not to create more "affluence," with trivial capitalist trinkets advertised to "create" a market for them, but less. One family does not need a bedroom for every person plus a guesthouse and a vacation home while others live in a single room shack. A single family does not need multiple autos when we already have too many traffic jams and too much pollution.

But this excessive wealth should only be used to alleviate very needy cases. An economically responsible ethic must not only be sustainable, but virtuous. The values of character should have priority over the values of desire and affluence, which no matter how neatly they are rationalized, are barely disguised greed. Limited social engineering of the Popperian type, which changes part of a culture but not all, has unintended consequences, for example overpopulation in the Third World and consequent starvation, based on the introduction of modern medicine, and other technologies; but not population control norms or methods. Although both government and corporations are bureaucratic in organization, government, representing the public interest, has to be strengthened in its regulative role. There can be no rights without enforcement: the role of government, as Thomas Hobbes so forcefully argued.

I realize that this consequence will be the most controversial for progressives. However, progressives generally recognize the need for population control. The point is that the present rate of consumption is not sustainable, especially if what was previously the "Third World" rises to levels of affluence in the First, as it is doing in Asia. The future involves a choice between quality of life and quantity, as the Chinese have realized. Humans can have equality and affluence, but not if they insist upon having too many children. When anyone travels through rural areas in the Third World and sees families in dire poverty with six or more children they cannot ignore the connection between freedom to breed and poverty.

We can interpret one goal of Planetary Management as normalization: similar agriculture everywhere, machines doing logging, and so on.[57] The "domination of nature" means a world totally organized according to reason, totally organized for exploitation: here a logging camp, there a designated wilderness, here a farm, there a suburban "development." In a word, "planning," where technology is the means. The effect on the wild of managerial totalism or normalization through organization would be the death of the wild. Rows of neat corn in the Midwest are not "nature," but rationalization: a form or pattern imposed on "substance," that is, the land. Organization as such is not inherently valuable but only a means whose value must be proven and shown. Previous values can be superceded. Organization is historically contingent, especially specific forms of it like managerialism.

Since the wild can do without managers, we can at least raise the issue of whether they are necessary for the rest of society? In logic a single counterexample contradicts and invalidates a universal. If the forests do not need managers, then their function in other institutions is contingent, not universally necessary (Daniel Simberloff, 2012). One standard for their effectiveness should be participation by employees and stockholders in decisions. Revaluing management would mean that managers serve, rather than rule; that they help, instead of control; that their incomes reflect actual contributions to society, not their power to set absurdly disproportionate salaries. This would be a true managerial revolution.

Another would be that managers refuse to enter where they are not needed, as in the forests. "Nature" is resilient and will gradually heal itself, *if it is left alone*. Good citizens of the world will respect the local customs of the wild, without trying to alter them. They are guests in the wild, not masters.

6. Limit or Outlaw Some Types of Technology

Should some types of technology be limited? Environmental impact and sustainability provide good tests for a technology. A technology that worsens the environment should be discarded in favor of those that are sustainable without polluting or harming other species. Thus fossil fuel use should be phased out in favor of wind and solar. Public transportation should be favored over private automobiles, and the cars heavily taxed, especially in cities, where they are more nuisances than conveniences.

A. Outlaw the use of acid-rain producing chemicals, such as sulfur-bearing coal. Such fuels poison the air and the lakes, and destroy the forests. The sooner we end such folly the better.

B. Get rid of the "damns" in the West. As Marc Reisner has noted, the "damns" have not actually changed the arid habitat of the West, but only allowed agriculture along narrow strips by the rivers.[58] But the price paid by fish and other riparian species has been catastrophic. Moreover, the agricultural payoff has been meager and superfluous in an era of bin-busting agricultural surpluses. A destructive waste of money representing political pork.

C. "Externalizing costs," such as the dumping of PCB's by GE in the Hudson and "coal ash" pits, should be banned, especially near waterways and wetlands. There is always someone downstream—some "other," whether humans or some other species. Waterways are home to fish and fowl consumed by humans and other species as well. The true costs should be calculated in terms that include disposal costs and borne by the consumer. This will correct the market. There should be governmental audits of where all "waste" goes, including chemicals and sewers (policing air and water). Much of this can be recycled or transformed from harmful to neutral by chemical means.

D. Also, get rid of factory farming.

As Matthew Scully (2002), hardly a radical, has written with respect to the suffering of animals,

> About 80 million of the 95 million hogs slaughtered each year in America, according to the National Pork Producers Council, are intensively reared in mass-confinement farms, never once in their time on earth feeling soil or sunshine. Scully, p. 29)
>
> As it is, I see in those factory farms a spirit not much better than that we observed in [the wild animal auctions in] Reno, a casual willingness to subject animals to suffering without end for the sake of little things we could do without. (Scully, p. 127-8)

There is a simple test we can apply to such practices, based on the golden rule and the principle of participation: would any human endure these concentration camp conditions willingly? I think not. In any case, the animals would escape if they could, practical evidence for how outrageous such treatment is. If we condemn the Nazis for treating humans this way, we should not put up with it for the sake of profits. These evils are a disgrace to the idea of humans as "moral agents" or the "rational animal."

Scully has also argued eloquently against superfluous animal research,[59]

> Many scientists and researchers themselves now advocate these methods, most prominently the Physicians Committee for Responsible Medicine. There is no longer any rational basis, they tell us, for the Draize test, dripping chemicals and personal care products into the eyes of immobilized rabbits. We can now test for eye irritancy by use of human tissue systems mimicking characteristics of the eye. We can stop pouring commercial and industrial chemicals into animals. Acute toxicity is determined more accurately by *in vitro* methods using human cell cultures obtained from cadavers. Damage to DNA can be studied in bacteria, as in the Ames assay developed thirty years ago, adopted slowly

by the EPA and yet now internationally accepted. Further experiments on animals for diseases of the heart, nicotine addiction, obesity, and many other disorders are unwarranted because we have already identified their primary causes by studying human populations.

E. Approach radically new technologies with extreme caution, especially those that are potentially catastrophic in their effects, such as genetic engineering. Ban any that constitute a serious threat to life or even to particular habitats.

Thus a vast reorganization of society is called for—away from consumer materialism and toward a thrifty, deep ecology life-style, in which religious and ascetic life-styles can be useful as a model. We should discipline ourselves to want less, not create more: true discipline. The condition of the fulfillment of our desires—which have been manipulated by those with narrowly economic values in the life of consumers—is a world in which desires can be fulfilled and objects of desire obtained. A dead world would end all desires. A life of virtue, not of desire is called for, an ethical life. We should revalue our lives in terms of a life worth living, whose measure is excellence—quality not quantity.

Epilogue

ENVIRONMENTAL PHILOSOPHY AS A SYSTEM: FROM ENVIRONMENTAL ETHICS TO ENVIRONMENTAL PHILOSOPHY

> The integrity of the species and the individual is a function of a field where fullness lies in interlocking predation and symbiosis, construction and destruction, aggradation and degradation. The inclusion of humans generates a philosophy, an ought, an intentionality, a transcendence. Yet for all their options, humans remain insiders. They are not spared environmental pressures; these precipitate the human uniqueness and define human integrity. Holmes Rolston III[1]

A revaluation of cosmopolitan ethics incorporates the insights of the environmental ethicists into ethics, thereby revaluing ethics.[2] Anthropocentric ethics is revalued as a part of ethics, and not the central part. Humans form one part of a larger whole, and the whole is the world of the environment as a whole, the biosphere. Thus environmental ethics is not "applied ethics" but ethics. Starting from the necessities of life is more universal: the necessity for all life of air, water, food, and other essentials. Calling these "physical" goods implies they are lower: a spiritual perspective. Thus cosmopolitanism should be revalued in accord with environmental ethics. Ethics is basic in evaluation and the environment is basic in practice. Environmental ethics is the new philosophical ethics.

1. Environmental Ethics and First Philosophy

The attempt to establish environmental ethics as a revolutionary change of outlook in ethics breaks with traditional Western philosophy, and presumably its reduction: of ethics to a branch of metaphysics, metaphysical ethics. Ethics is not confined to the metaphysical or the ontological as the ought goes beyond the is as its basis, while novel values transform what "is." Expansion of the moral sphere beyond the human to the non-human is based on what we "ought" to do and other ethical standards that regulate human action in terms of their impact on the environment. Environmental ethics expands the field of moral consideration beyond what is to what should be. Living in harmony with the environment is a valuable goal that is morally requisite, and by which the world can be transformed. Expansion of the moral sphere beyond the human to the non-human is based on the "ought" and other norms, standards, and ideals that regulate human action in terms of their impact on the environment.

Value does not require a basis in ontological metaphysics either, and indeed can be viewed on the contrary as the basis of what "is." Human goods are not "given," since they originated as a goal that was creatively actualized, or must be in the future. The reduction of ethics to metaphysics in the tradition makes ethics subordinate to a foundational theory of knowledge. The pragmatists argue for the primacy of practice and thus of value. The practical world reflects value, not theory, and the objects of theory are first made by practice. Moreover, we cannot separate theory entirely from a suitable environment.

What is the relation of Environmental Ethics to First Philosophy, evaluation of competing actualities? The condition of a good evaluation is that it be moral. An immoral evaluation is evil. Since the environment is a requirement of all values and morals, it is the first principle in any evaluation. All evaluation requires an environment even if bad decisions are made to destroy the environment. Environmental ethics, then, is the first principle of evaluation of competing outcomes, competing prospects of actualization. Environmental philosophy is basic to first philosophy.

If human life requires the environment and the goods of the environment as the most basic condition, then the value of the environment is basic to the life and quality of life of humans and other species. The condition of human values, goals and meaning is actualization within moral limits, and the most basic conditions of action are an environment in which to act and creatively actualize goods. As the most basic condition, the environment is the ultimate ground for all human and non-human endeavors; ultimate grounding is first philosophy. The environment, whose value was recognized last, is also the most basic good as the origin of all actual and potential goods. As the origin it is the principle of all good, that permeates all goods, including ourselves. As John Dewey argued, we are in and of the environment. Our value has as its condition the value of the environment. Thus moral limits in our relation to the environment are the most basic wisdom: a first principle of first philosophy.

If the environment is the practically required other, and universal ethics is ecocentric, then environmental ethics is revalued from conditioned to ultimate condition: the most basic principle of first philosophy. Environmental ethics is the other that constitutes first philosophy, since the biosphere is the framework for all other values. Ecocentrism is thus a new beginning for philosophy, in which humans can see their place in a much larger, greater whole, the non-human world. Giving ethics a primary role could provide a potent model for environmental philosophy. The scope of environmental ethics gives everything potential moral considerability.

But is environmental ethics actually a new ethic, or merely a progressive extension of previous ethics?[3] Alternatively, is it a new form of holism? I would reply that it is a new ethic based on its revaluation of traditional anthropocentric ethics and the principles and other factors in traditional ethics. Ecocentric cosmopolitanism is not a mere extension of traditional ethics, nor holistic, although it includes holistic value. The revaluation of the anthropocentric form of cosmopolitan ethics is made from within a larger, ecocentric framework. Also the new approach involves a revaluation of virtue, and other ethical principles, not an extension from anthropocentrism. Rather, anthropocentrism is revalued as within a larger whole.

Moral values are required for all others as their condition—but are revalued in ecocentric cosmopolitanism. Morality has a reciprocal relation with values, consistent with and in accord with moral principles, since they require actualization if we are to make them a part of the (dynamic) world. We can creatively actualize good or bad results. Bad results are, in general, self-destructive, either to agents or others. Good results are sustainable and meet standards set by moral principles: meliorism, reciprocity, and compossibility. In principle we evaluate different theories; in practice different outcomes—but evaluation has moral/environmental conditions, just as the environment requires moral treatment/actualization. Thus evaluation of different consequences in terms of values is the model for first philosophy, with consideration of moral consequences as the first condition and principle of good evaluation. Wisdom aims at creative actualization of a better world.

With this emendation I have not abandoned my earlier work on Radical Axiology, but expanded it while giving it a more concrete content. Ethical considerations are paramount in evaluation. Ethics has been revalued in accord with ecocentric cosmopolitanism. Moreover, the world has *moral* value in the *holistic* sense (compare Omnicide). Thus cosmopolitan ethics and Radical Axiology are compatible. Moreover, we could ask, what do we so evaluate as the better outcome? Surely it ought to be a healthier, greener world, not a polluted, dying one. If we argued that philosophy involves "evaluation of competing realities" in the earlier work, we can add here that a greener world is better than destroying the world for profit.

But does placing the environment as the ground mean or imply a form of realism? I would argue that it does not on several grounds. First, "the environment" is dynamic, both locally and in the long term. It is now well established that there have been several mass extinctions in the history of the earth. New eras in the history of life have superceded older ones. Moreover, habitats change as some species go extinct—whether locally or altogether. New species move in and others move out (Nelson, 2003). Such continual change argues against any fixed notion of a habitat, an ecosystem or a community. What is "real" today is gone tomorrow.

Second, organisms actively transform the "given," none so more than humans. Radical Axiology supercedes intrinsicality of "natures" with creative actualization—F_2 is created not "real" (actualized from possibilities). Reality is created and thus secondary. (But contrary to Richard Rorty, both external and internal relations are actual—relation of properties in any token (color, size, and so on; internal; spatial, temporal, causal, as external.) Because "reality" is historically contingent, it is secondary to wise evaluation of future actualities, that is, environmental ethics as the most general ethics. The movement of moral progress is the reverse of actual historical circumstances: life and culture dependent, not the source of value.

2. System

The relation of the parts of a system of philosophy can be gleaned from these considerations. Environmental ethics is the condition of human knowledge, since an environment is the condition of humans. It is also the condition of art, culture, the economy, religion, medicine, and all other kinds of goods. The philosophy of the future will examine how the environment limits and makes possible different forms of value and other kinds of goods. Environmental Ethics as a first philosophy is cosmopolitan more than cosmocentric. Ethics concerns how to live *in* the world, it is not simply neo-naturalism. The world is the source of human goods even as it is transformed to accord with human good. Creative actualization of knowledge, beauty, health, prosperity and other values is in the world. Environmental Philosophy is the resulting system, based on Environmental Ethics as the content of First Philosophy.

Human knowledge is within and a response to environments. Knowledge arises through human interaction with our environment; all knowledge is knowledge of parts of the environment in some respect. Once knowledge is reconfigured as within an environment, as responding to the environment and as requiring an environment for the life of the knower, the threat of solipsism and skepticism disappears. The theoretical stance in which we could somehow be locked up inside ourselves, is impossible in a world where practical action in an environment supercedes perception alone as the organ of knowledge (scientific experiments; "know how" instead of "knowing that"). The subject detached from an environment is at best an abstraction: with every breath we take we are attached to the world around us, in which we live. The problematic of subjective and objective is superceded by environmental philosophy, which grounds knowledge in the necessity for an environment and in the response of organisms to their habitat. Similarly, art arose as a copy of the surrounding world, although it has grown into something larger. Dewey has made a start at such an epistemology.[4]

In psychology, George Herbert Mead spoke for the other pragmatists in examining the role of the community in shaping the mind of the infant: social psychology. What unites the pragmatists, influenced as they were by Charles Darwin, is a functional view of mind. Psychological processes developed or emerged because they have survival value. Mind is not a distinct substance, nor identical with brain, but emerged over evolutionary time. Since mind is thereby continuous with the rest of the universe, it can interact with it.

Subjective values are only possible at a late stage of culture, when more basic, biological values for life have been taken care of. Subjectivists stranded in the wild in winter would not worry about pleasures or attitudes but survival. This value world is not distinct from the world described by physics, but the same world, contrary to the view of certain reductive "naturalists." Moreover, the "subject" is made by culture, not logically prior to a culture. Ethics and other norms are inculcated into individuals by their parents and other social agents. The model of individual moral subjects counterfactually ignores the social formation of individual character, a cultural universal that sociologists from the era of Emile Durkheim and G. H. Mead, and anthropologists have exhaustively documented. The moral self is a composite of individual impulses and social norms. This point was obvious to ancient ethicists, who would not have dreamed of ignoring moral education and development. Ancient ethics concentrated on the social consequences of action. Actions, not mental events are central. Virtue is manifested and exhibited in action, not feelings, desires and other internal states.

Ecocentrism is more than just an ethic. Ecocentrism also includes the idea that philosophy should center on environmental concerns and issues, not anthropocentric ones. The mainstream in philosophy, following in René Descartes footsteps, is still debating essentially sixteenth century ideas regarding knowledge: representationalism versus realism, subjectivism versus objectivism, and the like. As Richard Rorty (2007) remarked, the return to such issues by analytic and other philosophers tired of the "linguistic turn" appears reactionary to those who have moved past the Cartesian problematic. The Cartesian notion of detached subject—and thus all modern philosophy—seems at best an abstraction from the environment and from *practical* certainties, for example breathing. What is certain is that if we cannot breathe we will die. Anyone can prove this to themselves by trying to hold their breath for more than three minutes. These are the true starting point for "post-modern" philosophy.

Ecocentric cosmopolitanism views such epistemological philosophy as disguised anthropocentrism, since human knowledge is at the center of concern. Ecocentric philosophy would place knowledge of the relation of humans within their world at the forefront, and knowledge of the interaction of the various parts of a habitat functioning in a whole. But more, it would concentrate on moral wisdom, in working out the place of a destructive species in its own environment which it requires for survival.

Appendix:

Rare, Threatened and Endangered: Reflections upon the Categories of Botanical Scarcity

> Rarity is the attribute of a vast number of species of all classes, in all countries...
> Charles Darwin[1]

The California Native Plant Society has recently published a new edition of the *Inventory of Rare and Endangered Plants of California*. This book lists all the known plants in the state that are rare, threatened, endangered, presumed extinct, or in need of monitoring so that their populations do not decline to the point at which they will become rare.[2]

1. Rarity

What is a rare plant? How few individuals of a species must there be such that a plant should be considered rare? Where is the borderline at which a plant in decline moves from the threatened to the rare list? What is the borderline between threatened and endangered? Is rarity a human convention or a feature of certain species? Is it a standard or a concept?

Rarity is defined in the Inventory, following legal codes, as "when, although not presently threatened with extinction, it [a species] is in such small numbers throughout its range that it may become endangered if its present environment worsens." This is primarily a quantitative criterion or standard of rarity, as the number of plants of a species is the decisive factor in determining its rarity. However, qualitative factors are also recognized in the definition and quantitative factors complemented by qualitative in the definition of rarity adopted by the CNPS *Inventory*. That is, the "worsening" of the "present environment" of the plant is viewed as the direct cause of the decline in numbers of the plant. Each plant is linked, then to a specific environment or habitat, to which surveyors tie its survival as a species. If this habitat is altered, destroyed, or eliminated, then species dependent upon such an environment are put at risk.

The relation between specific environments and species varies. As Alan Chickering noted, *Calochortus weedii* is equally at home in soils with granite or sandstone as the "parent" materials of the soil. It is more adaptable, seemingly, than plants that are mostly confined to soils with a single parent material. An example of the latter is the commonly noted association of certain plants with "serpentine," which are soils with at least one of the parent materials consisting in ultramafic rock. (Ultramafic rock contains a high proportion of iron and magnesium, and is the product of geologic changes in which rock formed under the ocean is uplifted onto the land, often locally.) Some plants seem to be totally confined to "serpentine" soils, while others are more common on it than off of it. Another factor in this relation is rainfall, both with respect to quantity and to season. Plants which thrive in the low rainfall totals of the desert will not do well in wetlands or areas with larger quantities of rain, and vice versa. Similarly, plants which have adapted to seasonal rainfall, which includes by far the greater number of plants, do not do well in climates with year round rainfall, and vice versa. Most tropical climates have a wet and a dry season, although the "dry" season is not without some rainfall. This also holds for Mediterranean climates, with wet but mild winters and dry summers. Temperate and polar climates are "dry" during the winter, since water is largely unavailable to the plants, tied up in snow and ice.

Still another factor is heat and cold: the difference between temperate and tropical climates and the plants associated with each. The quality of an environment consists in a number of interrelated elements, then, all of which are important in determining whether a plant can be considered rare. As a desert wetland (sink, oasis) is much rarer than desert drylands, species that are confined to such wetlands are intrinsically rarer than dry desert species. This rarity is environmental and not the result of human intervention in the environment. The habitat suitable for *Calochortus striatus*, which grows in seasonal desert wetlands, would be rarer than that for *C. kennedyi*, a dry desert species, even if the human species had never evolved. The destruction of this habitat is more likely to endanger the plant than the destruction of nearby drylands is to endanger a species adopted to the latter environment, as the greater quantity of desert dryland allows for greater numbers, *ceteris paribus*.

Rarity of habitat is not the only limiting factor on plant distribution. There are other factors that are related to a specific species, rather than its environment, and may be grouped under botanical or biological limitations. Some species produce large numbers of non-viable seeds, such as the Coast Redwood. Annual species produce more seed than bulb species, as they must reproduce themselves annually to survive. They may outcompete bulb species, as they aggressively take over a habitat. Poison plants and plants with unpleasant odors, such as the *Allium* (onion) family, may be better able to keep predators away. None of these are environmental factors, for they are characteristics of the plant themselves. The quantity of a plant may reflect its biological ability to reproduce and compete, not the availability of habitat.

Another factor is endemism, that is, the localized distribution of a plant, often for unknown reasons. *Calochortus weedii* does not grow north of southern Orange and Riverside counties, while *C. albus*, which grows near *C. weedii* in San Diego County, gets as far north as San Francisco Bay, and even further north in the Sierras. The chaparral and sandstone habitat of *C. weedii* continues along the coast for hundreds of miles, and the rainfall does not vary significantly to the north. Yet *C. weedii* does not grow in these areas. The reason for this is not clear, and there are other localized endemics whose scarcity or limitations in range are also unexplained in terms of habitat.

Another factor is the age of the species as well as plant dynamics or succession. Relict species, such as some take the Giant Redwood to be, may be limited in distribution by habitat factors, but their primitiveness, an advantage in some respects, may limit their ability to compete. Other species may decline locally due to competition from taller species, as in the succession from meadows to forests, or pioneer to climax species.

The quantity of plants requisite to a judgement of scarcity is generally a relative number. That is, a comparison may be made with other plant species within a genus or even beyond it and a determination of scarcity can be made. If a plant occurs "in such small numbers throughout its range" relative to the numbers in other species and genera, then it can be adjudged rare. While *Calochortus albus* grows from San Diego to Butte and Plumas counties, *C. tiburonensis* grows in only one known location in the world. Thus the latter is intrinsically rare, even if it is not necessarily threatened. It is rare because the number of stands of the species is small, the numbers of plants at this stand are few, and thus by comparison with a more common species like *C. albus*, it is a rare species. *C. albus*, in turn, is relatively rare by comparison with *C. nuttallii*, which grows in parts of nine states, while *C. albus* confined to California. However, *C. albus* is not considered rare as it is fairly common in California, and does not occur in such small numbers throughout its range that it can be considered rare. On the contrary, it is fairly abundant where it does occur and its overall numbers are probably beyond counting.

In summary, rarity may reflect environmental conditions, independent of human intervention. These include environmental factors, such as rarity of habitat or suitable climate, and also limiting biological factors, such as an inability to compete with more aggressive plants. Further, there may be unknown factors limiting the spread of the plant, despite seemingly suitable and available habitat. Rarity can be adjudged not only from the actual rarity of numbers of a species, but also from lack of suitable habitat, climatic conditions or other favorable conditions for a species. Rarity is not an arbitrary concept then, a convention. It may be intrinsic to a species, given its conditions of life. Rarity can be established not only in relation to other species, in terms of relative quantity, but also in qualitative terms with respect to environmental and biological limitations.

2. Threatened and Endangered

The CNPS Inventory, as I noted, defines rarity as "although not presently threatened with extinction, it [a species] is in such small numbers throughout its range that it may become endangered if its present environment worsens."

Where or when does rarity devolve into threatened? One measure is (again) quantitative: there are too few plants to form a genetically viable population. Another is isolation and the corresponding decrease of each stand's genetic pools. Thus there may be a fairly large number of a species in isolated pockets, but the stands or herds may be so small in themselves and at so great a remove from others of their kind, that they are threatened with extinction. Human settlement compounds this problem, either by breaking up the continuity of the stand or by surrounding it, preventing genetic exchange. The CNPS Inventory regards endangerment as a higher or greater degree of threat. An extremely threatened species is endangered, that is, the species is in danger of extinction. Qualitative factors include erosion or poisoning of soils, decrease in rainfall or sunlight, due to overgrowth, and the invasion of a habitat of a rare plant by aggressive species, whether native or introduced, competitor or predator. A threatened species is one that is in danger of extinction; an endangered one is in immediate danger.

A species may be threatened or endangered due to many causes. These may be conveniently divided into wild or endemic threats, and human threats. The former includes competition and predation, threats from invasive species and habitat changes. Darwin noted that there were many species that have gone extinct in the past. Some of these disappeared due to competition from more aggressive species, or replacement species, that is, new varieties that outcompete their parent. New varieties may or may not have some quality that enables them to compete better. These may include more abundant seed production, more viable seed, more disease or pest resistance, and greater tolerance of variations in rainfall, soil or sunlight. More aggressive species, whether genetically related or not, adapt to a habitat better than less competitive, and this may threaten the survival of the less competitive forms.

Invasive species may also outcompete natives. Examples are the kudzu vine, introduced from Asia into the American southeast, the eucalyptus of California, and the bullfrog. In the absence of predators, the kudzu is so successful that it has virtually taken over many Southern United States wetlands. The bullfrog has expanded its range even into arid areas like Arizona, outcompeting native frogs in the few wetlands. Eucalyptus from Australia were introduced into California in the vain hope that this fast-growing and adaptable genus would be a cheap source of lumber. The lumber potential did not pan out, but the trees adapted well to California's mild, Mediterranean style climate and they killed plants around them. These examples also point to human intervention as factors in threatening species. Indeed, many common "American" weeds are actually natives of Europe, introduced inadvertently by migrating colonists, which have now "naturalized," in their new home. Wild pigs in North America are another example. Rabbits in Australia provide a more deliberate example of an introduced, invasive species. As many biologists and ecologists have pointed out, in the absence of their usual native predators, introduced species can expand unchecked. Humans are the cause of such changes.

Predators, widely defined to include pests and diseases, may also threaten a species to the point of endangerment. Often the predator is itself prey, and is thus kept in check to a degree. Herbivores must be on guard for carnivores while munching their meal. If carnivores are rare or absent in a particular area, the herbivores may graze unchecked and threaten or even extirpate a more delectable species of plant. In effect, as one student of the environment put it, the plants need carnivores to "protect" them. Otherwise, they will be grazed to death, and their numbers will decline. In the Eastern United States, the destruction of the timber wolf has meant unchecked grazing by deer, threatening many species, notably the wood lily, *Lilium philadelphicum*. Some predators are native, but not all are.

Pests may also endanger species. Some pests may be so successful that they kill their host, endangering the species. Both the chestnut and the American elm, once widespread, are now extremely endangered by a blight and a disease respectively. The paper birch, once common in northern woods is also in trouble. The white pine and the aspen are in threatened in the Rockies.[3] The trees may have little or no defense against pests if they are weakened by drought, global warming and other changes. Some of these pests are native, but some are introduced by humans, notably Dutch Elm disease, and the beetle that is destroying conifers throughout the American West. Cold air temperatures normally keep the beetles that threaten the white pine at lower altitudes, but global warming has allowed them to attack their prey at higher elevations. Thus dangers from pests that threaten species may come from changes ultimately induced by humans, whether from "hitchhikers"—species that are inadvertently introduced, like the Asian carp in the Mississippi River—or indirectly, from climate changes.

Habitat changes may result in the shrinking of the range of the plant, threatening its survival. Sudden geologic changes may introduce new parent soil materials that are relatively infertile, making it difficult for species to adapt. The species are used to the older soil and cannot adjust to the chemistry of the new soil. Serpentine endemics are such species. Areas may dry out, whether due to long term trends, as in the Mojave and the Sahara, or human induced climate change. Warming may threaten species that require pronounced cold to do well, or the results of cold. The polar bear is threatened by the reduction in ice near the shore in the Arctic, a by-product of Global Warming. Such changes can occur entirely apart from human intervention, for example in volcanic upheavals. Increasingly, however, they are the result of direct or indirect human activities. The orangutan and the highland gorilla face shrinking habitats as their jungle homes are cut down for timber and agriculture.

Human extirpation, whether deliberate or inadvertent, may also threaten or endanger a species. The expansion of human settlements, whether for agriculture or civilization, has proceeded to the point where it is estimated that fifty per-cent of the land surface is dedicated to human use. When the bulldozers move in, it is rare that anyone checks to see if threatened or endangered species will be destroyed. Such destruction is inadvertent, but careless. The burning of Amazonia for open grazing land, and other forests for lumber involves the destruction of countless species, whose rarity is unknown. Indeed, many rainforest species may not be known to science. Deliberate extirpation is generally of pest species, but may take other species with it, as in the use of general pesticides (Carson, 1958).

Global warming is an indirect threat. Global warming will raise temperatures everywhere, and so it constitutes a potential threat to every species. All the species on earth have evolved within certain temperature ranges. If these shift to the hot side, they may exceed the capacities of any species, regardless of its range. However, altitude increases create conditions that are relatively colder by comparison with those at lower elevations. The species at the tops of mountains differ from those at the bottom; the taller the mountain, the greater the difference. In effect, very tall mountains even near the equator have polar climates, with year-round glaciers and frigid conditions even on summer nights. Thus some species may adapt the way they might have during the ice ages, but in reverse, that is, they will climb to higher altitudes, as some did during the retreat of the glaciers in the past. The problem is that this may restrict their range, since there may be less area near the top of a cone than at the bottom. It may also increase genetic isolation.

Other species may move north or south, depending upon hemisphere, in search of cooler temperatures. The problem is that many will be unable to shift, due to differing conditions of soil, rainfall and other factors relevant to growth. Thus California species may be limited in their northward migration by the greater average rainfall in the Pacific Northwest. Low desert species may not be able to tolerate the colder temperatures of the medium or high altitude deserts to the north, which although hotter, may still have periods of freeze due to altitude, or periodic shifting of the Jet Stream.

With regard to our previous topic, rarity, and its relation to endangerment, a rare species may be threatened or endangered simply in virtue of being rare, whether from wild or human causes. If a species only occurs in one small locale, like *Calochortus persistens*, it is by that fact endangered, since it could easily go extinct with any drastic changes whether geological, meteorological, biological, ecological, or those caused by humans. A new local weather pattern could wipe this species out, as could a new disease or other pests, predators or competition.

NOTES

Introduction

1. In criticism of John Dewey, I would argue that his attempt to reduce the artificial to a type of nature cannot be justified.

2. As Sami Pihlstrom has pointed out, Iris Murdoch (1970/1985) argued that existential ethics also ends in solipsism (Pihlstrom p. 66).

3. What we do privately is our own affair—our freedom. Self-improvement is according to our own lights, at least to the extent of following our own interests. If it is seen as good for the society as a whole, then those who value it as such a good should do it.

4. For a critic of the notion of a philosophical ethics, see Richard Rorty, who defends cultural ethics. See Rorty, 2007, ch. 13.

5. For a counterargument, see Foucault, Michel, *Power/Knowledge: Selected Interviews and Other Writings, 1972-1977*, 1980, esp. p. 106 ff. on "normalization."

6. Held, Virginia, "Feminist Transformations of Moral Theory," orig. publ. in *Philosophy and Phenomenological Research*, Fall 1990, (Supplement), pp. 321-344.

7. For a counterargument see Karl Mannheim's *Ideology and Utopia*, 1936/1964.

8. See note 4.

9. For a parallel argument see Reiner Schürmann, *Heidegger on Being and Acting: From Principles to Anarchy*, 1990.

10. This point is echoed by Paul and Anne Ehrlich in *Extinction*, 1981; also by Jeremy Bentham, Charles Darwin, and Aldo Leopold.

11. Cf. empirical and juridical possession in Immanuel Kant.

12. Cf. Dewey on culture in *Freedom and Culture*.

13. Thus the importance of the Old Testament account of the flood and Noah's rescue of all species. Like Buddhism, I agree that one problem with contemporary civilization is that it is based on desires and even stokes desires through the use of advertising. Ethics means control of desires. Environmental ethics will include the moral evaluation of all desires. I will return to this point in ch. 6.

14. To argue God would not choose evil ignores the actuality of evil in the world, that must be in conformity with divine will, if the ethic is valid. Thomas Aquinas argued that we know the divine will through natural law, which, as a creation of God, reflects divine law (*Summa Theologica*). In that case, raping the environment is a violation of divine law, since it will ultimately result in the extinction of humans, to speak nothing of other species. Moreover, we have no way of knowing if our treatment of other species is a test of our compassion or any other transcendent test.

15. The notion of a "post-modern" philosophy, proposed by Jean Lyotard, includes some elements compatible with our revaluation and others that are not. Ecocentric Cosmopolitanism shares in the overthrow of the Cartesian subject, of ethics based on it, and a historical perspective. However, it rejects the skepticism about universal ethics, *inter alia*. Similarly, analyses of power by Foucault, may also apply to domination of non-human species. See Foucault, 1980. For Lyotard see, *inter alia, The PostModern Condition, A Report on Knowledge*, 1979, trans by Brian Massumi and Geoff Bennington, Univ. of Minnesota Press, (1979) 1985.

16. It is also a break with "phallocentrism," if the legitimacy of this concept is accepted. For it is not engendered, and replaces anthropocentrism with ecocentrism.

Anthropocentrism has been phallocentric in practice; feminists and others argue that logocentrism derives from masculine ways of thinking. Be that as it may, the environment includes both genders in a larger whole. I am indebted to Mr. Simon Høffding for raising these issues.

Chapter One

1. I am indebted to two anonymous referees who criticized an earlier version of this chapter, originally published in *Environmental Ethics*, 2001. However, I have altered the original for this volume. This chapter was inspired by the Anthony Weston-Eric Katz debate on intrinsic value (Katz and Light, 1996). However, I am arguing here for a view i.e. pragmatic but *not* subjective. For a pragmatic subjective view see Weston, Anthony, in *ibid.*, *Environmental Pragmatism*. Cf. Parker, Kelly, (*ibid.*, ch. 1). See also Fuller, R., 1992; Taylor, Bob P., 1990; and Hickman, Larry, 2007. For pragmatic ethics see John Dewey, Clarence I. Lewis, Sami Pihlström, Vernon Potter, and Henry Stuart in the Bibliography; for a critic see Morton White. See also William James, Mackey and Smith for introductions to pragmatism.

Intrinsic value theories in environmental ethics have been defended by, *inter alia*, Eric Katz, in *ibid.*, *Environmental Pragmatism*, ch. 15; J. Baird Callicott, 1989, ch. 8; Miller, P., 1982; Regan, T., 1980; and Holmes Rolston III (1988). For studies of intrinsic value in general see: Beardsley, M., 1965; Garnett, A.C., in Lepley, R., ed., 1949; Greenstein, H., 1971; Khatchadourian, H., 1969; Lemos, N.M., 1994; Nielsen, Kai, 1976; Richmond, S.A., 1976. See also the other authors in Lepley.

2. Dewey and Tufts, *Ethics*, 1932 rev. (Late Works 7, p. 270).

3. Value "monism" should not be confused with value theory. A value theory can argue that all values, instrumental and intrinsic, economic and moral, etc. have one, univocal character or "nature." Value monism is value theory which argues that all values derive their value from one source, e.g. God, the highest good, etc. The different kinds of good derive their value from the one source, and would not have value without this source. 'Monism' is perhaps an unfortunate term since it is primarily metaphysical (cf. Cartesian "dualism"). However, it is correct in locating value in a single source. I am indebted to an anonymous critic to my clarification of this point.

4. This hierarchy of values in the form of *summum bonum* theories is a commonplace in value theory. See e.g., Risieri Frondisi, 1971; Friedrich Nietzsche, 1968, B. IV sect. 1, "Order of rank"; Max Scheler, 1973, e.g. p. 22.

5. See Aristotle, *Nichomachean Ethics* I, 1 et al. for this relation. Although the instrumental value has its locus in a different being than the intrinsic value, the value relation is one of derivation as in *summum bonum* theories. For the value of the instrument refers back qua instrument, to that which it is valuable for, that which is valuable in itself, or intrinsically valuable.

6. Weston, A., "Beyond Intrinsic Value, Pragmatism in Environmental Ethics" in Katz and Light 1996. The substitution of the subject for substance in the relation in first philosophy is Cartesian. See *Meditations on the First Philosophy,* I-II. A critic has argued that there can be subjectively based views that nevertheless attribute intrinsic value to the object, e.g. Callicott's view. However, despite Callicott's arguments for intrinsic value, his is really a relational view, like that of C. I. Lewis. I would further argue that many extensionist arguments are also subjective, since they merely extend

subjective intrinsic value to non-human "subjects of a life." This is the basic pattern of the Animal Liberationists, e.g. Peter Singer and Tom Regan. Such arguments do not escape subjective grounding nor the metaphysics of the subject initiated by Rene Descartes. The ultimate ground of value lies in the relation to a valuing subject, not independent of the subject. Despite extension arguments within them, Utilitarian theories are also anthropocentric, e.g. Peter Singer's utilitarianism. For a pragmatic view, implicitly critical of these approaches see Light, A and McKenna, E., eds., 2004, esp. the Introduction. For Lewis's view see Lewis, C.I., 1946, (1962 ed.). See also his *Values and Imperatives*, 1969.

7. For articles on pragmatism and environmental philosophy see note 1 and McDonald, 2003. See also Weston, in Katz and Light, 1996, ch. 7. and Katz's, "Beyond Intrinsic Value, Pragmatism in Environmental Ethics," *Environmental Pragmatism,* ed. by Katz, E. and Light, A., New York: Routledge, 1996, ch. 14.

8. In Aristotle and Moore. The substance-attribute form: 'It *is* good.'

9. I am not denying personal values, e.g. preferring chocolate to vanilla. However, these reflect previous actualizations, e.g. the creative actualization of chocolate flavored sweets, and thus are derivative, not foundational. Action does not refer back to a subject as it is performed publicly in terms of an end. The achievement is not subjective as it is marked by actual achievement of the end in the world. I challenge those who believe that value is confined to the subject to choose a rotten, fungus-laden, necrotic fruit over a ripe, luscious one. On the same grounds, I would reject the notion of Moore (*Principia Ethica, I*) that value is a "non-natural" predicate, as it can often be perceived.

10. Descartes' mechanical metaphor in *Discourse on Method* is odd in any case, since mechanisms are artifices. Artificial is contrasted with natural for most thinkers.

11. For discussion of the issue of animal language see Olivia Judson, "'Leopard Behind You!'" *New York Times*, 2009; Nicholas Wade, "Boom! Hok! A Monkey Language Is Deciphered" *New York Times*, 2009 and "Sit. Stay. Parse. Good Girl!" *New York Times*, 2011, on recent views of animal language. Cf. also the controversial "speech" of Alex, the African Gray Parrot, who "learned more than 100 words," could pick out appropriate colors when asked and only performed when he was in the mood to. His epitaph was in *The New York Times*, 9/11/2007: Carey, Benedict, "Brainy Parrot Dies, Emotive to the End." See also the essays by Frey, R.G., Rollin, B., and Eckholm, E, in *Animal Rights and Human Obligations*, ed. T. Regan and P. Singer, 1989, P. II.

Dr. Grandin states that, "Animals have the ability to generalize, even though they do not use language…Autistic people like myself…would be denied the ability to think by scientists who maintain that language is essential for thinking…" (quoted from Temple Grandin, *Thinking in Pictures: and Other Reports from My Life with Autsim*, New York: Vintage 1995, p. 159, in Matthew Scully, *Dominion,* p. 244).

12. As Collier and Kalish (1981, p. 103) have observed, almost all the biological needs of humans heve an animal counterpart, even affection (the famous experiment by Harlow and Harlow using a wire and cloth surrogate mother).

13. Robert A. Hinde notes that methods of describing behavior include "description by consequence," that "lead to a specified result." Thus they describe behavior that aims at some future result, including "appetitive" behavior, motion, mating and others (Hinde, p. 19-23). Hinde also notes that some behaviors differ in intensity based on the object being acted upon. Thus "the hammering movement used

by great tits (*Parus major*) to open nuts is almost identical with the movement used in attacking rivals, being distinguished only by the speed of the movement" (*ibid.*, p. 12). The birds know the difference in the object and act accordingly, and this action is goal oriented, as is that of primates, especially chimps, using tools. For Hinde's view see Hinde, Robert A., 1970, ch. 1.

14. *Ibid.*, ch. 1, pp. 21 and 23. Scully also gives an example of animal intelligence and caring. "Witnesses told Mrs. Altsman [who had suffered a heart attack] that LuLu (a sow) waited until a car approached and then walked onto the road and lay down in front of it. Several times she returned to Mrs. Altsman only to leave again and try to get help…Finally a motorist stopped for the prone pig and got out. LuLu knew just what to do. She led the man to the house and the rescue…" (quoted Matthew Scully, 2002, p. 33, quoted from Fuoco, M., "Pig, Soooie! Weeping Porker Rescues Mistress," The *Washington Times*, 10/15/1998). See also Angier, Natalie, "Pigs Prove to Be Smart, if Not Vain," *New York Times*, 10/6/2009.

Brand Blanshard (1961, p. 352) and Mary A. Warren have argued for degrees of intelligence and thereby for degrees of moral considerability. But degrees of intelligence in humans do not affect the issue of their moral considerability. Retarded humans are still accorded basic rights. I do not see then why degrees of intelligence should affect moral considerability in animals. (Warren, Mary A., 1987, repr in Pojman, L., ed., 2001 46 ff.) Regan himself only includes the "higher" animals in his notion of animal rights. See his "The Case for Animal Rights," in *Animal Rights and Human Obligations*, ed. T. Regan and P. Singer, which has a whole section devoted to animal rights (P. IV), including a very good argument by Rachels, "Why Animals Have a Right to Liberty." See also Reagan, T., 1980, p. 99 ff. For a contrary view of intelligence see Hall, Matthew, 2009, 169-181.

15. For Animal Culture, see Yoon, Carol R., "Scientists Say Orangutans Can Exhibit 'Culture,'" *New York Times*, January 3, 2003.

16. My source on this point is my colleague, Prof. Frank Derringh. For other sources on Animal intelligence, learning and knowledge, see, *inter alia*, Gleitman, Henry, Fridlund, Alan J. and Reisburg, Daniel, 2000; Angier, Natalie, 2009, and 2010; Bhanoo, Sindya N., 2011; Fountain, Henry, 2010; Horowitz, Alexandra, and Shea, Ammon, 2011; and Zimmer, Carl, 2011.

17. Cf. Noam Chomsky on "innate linguistic capacity."

18. I am indebted to Marting Gorke on this point. Gorke argues that given the facts of evolution, and that humans evolved from other species, that the others have a similar anatomy, it is certainly plausible to believe, contrary to Descartes, that animals feel pain, hunger, etc. (Gorke, 2003, p. 252 ff.) Cf. also Rolin, B., "Animal Pain," and other articles in *Animal Rights and Human Obligations*, ed. T. Regan and P. Singer, on animal pain and suffering. See also Anton Frantzen, "Dissolving the Explanatory Gap: Do Fish Feel Pain?" for a pragmatic approach.

19. Scully, 2002, p. 196.

20. *Ibid.* He quotes a person with autism: "My experience as a visual thinker with autism makes it clear to me that thought does not have to be verbal or sequential to be real…I do believe that recognizing different capacities and kinds of thought and expression can lead to greater…understanding…" (Scully p. 239, quoted from Temple Grandin, *Thinking in Pictures: and Other Reports from My Life with Autsim*, New York: Vintage 1995, p. 164). Cf. Descartes definition of thinking as generic for mental content, including perception and

memory in *Meditations on the First Philosophy*, B. II; and the articles in T. Regan and P. Singer, Part II, on animal language.

21. Self-actualization is consistent with habitat changes of the sort described by Callicott in his more recent defense of the land ethic (Callicott, in Zimmerman, et al. 1998, 145 ff.) Plant succession is from pioneer to climax species. Amelioration *as* change cannot mean stasis, including a perpetual relation of species. But amelioration as bettering means flourishing. I will discuss meliorism below in section B.

Plant succession in terms of a climax has been challenged by Gorke and the authors he and Des Jardins cite; but this indirectly supports my view. For Gorke's view see 2003, ch. 11 esp. sect. 11b; for Des Jardins see 2001, p. 168 ff. See also Hettinger, Ned and Throop, Bill, "Refocusing Ecocentrism: De-emphasizing Stability and Defending Wildness," in Pojman, 2001. I will discuss this issue in ch. 14.

22. See Meadows, Donella H., "Biodiversity: The Key to Saving Life on Earth," from *The Land Steward Letter*, Summer 1990; repr., in Pojman, L., ed., 2001 206 ff. See also Wilson, Edward, O., "The Little Things that Run the World," *Conservation Biology*, 1: 344-46, repr. in Armstrong, S. J., and Botzler, R. G., eds., 2004.

23. I am indebted to J. Baird Callicott for a "community" concept as opposed to an organic form of holism (1989) But this does not affect the functional relation that different species may play as tokens. When I use terms like "the biosphere taken as a whole," I am not making a metaphor to the human body, but to the general relation of parts to wholes, or the resulting wholes themselves.

"Community" is of course only a metaphor for the human community; it has likenesses and differences to human communities, e.g. that in human communities people move in and out of an area; they die and are replaced (dynamism). Another similarity is that a community has different members—although biotic communities are much more varied in species. See also Hettinger and Throop in Pojman, 2001.

24. Cf. the adaptation of tuberculosis bacilli to penicillin. Personal values are by no means excluded from value as creative actualization, since actualization of personally valued states is a mode of value. What is excluded is the relation to such subjective states as the sole justification or basis of value.

25. Rolston argues a similar point in "Naturalizing Values: Organisms and Species," in Pojman, L., ed., 2001, 76 ff. For a critique of Rolston see Hettinger, Ned, "Comments on Holmes Rolston's 'Naturalizing Values'," repr. from conference presentation in *ibid.*, Pojman. Cf. Mill's arguments against nature as a model in *ibid.*

26. Thus Kant, Dewey, C. I. Lewis, Sir W. David Ross and Emmanuel Levinas, otherwise quite different in their philosophies, are united in their approach to ethics as concerned with the other person. Cf. the second statement of Kant's Categorical Imperative, the quote from Dewey at the beginning of this chapter, Lewis's separation of moral right from other kinds (prudential, technical) in that moral right involves relations to the other (Lewis, 1969, p. 34 et al.); Ross's critique of Utilitarian ethics (1930); and Levinas's arguments for ethics as a first philosophy as concerned with the irreducible other. (Levinas, (1982)/1985, ch. 1.) I am indebted to the work of all these philosophers in the development of my moral views, as readers of this chapter may realize.

27. In one sense, a culture is different: from other cultures. In another sense it gives the individual his/her identity. Thus cultural participation is difference in identity. The same considerations apply to identity in the sense of personal identity.

28. Kant agrees with this point, at least in moral deliberation. I will examine this point in ch. 5.

29. I am indebted to Jakob Liszka on this point, *Moral Competence*, p. 29. See also Gould, Stephen J., "The Golden Rule—A Proper Scale for Our Environmental Crisis," *Natural History*, Sept. 1990, repr., in Pojman, L., ed., 2001, 206 ff.

30. I am indebted to William Frankena, *Ethics*, on this point.

31. For an argument that we have obligations to the other even without reciprocity, see Levinas, 1982, ch. 8, p. 98 ff. I will echo this point in ch. 2 and 4.

32. A tangential approach to this issue is in Pihlström, 2005, which argues that for ethics to make sense, "justifiability can so much as be an issue…only for someone who already adopts a moral perspective, engages in moral discourses and practice. (34)" In my language, they already participate as a moral agent in their society. To opt out of moral responsibilities is freeloading on the morals of others.

33. Cf. Kant's example of the duty of charity in the *Groundwork*. R. M. Hare uses a case similar to my principle of participation as a principle. Could we universalize our act? If it would have bad consequences for us, we would not adopt a (hypothetical) principle such as put debtors in jail, as we might end up in jail. This ties the principle of participation to universality of moral principles, but also brings in consideration of consequences to a Kant-like principle. (Hare 1963 (1972).

34. With regard to economic values, reciprocity means equality in exchange: real choices not artificial scarcity, manipulated choices, or no choice. This rules out imposed choices imposed from above by either political or economic hierarchies: public standards based on prescribed "needs." I will cover this point in chapter 12.

35. Ladd, John, *Ethical Relativism*, p. 108-9

36. E.g. Vedanta, logical positivism, Heideggeriansim, Nietzscheanism, and others. For a critic of the "marriage" of pragmatism and environmental ethics see Samuelson, 32/2.

37. I have examined value as creative actualization in more detail in my book *Creative Actualization*, 2011. See also my "Creative Actualization as a Theory of Value," 2006.

38. Bentham, *Principles of Morals and Legislation*. If the greatest good for the greatest number results in inequalities, then the outcome is not equality. Later versions of Utilitarianism, that are influenced by economism, are impure forms of this ethic, but are more egoistic, and thereby even less ethical than the classic form, articulated by Bentham and Mill. I will briefly examine these forms in ch. 6.

39. Kant, *Groundwork of the Metaphysics of Morals*, P. II. For another reading of Kant from an environmental perspective see Martin Schönfeld's excellent article, "The Green Kant: Environmental Dynamics and Sustainable Policies," repr. in Pojman and Pojman, eds., 2008, p. 49 ff.

For classic treatments of duty see Plato's *Crito*; Cicero, M.T., *De Officiis/OnDuties,* 1974; and Ross, W. D., 1930 (1961); See also Gans, G. F., "What is Deontology?" *Journal of Value Inquiry*, 2001, 27-42; and Vallentyne, P., "The Teleological-Deontological Distinction," *ibid.*, Vol. 21, #1, 1987.

40. Duty is "objective" in Kant in the sense of the first critique: universally necessary *for us* as rational creatures. The role of universals and reason in deontological ethics is thereby tied to a specifically human capacity and intellectualist values in Kant.

41. The arbitrariness of such specific differences has been catalogued and dissected in more detail by the Routleys in "Against the Inevitability of Human Chauvanism" *Ethics and Problems of the 21st Century,* ed. Goodpaster and Sayre (1979), 36-59. See also, Ehrlich, P.R. and A.H., 1998.

42. For Tibor Machan's view see his "Why Human Beings May Use Animals," *Journal of Value Inquiry*, 2002, 9-14. Machan's fallacy is echoed in M. Goldman's "A Transcendental Defense of Speciesism," *JVI*, 35/1, 2001, 559-69. For an excellent critique of speciesism see Rachels, James, "Darwin, Species and Morality," in *Animal Rights and Human Obligations*, Regan and Singer, eds., part 3, sect. 3. Cf. n. 18 (Gorke).

43. Rolston has also used the term self-actualization with respect to the intrinsic value of other species in "Naturalizing Values: Organisms and Species," p. 83, although I thought of it independently (we published the same year). (Cf. Callicott, "On the Intrinsic Value of Non-human Species," ch. 8 of 1989, p. 131). I think that I have developed it more, since it is a type of creative actualization, my theory of value. Although my position is close to his and I admire his work, I believe he, like Callicott, does not quite wish to break with Cartesian inspired categories, e.g. the "objectivity of value," which presumes a world of subjects and objects. My argument here is that environmental philosophy is a fundamental break with Cartesian categories. An organism in an environment cannot be characterized in terms of subjects and objects.

Cf. Naess, Arne, "Ecosophy T: Deep Versus Shallow Ecology," repr. in Pojman and Pojman, 2008 and his notion of "self-realization," in relation to deep ecology.

44. Cf. ch. 4 and the critical evaluation of self-interest; and of indiv. relativism in ch. 6.

45. The notion of moral considerability in environmental ethics probably originated with the article by Kenneth Goodpaster, "On Being Morally Considerable," *Journ. Phil.* 75, 1978.

46. See above p. 9. Nietzsche, *inter alia*, argues on the contrary, that organs are instruments. Others argue by analogy with biological needs, e.g. the need for all organisms for water. Function as needs no more persuasive than functioning well or not—to show functional value. For a treatment of functions see Emmet, Dorothy, 1958 (1972). Cf. Taylor, R., 1966.

47. The relation of wholes to parts in biosphere is environmental not normative, although it may be that one can derive a norm of a healthy whole based on studies of this relation. I am indebted to Gorke on this point.

48. There is also a holistic method, associated with Hegel and perhaps Dewey. We are not invoking such a method here, but may appeal to analysis of environmental wholes in making holistic arguments.

49. For parallel arguments concerning duties to species and the biosphere see Rolston, 1991 (repr. in Zimmerman, et al., 1998, p. 124 ff.) and 1988. See also the articles in P. VIII of *Animal Rights and Human Obligations*, Regan and Singer eds., 1989 (2nd ed.) especially Russow, L-M., and Gunn, A., 1980. Cf. Godfrey-Smith, W., 1979, p. 309; and Parker, K., 1990.

50. Heidegger emphasized this relation in *Being and Time*. The modern form of "worldlessness" is a subject outside an ecosystem.

51. Singer has raised the issue of the duty of beneficence in the context of overpopulation. Is there perhaps a duty of benevolence toward animals and the biosphere? Benevolence may be a virtue, but has a anthropocentric connotations. Yet it

may be important for moral agents. For Singer's view see "A Utilitarian Population Principle," in Bayles, 1976, p. 85, 92. I will address population in ch. 8.

52. I am indebted to an anonymous referee for my clarification of this point. Gorke, in 2003, ch. 28, p 279 ff. also considers this point. Cf. symbiosis.

53. Cf. the second and third formulas of the Categorical Imperative in Kant, *Groundwork of the Metaphysics of Morals*, P. II, p. 100 ff.

54. E.g. in "The Field of Value" in Lepley, ed., 1949, p. 64 ff.

55. I am addressing a point raised by Rosenthal and Buchholz in "How Pragmatism is an Environmental Ethic," in *Environmental Pragmatism*, p. 44.

56. The "right to self-defense" arises in civil society, not the world as a whole. I will cover this point in ch. four.

57. For another view arguing that species can be value bearers, see Rolston, "Naturalizing Values: Organisms and Species," in Pojman, ed., 2001, 76 ff.

58. For arguments in favor of extending the notion of rights to animals see Charles Hartshorne, "The Rights of the Subhuman World," *Env. Ethics*, I/1, 1979, p. 49; Roderick Nash, 1989; and the articles in P. IV of Regan and Singer, 1989, especially Regan and Rachels.

59. I will discuss the smallpox virus and similar hard cases in ch. 7.

60. Gorke (2003, p. 201 ff.) argues that one problem with what he terms biocentrism, in a sense opposed to ecocentrism, is that the former is still individualistic.

61. For Gorke's view see Gorke, 2003, p 60 ff. See also Robbins, Jim, "Hunting Habits of Wolves Change Ecological Balance in Yellowstone, *New York Times*, 10/18/2005 and Judson, Olivia, "Where Tasty Morsels Fear to Tread," *New York Times*, 9/29/2009.

62. I will discuss this relation in more detail in ch. 5.

63. Kant does not ground obligation in intrinsic value, although intrinsic value plays a role in moral deliberation and especially the scope of intrinsic value. Other critics of this model include the pragmatists, especially Dewey, who argues for a more holistic notion of moral considerability.

64. One version of ecofeminism calls for "caring for" nature, but this would result in the death of the wild and its unique values.

65. I will discuss this issue in more detail in ch. 4.

66. These include the prevention global warming; resulting in rising ocean levels, and of salinization of soils and erosion: as threats to future generations, among others. I will cover these topics in ch. 4 and 13.

67. A species is a universal of a type, but not completely universal, for it is limited to one species, as I discussed above.

68. Cf. Garrett Hardin's notion of the rights of the community in 1993, p. 109. Harding notes that the British political tradition places strong emphasis on individual rights, e.g. in the philosophy of Locke. He has accused this tradition of being "community blind," and has emphasized "community-oriented value systems" by contrast with "individual-oriented" ones. Cf. ch. 8 below and the issue of the "right" to reproduce.

69. Thus a revaluation includes a rejection of the Delphic problematic of Socrates. I have discussed this point in my *Radical Axiology,* 2004.

Chapter Two

1. I am indebted to Dr. Larry Nannery for an exchange that improved this chapter.

2. Lynton Caldwell, International Environmental Policy, quoted in Manes, C., 1990, p. 141.

3. Gorke, Martin, *The Death of Our Planet's Species,* 2003, p 196. Gorke also argues against the convergence hypothesis.

4. I argued the last point in ch. 1. I will cover this issue again in the last section. Other authors who have called for a (new) environmental ethic include, J. Baird Callicott, 1989; Martin Gorke, 2003; Regan, Tom, 1981, (repr. in Regan & Singer, eds., 1976/1989, p. 153 ff.); Carolyn Merchant, 1990; E. B. Phihar, "The Justification of an Environmental Ethic," *Environmental Ethics,* Vol. 5, #1, pp. 47-61; James Rachels, 1983; Holmes Rolston, 1988 and 1989; Richard Routley (Sylvan), in Zimmermann, et al., 1998); Peter Singer, 1975; and Paul W. Taylor, 1981.

5. For traditional anthropocentric ethics see: Vernon J. Bourke, 1968 (1970); R. B. Brandt, 1979; J. N. Findlay, 1970; William Frankena, 1963; James J. Liszka, 2002; Hillary Putnam, 2004; Phillip Rice, in Ray Lepley, ed., 1949, ch. 14; Henry Sidgewick, 1966; Stohr and Wellman, "Recent Work in Virtue Ethics," *American Philosophical Quarterly*, Jan. 2002, 49 ff.; Stuart, H.W., "Dewey's Ethical Theory," in Schilpp, P.A., ed. 1939 (1951), pp. 293-333 (sect. 10); William G. Sumner, "Folkways," repr. in Ladd, John ed., 1973, p. 23 ff.; J. H. Wellbank, "Is a New Definition of Ethical Naturalism Needed?" *Jour Val Inq*, Vol. 8, #1, 1974, p. 46 ff.; as well as Plato, Aristotle, Cicero, Hume, Kant, J. S. Mill, T. H. Green and other ethical classics in the Bibliography.

6. This is the gist of the debate between the Sophists and Socrates, e.g. whether justice as the interests of the stronger is "natural" in *Rep.* I. Cf. Callicles arguments in *Gorgias*.

7. Our best source on the Hellenic Cosmopolitans remains Diogenes Laertius' *Lives and Opinions of Eminent Philosophers*.

8. I examined philosophical ethics and the relation of it to culture in ch. 1 C.

9. Callicott discusses the extension of "moral considerability" to larger and larger numbers in 1989). The issue was first treated in Jeremy Bentham, and later by Charles Darwin, by Aldo Leopold in his formulation of the land ethic, by Singer and Regan for Animal Liberation, and by the Ehrlichs for endangered or threatened species. For a critical view see Gorke (2003) who critically evaluates linear model of moral progress as only roughly true.

10. The phrase is Herbert Marcuse's (1964). The totalitarian domination of the earth by technology and other human intrusions has been noted by both left (Marcuse's "technological rationality") and the right (Heidegger in *The Question Concerning Technology,* and Ernst Junger's notion of "planetary mobilization").

11. I am indebted to Regan (1981, pp. 19-34) for the notion of a "management ethic."

12. For this reason, the concept of moral progress rejects the notion of "popular opinion" as decisive in ethics. Integration was not popular in 1950. Popular opinion can be and has been manipulated. Moral progress often requires extensive persuasion before it can be actualized. Popular opinion has to change. I am indebted to John Ladd (1973) for raising this issue in another context. 1973. Gorke paraphrases Meyer-Abich, who calls "nepotism" the ethical position that considers only "one's own tribe"

(reference group); and "nationalism: ones own people or race." (Gorke, 2003, pp. 22 ff).

13. Rolston (1988) anticipated the cosmopolitanism of ecological ethics. In general, Rolston's approach to environmental ethics goes into much richer detail on specific policies. Rolston seems to think that environmental ethics should be confined to the wild; our approach is that obligations are mutual for both humans and animals, i.e., value pluralism but moral monism, or rather, moral consistency. However, Rolston does note some impact of the wild environment on human "urban" areas in the form of air and water quality, etc.

14. The first philosopher to combine a belief in moral progress and cosmopolitanism was Kant, who also used universality as a test for moral consistency and thus as a Categorical Imperative. See note 2 for Kant's view.

15. For a recent view of anthropocentrism see W. H. Murdy, 1975; repr. in Armstrong and Botzler, eds., 2004. For a critical study see David Ehrenfeld, 1978. Cf. Zimmerman, M., 1985.

16. E.g. Bryan Norton. (1987) and Anthony Weston, 1985.

17. Callicott, 1989; Rolston, 1988; Paul Taylor and many others.

18. Regan, 1976. See also his "The Nature and Possibility of an Environmental Ethic," pp. 19-34.

19. A late version of this theory is Richard Rorty's theory of human solidarity, in which we only have to justify ourselves to other humans. He argues that justification in terms of "nature" substitutes nature for God as an impartial standard. He ignores our place in the world and the requirement of an environment for human survival and flourishing. For Rorty's view see Brandom, ed., 2000, esp. the first chapter by Rorty.

20. Some modern humanists argue for a more naturalistic humanism, in which divine intervention is ruled out. However, it is difficult to fathom why this is referred to as humanism and not naturalism. See Radest, Howard, "Modern Humanism," paper delivered to the Society for Advancement of American Philosophy convention, Michigan State Univ. 3/14/08. See also Shook, John 2008.

21. Similarly, those who uphold the Old Testament view of human dominion are attempting to preserve an archaic anthropocentric worldview. If we don't believe in the literal account of creation in the Old Testament on scientific grounds why believe the bit of human arrogance contained in the idea of human dominion? This is faith not philosophy. But religious texts do not actually place humans first in any case.

22. But more, the "subject" may be a myth. "Post-modern" philosophy, and Michel Foucault in particular recognize that the disappearance of the subject requires a new approach. If the subject is denied a ground by contemporary philosophy, the attempt to base value theory on subjective states of humans like feeling or reason is undermined. For a critique of the significance of post-modern philosophy for environmentalism see Callicott, 1989. For the view that the environment is socially constructed see *ibid.*; Anna L. Peterson, "Environmental Ethics and the Social Construction of Nature," repr. in Armstrong and Botzler, eds., 2004. Foucault can be read as a social constructionist.

23. There is something in humans that makes us avoid seeing the truth. We want to cloak our animality in finery, and create myths to avoid seeing the obvious. We want to tell our children that came by way of storks, not sex—a denial of our "beastial" sexuality. But such perspectives belong with Santa Claus and the Easter Bunny as self-delusions. Our need for myths is one source of "humanism," the

delusion that we are something more than, in Rorty's words, "clever animals." Cf. Sorel, G., 1941 where he discusses the role of myths in moving the masses to action.

24. By variant I mean cultural versions, e.g. Kung fu-tse's "Treat others as you would like to be treated." There are versions in Africa as well. Cf. Gould, Stephen J., 1990, repr., in Pojman, L., ed., 2001, 206 ff.; and Wade, Nicholas, "Is 'Do Unto Others' Written Into Our Genes?" *NY Times*, 9/18/2007.

25. Including such figures as Plato, Cicero, Confucius, Mo-tse, Aristotle, the Stoics, Meng-tse, Kant, J. S. Mill, etc. By contrast Hegel, Marx, Nietzsche, Spencer and Heidegger called for a class or race ethic, tribalism in one form or another.

Religious thinkers like Augustine and Aquinas also elevate humans, but of course place humans below the divine. Nevertheless, humans are flattered as at the center of creation.

26. I addressed this point in ch. 1.

27. The late Prof. Reiner Schürmann argued that the movement in European philosophy known as "philosophical anthropology" (Scheler, Gehlen, Plessner, et al.) was a modern form or development of the humanist thesis. The latter included rationality as the specific difference of humans, a hierarchy of values and faith in unlimited progress.

28. The Ehrlichs (1996) have compiled a list of such arbitrary species differences that have been used to defend the indefensible.

29. Matthew Scully (2002, p. 118) criticizes hunting in *Dominion*: "As with so much modern 'conservation,' Mr. Scruton's entire scheme is nothing but an elaborate evasion of duty, leaving us to make hard judgments about the creatures and none about ourselves, to deny them their lives and habitats while denying ourselves nothing." (Scruton is a British Tory who defended fox hunting.)

30. *Ibid.*

31. Humans as privileged because they are the "rational animal" is actually a parasite ethic. A parasite ethic is a freeloader ethic.

32. E.g. Carl Cohen (Reprinted in Cahn and Markie, 1998, pp. 78-80). Callicott discusses similar claims (1989, ch. 8 et al.). Cf. Tibor Machan (2002), a fallacy echoed in M. Goldman's "A Transcendental Defense of Speciesism," *JVI*, 35/1, 2001, 559-69.

The arbitrariness of such specific differences has been catalogued and dissected in more detail by the Routleys (Goodpaster and Sayre eds., 1979, 36-59). See also, Ehrlich, P.R. and A.H., 1998. For an article critical of Cohen see Nathan Nobis, "Carl Cohen's 'Kind' Arguments *FOR* Animal Rights and *AGAINST* Human Rights," paper presented at the Eastern APA.

33. See last note.

34. The only one with the elan to do so was Nietzsche. The rest were prudent but disingenuous.

35. Albert Schweitzer, (in Regan and Singer, eds., 1976, sect. 10; part reprinted in Pojman, L., ed., 2001, 95 ff.

36. My source on this point is Callicott (1989).

37. Wade, Nicholas, "Darwin, Ahead of His Time, Is Still Influential," *New York Times*, 2/9/09.

38. I take it that I do not need to argue against meat industry propaganda that somehow humans "require" meat in their diet. Such lies belong in fiction.

Note for academics: show evidence to your class that humans can survive perfectly well on vegetable proteins and then ask how many will give up meat. Few will. Ask why: they will respond that "it tastes good." In other words, pleasure is the ground to kill other species. And Cohen wants to argue that humans alone are moral agents?!?

39. For an argument that we have obligations to the other even without reciprocity, see Levinas, Emmanuel (1982/1985) ch. 8, p. 98 ff.

40. Cohen also fails to distinguish moral rights from moral goods. "Moral goods are things like benevolence and charity that we have a duty to perform, with no corresponding right on the part of others to expect it of us. We owe these to animals, too, says Prof. Oderberg, but they have no more a moral right to kindness and charity, much less a legal right, than we have toward one another." (Scully, 2002, p. 342, precis of David Oderberg, *Applied Ethics*, who argues animals do not have rights, but that humans nevertheless have duties toward them).

41. Cf. ch. 1: Individual Rights are reciprocal, a relation excluded by anthrofascism.

42. The argument has also been made that, "humans do not exploit nature any more than other species" (Lawrence Nannery, pers. communication). But then they are just another species! This is a statement i.e. hard to take seriously in any case: while other animals kill prey for food, the analogy ends there. Humans do indeed "exploit nature" more than any other species.

43. E.g., Max Scheler 1928/1961, esp. ch. 2; and Walter Weisskopf's comments on humanist ethics. in Maslow, A., ed., 1959/1970, p. 214.

44. Even Rolston (1988) argues this, although he is aware of the emergence of prized "subjective values" from nature.

45. See the article by L. Gerber, 2002, 41-56.

46. I am indebted to C. Merchant, 1990, on this point.

47. As for "human rights," the moral considerability of animals and plants somehow are overlooked: they fail to get considered by anthropocentrism and the duties to animals without which rights are impossible. However, Callicott has argued that the idea of extending human "rights" to the wild is untenable (1989, ch. 1) although he later revised the views expressed in this article. I will discuss rights in chapter four.

48. Dewey and Tufts, *Ethics*, 1932 rev. Cf. p. 74 with ch. 1, p. 7. Cf. p. 132 bottom.

49. Montesquieu, Baron de C.S., 1966. Cf. Darwin on the effect of climate on the Origin of Species, p. 106 ff also ch. 11-13. Cf. p. 158 on the environment as the cond. of life. See also Jared Diamond, 1992, esp. Part IV, and his later work, *Guns, Germs and Steel.* I will examine this issue in more detail in chapter thirteen.

50. See e.g., M. Zimmerman, in Gottlieb, R., ed., 1997, ch. 12. Cf. Wenz, Peter, in *ibid.*, Gottlieb, R.S.; and his "Just Garbage: Environmental Injustice" (repr. in Armstrong, and Botzler, 2004.)

Following Callicott, I have no wish to pick a fight with animal liberationists and want to be "back together again" since anthropocentrism is a far greater threat, and both camps agree on this point. Further, I remain a vegetarian! The "ecofascist" charge may also stem from the fact that Heidegger wrote against technological subversion of the environment and that he also, however briefly or misguidedly, endorsed the Nazi regime in the early thirties. However, the American environmental movement stems from native pioneers such as John Muir and Aldo Leopold, who almost certainly were unfamiliar with Heidegger.

51. Zimmerman, 1997, p. 231. Another tack has been that ecocentrism denies individual rights. I argued in ch. 1 that the notion of rights is confined to civil society. Also, see below.

52. Nor is modern environmentalism romantic, since it traces back more to Darwin than Rousseau, as Callicott explained (1989, passim).

53. There is also the question of whether we should follow nature. Cf. Rolston, 1989, ch. 2. Cf. John Stuart Mill, 1874 (in Pojman, 2001, 89 ff.); and Nietzsche, F., *Beyond Good and Evil*, I, 9.

54. George E. Moore, *Principia Ethica*.

55. Philosophers not willing to take this step are either inconsistent—willing to make such judgments in the case of science but not ethics—or relativists. But scientistic relativism is incomprehensible.

56. Des Jardins, Joseph, R., p. 18.

57. I am indebted to Christopher Manes (1990) for bringing Bradford's comments to my attention.

58. Bradford, George., 1987. Quoted in Manes, C., 1990, p. 157. Bradford's views are shared by H. Smolinski, M. Bookchin and other "social" environmentalists. In my judgment, these individuals are watermelons: green on the outside, red on the inside. Gorky calls these views the "Epistemological argument" for anthropocentrism: "it is always humans who think and pass judgment." We cannot escape human values, categories, etc. (Gorke, 2003, pp. 215-6) and notes that it is a version of anthropocentrism.

59. *Ibid.*, Gorke, p. 218. He notes that it becomes even more difficult in the case of plants, landscapes, rivers, etc.—the more distance there is between humans and other organisms or formations (*ibid.*). In effect, Gorke argues for meliorism in evaluating competing theories, world-views, etc. (p. 241 ff.). For Singer's and Regan's views see Regan, T., and Singer, P., 1976.

60. I am not confusing the ecological notion of balance with the normative notion (balance as justice). Humans are unjust to the environment apart from a balance within a habitat.

61. The "humane" treatment of animals little more than a joke since it consists of politely killing them.

62. The universality requirement is not confined to Kantian ethics, but is included in Dewey's (1932) pragmatic consequentialism and Singer's utilitarianism.

63. Other species are the "other" of reciprocal relations. Recognizing otherness is a part of cosmopolitanism, including other ways of life, knowing and associating. Cf. Hall, Matthew, "Plant Autonomy and Human-Plant Ethics, (2009, pp. 169-181).

64. Of course it wouldn't be a bad idea to adopt this practice universally and extend it to other "domesticated" animals, i.e., adopt a vegetarian diet. It is certainly better if humans do not eat meat; there would then be less impact on the wild, especially in delicate habitats, like the deserts of the West. Meat is not needed for nutrition and certain forms are deleterious to health.

65. My arguments for this point are in ch. 1. See n. 2.

66. However, I have revalued it as an ecocentric principle.

67. The ethic of Ecocentric Cosmopolitanism includes several principles, whose relation is internally reciprocal. This point brings in technical issues in theory of value beyond the scope of this book, which I have discussed in McDonald, 2011.

68. "Harmony" of values and virtues is a requirement of justice to each, just as living in harmony with the biosphere and other species requires doing justice to them all. Harmony is a relation of the parts to the whole, as Plato realized. Justice as fulfilling our duty to the other is a reciprocal relation in which duty is required in moral evaluation, but we should consider environmental consequences. Our obligation includes consequences to the whole and each of the parts, not just the human part, in environmental ethics.

69. Thus some self-interest lies even in justice, if only an abstract self-interest.

70. Such issues are not addressed by science as they are not in the realm of knowledge. However, scientific knowledge of the environment overlaps environmental ethics. Nor am I drawing a fact-value distinction. That "food is good for you" is a "fact-value." Knowledge may be normative as well as descriptive: "one must eat to live." Science may be descriptive but reach normative conclusions-"one should not smoke, because it causes cancer." The fact-value distinction is attenuated by action, since actions manifest values. The distinction is only valid in some cases involving time reference: present is, future ought which becomes realized later and thus may be a future fact.

71. See McDonald 2003, esp. ch. 3 for an examination of Dewey's moral holism. For a critic of the "marriage" of pragmatism and environmental ethics see Samuelson, 32/2.

72. I am indebted to Prof. S. Fesmire on this point. See Dewey's essay (July 1966). Originally published in French.

73. I will address this point in chapter five. Another criticism of Kant is his rationalism. As I argued above, morality cannot be ruled by logic, it can only be served by it—or it becomes corrupted. The reasoning of a thief is a morally corrupt use of logic. Note against both intellectualism and "rationality in ethics." Moral values have priority in evaluation over rationality, as greater.

74. The proof is pragmatic: we reject past technology and political forms.

75. *Republic*, I-II. Cf. Prof. R. Neville (1981) for whom value is a harmony.

76. Learning from our mistakes: failures contribute to a larger whole of success.

77. E.g. in Frankena.

78. Time impinges upon our choice of ends.

79. Moreover, as I will argue in chapter six, self-interest is not an ethical stance at all, as it does not reciprocate with the other.

80. See ch. 3.

81. Should human population be limited? I will address this issue in ch. 8.

82. "Queering" environmental philosophy or "ecofeminism" are neo-anthropocentric insofar as they attempt to project environmental philosophy as subordinate to, or adjuncts to gay or feminist philosophy. Gays and women are in the environment and dependent upon the environment. This point should not, of course, be taken as an argument against feminism or gay lib. Rather, it is to put things in perspective: human concerns, including those of female and gay humans, are only a part of a larger whole. The agenda of liberating human potentialities will not be possible if global warming, overpopulation and destruction of habitats continues. Such views are a retreat to a personal, perspectival ethic, a regression not an advance.

Femimism and gay liberation are parts of a part, and their perspectives cannot form the whole.

83. Bookchin, Murray, (1988); repr. in Pojman, L., ed., 2001, 168 ff. For a critical view of Bookchin see Des Jardins, 2001 p. 246 ff. Cf. n. 58.

84. Cf. animal quality of life.

Chapter Three

1. *Principles of Political Economy*, 1848.

2. I am indebted to Stohr and Wellman, 2002, for a survey of recent work in this area. Also, a Society for the Advancement of American Philosophy conference in Las Vegas in 2001, at which one session was organized around the application of the ethics of character to environmental issues.

3. John Dewey and James Tufts (1932 revision) have emphasized the role of praise and blame in formation of character, and thus the social direction of character development. Virtue is creatively actualized in virtuous actions, and not part of human "nature" as discipline instilled. However, it involves development of human capacities. See also Dewey, 1922, [M14]; and 1930/ 1966.

4. Cf. Sigmund Freud, *Civilization and its Discontents*, whose thesis is that repression is necessary for the development of civilization.

5. The critics of acculturation may also have included the Cyrenaics, and other hedonists. See Windelband, Wilhelm, (1901) 1958, P. I.

6. The term 'behavior' originally implied character, i.e., one who behaves well. However, it was kidnapped by a mechanist school in psychology that was attractive to materialists in philosophy, although its influence in the latter seems to be waning. Functionalists in philosophy of mind sometimes posit an agreement between their views and those of Aristotle, without realizing that Aristotle's view includes the connection of functioning and *telos*, to speak nothing of excellence and deliberation.

7. Here I am outlining Aristotle's view, as opposed to Plato's without endorsing the former. Plato's Socrates argued that if all the virtues had a common name, they must have a common element (*Meno*). Aristotle argued that, e.g., intellectual virtue was of a different sort than moral, since moral virtue was a mean, but intellectual virtues were not.

8. Stohr and Wellman, Jan. 2002, 49 ff.

9. Cf. Talcott Parsons, who argues for the "social structure" of actions in 1937/1968.

10. This point has been made by a number of authors, from G.H. Mead to V. Held.

11. On this view, suicide is permissible, not morally wrong, even if it should be discouraged. However, the consequences for family and friends are a moral consideration. Also, acts of excellence and to the good should always be aimed at.

12. It could be argued that self-defense is not murder but justified killing, and that there are thus no exceptions to the law against murder.

13. On a formal level, harming another treats the other inconsistently with how one treats oneself, violating the identity relation of a culture, equality, as well as the human bond. It is "inhumane."

14. There are rare cases where this might result in the happiness of only one, but this is still not the principle of action. Also, I am ignoring the differences between the average, total and median happiness (Frankena, 1963).

15. Cf. honor among thieves. If even thieves require a virtue to cooperate, the case for virtue is stronger.

16. Also, as Sir W. David Ross (1930) points out, development of character is good in itself, apart from any pleasure or other consequences we may obtain from it, including aesthetic models of ideal human beings.

17. Asceticism may have a relation to self-sufficiency, the Stoic ideal. The "ascetic ideal" of Friedrich Nietzsche (*Genealogy of Morals*) is based on false ideal of self-sufficiency. The materialist version comes from Stoicism (and originally the Sophists and Epicurus?). The idealist version is Christian monasticism, and later, Calvinism, but these derived from Stoicism. Stoicism was rightly criticized by Hegel as based on extremely restricted sense of freedom (free will), not an expansive sense that includes action, which is the locus of value in the original Greek version of the ethics of character.

18. As I noted in another chapter, subjectivity as the inward turn cannot generate duty. Duty is to others, an external, not an internal relation, like that of the self-regarding or self-interested subject.

19. Cf. the fallacy of induction. Bertrand Russell illustrated this fallacy with the example of a turkey i.e. treated very well, and believed itself privileged up to Thanksgiving Day.

20. Aristotle argues this point in *N. E.* 1.

21. Alasdair MacIntyre, 1984.

22. Thus Cooper (1975) argues that Aristotle's account of deliberation includes what have been called deontological principles. Dewey is another author who argues against the separation of duty and virtue, and for that matter of consequences.

23. Cf. the book of Job. I am indebted to James Tufts on this issue, which I addressed in a previous chapter.

24. Prof. Critchley raised these points at the weekly presentation given by guests, faculty and students at the New School, in fall 2005. I briefly discussed his views with him. He has stated (personal communication) that his reservations are largely about MacIntyre, rather than the ethics of character development as such.

25. Emmet, Dorothy, 1958 (1972); cf. *inter alia*, Smith, P. and Jones, O.R., 1986, on functions in philosophy of mind.

26. This is particularly true of those influenced by the work of psychologist Abraham Maslow. Maslow has argued for the value of certain human capacities, thus against the "value-free" model in psychology. See also his theory of values and of human self-realization in (1959/1970). Cf. Drews, Elizabeth and Lipson, Leslie, 1971.

27. This distinction has been noted by others, viz., Emmet 1958 (1972); and also Henry Wang, 2005.

28. Just as knowledge is made, (factum), and reality is made (techne), so are humans are made (character). Thus they transcend subject-object distinction.

29. George Graham raises the issue of whether there is a "self," or whether this is a fiction, as Dennett has argued. This issue is beyond the scope of this chapter, which concerns the moral self. I will cover this point in more detail in a projected future work. In brief, the self has overlapping layers, sometimes referring to the inner self,

sometimes to the entire organism. For Graham's treatment see 2002, 109-134. Cf. William James famous chapter on the self in 1890.

30. I am indebted to Dewey and Tufts (1932) (and George Mead) on this point; and also Virginia Held, 1990, which agrees with pragmatism on this point.

31. The advantage of the "moral self" is that it does without the problematic notion "free will" and gives personality to agency, but is derivative, as a posteriori synthesis of imagination and social norms/cultural values.

I am indebted to the late Karin Stokkink for basic insight that human psychology centers on the imagination. The application of it to issues of character is my own. I will examine the imagination in a future work on the imagination written in collaboration with her, including issues of the development of the self.

32. Cf. Descartes on the role of will in judgments of truth.

33. The value of the "normative self" involves two distinct issues between idealism and nominalism: is there a unique, irreducible self i.e. of value (Windelband) or is self an instance of an *eide* (Duns Scotus and Plato by inference. However, Plato also speaks of the soul as the form of the body; the individuality of the soul is thus problematic). The developmental view lies somewhere in the middle.

34. Pappas, G.F., 1997, 447-472.

35. Conscience as an inner creation both (a) within the Cartesian paradigm, originating from the Neo-Platonic inward turn; and also (b) an aspect of self development of moral autonomy—moral maturity or wisdom.

36. Cf. Putnam, H., 2004.

37. Reciprocal with a moral condition.

38. Cf. Plato's largest and most famous texts: the *Republic* and *Laws*.

39. Thus the agent is not a substance with action as an attribute. Substance cannot be separated ontically from any attribute—they are a unity. There is no "substance" that does not have a color, e.g.. The distinction is an abstraction, at best, but also an attempt to match actuality to our language. Here language is inadequate, for it separates what is a unity. However, the substance and attribute metaphysics itself presumes hierarchy, i.e., a value relation. The *pros hen* relation of Aristotle is an evaluation disguised as ontic analysis.

40. However, the Enlightenment's view of the perfectibility of humans had an origin in Greek philosophy. Descartes and Spinoza followed the Stoics, and Hobbes, Epicurus, while Rousseau admired Plato. Rousseau wrote his "Discourse on the Arts and Sciences" as a critique of modern humans from the perspective of ancient notions of virtue.

41. See, e.g., the section on Condorcet in Gardiner, Patrick, ed., 1959, p. 49 ff..

42. Cf. the *Autobiography* of Ben Franklin, who tried to improve himself in each virtue.

43. Kant argued for a duty to develop our talents, and thus seemingly for a moral requirement to develop them. But I doubt he would argue that a tone-deaf person should develop their musical talents. We should develop our talents, but only if we have a talent.

44. I am arguing here for the pragmatic theory of value, contribution to life as a whole, not "naturalism" in sensu human nature. Albert Camus in "The Myth of Sisyphus" (1991) argues that the most important issue is suicide, or the value of life.

45. Nietzsche thought that "naturalization of morality" (restoring innocence to becoming) would tend to the primacy of the overman, yet recognized that "higher

types" needed requisite conditions (Sparta)—that the emergence of virtue was rare, as the "higher types" were weaker in terms of quantity (anti-Darwin). Thus the overman, whatever his virtues, is an artificial creation of the "philosophers of the future."

46. However, Plato's account is complicated. His rhetorical opponents, the Sophists, argued that the interest of the stronger is "natural" (*Rep.* I, *Gorgias*) and Plato argued against this. But he also argued for nature in the sense of "essence."

47. One counter-argument to infinite perfectibility is the Judaeo-Christian view of the human condition, in which human imperfections are paramount and improvement of the human condition without divine grace is impossible. Perfection belongs to the Gods. However, we can improve ourselves with the grace of God.

48. See Lucretius, *De Rerum Natura*, esp. book V. I have argued for the materialistic origin of the notion of a state of nature elsewhere.

49. Plato argued for the virtue of both the rulers and the ruled against the realism of the Sophists, who argued for self-interest, e.g. in *Rep.* I; Aristotle for *telos* as the fullest nature, thus a model for state in the highest development of human nature.

50. This view is shared by Hobbes, Hume and Mill as well as some feminists, in different forms. Thus it is modern, attached to the Cartesian philosophy of the subject, and the internalization of ethics.

51. My friend Prof. Jim Liszka, a good Humean, argues that "those who have acquired moral sentiment have successfully internalized some moral norms" (2002, p. 39 ff.). But are they sentiments at all as normative? Does the Golden Rule require feelings or reason? Doing what we should do cannot be based on our feelings, as Kant argued, for norms are not feelings. The same goes for his definition of duty in terms of sentiment.

52. For their views see King, M.L., "Letter from a Birmingham Jail," reprinted in Estey, G and Hunter, D., eds., 1971, which also has an essay by Gandhi.

53. Pihlström (2005, p. 34) argues that ethics makes sense, "only for someone who already adopts a moral perspective, engages in moral discourses and practice." They already lead a moral life. Cf. his discussion of Peter Winch's work p. 55 ff.

54. It may refer to the future (telos, ought) or may not ("don't __," "this is good"). Thus ways of life can become a telos, a project or "way."

55. Cf. the ways of life outlined by Aristotle, NE 10; and those of Kierkegaard.

56. Pihlström, 2005, argues this point in a tangential way.

57. Cf. Nietzsche's overall project from *The Birth of Tragedy* onward, the "metaphysics of the artist."

58. However, as I will note below, there are some universal virtues, recognized by all cultures, e.g. courage.

59. For a somewhat different take, see John McCumber, (2000), who argues that the self-censorship was in reaction to political pressure.

60. But cf. Cicero on the virtues of animals, in his *Academica.*

61. Utilitarianism as majoritarian (greatest number) can be tribal but as democratic can be a pragmatic form of universal cosmopolitan ethics. I will discuss Utilitarianism in a later chapter.

62. Cicero (*Academica*) recognized the virtue of animals and his view reflected a consensus among many ancient students of virtue.

63. I.e., presuming for the moment that loyalty is a virtue, following Royce (*Philosophy of Loyalty).* One could argue, like Sahakian, that at best loyalty is a minor virtue and at worst a vice (*Ethics*, 1974).

64. I am indebted to J. Baird Callicott on this point.

65. I will also address the issue of ecocentric duties and virtues in ch. 4, and 9-11. For an argument for the environmental value of humility see Ian Smith, 2010.

66. For deep ecology see Naess, Arne, "The Deep Ecological Movement: Some Philosophical Aspects" (1986); and Glasser, Harold, "Demystifying the Critiques of Deep Ecology," 1996; both repr. in Zimmermann, et al., 1998. See also diZerega, Gus, "Empathy, Society, Nature and the Relational Self: Deep Ecology and Liberal Modernity," (1995), repr. in Gottlieb, R.S., ed., 1997, p. 56 ff.; and Fox, Warwick, "The Deep Ecology–Ecofeminism Debate and Its Parallels," in Zimmermann, et al. For a critical view see Bookchin, Murray, "Social Ecology Versus Deep Ecology," (1988), 11-29; repr. in Pojman, L., ed., 2001, 168 ff. The earlier versions of Zimmernan's, Callicott's, Sessions's, Warren's and Clark's *Environmental Philosophy* contained readings in deep ecology. Unfortunately, the editors deleted these authors and this entire section from the latest edition. One would think that they could have retained at least some of these seminal essays, especially since they have replaced them with trendy texts in feckless "ecophenomenology," whatever that is. They also have excluded the work of environmentally minded pragmatists such as Light, Norton, Weston, Katz and others. For the latter views see Katz, E., and Light, A. 1996.

Deep Ecologists have also tied their views to the ethics of self-realization, which can be traced to T. H. Green (*Prolegomena to Ethics*, 1969 ed). For a discussion see Sahakian, W.S., 1974; and Des Jardins, 2001, 225 ff. See also the Maslowian version, n. 26. Self-realizationism, which tried to reincorporate the ethics of character into modern philosophy is the subjective view of development of character, thus too self-centered. However, Green influenced the pragmatists, particularly Dewey. The pragmatists and deep ecologists have moved self-realization away from subjectivism. From another quarter, Michel Foucault spoke of "self-creation" and this could be taken as a form of self-realization. But the relation of "self-creation" and self-realization is not clear. It is close to self-realizationism in respect of its individualism, and anthropocentrism but not in respect of its aesthetic grounding ("creation" of the self), its emphasis on creating a new way of life, as opposed to standard anthropocentric virtues, and its revaluation of asceticism as "self transformation" (Foucault, 1994 (1997), esp. P. II, sect. 2-3).

67. Plato's *Republic* harmonizes the different virtues with the virtue of the whole: justice, both in the individual and the state.

68. For the ethics of duty, there is a duty to be virtuous (Plato, Kant). But according to some scholars, the ethics of virtue does not include duties. Thus the ethics of duty has a larger scope with more elements. But according to John Cooper (1975), this view is an exaggeration. He argues that in the classic statement of the ethics of character, Aristotle's *Nichomachean Ethics*, Aristotle is closer to Kant than at first appears, since the good man, the virtuous man, acts rightly.

69. I will argue this point in more detail in ch. 8.

70. Aristotle and Aquinas, e.g., realized that humans were biological creatures, albeit ones capable of higher reasoning.

71. Norton has discussed the "transformative value" of the wilderness in a life. In other words, the environment can have a direct impact on one's character, transforming for the better. As Leopold noted, even a few flowers in a weed-strewn lot can give lessons to urban dwellers, a point emphasized by Andrew Light.

72. I will argue for the virtue of thrift in more detail in ch. 9.

73. C.I. Lewis's "contributory value" (1969, p. 92) argues for deferred gratification thus control over desire. Cf. "prudence" p. 108 ff , and 117 ff. on intelligence.

74. My source for this statistic is the *New York Times*. Cf. closed unions. In the thirties, the unions favored full employment, not closed shops that minimize the distribution of wealth.

75. As Thorstein Veblen has pointed out, emulation of the wealthy is both the cement of society and its motivating force.

76. Freedom from economics—from the rule of the necessity of laboring—would be universal prosperity based on complete machine labor. But this entails reliance on (1) technology and (2) bureaucracy (for distribution). Any "return to nature" would precisely destroy the possibility of freedom from economics. Yet the two conditions of economic freedom are often despised precisely by those who desire economic freedom. (Complete freedom would here be unlikely also—engineers and bureaucrats would also be needed to keep things running—as in politics, a minimum state is always needed—even in the most free society. Anti-technologists, such as Heidegger and Marcuse, want a return to *slave labor*, without knowing it. They themselves do not volunteer to do farm labor by hand, without machines. But they advocate such a situation by more or less condemning technology.

77. Despite the availability of condoms and other forms of birth control, the boomers have used abortion as a form of birth control, rather than for accidents or crises (rape, life of the mother), where it would remain the exception. While abortion could aid in the population problem, it should be used as a last resort, given the wide availability of birth control devices.

78. Sue Stafford (2010) argues, like Plato, for treating intellectual virtues like wisdom along the lines of moral virtues (341). Aristotle argued that intellectual virtues are distinct from moral virtues on the basis of the place of moderation in the moral virtues. While moderation is essential for virtues like generosity, there can be no such thing as too much wisdom. The key to the intellectual virtues is different than the rule of moderation.

78. E.g., Rush Limbaugh and Ayn Rand. Rand is the model of the want-to-be princess; Limbaugh, the druggie couch potato.

79. There is no theory of responsibility or of moral value in Karl Marx. Thus he never raises the issue of the point where another's need (or others' need) become exploitation. At what point do others' self-perceived needs stand in need of a value critique? What needs have legitimate value?

80. This point has been argued by others, e.g. Paul Taylor. See also DesJardins, J., 2001, p. 144 ff.

81. Note for the religious: in the religious view, "we are all equal in the eyes of God." Have religions fought for this principle in the world, as Christ and others enjoined? Or have they endorsed the status quo, so that they will get a share of the wealth?

82. As I noted in ch. 2, "Eco-feminism" subordinates the whole, the environment, to one part, human females. Similarly, "queering" environmentalism puts a small part before a much larger and more valuable whole, precisely because it is the condition of all life-styles, including gay ones. Both are anthropocentric, i.e., not really environmental so much as an attempt to make some form of identity politics relevant to other issues of far greater purport.

No doubt I will be criticized as anti-feminine or anti-gay. But such accusations are tactics, which only disguise the fact that the movements for liberation have, with some notable exceptions, failed to come to grips with the very real problem that affluence and self-indulgence have deleterious effects on the environment; that "liberation" has been interpreted to mean raping the environment and thus their own calls for liberation should be within an ecocentric ethic, not its source. It is hypocrisy to believe in liberation for one group, without respect to others.

83. Thus Rabbi Bleich, despite his personal detestation of being lied to, argues that Judaism accepts such white lies to save lives: in any conflict between truth and life, it is vitalist (Bleich, J. David, 1998, esp. ch. 2, 27 ff.; and 3, 61 ff.) Nor is Judaism alone in this evaluation of truth-telling, which is probably the majority view in ethics. Thus Utilitarianism would accept lies that served the greatest good. Kant, as is well known, argued that truth telling did not bear exception. Cf. n. 87 below.

84. However, we should not ignore the environmental conditions of truth, i.e., that the condition of truth is having an environment. We are capable of being truthful because we are a certain kind of creature that arose in a certain environment, upon which we depend. Our lives are dependent in many cases upon recognizing and acting upon the truth. There is another sense of the environmental conditions of truth, viz., a certain causal relation between features of the environment and the holding true of sentences. This is the theory Rorty attributed to Davidson in. Rorty, 1999, p. 33. The environment is the cause of the truth of the sentences.

85. There is a whole ethical literature that argues we cannot be self-deceived and counterarguments. It is enough to argue that humans ignore the issue; formal self-deception is not required.

86. Gorke (2003, p. 183) notes that predictions of the death of seals on the North sea coast, and the "death of German forests" had to be revised later. Thus it is not wise to "encourage expectations too great for science to fulfill."

87. Even here, our first duty is to prevent future Hitlers. Moreover, I do not agree with Bleich (1998) that lying to save a life is justified. For it conflicts with another duty to tell the truth (cf. the commandment to "not bear false witness") and I do not agree that telling the truth to patients will result in their death. On the contrary, withholding the truth from them may make their life worse by preventing them from settling their affairs in a way they would have liked.

88. Howard, Ross and Perly, Michael, *Acid Rain*, 1980.

89. I agree with K. O. Apel's ethics (the "transcendental communication community") on this point: it is a good argument for truth telling; that by being a member of a communication community (society), we tacitly accept this as normative over our use of language. Universalized lying would destroy communication.

90. For the principle of benevolence see Liszka, 2002, p. 376 ff.

91. I examine this virtue in detail in ch. 9.

92. For defenders of loyalty see J. Royce, *The Philosophy of Loyalty*. Cf. Rorty, R., 2007, ch. 3. For a critical treatment see Sahakian 1974.

Chapter Four

1. See especially Tom Regan, "The Case for Animal Rights," in T. Regan and Peter Singer, eds. 1976/1989, which has a section devoted to animal rights, including a very good argument by James Rachels, "Why Animals Have a Right to Liberty." See

also Reagan, T., 1980, p. 99 ff. However, Regan only includes the "higher" animals in his notion of animal rights. For other arguments in favor of extending the notion of rights to animals see Charles Hartshorne, 1979; Roderick Nash, 1989; A. J. Povilitis, 1980; and the articles in P. IV of *ibid.*, Regan and Singer. For a view critical of Regan, see Albrecht, J. M., in Andrew Light and Erin McKenna, eds., 2004; and Warren, Mary A., 1987 (in Pojman, L., ed., 2001, 46 ff.). See also Christian Diehm's (2010) article on Val Plumwood, who is critical of extensionism.

For degrees of reason/intelligence in animals see Blanshard, Brand, 1961, p. 352 ff. For a contrary view of intelligence see Hall, Matthew, 2009, 169-181. For a view that all life has value, and thus an implicit critique of limiting rights to higher animals, see Albert Schweitzer, in *ibid.*, Regan/Singer, eds., P. I, p. 32 ff.

For a good summary of the issues surrounding "wildness," and the wilderness debate see Michael Nelson, "The Great New Wilderness Debate: an Overview," in Pojman, 5th ed., p. 200 ff. See ibid., Hettinger and Throop, p. 186 ff. (esp. the source listed on p. 197, n 5) for arguments against ecological stability (stable habitats).

2. *A Sand County Almanac*, p. 204. I am indebted to J. Baird Callicott (1989) for his reading of Leopold's Land Ethic. However, my comments are my own.

3. *Ibid*, p. 224-5. This notion of a biotic community places Leopold squarely in the tradition of American philosophy. Peirce, e.g., argues that truth and knowledge require a community of inquirers; Royce extended the community notion to ethics. Leopold gave a more ecocentric reading to community, extending it to include the elements of the whole excluded by more anthropocentric interpretations.

4. *Ibid.*, p. 204.

5. For a holistic view, see McDonald, 2003. Cf. Light and McKenna, 2004.

6. Callicott, 1989, ch. 1.

7. In logical terms, it would be contradictory

8. For a slightly different view of rights see Feinberg, J., 1970, in which rights are treated as "claims." I would argue that this characterization applies more to "positive" than to "negative" rights.

10. My colleague Prof. Derringh has made this objection.

11. I argued this point in more detail in McDonald, 2011, ch. 4.

12. Cf. Levinas. The wild as other is a fortiori an example of the irreducible otherness of the other.

13. I argued this point in more detail in ch. 1. The point about finite value is in response to Kant and Hugo Münsterberg (1909) who argued in particular that intrinsic value could not be relative and therefore was absolute. However, animal life is of finite value (even if we grant the "infinite worth" of the individual human, which is at best dubious). If it is of finite value it cannot be of absolute value.

14. Kong fu-tse (*Analects*) agreed: it is our duty to be virtuous; virtuous to do our duty.

15. See the "Doctrine of Virtue", the second part of the 1797b (1994). Cf. *Lectures on Ethics*, 1775-80 (1930, 1963 ed.)

16. "Three Independent Factors in Morals," (July 1966): 198-209, I am indebted to Prof. S. Fesmire for bringing this essay to my attention.

17. Cf. Dr. Martin Luther King Jr., (1971) who argued for civil rights that transcended the law; and the tradition of natural rights in Jefferson.

18. For an argument that defensive functions in plant argue for some form of intelligence see Hall, Matthew, 2009, 169-181.

19. Emmet, Dorothy, 1958 (1972)

20. Friskics, Scott, 2008, 381-399. Cf. Godfrey-Smith, W., 1979, p. 309; Miller, Peter, 1982; and Sessions, George, in Zimmermann, et al., 1998.

21. Stone, Christopher D., 1987. Callicott, 1989 has criticized Stone and argued for moral "monism".

22. *Ibid.*, p. 120.

23. *Ibid.*, p. 75 ff. More strictly, he argues that critics of environmental ethics have argued that trees and other non-human living things do not have interests and therefore their interests cannot be protected. This approach sounds more like an application for a loan than an ethical principle. As I argued, interests can no more generate ethics than greed can.

24. King, 1971.

25. This point was emphasized by sociologist Robert Merton (1968) in his justification of "middle-level theories."

26. 1987, p. 122 ff. The arguments against self-interest apply equally well against hedonism. We may derive pleasure from torture, killing, pederasty and many other practices. But these hardly constitute moral values.

27. *Ibid.,* p. 142 and 148. Cf. cultural relativism. I will examine cultural relativism in chapter eight.

28. Merton, 1968. Cf. maxims in Kant, which formulate rules for particular circumstances and cases.

29. This view is urged by Rabbi J. David Bleich, 1998.

30. Thus Christopher Stone's contention that we might apply different ethics in different areas of our lives (1987, p. 118 ff.) would be morally hypocritical. We call on others to do their duty but are ourselves freeloaders.

31. Since, according to anthropologist and Professor D. Lee, both of these norms are cultural universals, the problem of cultural variation does not arise for them. ("Culture and the Experience of Value," in Maslow, A., ed., 1959/1970, p. 165 ff.)

32. See J. Baird Callicott 1990. Callicott was responding to Stone 1987.

33. The inability to form a contract with at least some wild animals (cf. domestication) is of subordinate importance from a moral point of view. The "contract" of modern liberalism, e.g. Rawls, is based for Kant on moral principles independent of the contract (Kant 1965). The contract is derivative from morals, a special case in which moral universals should be regulative. We cannot form a contract with violent criminals or the Nazis either, but extend them moral consideration. Duty to others is superordinate. (Thus contractualism is not a moral theory, contrary to its supporters.)

34. For situation ethics see especially Kockelmans, Joseph, 1972, Part IV.

35. Identity and difference: this is complicated by what I have elsewhere labeled the correlativity of identity and difference. Each identity is different from other identities; each difference is identified as such. Thus each situation, even if it were unique in some sense, is both an identity and different. I have tried to untangle this complicated relation in McDonald, 2011, ch. 1.

36. Just as moral virtues are similar to scientific predictions, so moral rules are similar to mathematical axioms. Ethical theory and principles can be applied, by analogy with math, although both can also be viewed at the abstract level. The violation of ethical rules in society leads, or would potentially lead, to such and such

results, as Hobbes argued, just as violation of mathematical rules in engineering will cause of result in the collapse of a bridge.

37. Cultural relativism also fails on this ground. I will discuss the latter in more detail in chapter eight. "Act" theories also fail on this ground. They are too atomistic to be moral. To define an act as moral is to appeal to a priori or a posteriori rules (resp. Kant and Mill).

38. There is also the formal rule of how to bring a situation under a concept in Kant, i.e. the practical manifold. Form unites situations of a kind, which then come under a rule. Thus situations are classified as kinds, identities. This is covered in the *Critique of Practical Reason.*

39. The role of universals and reason in deontological ethics ties Kant's morals to intellectualist values. However, Kant argued against reason qua human nature as determining the moral principle, although humans are rational agents. Moreover, morals also require an end of willing: that moral values be willed by the moral agent. Kant identifies the moral agent as a person, who thereby forms the content for the Categorical Imperative i.e. willed, the second statement of the Imperative.

40. C. I. Lewis, following Kant, wrote of moral imperatives, e.g. in his late work *Values and Imperatives*. (See also Murphey, M., 2005). Since morals are voluntary, it would be better to refer to moral principles rather than "imperatives."

41. I examined the role of situations in Dewey's ethics in McDonald, 2004, ch. 4.

42. Formalism vs. rule decisionism: should the rule always govern the situation? Or are we free to decide which rule to apply? The latter would provide more flexibility, since we might invoke a different rule to cover a similar situation that we normally cover with another rule. This would allow us to act differently in certain situations. Ross discusses such situations (in 1930/1961). There is also the problem of unique ends, where there is no covering principle, but inference and extension can be used to formulate such rules.

43. I am indebted on this point to Frankena, 1963, ch. 3, p. 30 ff. where the "rule-utilitarian" critique of "act-utilitarianism," is discussed.

44. This criticism applies as well to the European situation ethicists. See Kockelmans, 1974 for selections from these authors.

45. *Earth and Other Ethics,* p. 115 ff.

46. Steve Fesmire has argued that Dewey was a moral pluralist in Light, A and McKenna, E., eds., 2004. However, Fesmire contrasts pluralism with foundationalism, which is a slightly different distinction than the one Stone makes.

47. Cf. in my *Creative Actualization*, ch. 4.

48. I have argued this point in more detail in McDonald, 1997ch. 3.

49. See Cronin, John, and Kennedy, Robert F. Jr., *The Riverkeepers*, 1999, pp. 149, 235 and 270. For parallel criticisms of the right to private property see Des Jardins, 2001, p. 33-4. Cf. Rowe, Jonathan, (July/August 2005), 54 ff. where he argues that "courts generally have agreed that the assertion of public trust is not a 'taking' that requires compensation."

50. For an argument that we have obligations to the other even without reciprocity, see Levinas, Emmanuel, (1982/1985) ch. 8, p. 98 ff.

51. For another approach to duties to future generations see Dewey and Tufts, 1932, esp. ch. 23.

52. For rights of future generations see, *inter alia*, O'Neill, John, in Gottlieb, R.S., ed., 1997, p. 22 ff.; and Hardin, G., and Baden, J., 1977, in Bayles, Michael D., 1976. For a Utilitarian view see Narveson, J., in *ibid.*, Bayles, ch. 4, esp. p. 62 ff.

In an interesting column, David Brooks asks how life would change if some cataclysmic event sterilized half the planet. He notes that society as we know it would break down for that half, since there would be no future to inspire us. See Brooks, *New York Times*, 7/28/09.

53. Blackstone, W., ed., 1974;. Quoted in Des Jardins, 2001, pp. 100-1.

54. Thus Stone (1987, p. 74) follows those, like Tooley, and Mary Ann Warren, who argue for a subjective notion of intrinsic value. If only "moral persons" have rights there is an implication that helpless dependents have no rights, and Tooley draws this conclusion, arguing that infanticide is allowable. But is the formula that helpless dependents have no rights any more than might makes right? Cf. victims of human violence.

55. Ben Bradley (2001, 43-56) notes the "contributory value" of other species i.e. distinct from the instrumental-intrinsic distinction, as a part of a larger whole as an argument for endangered species.

I will examine this issue from another angle, the actuality of species, in ch. 7.

56. On the relation of rights and duties, see Clement, R.C., 1979.

57. This point has been made by others, including Leopold and Callicott.

58. I will address this point in more detail in ch. 7.

59. Cf. Godfrey-Smith, W., 1979, p. 309; Miller, P., 1982; Sessions, George, in Zimmermann, et al., 1998; Wilson, Edward, O., "The Little Things that Run the World," repr. in Armstrong and Botzler, 2004.

60. There are exceptions. Apparently, *Tigridia multiflora* has been collected at widely separated stands in Mexico, but there are only a few such stands. It could be that this species is actually much less rare, and that other occurrences of it have been overlooked by field botanists. But until such stands are discovered, this species must be accounted fairly rare, as far as is known. My source for this species is E. Molseed's doctoral dissertation on *Tigridia*, 1970.

61. I will examine population in ch. 8.

62. I am indebted to Callicott for the underlying axiological argument on this point. Cf. Guevara, Sergio, and LaBorde, Javier, 2008.

63. Cf. Brian Norton (1987) on the attempt to apply cost-benefit analysis to the value of such beauty. Beauty is not an economic commodity except in a whorehouse. Economists who would engage in such analyses must have learned value theory at the Mustang Ranch.

64. Charles S. Peirce and Samuel Alexander (1933) argue for this standard.

65. Cf. the hyperinflation in Germany during the nineteen-twenties.

66. The UN has identified certain botanical "hot-spots" that contain unique flora and fauna. For a map and discussion of these see Al Gore, 2006, p. 130 ff. and map pp. 134-5. See also Pimm, Stuart, and Jenkins, C., 2005, and map, "Special Places," p. 70.

67. Friskics, Scott, 2008, 381-399.

68. For a discussion of the management ethic approach to dealing with animals see Regan, T., "The Case for Animal Rights," in T. Regan and P. Singer, 1976. I will critically evaluate managerialism in ch. 11.

69. Cf. Gordon Allport's notion of "inclusion" and "exclusion" in Maslow, A., 1959, p. 137 ff.

Chapter Five

1. Lewis, Clarence Irving, in 1969, p. 112. Cf. p. 182.

2. Ross, Sir W. David, 1930 (1961) and also Prichard and other deontologists See also Gans, G. F., "What is Deontology?" *Jour Val Inq*, 35/1, March 2001, 27-42. Consequentialist ethics has also been referred to as "teleological" (by Frederick Copleston in the treatment of Aristotle's Ethics in his well-known *A History of Philosophy,* (in 9 Volumes, 1946-74.) perhaps following Ross. It is now a commonplace in the literature of ethical theory. However, there are some recent works that contest the distinction. See e.g. John Cooper, John, 1975, who favorably compares Aristotle's and Kant's ethics. See also Vallentyne, P., "The Teleological-Deontological Distinction," 1987. Martin Gorke (2003, p. 116 ff.) quotes Zimmerli to the effect that deontological ethics is inadequate, since environmental ethics must include consideration of consequences to the environment. I would argue that the separation of deontological ethics from consequences has been exaggerated.

I will argue against the separation, at least in Kant, in more detail in my upcoming *Essays in the History of Value* (ined.)

3. I have examined the moral theory of value in my paper "Kant's Theory of Value," ined. I hope to publish it shortly in *ibid.*

4. Duty is objective in Kant in the sense of the first critique: universally necessary for us.

5. Kant, I, 1785 (1873) and The "Doctrine of Virtue", the second part of the 1797b (1994). For a classic study of Kant's ethics, see Paton, 1947.

6. Lewis examines Kantian ethics in depth in *Values and Imperatives*; a comparison of Lewis and Kant can be found in Saydah, R., 1969.

7. I examine reasoning in ch. 6.

8. And one with so many problems that it is untenable, as I will argue in ch. 6.

9. See Paul Guyer, "Ends of Reason and Ends of Nature: The Place of Teleology in Kant's Ethics," 2002, 161-186.

10. A pragmaticist criterion of values includes an implicit critique of intentionalist ethics, since the meaning lies in the practical consequences of acts of various possible intents.

11. See Lewis, C.I., 1946, (1962 ed.), esp. P. III; and also 1969. In the latter work, Lewis is largely in agreement with this work, differing mainly in his theory of value. In his essay "The Meaning of Liberty," e.g., he brings in versions of the principles of reciprocity and participation as well as consideration of consequences in practical reason. He also argues for a separation of value and obligation and the primacy of the latter over the former. See also Saydah, R., 1969; and Murphey, M., 2005, esp. ch. 10. Robert Louden (1992, cited in Pihlström, Sami, 2005, p. 138) argues that the priority of the practical in Kant connects his philosophy to the pragmatists.

12. Both George E. Moore and Dewey stated this objection, despite the consequentialism of both.

13. Cf. ch. 1. See also, Gould, Stephen J., "The Golden Rule—A Proper Scale for Our Environmental Crisis," repr., in Pojman, L., ed., 2001, 206 ff.

14. For a discussion in the context of Kant's ethics, see Guyer, 2002, 161-186

15. *Groundwork*, Paton transl.. p. 91. Note that consideration of "sympathy" in moral deliberation, a feeling, is also included, although Kant correctly argued that

individual feelings are not morally relevant—morals arise where others are involved and involve acts, not feelings toward others. Kurt Baier (1954,; repr in Cahn, S., and Markie, P., 1998, pp 104-135) agrees with Kant.

The problem with "moral sentiments," "caring for," and other appeals to emotion, especially as a foundation, is that these have nothing to do with ethics. Ethics is concerned with right and wrong, (a) good and bad (life): actions, not feelings. Feeling is irrelevant—a throwback to Cartesian metaphysics in which mind and matter are in separate universes as subject and object, and mind cannot influence reality (reductive materialism). But good feelings, as Kant noted, will not make a wrong act right. Thus ecocentrism decenters ethics from a subject who sets value—the egoism at the center of modern ethics. Value is creative actualization and is tied to action and consequences, not feelings. Even moral control of feeling is not a feeling but control over feelings. See also ch. 3 on control of feeling as a virtue.

Caring for belongs to families, not others, the latter including both strangers and wild animals and plants. Caring confuses intimate association with casual. Indeed, as I argued in chapter one, we have no obligation to tend each wild plant such that it comes to maturity—on the contrary, we have a duty not to. For this would result in domestication, ending wild value as different. However, if caring means acts, as in Dewey, then caring is more relevant. Dewey and Tufts (1932, p. 297-8) "revalue" sympathy as a virtue or perhaps a virtuous feeling. While sympathy cannot be the source of moral principles, if it floats your boat, pragmatists have no problem with it, especially if it means sympathy for the plight of other species.

16. In "The Meaning of Liberty," 1969, p. 146. Compare the quote at the beginning of this chapter. Kant would say that consideration of consequences is the content or 'matter' of rational consideration, not the rational element as such.

17. However, Dewey does not eliminate intent as a factor in ethical decisions. Indeed, there is a very interesting discussion of the intent vs. act/consequences issue as different degrees in Dewey and Tufts, 1932, pp. 184 ff.

18. For Kant reason is the faculty of principles. Prichard and Sidgwick interpret Kant to use "moral intuition," but as Rashdall has argued, Kant uses deontic logic, not "intuition," and pure practical reason does not appeal to intuition. See above p. 9. For Rashdall, see 1907.See also Gans, 2001. For Pritchard, see 1912. Reprinted in Margolis, J., 1966. For Sidgwick's view see 1966.

19. See McDonald 1997 for a discussion of this point.

20. Kant of course addressed this point in an essay. Whether his argument succeeds is another question.

21. "Prescriptivism" is sometimes lumped with other non-cognitive positions in ethics, but it is not necessarily non-rational. (1) Ethical imperatives are less "commands" than moral requirements (oughts). (2) In deontic ethical logic: One ought to do x in situation y, this is (like or close to in morally relevant respects) situation y, so I ought to do x. The reasoning is of the rule-instance sort. This point also applies to R. M. Hare, 1952, and 1963. For a critique, see Brand Blanshard, 1961, ch. 9.

22. In *Academica*. This view was endorsed earlier in Aristotle's distinction of moral, intellectual and technical excellence or virtue in *N. E.* VI.

23. Kant of course recognized the moral limitations of virtue in the opening paragraph of the *Groundwork.*

24. See McDonald 2003, ch. 3. C. I. Lewis, G.E. Moore and even J. S. Mill incorporate holistic elements into their theories of value, but are consequentialists in ethics. For Lewis, value in a life as a whole, see 1969, p. 35 ff.

25. Does using arguments from conditions of possibility harmonize with anti-foundationalism? Moral evaluation (Dewey's 'deliberation') requires consideration of plural elements or values and norms, including duties, consequences, etc., by a moral evaluator. Among these may be the organic relation of parts to a whole, which is especially useful in consideration of environmental relations. (Not all relations are holistic, as Kant noted in his table of categories: holism is a form of either the substance attribute or the community relation. Cause and effect is another relation. Cf. the community concept vs. the organic model of the environment as a whole in ch. 1. Callicott argues for a "community" concept as opposed to an organic form of holism in Callicott, 1989) But relations in a pluralist system are actually reciprocal, from which the compossibility of intrinsic value follows. For they are conditions of each other and thus of equal value. Moreover, their relation follows moral relations of social equals. For moral relations are conditions of the others. Thus pluralism is more general than holism. It implies the latter as a method of consideration of all relevant elements in (moral) evaluation. Holism is non-foundational; pluralism as reciprocal is also.

26. Cf. Iris Murdoch's (1970) critique of existentialism as solipsistic.

27. I am not arguing that ethics is the foundation, since I reject the foundational model.

28. For another treatment of ethical rules, see Kurt Baier, 1954.

29. I am indebted to James Liszka on this point, 2002, ch. 4.

30. Saydah compared C. I. Lewis to Hare in his book (1969). But Lewis brought in consequences in a way ignored by Hare, as well as moral actions as facts. Hare is a closet analyst, separating moral and factual.

31. Note for contractualists: By tying duties to abstractions such as life and society, the obligation is not contractual but absolute: if x lies to y, Y may not in turn lie. The reciprocity is abstract not concrete.

32. Cf. the film *Sophie's Choice*, one of the most tragic stories in recent cinema. Such choices dwell in the realm of tragedy, since either choice would be horrendous, and the situation she was placed in amoral to begin with.

33. Dewey and Tufts agree that moral principles as general, not absolute (1932, p. 304).

34. It is also distinct from the subordinate (hierarchical) types of reciprocity involved in moral evaluation. Hierarchical subordination of ends in themselves means superordinate (moral) values are not a ground but a limit: subordinate values are reciprocal with superordinate in their place as both are needed: there can be no categorical ends in themselves without moral limits. This topic will be discussed more fully in a projected future work on evaluation.

35. Justification in Dewey follows an organic model in which a whole of the problematic situation is analyzed in terms of its parts as elements. I have discussed this relation in more detail in McDonald, 2003, ch. 4. However, in our own view, holism is a type of pluralism. See below.

36. Am I a pragmatist in ethics? Do I reject a priori principles, including the Categorical Imperative? I am a Lewisian pragmatist, which is Kantian but with a pragmatic bent. The environment is neither a priori nor a posteriori. The air and soil are creatively actualized by (IV) life, but also the condition of the possibility of life.

The relation is reciprocal and holistic. For another type of pragmatic approach, see Light, A., 1995, repr. in Gottlieb, R.S., 1997, p. 255 ff.

Chapter Six

1. *Ethics*, 1932 p. 86.

2. Light, A and McKenna, E., eds., *Animal Pragmatism,* 2004, esp. Introduction. For Peter Singer, see "Not for Humans Only," repr. in Goodpaster, K. E., and Sayre, K. M., eds., 1979. Cf. "All Animals are Equal" reprinted in Zimmerman, M., et al., 1998 p. 26 ff. and J. Baird Callicott's introduction, where he uses the term. See also DeGrazia, D., 1999, 23-24; reprinted in Beauchamp and Walters, eds., 1999, pp. 418.

3. Norton, B., 1987.

4. Feinberg, Joel, 1974, pp. 43-68; Rescher, Nicholas, 1980, p. 83. My source for these citations is Holmes Rolston, 1989, ch. 10, p. 213 and 219. Rolston notes that the arguments for interests are used to deny species interests, since only individuals can have interests. I will address the notion of species in ch. 8. Here I will only note that "nations," "corporations," and other collective formations are frequently assigned interests and also act on them. See also Rolston, "The Value of Species," in T. Regan and P. Singer eds., 1976, p. 252 ff.; and Kenneth Goodpaster, 1978, in Pojman and Pojman, 2008, p. 154 ff., esp. 158 on Feinberg. For a critique of "biological interests" see Aaron Simmons, 2010.

5. Joseph Des Jardins calls this "second" version of Utilitarianism "preference Utilitarianism" (2001, p. 26 ff.). This version conceives the good as "the happiness that results from satisfaction of our desires." Note the subtle shift here from affective to conative, lacking from classical Utilitarianism. I will address this point below.

6. For a critique of this concept by an economist see Veblen, Thorstein, "Why is Economics Not an Evolutionary Science," and "The Preconceptions of the Classical Economists," both reprinted in Max Lerner, ed., 1948/1969.

7. I am indebted to Prof. Thomas Avery for a critical review of this section, which, I hope, helped me to improve it.

8. This view may be a version of "psychological egoism." See below.

9. For a treatment of the latter see Joel Feinberg, "Psychological Egoism," 1978; repr. in Cahn, S., and Markie, P., 1998. Des Jardins gives a critical treatment in 2001, esp. p. 84 ff.

There is also an evolutionary version of self-interest, bound up with some readings of genetics, but Darwin argued for "genuinely altruistic" behavior in, e.g. birds. Darwin believed in a continuity between humans and other species, which led him to think of human morality as related to the sympathy seen among social animals. This long-disdained idea was resurrected only recently by researchers like the primatologist Frans de Waal. Darwin "never felt that morality was our own invention, but was a product of evolution, a position we are now seeing grow in popularity under the influence of what we know about animal behavior," Dr. de Waal says. "In fact, we've now returned to the original Darwinian position" (*New York Times*, 2/10/09).

10. I have been greatly aided in my discussion of practical reason by previous treatments including Blanshard, Brand, 1961; R. M. Hare, 1963; Arthur Murphy, 1964; Steven Toulmin, 1970, John Cooper, 1975, John Searle, 2001; as well as C.I. Lewis.

11. Cf. Aristotle on intuition in NE 6; is this also an inference from instance to kind? Cf. fallacy of induction from single instance.

12. Hare has proposed another model with an evaluative major, and a factual minor (1952). This model was criticized by Philippa Foot (1958). The issues this raises are too wide and complex for this chapter; I hope to address them in a projected book on Evaluation and Norms. Foot's view was repr. in Margolis, J., 1966. Cf. Hampshire, S., 1949; repr. in *ibid.*

13. Aristotle examines the first model, although Plato anticipated his analysis. The origin of the rule-instance model is probably Kantian. I am indebted to Cooper (1975) on these points.

R. M. Hare uses case similar to my principle of participation as example (in *Freedom and Reason).* Could we universalize our act? If it would have bad consequences for us, we would not adopt a (hypothetical) principle such as put debtors in jail. or we might end up in jail. Hare (1963) argues that ethical reasoning utilizes principles in the light of facts. He argues (*ibid.*, p. 95 ff.) that ethical reasoning is deductive not inductive. This view compares with Popper's, who is close to Peirce, on scientific hypotheses, but it concerns ought not 'is.' Ethical reasoning is prescriptive and universal. (I read Hare as an analytic Kantian). C. I. Lewis (1969) also discusses commands or imperatives in ethics.

14. Decision theory utilizes reason in the attempt to evaluate various alternatives in order to choose optimum outcomes or consequences. Whether evaluation can be thought of as reasoning or is distinct is at least not a settled issue. Moreover, deciding optimum outcomes brings in non-rational considerations, including notions like 'desirable' that, however normative, go beyond reason as such. Game theory may be an offshoot of this. Also, game theory may be based on self-interest. See Jakob J. Liszka, 2002, p. 360 ff.

15. Thus Dewey (1939) criticized Perry's form of conativism as not distinguishing desire from a warranted desire, the desirable, which has a normative element. Searle (2001) also has criticized desire as a "reason."

16. It is probably here that Apel and Habermas's notion of communicative rationality belongs, although it is not innate. Cf. also Richard Rorty, 2007, ch. 12, 13, where he analyzes Habermas to mean that rationality is social, not subjective. For a critic of reason as a faculty see Searle, 2001.

17. Indeed, some have questioned whether reason can be practical. Kant struggled with this point, in the light of Hume's dictum that reason is "the slave of the passions." Arthur Murphy covers this issue in 1964, p. 32 ff.

18. In practical reasoning, a case is classified morally then evaluated or brought under a rule: this is/would be (a case of) a lie, (all) lies are wrong, I shouldn't do what is wrong, so I won't lie in this situation. Deduction using normative practical reasoning of the deductive type: not strict induction despite elements of instances used didactically to teach what lies are. Lies are normative identities, thus universal as including all possible instances of deliberate untruths, however defined (whether evasions, not answering, etc. are included as 'lies') including past and future instances under a normative identity.

19. However, as Karin Stokkink argued, imagination may actually connect means and ends. I am working on a book presenting her ideas on this point.

20. See n. 14. Arthur Murphy (1964) examines practical reason in terms of justification with his notion of "worthy" ends, esp. in ch. 2. He notes that grounds

must meet certain norms, such as whether they are suitable, appropriate and effective; and ends should meet the standard of being worthwhile. Justification means having "good and sufficient reasons" (p., 28). Thus there is a critical sense built into the notion of practical reason, that includes certain normative standards.

21. Similarly, Liszka notes means and execution must be "morally satisfactory," i.e., within moral limits (2002 sect. 5.6).

22. Cf. judgment in Kant. (One example of "giving reasons" in morals: universal rules).

23. Cf. Aristotle's comparison of argumentation in math and rhetoric in NE I. Cf. Toulmin, 1970, 114 ff.

24. I am indebted to James Tufts on this point (Tufts/Dewey, 1932, ch 7).

25. See note 20; Liszka (2002) argues that we must consider morals first in evaluation decisions.

26. *Groundwork of the Metaphysics of Morals*.

27. Aristotle, *Meta*, I, 5, $_{985}$b23 ff. Cf. his argument in II, 2 and IV, 3 that there is a limit to the reasons or "causes" that one can give if we are to avoid infinite regress.

28. Cf. the "paradox of rising expectations," in which if a certain level of satisfaction of desires are satisfied, new ones arise. This paradox was formulated by Mihaly Csikszentmihalyi in *Flow*, 1990. Quoted in Liszka, 2002, p. 246.

29. Plato, *Phaedo*. Cf. his treatment of the relation of reason and desire in *Republic*, esp. B. VIII and IX. 2. Cf. also Plato's *Crito* and *Gorgias*, in which the life of reason is contrasted with the life of desire, advocated by Callicles, with Aristotle, *Nichomachean Ethics* 10 which defends the life of contemplation, Epicurus' "Letter to Menoeceus," (from Diogenes Laertius' *Lives of the Eminent Philosophers*, X, 27) in which the life of tranquil pleasure is outlined, and *Lives*, VII, 1 "Zeno," in which the life of reason is judged best by the Stoics. This covers the main schools of ancient philosophy.

30. I criticized this anthropocentric view in chapter one. The point here is not to defend it, but to shed light on the ancient view of reason by contrast with desire.

31. I am following Kant here.

32. Personal communication.

33. Peirce argued that logic is a normative science. Cf. Sellars, Wilfrid, *Empiricism and the Philosophy of Mind*, 1956/1997.

34. This is a failing of Prof. Searle's otherwise excellent book, *Rationality in Action*, where he wants to "map" formal inference onto the psychology of desire (ch. 8). Reason is normative, not descriptive of how people think.

35. For a critique of such models see Veblen, T., "Why is Economics Not an Evolutionary Science," and "The Preconceptions of the Classical Economists," in Max Lerner, ed., 1969, p. 215 ff. Avery argues that ethical egoism is normative. But again, ethical egoism requires that we go beyond the singular, using reason to argue that one case, my self-interest, is a universal model.

36. Cf. Feinberg's critique of ethical egoism, in Cahn, and Markie, 1998.

37. Dewey argued this point in 1939, and its critique of conativism.

38. The phrase is Arthur Murphy's.(1964, p. 28).

39. Thompson, Jenna, "A Refutation of Environmental Ethics," reprinted in Hinman, L., ed., 2000, pp. 576-586. For a critical view of this article, see Lars Samuelson, 32/3, 2010.

40. *Ibid.*, Thompson, p. 583.

41. I have critically evaluated "needs" as a substitute for a theory of value in McDonald, 2011. In general, needs other than biological needs of the organism are actually values. What people say they "need," is actually not a need at all, but some valued goal. A person who claims they "need" a car no more needs one than they need a flying machine. Humans survived for centuries without cars.

42. I have documented how four environmental ethicists, Regan, Callicott, Rolston and Norton, incorporate the distinction of instrumental and intrinsic value in McDonald, 2003.

43. I am indebted to Callicott for this point.

44. I have argued for a theory of intrinsic value in ch. 1, which meets Thompson's substantive criticisms.

45. The relation of self-interest and action is also problematic. Actions are public, but self-interest is not Are "self-interested" actions an oxymoron? I examined this issue more fully in McDonald, 2011, ch. 6.

46. Norton, B., 1987. Cf. Singer and the classical Utilitarians like Bentham.

47. Cf. the Routleys' "last man" argument, repr. in Zimmermann, M., et al., 1998.

48. Sam Harris, 2004, p. 186. The quote is from J. Glover, *Humanity: A Moral History of the Twentieth Century* (New Haven: Yale Univ. Press, 1999, p. 24.

49. I am speaking of economic self-interest or self-interest in the wider sense, e.g., Feinberg's "realist" model for morals. These theories are distinct from the Perry-Prall interest theory, but are related as subjective. I am not arguing that ethical egoism and psychological egoism are identical, but that the former uses self-interest as a model or standard for human behavior.

50. Cf. the interrelation of species and habitats: symbiosis and holistic relations (the role of predators in healthy environment, e.g. birds of prey, wolves, etc.) Cf. Also Virginia Held's (1990) critique of the "artificial" autonomous individual.

51. Thus Kant, C. I. Lewis, W. D. Ross and E. Levinas, otherwise quite different in their philosophies, are united in their approach to ethics as concerned with the other person. This is reflected in the second statement of Kant's Categorical Imperative, Lewis's separation of moral right from other kinds (prudential, technical) in that moral right involves relations to the other (1969, p. 34 et al.), Ross's critique of Utilitarian ethics (1930), and Levinas's arguments for ethics as a first philosophy as concerned with the irreducible other. (Levinas, 1982, ch. 1.)

52. Deontological ethics meets this test.

53. Cf. the story of the invisible man in Plato's Republic, Book I, the "ring of Gyges."

54. For a detailed look at this problems see Dillingham, Catherine K., Newton, Lisa H., 2002, esp. ch. 8.

55. These arguments also apply to the closely related "preference Utilitarianism," in which "satisfaction of desires" is the highest good. Both Prall and Perry linked conativism to desire and interest. Cf. n. 5.

Cf. Joel Feinburg's distinction of psychological and ethical hedonism, and his critique of the former in "Psychological Egoism," in Cahn, S., and Markie, P., 1998.

56. This is the problem with Gauthier's project of incorporating morals into a decision theory based on self-interest. Gauthier must appeal to reason to justify taking long-term and community interests into account over short-term and selfish ones. But this defeats the theory. See Gauthier, David, 1986; quoted in Liszka, 2002, p. 256 ff. and 303 ff.

57. See McDonald, 2003, in which I have argued that Dewey's philosophy could be used for a holistic approach.

58. This point has been emphasized by both the pragmatists, especially Mead, who argue for the social formation of the individual; and sociologists like Durkheim.

59. I will mostly, if not completely ignore the subtle distinctions between net, total, average, and other Utilities, (Frankena 1963) since they are not relevant to my critique. From a moral perspective, it hardly matters whether we have a net increase in happiness a total or an average increase, if the happiness achieved in a particular context is immoral.

60. The theory confuses value and evaluation, denies the validity of non-hedonistic value judgments and thus reduces all value to hierarchies based on solipsistic pleasures. However, since our topic is the relation of hedonism to more contemporary theories based on self-interest, I will not cover these topics. I have made an extensive critique of hedonism in McDonald, 2011, ch. 6.

61. I am indebted to W. D. Ross for this example.

62. In Mill's version of the theory it is better to be Socrates dissatisfied than a satisfied pig. Bentham, however, recognizes that other species feel pleasure and pain. Perhaps they have some intrinsic value, then, but not all species can be considered inherently valuable on this view.

63. The connecting link is Hobbesian, in that it is in my self-interest to acknowledge the rights of others in an implicit contract, since the alternative is a life i.e. "nasty, brutish and short" (*Leviathan*). However, Utilitarianism itself cannot establish the connection, but must resort to contract.

64. See Mill, 1973, ch. 5, p. 446.

65. *Ibid.*, p. 419. Mill does not seem to distinguish ethics and morality although he may limit ethics to obligation, and morality as the larger whole that includes both obligation and theory of value.

66. *Ibid.*, p. 419. This is an implicit critique of Kant. That which is desired but not a duty is not punishable, according to Mill (ch. 5). Thus his theory of duty requires a sanction, which Mill discusses in detail (ch. 3).

67. This theory is defended, inter alia, by Joel Feinberg, in "The Nature and Value of Rights," 1970. The theory works better as a characterization of positive rights than negative rights, but I will not argue this point here.

68. John Rawls argues for this principle in an implicit critique of Utilitarianism.

69. Notably, Singer and, based on his work, other animal rights advocates.

70. Kantians will note that Kant would agree that moral good is not simply what is willed but determines what ought to be willed. But then Kant is not a voluntarist in value theory, a point I have argued in my paper, "Kant's Theory of Value" (ined.) The question becomes why the psychological element is brought in to Kantian ethics: the answer is that Kant was a subjectivist and had to both argue for the subjective elements of agency and how the moral law determines the agency to action. Once we purge ethics of subjectivity we can jettison the "will" as well as a superfluity.

71. Perry, Ralph.B., 1954. Prall, David, W., 1921.

72. The connection of self-interest and conativism lies in Hobbes, who argued both for a conative theory of value and for self-interest in the Leviathan. He was preceded in ancient times by Plato's Callicles in the *Gorgias*.

73. Moore, 1903, ch. 3, p. 65 ff.; Dewey 1939 passim.

74. Qualitative hedonists recognize this problem. Indeed, this is the gist of both Epicurus' critique of the Cyrenaics and J.S. Mill's critique of Bentham.

75. For a discussion of the *pros hen* relation see ch. 1.

76. I am treating self-interest and Utility in this chapter, not conativism in its various incarnations as conation, desire, willing, liking, taking an interest in and other variations. Thus I will avoid any critique of desire, demand or other variations of conativism, especially interests as such. I have critically evaluated desire in esp. ch. 10, sect. B, as well as in ch. 2, 3, and 5; demand value esp. in ch. 12. See also McDonald 2011, ch. 7 where I have critically evaluated conativism in detail. However, the irrationality of satisfaction of interests applies to desire as well.

77. See Veblen's *The Engineers and the Price System*, 1921, ch. 1.

78. Documented in Engerman, Stanley L. and Fogel, Robert W., *Time on the Cross* 1974/1995.

79. Quoted in Gorke, 2003, p. 194.

80. As I noted at the beginning, these concepts have even invaded animal rights discussions (Singer, Attfield, and all the literature of "animal interests," e.g. in De Grazia, Brody in Medical Ethics and others. But we cannot either speak for animals or calculate their ultimate interests. Discussions of whether animals can have interests are like speculating on whether there are alien life forms, since we cannot know their form of consciousness. Of course treating them as we do is slavery, in any case.

81. Cf. Dewey's arguments (1939) against the interest theory: that it confuses simple desire with the desirable. Dewey also provides a non-foundational notion of reasoning, including practical reasoning. I have examined his more holistic model of practical deliberation in McDonald, 2003, esp. ch. 4.

Chapter Seven

1. I am indebted to Prof. Frank Derringh, whose critical comments, incisive as usual, helped me to improve this chapter. See also Dillingham, Catherine K., Newton, Lisa H., 2002, ch. 2. On immigration see Cafaro, P. and Staples, W., "The Environmental Argument for Reducing Immigration into the United States," *Env Eth*, 31 (Spr. 09). This fine article presents the facts and statistics showing that immigration is the main cause of recent population increases in the United States. Cf. John Dewey and James Tufts, 1932 [L7], p. 481 ff.; and Revkin, A., "What's the Right Number of Americans?" *NewYork Times*, 8/4/2009.

2. Jared Diamond, 1992, p. 190.

3. Ehrlich, Paul R. and Anne H., 1998, p. 68. They also note the decline in the amount of good farmland, soil, fresh water, and biodiversity—all elements necessary for sustainability. (*ibid.*)

4. I will outline possible enforcement tools in chapter twelve.

5. For articles that examine the relation of culture to the environment see Guha, Ramachandra, 1989, (repr. in Pojman, L., 2001); Moncrief, Lewis W., 1970, (repr. in Pojman, 2001); and Ogungbemi, Segun, "An African Perspective on the Environmental Crisis," (in Pojman, 2001). Culture is often confused with religion.

6. I am indebted on this point to Virginia Held, 1990.

7. I am indebted to Garrett Hardin 1977 and 1993, *inter alia*, for the notion of "carrying capacity." For arguments against this notion see Bill McKibben, "A Special Moment in History," 1998, repr. in Pojman and Pojman, 2008, p. 378 ff., esp. 381 ff.

8. Speth, James G., *Red Sky at Morning*, 2005. (cf. the Sahara as non-arable).

9. Cf. Donella Meadows, et al., 1972. Cf. Rosenthal, Elizabeth, "Likely Spread of Deserts to Fertile Land Requires Quick Response, U.N. Report Says," *New York Times*, 6/28/07. Even Norman Borlaug, who developed many of the new strains of grains, which have again allowed expanded populations, argued that overpopulation is the actual problem, not developing new food strains. See Gillis, Justin, et al., "Norman Borlaug, Plant Scientist Who Fought Famine, Dies at 95," *New York Times*, 9/14/09.

10. Revkin, Andrew C., *New York Times*, 5/15/03.

11. I examined the intrinsic value of animals in ch. 1 and 4; the symbiosis with other species in ch. 7.

12. I am indebted to Prof. Derringh for raising this point. I addressed this issue in ch. seven in more detail.

13. For information on the Sahel see, *inter alia*, www.eden-foundation.org/project/desertif.html. Cf. n. 11 and also Al Gore, 1992/2006, pp. 75-6, 123-4 and 309.

14. I am indebted to Prof. Derringh for raising this point.

15. I am indebted to Prof. Derringh for raising this point also.

16. See "An Even Poorer World," *New York Times*, 9/2/08.

17. The case of California Court of Appeals, "*Bouvia v. Superior Court*," (repr. in Beauchamp and Walters, eds., 1999) provides many arguments for the value of quality of life over the value of life alone. Bouvia judged that her life was hopeless and useless, since she was confined to bed, had to be fed by tube, suffered from numerous ailments etc. Also if the quality of life of children is poor, we are justified in limiting births on Utilitarian Grounds. See Singer Peter, in Bayles, 1976, p. 68 ff.

18. I am paraphrasing John Dewey 1939, sect. 6, p. 427 here [LW 13: 235].

19. For two views that even on Utilitarian grounds population control would be better see J. Narveson, J., and Peter Singer, in Bayles, M., 1976; as well as the quote from Mill beginning this chapter. Singer is critical of both the "Average" and the "Total" views of Utilitarianism as applied to population; Parfit (*ibid.*) thinks Utilitarianism is an impossible view.

20. The issue of the rights of collectivities has been raised by several authors including Regan, Callicott, and Stone. Cf. ch. 1 above on the (intrinsic) value of collectivities. See also ch. 6, n. 4 on the "interests" of collectivities.

21. Economists are the main champions of this view. See e.g., Wilfred Beckerman, *Two Cheers for the Affluent Society*, 1974. Of course increased production with a stable population or a decrease in population would have a different outcome.

22. See Revkin, Andrew C., and Seelye, Katherine Q., "Panel Tells Bush Global Warming is Getting Worse" *New York Times*, June 2001. See also Blakemore, Bill, "North Pole Meets South Pole: Earth is Melting at Both Ends," ABC News.com, 3/3/06; Dreifus, Claudia, *NY Times*, 12/19/03; Gorman, James, "Scientists Predict Widespread Extinction by Global Warming," *NY Times*, 1/9/04; Kanter, James and Revkin, Andrew, "Scientists Detail Climate Changes, Poles to Tropics," *NY Times*, 4/7/07; Kelley, Katie, "Emblem of the West is Dying and No One Can Figure Out Why," *NY Times*, 9/27/06; Revkin, Andrew C., and Seelye, Katherine Q., "Report by the E.P.A. Leaves Out Data on Climate Change," *NY Times*, 6/19/03; Revkin, Andrew C., "Climate Talks Shift Focus to How to Deal with Changes," *NY Times*, 11/4/02; Revkin, Andrew C., "Global Warming is Expected to Raise Hurricane Intensity," *NY Times*, 10/1/04; Revkin, Andrew C., "Big Arctic Perils Seen in Warming, Survey

Finds," *NY Times*, 11/1/04; Revkin, Andrew C., "Yelling 'Fire' on a Hot Planet," *NY Times*, 4/24/2006; Revkin, Andrew C., "Federal Study Finds Accord on Warming," *NY Times*, 5/9/06; Revkin, Andrew C., "Emissions Already Affecting Climate, Report Finds," *NY Times*, 4/6/07; Rosenthal, Elizabeth, "As the Climate Changes, Bits of England's Coast Crumble," *NY Times*, 5/4/07; Than, Ker and Thompson, Andrea, "Frightening Forecast," Live Science/Imaginova Corp., 4/22/07; Yardley, William, "Engulfed by Climate Change, Town seeks Lifeline," *NY Times*, 5/27/07.

23. Cronin and Kennedy also criticize the use of the Old Testament to justify overpopulation. See Cronin, John, and Kennedy, Robert F. Jr., 1999, p. 149, 235 and 270. Cf. Rowe, Jonathan, "The Common Good," *Sierra*, (July/August 2005), 54 ff. where he argues that "courts generally have agreed that the assertion of public trust is not a 'taking' that requires compensation." Also cf. arguments on property in Joseph Des Jardins, p. 33-4. One might add that there should be a right not to be a victim of the decisions of others to have too many children. There are others within the rights tradition who argue for restrictions of the right to property, even, in Prof. Derringh's reading, John Locke.

24. Hardin, G., 1976, p. 297 ff.

25. Cf. Hardin's notion of the rights of the community in *Living Within Limits,* p. 109, 169 ff.

26. I am indebted to a PBS documentary on this point.

27. For a fairly complete treatment, see John Ladd, ed., 1973. Ladd defines relativism in terms of the "society" one belongs to rather than the culture, but later discusses them such that they are more or less coextensive with culture in our sense. Ladd also notes that the fact of different societies, cultures and customs can also lead to ethical skepticism or cynicism (Introduction.)

28. In particular, I will not examine epistemological forms of relativism except as they are germane to our central, ethical project. Epistemological or perceptual relativism arises from a nominalist epistemology: of contrary perceptions, conceptions or ideas of different individuals, each is equally valid or equally false (skepticism), or different and/but incomparable. Peirce used the notion of community to overcome this model. Historical relativism can be interpreted as a species of this (cf. "historicism" in Leo Strauss): each historical period is distinct and not comparable to others. It must be "understood" in its own terms. W. Dilthey may have advocated this position. Finally, there is hermeneutic relativism: different interpretations are possible of the same data, text (Derrida), words, "formations" (Foucault), etc. (I am uncertain whether Foucault is a relativist, but some have interpreted him as such).

29. Sam Harris, 2004, p 178. Harris himself argues against relativism.

30. I am referring to Ayer, other relativists and irrationalists in ethics. I have critically examined Ayer's views in McDonald, 2011, ch. 6.

31. Sumner, William G., "Folkways," repr. in Ladd, 1973, p. 23 ff. Ruth Benedict (1934) has also been interpreted as arguing for cultural relativism although one could contest this interpretation.

32. In "The Issue of Relativism," in *ibid.*, Ladd, 1973, p. 109 ff. The diversity thesis infers that there is no consensus on moral issues between different cultures. This is weaker than the dependency thesis. Sumner upholds both.

33. Categorial standards are standards attached to a kind of value, whether economic, social, religious, aesthetic or some other category of value (McDonald, 1997). They are standards of cultural viability—they assume pluralism—not formal

standards. They only presume that a cultural mode is viable for the participants not according to arbitrary and capricious sophiacratic public standards.

34. I am indebted to my late first wife on this point. See Karin Stokkink, *Cultural Relativism*, unpublished essay submitted to and accepted by the philosophy faculty of Lehman College, CUNY, and winner of the first Weinstock Award.

35. Again, there are limits to tolerance in practice that belie theoretical tolerance.

35. See Karin Stokkink, "Ethical Relativism," where she argues that at every step, cultural relativism is a disguised absolute.

37. Thus in the ethics of Christianity as in that of Chinese Moism, universal love is the duty and goal. Love in principle is a "doing unto others..." i.e. as an ethical principle, apart from the subjective feelings involved. Because others are treated as is self, equality is the only truly moral relationship for this ethic. As life is bearer of value, "doing unto others" can never be destructive of life, only supportive.

38. This point doubles as a critique of perspectivism and historicism.

39. Rachels, James, "The Challenge of Cultural Relativism," 1978; repr. in Cahn, and Markie, 1998.

40. Kluckholn, Clyde, "Ethical Relativity, Sic et Non," in Ladd, , 1973.

41. "Historicism" has received different definitions from different authors, e.g. Karl Popper 1963, and Leo Strauss 1974. I am following Strauss here on the whole, with reservations. See also Popper, Karl, *The Poverty of Historicism.*

42. I will examine historicism in a projected future work.

43. Individual relativism was the preferred ancient variety, based on the truth of individual perception in Epicurus, et al. Knowledge claims led the way in this as in other areas of philosophy.

44. Cf. normalization in Foucault, Michel, *Power/Knowledge*1980, p. 106 ff.

45. Technology as science creating values: active not reactive science, in the form of transuranic elements, etc.

46. Behind all the differentiations, cultural assertiveness, elitism, and other attacks on the universality of obligations lies Nietzschean thought with its petty elitism. The "overman" is a cop-out, totally without value to humans—it tells us nothing about the human condition, which it seeks to transcend. Nietzsche wants the ingrown ethic of the tribe, the special "overman" which excludes the human.

Nietzsche cannot find his great men, his "higher type," for the seeming great have their foibles: he constantly qualifies his definition of greatness in a vain attempt to get past "human, all too human." Great men are as rare among an aristocracy as among the masses; most members of the "aristocracy" spend their time in decadence and dissolute pursuits; not renaissance men but alcoholics. To sacrifice hundreds of others for the sake of such is truly a "revaluation of all values"—slavery as decadent: force over value.

Nietzsche's equation of moral and religious values is the opposite of their true genealogy. Religion is tribal, ideological: our gods support us in battle and our conquests divinely ordained and justified ("holy" war). Judaism, Christianity and Islam represent hierarchy ("Lord," "Father" etc.). Thus their value theory is not a recognition of the intrinsic value of each, but hierarchical value. Value in this view is a reflection of supreme value: theocentric emanations of value, despite subtle dissemblings: love, the infinite worth of the individual (as per Windelband inter alia, see his Rectorial Address), and the like. These are values because they are commanded by God, i.e., by divine command, in a hierarchy from above, not intrinsically valuable.

Thus Nietzsche's "genealogy" of morals is erroneous. Tribal ethics, including his own, are hierarchical and do not recognize intrinsic value. Cosmopolitan morality, i.e. duty, is an invention of Greek philosophy, not the Judaeo-Christian-Islamic religious tradition, and is anti-hierarchical in cosmopolitan recognition of non-tribal values. The "exclusive truth" claimed by religious creeds is tribal. Thus Nietzsche is the true Jew, contrary to his claims—us versus them (the Canaanites or Philistines)=man versus overman, with the same notion of a "chosen race." Violence is justified by a tribal "lord." Cosmopolitanism is Greek, as was democracy and universal ethics.

To be fair, Nietzsche also thought of himself as a "good European." Thus he was not tribal in the ethnic sense. But then his elitism is even less excusable.

47. Thus false identities can bring strife also (Catholic vs. Protestant "Christians"). Only identification with more universal values, i.e. life in the form of the biosphere, and universal morals, can avoid such strife. Again, identity as a "worker" or "Marxist" differentiates from "fascist" or "bourgeois," the root of political strife. Only recognition of the moral condition that makes for the best society will allow that "cities see an end to their evil…" I have examined the political aspects of such false identities in McDonald, 1997.

48. Cf. the notion of "inclusion" and "exclusion" in Alport in (Maslow, ed.) *New Knowledge of Human Values*.

49. One factor may be the trivialization of ethics by positivism (Russell: a philosopher moralizes on his own time—despite his many books on morals, e.g. *Marriage and Morals*). Despite the abundance of material on ethics, this has not yet been overcome.

50. Cf. Aristotle, *N. E.*, X.

51. Cf. Riesman's (1950) categories of "other-directed" and "inner directed."

52. See e.g., LaFraniere, Sharon, (*NY Times*, 10/29/06), which alludes to forced prostitution in Asia and other use of coerced child labor in Latin America.

53. For articles that examine the relation of religion to the environment see White, Lynn, "The Historical Roots of our Ecologic Crisis," *Science*, 155: 1203-7; repr. in Armstrong, and Botzler, 2004; De Silva, Lily, "The Buddhist Attitude Toward Nature," from *The Buddhist Attitude Toward Nature*, 1987; repr. in Pojman, 2001, 256 ff.; Dwivedi, O.P., "*Satyagraha* for Conservation: Awakening the Spirit of Hinduism," *Ethics of Environment and Development*, Engel, J.R. and Engel, J.G. eds., 1990; repr. in Pojman, 2001, 250 ff.; Izzi Deen, Mawil Y., "Islamic Environmental Ethics, Law and Society," *Ethics of Environment and Development*, Engel, J.R. and Engel, J.G. eds., 1990; repr. in Pojman, 2001; and Kim, Sung-Hae, "The Immortal World: The *Telos* of Daoist Environmental Ethics," *Env Eth*, 30/2 (Summer 2008) 135-157.

54. Tribal attitudes: define religion exclusively, e.g. by the elect. Cosmopolitan attitudes in religion: St. Francis, Gandhi, Moism.

55. "Ethics and Population Limitation," in Bayles, 1976, ch. 2, p. 27. Bayles himself concurs in another article in the same book. Cf. Justice Oliver Wendell Holmes: "It may be said in a general way that the police power extends to all the great public needs….It may be put forth in aid of what is sanctioned by usage or held by the prevailing morality or strong and preponderant opinion to be greatly and immediately necessary to the public welfare." Quoted in Dewey and Tufts, 1932, p. 462.

56. For the idea that use of common grazing land will result in "weed dominance," see Leopold, 1949; and Hardin, 1977.

57. In an excellent dissection of the so-called right to procreation, Sarah Conly argues that government "limits on reproduction are morally acceptable" (2005, p.105). She argues that such limits do not violate rights since the right to have as many children as you want is confused with other rights, including the right to procreate. The former does not follow from the latter and the latter is not absolute, any more than a right to work at any job you want follows from a right to work.

Conly distinguishes the axiological argument for the right to reproduce, that from it follow important values. But even if the family is valuable, it does not require an unlimited number of children. Although she recognizes a "right to non-interference," again, it is not absolute: too many children would interfere with others. We also restrict the right to what happens to one's body, e.g. in public vaccinations. She notes that many problems are aggravated by overpopulation, not just starvation, but poverty, global warming and others.

58. I addressed this point in more detail in McDonald, 1997.

59. I argued this point in principle in ch. 4. (Cf. Hardin, 1993, ch. 27, p. 307).

60. Some have argued that a right is more of a claim than a freedom, e.g. Joel Feinberg (1970). This point is valid for positive rights, but less so for negative rights. Insofar as a negative right is a claim on others, it is a claim of non-interference and thus a duty for others in our sense.

61. Note for libertarians as Hardin pointed out, loss of freedom will be the inevitable consequence of unlimited population growth. Thus arguments for freedom to reproduce are self-defeating. (1993, ch. 27, p. 297; cf. Hardin 1974.

62. Gorman, James, "Scientists Predict Widespread Extinction by Global Warming," *NY Times*, 1/9/04. Lions in Africa and tigers in India are species threatened, among many others, by overpopulation.

63. Cf. de-Shalit, A., "Is Liberalism Environment-Friendly?" repr in Zimmerman, et al., 1998.

64. Than, Ker and Thompson, Andrea, "Frightening Forecast," Live Science/Imaginova Corp., 4/22/07.

Chapter Eight

1. *Origin of Species*, 1991 ed., ch. 5, p. 118.

2. *Confessions of an Ecowarrior*, 1991, p. 133.

3. Daniel Elstein's paper was presented to the Society for the Study of Ethics and Animals, at the Winter 2003 APA Conference held in Washington, DC. Elstein has since published a slightly altered version of this paper, "Species as a Social Construction: Is Species Morally Relevant?" in *Animal Liberation Philosophy and Policy Journal*, I, 1, 2003 (see Appendix). I will only address some of Elstein's specific arguments in this chapter. For another paper advocating the social construction of species see Peterson, Anna L., "Environmental Ethics and the Social Construction of Nature," *Env Eth,* 21/ 4, pp. 339-57; repr. in Armstrong, Botzler, eds., 2004. Cf. Sagoff, Mark, "The Ecological Critique of Agricultural Biodiversity," 2001, in *ibid.*, Armstrong *et al*, 2004.

4. I argued for the value of species in ch. 1 and the notion of rights in ch. 4.

5. A useful examination of social constructivism is in Hacking, 2000.

6. See the *Stanford Encyclopedia of Philosophy*, online, for an account of social constructivism in science, of which this is a precis, under the topic "Social Epistemology."

7. An early version of this thesis is contained in Nelson Goodman's *Ways of Worldmaking* (1978). The source of the social constructivist thesis, however, is Berger's and Luckmann's *The Social Construction of Reality*, 1966.

8. This thesis was anticipated by the English pragmatist F. C. S. Schiller. See his *Humanism*, 1903.

9. Stamos and Dobson have noted that there are many biologists who do not accept the reality of species, although they are not social constructivists. Some are nominalists; some demur on other grounds (Stamos, David N., *The Species Problem*, 2003; Dobson, A., "Biocentrism and Genetic Engineering," *Env Values*, 4, 227-239, 1995.

10. I have used a dictionary definition to avoid having to deal with the considerable problems defining species, by using a consensus definition. For an excellent, balanced and thorough treatment of the problems of defining species and the attendant metaphysical connection, see Stamos, 2003.

11. This claim may be made by certain followers of Derrida, who argue from the latter's notion of the "logocentric" nature of Western philosophy.

12. Elstein specifically addresses this point in his paper. Color may also be a "secondary" characteristic, i.e., added by the mind (an argument made by Descartes and Locke and still debated in contemporary philosophy of mind). Thus it may not be essential to species identification in the taxonomic or botanical sense, but only for practical identification.

13. George Lakoff and Mark Johnson (1999, ch. 3, p. 17) reinforce this point when they note that all species categorize the things they encounter in their environment. Even the amoeba must classify food and not food. Lakoff and Johnson also argue that our perceptual systems "have no problem distinguishing cows from horses, goats from cats, or elephants from giraffes" (p. 27) and "since color is created jointly by our biology and the world, not by our culture…" social constructivism "fails to explain color" (p. 25).

14. Stamos (2003, ch. 2, sect. 3) agrees with me on this point: that reading Darwin as a nominalist misses Darwin's actual view.

15. Darwin's views on the issue of the validity, true to the nominalism of the British tradition, argue foundationally from individuals to varieties to species. However, Darwin ignored the genetic tie based on interspecific breeding, although he mentions it. For his treatment of these issues see 1859/1991, esp. chapter two.

16. *Ibid.*, 1991, p. 82.

17. *Ibid.*, p. 75 ff. et al.

18. *Hymenocallis* is genus of bulbous perennials, primarily of sub-tropical and tropical distribution in the New World, with a few sp. in Africa. There are approximately 50 species, although splitters claim as many as 80. See next note.

19. I am indebted to the work of Rogers McVaugh for the key and description of these taxa, based on the work of Bauml. See McVaugh, R., 1989. For *Hymenocallis* sp. see p. 203 ff.

20. However, this test for species differentiation should not be confused with the definition of species. See n. 10.

21. Cf. Yoon, C. K., "Reviving the Lost Art of Naming the World," *New York Times*, 8/13/09, who argues for common principles and outcomes in naming species among cultures. Cf. Hunn, E., *Tzeltal Folk Zoology: The Classification of Discontinuities in Nature*, New York: Academic Press, 1977.

22. This Paiute word is the origin of the "sego" lily, the common name of *Calochortus nuttallii*, the Utah State flower. The term "sego" is frequently applied to several other species in the Southwest area, notably *C. gunnisoni*.

23. Lakoff and Johnson, 1999, pp. 90-1. The references are to Berlin, B., Breedlove, D. and Raven, P., *Principles of Tzeltal Plant Classification*, New York: Academic Press, 1974; and Hunn, 1977. Lakoff and Johnson also defend embodied realism: "at the heart of embodied realism is our physical engagement with an environment in an ongoing series of interactions." (*ibid.*)

24. Because this work is primarily about moral theory and moral principles, the question of cultural limitations and of the application of Western values to other cultures is irrelevant, since this work is not an anthropological treatise. In fact, Western values are now in many cases universal for the entire world, willy-nilly. Technology knows no borders. I will discuss cultural relativism in the next chapter.

25. Kinsey's famous studies have provided evidence, however unscientifically gathered, of human bestial behaviors.

26. See Stamos, 2003, ch. 2. One woman at Elstein's talk raised this issue, beside philosophers of biology and metaphysical nominalists.

27. One corollary to Windelband, Rickert and the distinction of the human and natural sciences is the meaningfulness of the individuality of the individual and therefore that the individuality of a culture is not merely an instance of a general. (But we may raise the question of whether it is an instance of another general—culturally mediated?) See Windelband, W. (1894) trans. G. Oakes, in *History and Theory*, 1980.

28. For several papers discussing the metaphysics of tropes, see Laurence, Stephen, and Macdonald, Cynthia, 1998.

29. Our pragmatic pedigree here is that William James argued for the actuality of relations in *A Pluralistic Universe*, *inter alia*, and C. S. Peirce for a type of realism closely allied to his interpretation of scientific laws as generals or "thirds."

30. I am arguing against reduction of events to things by some contemporary philosophers, e.g. Lombard in Laurence and Macdonald, 1998.

31. Lynn Margulis, "Symbiosis in Cell Evolution," 1993, quoted in Stamos, 2003, ch. 5, p. 333-4.

32. It was also a denial of any supernatural model of understanding. Unfortunately, Descartes did not distinguish supernatural design from immanent functioning. However, there are signs that the latter is now being given fresh life.

33. Especially in *Philosophical Investigations*.

34. Davidson has argued against any scheme-content distinction. Is kind and instance a scheme-content distinction? Our argument is that kinds are normative, not simply conceptual, i.e., actual norms.

35. There is an element of conventionalism in Elstein's position as in all social constructivism. Hacking (2000) ties social conventionalism to nominalism.

36. I would argue against any metaphysics of essence-nature that the essence is not a *per se* being, but an abstraction of actual characteristics from a type-token. There is no distinct actuality, the essence, but only composites. I have argued this point in more detail in McDonald, 2004, ch. 5 and 2012, ch. 1.

37. Which principle it "comes under" at any one time is the trick, which is why Plato devotes so much space to classification (*Sophist* and *Statesman*). Plato's forms become "natural kinds" in Aristotle: species. Form is also connected to rationality, since form limits an instance to a finite size and shape, and only the finite is rational.

38. "Norm" means a range with limits in this context, like a species: not a standard, although it becomes one once the norm is established. It holds equally for all instances, although it may hold in general rather than invariably. Thus norm is not defined hierarchically. I argue in more detail for the generality of norms in ch. 6 of my *Speculative Evaluations*, 2012.

39. I am indebted to Stamos for bringing this point, dimorphism, to my attention. See 2003, passim. Stamos also discussed metamorphic species, in *ibid.* p. 306: while complicating the morphological criteria I defended such metamorphoses are normal for these species, just as changes in morphology during gestation and maturation are for most species, including humans.

40. Darwin recognized such relations as essential to his thesis. See 1859/1991, p. 87, 95-97, and 100, where it is linked to variations in the form of causal relations.

41. I am indebted to J. Baird Callicott, 1989 for many of the insights I have applied here.

42. Cf. the distinction of primary and secondary qualities in Descartes and Locke. But this distinction was later undermined, when Berkeley realized that primary qualities were also subjective. I have discussed nominalism and quantification in McDonald, 2012, ch. 6.

43. In this also it meshes with pragmatism. See William James' *A Pluralistic Universe*, 1971. Cf. ch. 5, where I argue for holism as a subset of pluralism.

44. I examine rare, threatened and endangered in Appendix 1.

45. Ben Bradley notes the "contributory value" of other species i.e. distinct from the instrumental-intrinsic distinction, as a part of a larger whole as an argument for endangered species. See "The Value of Endangered Species," *Jour Val Inq*, 35/1, 2001, pp. 43-56. (Gorke, 2003 at least qualifies this argument.) For arguments tying species to intrinsic value see Rolston, Holmes III, "Naturalizing Values: Organisms and Species," in Pojman, 2001. See also Gunn, A., "Why Should We Care About Rare Species," *Env Eth*, 1980; Hartshorne, C., "The Rights of the Subhuman World," *Env Eth*, 1979; and Gould, Stephen, "The Golden Rule—A Proper Scale for Our Environmental Crisis," *Nat Hist*, Sept. 1990, repr., in Pojman, 2001; and Gorke, *ibid.* For a counterargument see Goldman, M., "A Transcendental Defense of Speciesism," *Jour Val Inq*, 2001; and Russow, L.-M., "Why Do Species Matter?" *Env Eth,* V3, #2, pp. 101-112. Also, see ch. 1.

Gorke also considers aesthetic arguments, but judges that these are considered inadequate to justify species protection, since they are anthropocentric. (178 ff.) Aesthetic arguments are contradictory if value is anthropogenic, since beautiful landscapes etc. only derive beauty from humans. I would add that not all species are beautiful. (However, I would argue against the subjectivity of beauty. See McDonald 2011, ch. 1, where I argue that the presence of the object of beauty is a condition of the experience. Moreover, artists and others creatively actualize beauty in the world, not just in their minds.)

46. Callicott stressed this point as did Al Gore (1992 p. 50 ff). Some have argued that removing one species may have no impact on the environment at all.

Gorke (2003) calls the idea that ecology cannot tell us the impact of the removal of any one species the "insufficient knowledge argument." But this is irrelevant to anthropocentrism, based as it is on human interests. The environmental issue is, what is the tipping point? How many species can we eliminate before we destroy a habitat? Since we do not know, gambling is dangerous, violating the precautionary principle. Moreover, as I argued in ch. 1, a particular species may serve an essential function in a habitat, e.g. the predator role, and its removal may cause devastation.

47. Robbins, Jim, "Hunting Habits of Wolves Change Ecological Balance in Yellowstone, *NY Times*, 10/18/2005.

48. See Meadows, Donella H., "Biodiversity: The Key to Saving Life on Earth," 1990; repr., in Pojman, 2001. See also Wilson, Edward, O., "The Little Things that Run the World, repr. in Armstrong, et al., 2004. For a somewhat critical counterargument see Sagoff, Mark, "The Ecological Critique of Agricultural Biodiversity," 2001; in Armstrong, 2004.

49. Heywood, V. ed., *Global Biodiversity Assessment* 1995, from forward by E. Dowdeswell, p. vii. Quoted in Ehrlich, Paul R. and Anne H., 1998, p. 110.

50. For parallel arguments concerning duties to species and the biosphere see Rolston, "Challenges in Environmental Ethics," 1991; repr. in Zimmerman, et al., 1998; and his *Environmental Ethics,* 1988. See also the articles in P. VIII of *Animal Rights and Human Obligations*, ed. T. Regan and P. Singer, 1989 especially Russow, L-M., "Why Do Species Matter?"; and Rolston's "The Value of Species." See also Gunn, A., "Why Should We Care About Rare Species," *Env Eth* 1980; Godfrey-Smith, W., "The Value of Wilderness," *Env Eth* 1979; Parker, K., "The Values of a Habitat," *Env Eth* 1990.

51. I examined animal intelligence in ch. 1.

52. The issue also is, should the wilderness be preserved for its own sake—just so that there will be some unspoiled nature? I have examined the issue of the intrinsic value of landscapes in ch. 4.

53. I have given an outline of these in chapters. 3, 4, 11 and 12.

54. Cf. Tamerlane who is reputed to have wiped out an entire district in China.

55. I was stimulated by my reading of Richard Bernstein's *Radical Evil,* 2002.

56. Cf. Wilshire, Bruce, *Get 'Em All! Kill 'Em!: Genocide, Terrorism, Righteous Communities*, 2004. For another view of evil, tied to reverence for all life see Schweitzer, A., *Civilization and Ethics*, trans. J. Naish, (part repr. in Regan and Singer, eds., 1976; and as "Reverence for Life," in Pojman, 2001).

57. There is also the point of asking in whose interest is medicalizing evil? As Foucault argued, it aids the power of the medical profession and psychiatric technocracy who prescribe sanity. Social maladjustment is a curable "illness." Economic circumstances responsible for all social ills, never personal failings or organized violence.

58. Cf. bad will in Kant as willfulness without a moral end; on superficiality cf. the "banality of evil" in Arendt.

59. Cf. Rorty on Human Rights in ch. 9 of *Truth and Progress*, 1998.

Chapter Nine

1. Benjamin R. Barber, *Jihad vs. McWorld*, 1995. p. 93.

2. "The Viable Human," originally publ. in *Revision*, 9/2, 1987; reprinted in Zimmerman, et al., *Environmental Philosophy,* 1998, p. 183 ff.

3. Cf. Thorstein Veblen (1948/1969, p. 215 ff.) on whether economics is a science, with the pretensions to scientific rank in economics today.

4. The literature on Marx contains extensive debates between those who interpret Marx as an economic determinist and those who argue for a more nuanced view. One of the more recent is between L. Althusser and his critics, who include the "humanist" Marxists.

5. Against vulgar economism we can argue: do its advocates choose their spouses on economic grounds? Is their taste in food and art economically determined? Friedan has raised the issue of whether marriage is more of an economic issue and sex within marriage "economic sex." "Golddiggers" represent the best counterexample, but we would hope that not all romances disguise economic motives.

6. See Engels' "Letter to Joseph Bloch," 9/21/80. Reprinted in Feuer, L.S., ed., 1959. p. 397 ff.

7. Cf. Marshall Sahlins in *Stone Age Economics*. The distinction of necessities from luxuries is as old as philosophy and has been noted by economists. Some have noted that standard of living as a motive is culture bound, and is not accepted by all cultures.

8. This point is in part a response to Althusser's notion of invention as well as to historical determinism in general.

9. Cf. the pithy feminist phrase, "anatomy as destiny," in their critique of the "mommy" track.

10. Veblen (1968) was an acute critic of such simplified economic models.

11. Apart from the issue of the validity of a *summum bonum*. I criticized *summum bonum* theories in chapter one.

12. R. Michels (1968) and V. Pareto (1968) have stressed this point in the political context, where the unwillingness or inability of the masses to participate in politics allows an elite to rule in their name. Also cf. James Burnham's (1943/1963) interpretation of these figures.

13. I have discussed this point in more detail in my *Creative Actualization*, 2011.

14. Kantians will note that Kant would agree that moral good is not simply what is willed but determines what ought to be willed. But then Kant is not a voluntarist in value theory, a point I have argued in my paper, "Kant's Theory of Value" (ined.) The question becomes why the psychological element is brought in to Kantian ethics: the answer is that Kant was a subjectivist and had to both argue for the subjective elements of agency and how the moral law determines the agency to action. Once we purge ethics of subjectivity we can jettison the "will" as well as a superfluity.

15. E.g., some socialists use this language. They argue that economics should serve certain "needs" that are universal. Prescribing needs on some *ad hoc* model of human "nature" is itself repressive, for it separates needs from other possibilities and excludes freedom to choose between them.

There is also the use of needs by the psychological school of A. Maslow (1962) to accompany self-realization.

16. For an analysis of the use of sex in advertising see Goodman Paul, "Can Technology be Humane?" from *The New Reformation*, repr. in Teich, A., ed. 2000, pp. 90-103. See also Vance Packard, *The Hidden Persuaders*, 1976. Some feminists have used these examples to argue that advertising is "sexist."

17. Socialism should no more be exempt from morals than *laissez-faire*. The problem here is the notorious ambiguity of the word "socialism," which may mean anything from mild forms of the welfare state to Stalinist totalitarianism. In view of this ambiguity, the immorality of the latter cannot be used as an argument against the former.

18. Prices are also prescribed by the largest firm.

19. See *The Affluent Society* and *The New Industrial State.*

20. *N.E.*, IV, 1. Cf. Kong fu-tse, who like Aristotle argued that moderation, the mean, is the key to the virtues.

21. Benjamin Franklin treats thrift as a virtue. Cf. Alasdair MacIntyre, *After Virtue.* Thrift probably goes against the aristocratic Greek spirit on the whole, although moderation or temperance is a virtue for both Plato and Aristotle, and this cannot encompass wastefulness.

22. Donella H. Meadows, et al., *The Limits to Growth*, 1972. Cf. Des Jardins, Joseph, 2001, p. 60 and 84 on a sustainable economy.

23. I am interpreting duty such that it imposes stronger demands than virtue. However, there may be cases where virtue makes greater demands. See e.g., Edmund D. Pellegrino, "The Virtuous Physician and the Ethics of Medicine," in Shelp, E. ed., 1985.

24. An example is Vice-president Cheney's recent remark that energy conservation should be considered a "virtue," but that increased supply is still required. Thus drilling in the Arctic Wildlife Preserve is "required," QED. Apart from oilman Cheney's disinterested judgments, and his inability to grasp the concept "wilderness," the question arises of whether thrift, as a virtue, is required. For the view that virtues can be a duty see Plato's *Crito*, Kant's *Metaphysics of Morals*, P. II., and Kung fu-tse's *Analects*.

25. Dr. H. Daly (and my colleague Prof. F. Derringh) regard "sustainable growth" as an oxymoron. However, Daly argues there can be sustainable development, i.e. qualitative changes in the lives of the average person. For Daly's view see "Sustainable Growth? No Thank You," in Mander and Goldsmith, eds., 1996. Cf. R. Goodland's essay in the same vol., "Growth Has Reached its Limit." Cf. Meadows, Donella H. et al., 1972.

26. Freedom *from* economics—from the rule of the necessity of laboring—would be universal prosperity based on complete machine labor. But this entails reliance on (1) technology (2) bureaucracy (for distribution). Any "return to nature" would precisely destroy the possibility of freedom from economics. Yet the two conditions of economic freedom are often despised precisely by those who desire economic freedom. Complete freedom would here be unlikely also—engineers and bureaucrats would also be needed to keep things running—as in politics, a minimum state is always needed—even in the freest society. Anti-technologists, such as Heidegger and Marcuse, want a return to *slave labor*, without knowing it. They themselves do not volunteer to do farm labor by hand, without machines. But they advocate such a situation by condemning technology.

27. Purchases by consumers only occasionally reflect desire, of course, which is perhaps why economists switched to "demand" as the model. Purchases reflect an evaluation of many factors by the purchasers, including perceived "need," e.g., for food, cost, family and cultural values, economic and social trends and whether the

good will fit the purpose "suitability." Most purchases do not reflect desire simple. Cf. Veblen (1969, p. 215 ff.) on why economics is not a science.

28. See "Scientists Detail Climate Changes, Poles to Tropics," *NY Times*, 4/7/2007.

29. I will address the latter point in chapters 10 and 11.

30. Revkin, Andrew C., "Commercial Fleets Reduced Big Fish by 90%, Study Says," *New York Times*, 5/15/03.

31. Cf. Garrett Hardin's essay, "The Tragedy of the Commons" for the notion that conscience is "self-eliminating."

Chapter Ten

1. Cf. the title of ch. 2 of Christopher Manes's *Green Rage*, 1990: "The Culture of Extinction."

2. *Principles of Political Economy*, 1848.

3. "The Viable Human," originally publ. in *Revision*, 9/2, 1987.

4. Jared Diamond, *The Third Chimpanzee,* 1992, p. 318 ff. and 335. I am indebted to Prof. Frank Derringh for bringing this text to my attention.

5. However, see Diamond, 1992 on the disappearance of the Anasazi of the Southwest.

6. This point has been documented at length by Alfred Crosby, in his book *Ecological Imperialism* (1986) who argues that Europeans created "neo-Europes" in the Americas, *inter alia.* I am indebted to Dave Foreman (1991), who brought this book to my attention.

7. Diamond, 1992, p. 218.

8. *Ibid.*, p. 224. Cf. Darwin on the effect of climate in the *Origin of Species*, 1859 (1991 ed., based on the sixth, final edition) p. 106 ff., also ch. 11-13. Cf. p. 158: the environment as the condition of life.

9. I am indebted to Prof. A. Light on this last point.

10. Cf. Kaufman, Frederick, "Wasteland: A Journey Through the American Cloaca," *Harper's Magazine*, February 2008.

11. Cf. Packard, Vance, *The Status Seekers*, 1969.

12. For a detailed look at these problems see Dillingham, Catherine, and Newton, Lisa, *Watersheds 3*, 2002, esp. ch. 3, 5, 6, 8 and 9.

13. Manes (1990) notes that restoring the environment will be a difficult process, one in which calls for a "more rational regime of technology" have a superficial appeal (p. 240). But is any problem its own solution?

14. However, Prof. Krugman argues that the costs of switching to a green economy have been exaggerated by the right, and is actually rather small and manageable. See Krugman, Paul, "It's Easy Being Green," *New York Times*, 9/25/09

15. I presume that I do not have to take on the deniers of Global Warming, who either are paid by the fossil fuel industry, directly or indirectly, or who deny Global Warming on ideological grounds (cf. next two notes). As Nate Lewis, Caltech energy chemist noted: "Many climate change deniers argue that the variability of the earth's orbit and the different solar radiation impacts that this has on the planet—*and only those factors*—are what account for the big temperature fluctuations. They claim that human activities have no real effect. There is just one problem with that argument, says Lewis: the 6-degree-Celsius change in global average temperature that

distinguishes warm periods from glacial periods is too big to be accounted for simply by the changes in sunlight. The large observed temperature differences cannot be explained solely by the small differences in the amount of the sun's radiation hitting the earth as a result of the small variations in the earth's orbit." Quoted in Thomas Friedman, *Hot, Flat, and Crowded*, 2008, p. 118. Cf. Dillingham and Newton, 2002, ch. 3, as well as the current IPCC report, available on-line.

16. This and other consequences of global warming are drawn from the 2007 IPCC report. See www.ipcc.ch. See also Blakemore, Bill, "North Pole Meets South Pole: Earth is Melting at Both Ends," ABC News.com, 3/3/06; Dreifus, Claudia, "Listening to the Climate Models, and Trying to Wake Up the World: A Conversation with Jerry Mahlman," *New York Times*, 12/19/03; Kanter, James and Revkin, Andrew, "Scientists Detail Climate Changes, Poles to Tropics," *New York Times*, 4/7/07; Revkin, Andrew C., and Seelye, Katherine Q., "Report by the E.P.A. Leaves Out Data on Climate Change," *New York Times*, 6/19/03; Revkin, Andrew C., "Climate Talks Shift Focus to How to Deal with Changes," *New York Times*, 11/4/02; Revkin, Andrew C., "Global Warming is Expected to Raise Hurricane Intensity," *New York Times*, 10/1/04; Revkin, Andrew C., "Big Arctic Perils Seen in Warming, Survey Finds," *New York Times*, 11/1/04; Revkin, Andrew, "Steps to Limit Global-Warming Gas," *New York Times*, 9/25/05, Revkin, Andrew C., "Yelling 'Fire' on a Hot Planet," *New York Times*, 4/24/2006; Revkin, Andrew C., "Federal Study Finds Accord on Warming," *New York Times*, 5/9/06; Revkin, Andrew C., "Emissions Already Affecting Climate, Report Finds," *New York Times*, 4/6/07; Than, Ker and Thompson, Andrea, "Frightening Forecast," Live Science/Imaginova Corp., 4/22/07. Schell, Orville, "The Thaw at the Roof of the World," *New York Times*, 9/26/09 [Asian Glaciers melting], Gettleman, Jeffrey, "Even the Camels Are Dying," *New York Times*, 10/8/09 [Somalia Drought]

17. This was documented in Al Gore's famous film, "An Inconvenient Truth." See also the photographs accompanying McKibben, Bill, "Year One: Climate Chaos Has Arrived," *Sierra*, Jan-Feb. 2006, pp. 30-35 and Schell, Orville, "The Thaw at the Roof of the World," *New York Times*, 9/26/09 [Asian Glaciers melting].

18. See MacFarquhar, Neil, "Refugees Join List of Climate-Changing Issues, *New York Times*, 5/29/09, Revkin, A., "Global Warming is Expected to Raise Hurricane Intensity," *New York Times*, 10/1/04, Rosenthal, Elizabeth, "As the Climate Changes, Bits of England's Coast Crumble," *New York Times*, 5/4/07, Rosenthal, E., "Likely Spread of Deserts to Fertile Land Requires Quick Response, U.N. Report Says." *New York Times*, 6/28/07, and Saulny, Susan, "Holding Out, to Last Tiny Isle, as Cajun Land Sinks Into Gulf "*New York Times*, 8/24/08.

19. Gorman, James, "Scientists Predict Widespread Extinction by Global Warming," *New York Times*, 1/9/04.

20. *Ibid.*, cf. Kelley, Katie, "Emblem of the West is Dying and No One Can Figure Out Why," *New York Times*, 9/27/06; and Dillingham and Newton, 2002, ch. 4.

21. I argued against such unlimited population expansion in ch. 8 and against threatening technologies in ch. 9 and 10. Another consequence has been prolonged drought in the Sahel and elsewhere. See Gettleman, Jeffrey, "Even the Camels Are Dying," *New York Times*, 10/8/09 [Somalia Drought]

22. Judson, Olivia, "Where Tasty Morsels Fear to Tread," *New York Times*, 9/29/09

23. Howard, Ross and Perly, Michael, *Acid Rain*, 1980.

24. See also Cox, Stan, "Turf Wars," http://www.alternet.org/envirohealth/28361/ AlterNet. 11/17/05 [use of pesticides and fertilizers in lawns]; and Ritchie, John, "Toxic Legacies," a Canadian TV Documentary on pesticide damage to human health, including estrogen mimicry.

25. But cf. Arnold Toynbee on the relativity of technology. He notes in *A Study of History* (1946) that other cultures also had their technology.

26. Extremist anti-technology groups have tried to overthrow human's increasing technical mastery of and liberation from nature. For Popper's view see *The Open Society and Its Enemies*, (1963) P. I; cf. R. Pirsig in *Zen and the Art of Motorcycle Maintenance*, 1974.

27. See e.g. Dodds, E. R., *The Greeks and the Irrational* (1951, 1973) on the "fifth century (B.C.) enlightenment."

28. I discussed the value of technology in chapter seven.

29. Cf. Jacques Ellul, *The Technological Society*, 1964. There is a whole history of technocracy as a movement. See Akin, W.E., 1977.

30. The Heideggerian and Frankfurt School philosophers have drawn attention to the dangers of technological domination. See Heidegger, Martin, *The Question Concerning Technology and Other Essays*, 1977; and Schroyer, T., *The Critique of Domination*, 1973.

31. I have discussed the notion of sophiacracy in my McDonald, 1997, esp. ch. 3.

32. Marx, Leo, "Does Improved Technology Mean Progress?" first published in *Technology Review,* January 1987.

33. I discussed the political side of technocracy in McDonald 1997.

34. Winner, Langdon, "Technologies as Forms of Life" From *Epistemology, Methodology and the Social Sciences*, Cohen and Wartofsky, eds., Kluwer Academic Publishers, pp. 249-263.

35. The "needs" paradigm is technocratic: man has needs, technology alone can fulfill them, therefore, progress=technology. Imposed needs are a public standard of technocrats: technocracy. I examined the fallacy of "needs" in McDonald, 2011.

36. For a counterargument see McDermott, 1969.

37. Compare Marx on technology as the determining factor: but technology interpreted economically, not intellectually.

38. It is in military strategy that its worse effects are to be feared. Nuclear weapons are sought to offset those of other nations creating an arms race. I have discussed this issue in more detail in McDonald, 2012, ch. 3.

39. Cf. H. Kahn on thermonuclear war in *Thinking the Unthinkable*, 1962, where he argued that the United States could fight and win such a war. A nuclear war would probably end all life on earth, but would in any case be a form of genocide as cities are destroyed along with civilians. The arms race as a theoretical, even metaphysical war. Closed techno-logic of this type is comparable to closed ideologies of the sort analyzed by Karl Popper, *inter alia.*

40. Harris, Sheldon, "Factories of Death," repr. in Beauchamp and Walters, eds., 1999; excerpt taken from *Factories of Death*, 1994.

41. "Pax Atomica" in his column in *National Review*, 1969.

42. Or, for that matter, more freedom from government (unregulated capitalism). The latter would involve reduction of one freedom-power and the enhancement of another. Thus it would not create more freedom, only a new master (McDonald, 1997).

Chapter Eleven

1. Dewey, John, and Tufts, James, *Ethics*, 1932 revision [M5/L7], P. III, ch. 20, p. 452. Cf. p. 481.

2. Cf. Godfrey-Smith, W., "The Value of Wilderness," *Env Eth,* Vol. I, #4, 1979, p. 309; Miller, P., "Value as Richness: Toward a Value Theory for an Expanded Naturalism in Environmental Ethics," *Env Eth,* Vol. 4, #2, 1982; Sessions, George, "Ecocentrism, Wilderness and Global Ecosystem Protection," in Zimmerman, et al., 1998; Schönfeld, Martin, "The Green Kant: Environmental Dynamics and Sustainable Policies," repr. in Pojman and Pojman, eds., 2008, p. 49 ff. esp. p. 56 ff.; Wilson, Edward, O., "The Little Things that Run the World," *Conservation Biology*, 1: 344-46, repr. in Armstrong, S. J., and Botzler, R. G., eds., 2004, ch. 5.

3. See ch. 9.

4. I am indebted to an article by J. Baird Callicott, "Do Deconstructive Ecology and Sociobiology Undermine Leopold's Land Ethic?" which he wrote in answer to his "post-modern" critics, for bringing this issue to my attention. Reprinted in Zimmermann, et al., 1998. The critique of ecosystem stability was also raised by Joseph Des Jardins (2001, p. 175 ff.); and Martin Gorke, 2003.

5. Gorke (in *ibid.*, p. 68 ff.) notes that a "mosaic cycle" theory is slowly replacing the climax theory. Cf. also *ibid.*, p. 164 where he criticizes ecological equilibrium. A species may have different functions in different habitats. While true, this does not affect the point that each function at a generic level may be required for a healthy habitat, as I argued in ch. 1.

6. Norton makes a similar argument for prudence. Gorke (2003, p. 148 ff., cf. 164 ff.) argues that some ecologists believe that only some species are vital ("keystone" spp.), while others are "free riders," and do not serve an essential function. Thus for the environment, there is a hierarchy of species in terms of importance (cf. Callicott, 1989) on the greater importance of endangered species.) Des Jardins also addresses this point (2001, p. 168 ff). However, we do not really know whether they are essential in advance. The prudence argument still holds. Moreover, as Gorke's arguments indicate, the habitat may change and bring changes in keystone species with it. If a species goes extinct its future services as a keystone species may be lost. Cf. wolves in Yellowstone National Park, whose reintroduction rescued the habitat. Cf. also his arguments against "interventions" in the environment that might lead to irreversible consequences, or interfere with the "trial and error" processes of evolution.

7. In previous chapters, I noted the role of predators in healthy environment, e.g. birds of prey, wolves, etc.

8. E.g. Guha, Ramachandra, "Radical Environmentalism and Wilderness Preservation: A Third World Critique," *Env Eth* 11, Spring 1989, (repr. in Pojman, L., 2001.) By contrast, Prof. Workineh Kelbessa calls for learning from indigenous peoples how to live within a habitat and not destroy it, but also recognizing which of their practices are not ecofriendly, a more balanced and sensible approach (Kelbessa, "The Rehabilitation of Indigenous Environmental Ethics in Africa," D*iogenes.*2005; 52: 17-34, and *The Utility of Ethical Dialogue for Marginalized Voices in Africa.* Available on-line.

9. Cf. the "Tu, quoque" fallacy.

10. Foreman's list:

Area	State(s)	Area
North Cascades	Washington	Olympic Mountains
Kalmiopsis/Siskiyous	Oregon/California	Hells Canyon/Eagle Cap
Selway/River of No Return	Idaho /Montana	Great Rift
Owyhee	Idaho /Oregon / Nevada	Oregon Desert
Bob Marshall	Montana	Beartooth
North Absaroka	Wyoming	Upper Yellowstone/ South Absaroka
Tetons/SW Yellowstone	Wyoming/Idaho	Wind Rivers
Red Desert	Wyoming	Maroon Bells
San Juan Mountains	Colorado	Desolation Canyon
High Uintas	Utah	Canyonlands
San Rafael/Wayne	Utah	Escalante/ Kaiparowits/Henry Mts,
Desert Game Range	Nevada	Black Rock Desert
Smoke Creek Desert	Nevada/California	High Sierra
Yosemite North	California	Los Padres
Inyo/Saline/Cottonwood	California	Panamint Mountains
Mojave Desert	California	Bill Williams River
Kofa	Arizona	Cabeza Prieta
Galiuro/Pinaleno	Arizona	Grand Canyon/Kaibab
Gila/Black Range	New Mexico	Guadalupe Escarpment

(adopted from Foreman, D., 1991, ch. 17.)

11. "Stochastic is used by ecologists to indicate a random alteration in an ecosystem, brought about by fire, hurricane, or other natural perturbation." (*Ibid.*).

12. *Ibid.*

13. *Ibid.*

14. See Throop, William, "The Rationale for Environmental Restoration," in Gottlieb, R.S., ed., 1997, p. 39 ff. The task of undoing the damage was also raised by Joseph Des Jardins (2001, p. 121. Cf. also p. 88). However, restoration remains controversial among environmentalists, since the restored area is never quite the same as the original. Cf. Guevara, Sergio, and LaBorde, Javier, "The Landscape Approach: Designing New Reserves for the Protection of Biological and Cultural Diversity in Latin America," *Env Eth*, 30/3 (Fall 2008), 251-262. Cf. also ch. 4 above on wilderness preservation.

15. Such purchases could be financed by a steeply progressive tax on the inflationary salaries of CEO's. However, Rowe argues that "courts generally have agreed that the assertion of public trust is not a 'taking' that requires compensation." See Rowe, Jonathan, "The Common Good," *Sierra*, 90/4 (July/August 2005), 54 ff.

16. Foreman agrees with this point arguing we should "reestablish native species," since they cannot survive in puny reserves. "The Grizzly will not survive restricted to the dwindling Yellowstone and Bob Marshall /Glacier ecosystems." He makes similar points about the Gray Wolf, and other species. "In suitable areas of southern New Mexico, Arizona, and Texas, the Jaguar, Ocelot, and Jaguarundi should be reintroduced. Bighorn Sheep, Bison, Pronghorn, River Otter, Woodland Caribou, and other once-widespread species should be widely propagated in their former habitats." In some cases a wilderness might have to be completely closed to human use to preserve a species. Thus he argues that "A million and a half acres in the Los Padres

National Forest, northwest of Los Angeles, should be totally closed to human use or entry in order to protect the California Condor after reintroduction." (Foreman, 1991)

17. Judson, Olivia, "Where Tasty Morsels Fear to Tread," *NY Times*, 9/29/09 [role of predators]

18. Robbins, Jim, "Hunting Habits of Wolves Change Ecological Balance in Yellowstone," *NY Times*, 10/18/2005.

19. Cornelia Dean, "Coral Reefs and What Ruins Them," *NY Times*, Feb. 26,2008 and Judson, n. 21

20. Thomas Friedman, *Hot, Flat, and Crowded: Why We Need a Green Revolution—and How It Can Renew America*, 2008, p. 252-3.

21. Al Gore, *Earth in the Balance: Ecology and the Human Spirit*, 1992/2006, p. 184 ff.

22. The simple expedient of changing the color of roofs to reflect more sunlight has resulting energy savings. See Barringer, Felicity, "White Roofs Catch on as Energy Cost Cutters," New York Times, 7/30/09.

23. According to Al Gore, half of the topsoil in Iowa has been lost to erosion, some of the richest farmland in the world (1992/2006, p. 184).

24. For other arguments on sustainable economies, see Des Jardins, 2001, p. 60 ff.

25. Galbraith, Kate, "Efficiency Drive Could Cut Energy Use 23% by 2020, Study Finds," *NY Times*, 7/30/09. The issue of efficiency has been raised by other authors as well, including Horn, Miriam and Krupp, Fred, in *Earth: the Sequel,* 2008; and by Al Gore.

26. Economists might argue that this might result in a decline in demand, and thereby end economic growth. (For a counterargument, see Meadows, Donella H. and the Club of Rome Staff, *The Limits to Growth*, 1972.) The gradual improvement in the quality of life would more than offset the temporary decline in demand, as costs for basics slowly dropped with a relative increase in supplies. Moreover, it might also result in lower prices as marginal costs of production become reduced.

As I noted in the last chapter, even Norman Borlaug argued for population control; the development of new plant strains provides only a temporary breathing space for dealing with the main problem of overpopulation. See Gillis, Justin, Jonas, G. and Wheaton, S., "Norman Borlaug, Plant Scientist Who Fought Famine, Dies at 95," *NY Times*, 9/14/09.

27. This point has been made especially by Cafaro, P and Staples, W. III, "The Environmental Argument for Reducing Immigration into the United States," *Env Eth*, 31 (Spr. 09).

28. See Revkin, Andrew, "What's the Right Number of Americans?" *NY Times*, 8/4/2009, where evidence is presented that without immigration, the population of the United States would have already leveled off.

29. For a counterargument see Singer, P., "Famine, Affluence and Morality," (reprinted in Cahn and Markie, eds., 1998). I hardly think that it is in the "interest" of the "greatest number" to be crowded and starving.

30. For another take on enablers see Al Gore (1992/2006) p. 223 ff. Cf. the notion of "responsibility" of rulers for their own people in Hardin, 1993, p. 307.

31. Gettleman, Jeffrey, "Even the Camels Are Dying," *NY Times*, 10/8/09 [Somalia Drought]

32. I argued this point in ch. 10. For Dewey's and Tuft's view on population see their *Ethics* (1932 revision) [L7], esp. 508 ff.

33. 1992/2006, p. 104. For the evidence that the global warming is caused by human use of fossil fuels (and livestock) see *ibid*, p. 92 ff. Also see the IPCC report at http://www.ipcc.ch/ and Blakemore, Bill, "North Pole Meets South Pole: Earth is Melting at Both Ends," ABC News.com, 3/3/06; Dreifus, Claudia, "Listening to the Climate Models, and Trying to Wake Up the World: A Conversation with Jerry Mahlman," *NY Times*, 12/19/03; Gorman, James, "Scientists Predict Widespread Extinction by Global Warming," *NY Times*, 1/9/04; Kanter, James and Revkin, Andrew, "Scientists Detail Climate Changes, Poles to Tropics," *NY Times*, 4/7/07; Kelley, Katie, "Emblem of the West is Dying and No One Can Figure Out Why," *NY Times*, 9/27/06; Revkin, Andrew C., and Seelye, Katherine Q., "Report by the E.P.A. Leaves Out Data on Climate Change," *NY Times*, 6/19/03; Revkin, Andrew C., "Climate Talks Shift Focus to How to Deal with Changes," *NY Times*, 11/4/02; Revkin, Andrew C., "Global Warming is Expected to Raise Hurricane Intensity," *NY Times*, 10/1/04; Revkin, Andrew C., "Big Arctic Perils Seen in Warming, Survey Finds," *NY Times*, 11/1/04; Revkin, Andrew, "Steps to Limit Global-Warming Gas," *NY Times*, 9/25/05; Revkin, Andrew C., "Yelling 'Fire' on a Hot Planet," *NY Times*, 4/24/2006; Revkin, Andrew C., "Federal Study Finds Accord on Warming," *NY Times*, 5/9/06; Revkin, Andrew C., "Emissions Already Affecting Climate, Report Finds," *NY Times*, 4/6/07; Rosenthal, Elizabeth, "As the Climate Changes, Bits of England's Coast Crumble," *NY Times*, 5/4/07; Than, Ker and Thompson, Andrea, "Frightening Forecast," Live Science/Imaginova Corp., 4/22/07; Yardley, William, "Engulfed by Climate Change, Town seeks Lifeline," *NY Times*, 5/27/07; and Schell, Orville, "The Thaw at the Roof of the World," *NY Times*, 9/26/09.

34. *Ibid.*

35. Cf. Packard, Vance, *The Waste Makers*, 1978. Also, Kaufman, Frederick, "Wasteland: A Journey Through the American Cloaca," *Harper's Magazine*, February 2008; and Dewey and Tufts, 1932, p. 481.

36. Cf. Prof. Owen Gutfreund, "Pick on the Big Guys," *NY Times*, 4/9/08

37. I am indebted to Prof. Andrew Light on this point, who argued that environmental philosophers concentrate on wilderness issues, ignoring the impact of patterns of human settlement.

38. I am indebted to Prof. Andrew Light on this point.

39. By contrast, private motor vehicle use, wasteful though it sometimes is, accounts for only about 20%. Source: Prof. M. Oppenheimer, Prof. of Geosciences, Princeton University, quoted in the *NY Times*, 10/24/06

40. I am indebted to Garett Hardin on this point and the rest of the paragraph.

41. For Horn and Krupp, see *Earth: the Sequel,* 2008. Also, see Belson, Ken, "Air Storage Is Explored for Energy," *NY Times*, 8/26/08

42. See Rauber, Paul, "Why Not Nukes?" *Sierra* Magazine, 2007, p. 37.

43. Thomas Friedman, 2008, p. 193. He also argues that "we should not dismiss calls to simply consume less by the "back to nature" wing of the environmental movement" (*ibid.*).

44. *Ibid.*, (Friedman) p. 195 and pp. 282-3.

45. "A king of Lacedemonia gave a very wise reply to those who praised in his presence the extreme goodness of his colleague of Charillus. 'And how can he be

good,' he said to them, 'if he doesn't know how to be terrible to the wicked?'" Rousseau, "On the Sciences and Arts," in *Collected Writings*, 1992, pp. 110-99, n.4.

46. See "Scientists Detail Climate Changes, Poles to Tropics," *NY Times*, 4/7/2007, p. 1.

47. I am indebted for this proposal to Prof. James Speth, 2004.

48. Some may argue that other species dirty their own habitat, e.g. pigeons foul everywhere. However, this cannot be a norm for humans; and animals do not generally live in their own mess.

49. I am indebted to Garrett Hardin for this suggestion, 1993, p. 201.

50. E.g. the chief executives and lobbyists of Exxon-Mobil, who used their money to deceive the public about global warming. Since their ecocrimes are anti-social, solitary confinement would be a suitable punishment, until such ecocriminals have reformed and are fit to re-enter human society.

51. These include Huxley's *Brave New World* (1946/1956), and Orwell's *1984*, among others. Both of these works are critiques of managerialism.

52. *Confessions of an Ecowarrior*, 1991.p. 95. Foreman notes that it costs more to "administer" these lands than the total fees the two bureaucracies receive as payment. The polices are not justified economically, much less environmentally. (p. 100)

53. 2003, Part A, ch. 1, sect. 2-4. Cf. p. 30 and 259.

54. For a parallel critique of the managerial "ethic," see Regan, Tom, 1981, pp. 19-34. See also his "The Case for Animal Rights," in Regan and Singer, eds., 1976.

55. For a counterargument, see Guevara, Sergio, and LaBorde, Javier, "The Landscape Approach," *Envir Eth*, 30/3 (Fall 2008), p. 251.

56. Thus Bookchin's (1988) "social ecology," is an oxymoron. The environment cannot be saved by redividing the pie. This is the outlook of a watermelon.

57. Cf. E. Junger and his concept of total mobilization, which I am adapting here.

58. Marc Reisner, *Cadillac Desert,* 1986/1993.

59. Matthew Scully, *Dominion,* 2002, p. 384-5.

Epilogue

1. Rolston, *Philosophy Gone Wild*, 1989.

2. As Joseph Des Jardins (2001, p. 167 ff.) has noted, "philosophers disagree about the lessons to be drawn from ecology…." I am arguing that the lessons are profound and should alter the direction of future philosophy.

3. I am indebted to Martin Gorke for raising this point (2003, p. 116 ff.).

4 Cf. also, Cheney, Jim, and Weston, A., "Environmental Ethics as Environmental Etiquette: Towards an Ethics-Based Epistemology," *Env Eth,* 21, 2, 115-34, repr. in Armstrong and Botzler, 2004. Cf. also Francis Bacon who thought of nature in terms of knowledge of how to do things.

Appendix

1. *Origin of Species*, 1991 edition, ch. 11, p. 280.

2. I covered the ethics of endangered species in a previous chapter. This appendix is supplemental.

3. I am indebted to articles in the *New York Times*, for information on these taxa. See Kelley, Katie, "Emblem of the West is Dying and No One Can Figure Out Why," *NY Times*, 9/27/06.

BIBLIOGRAPHY

Akin, William E., *Technocracy and the American Dream: the Technocrat Movement, 1900-1941*, Univ. of Calif. Press, 1977.

Alexander, Samuel, *Beauty and Other Forms of Value,* New York: Thomas Crowell Co., 1933/1968 ed.

Alland, Alexander. *Race in Mind: Race, IQ and Other Racisms,* Palgrave Macmillan, Pub., 2002.

Allport, Gordon. "Normative Compatibility in the Light of Social Science," in Maslow, A., ed. *New Knowledge of Human Values* (Chicago: Henry Regnery Co. 1959), p. 137 ff.

Althusser, Louis, *For Marx*, trans. by Brewster, B., New York: Vintage/Random House, 1970.

Anderegg, William R.L., "Diagnosis Earth: the Climate Change Debate," *Thought and Action*, Fall 2010.

Anderson, Terry, and Leal, Donald, "Free Market versus Political Environmentalism," repr in Zimmerman, et al., eds., *Environmental Philosophy*, 1998, p. 364 ff.

Anselm, *Proslogion.*

Arendt, Hannah. *The Origins of Modern Totalitarianism*, 3 vols., NY: Harcourt, Brace and World, 1951/1968 P. I, "Imperialism."

Arias-Maldonado, Manuel, "Let's Make It Real: In Defense of a Realistic Constructivism," *Environmental Ethics*, 33/4 (Winter 2011), p. 377 ff.

Aristotle, *Nichomachean Ethics, Politics, Metaphysics,* in Basic Works of Aristotle, ed. by McKeon, R., NewYork: Random House, 1941/1970.

Armstrong, Susan J., and Botzler, Richard G., *Environmental Ethics: Divergence and Convergence*, McGraw Hill, 2004.

Aquinas, Thomas, *Summa Theologicae. Commentary on the Nicomachean Ethics.*

Attfield, Robin. *The Ethics of Environmental Concern*, New York: Columbia University Press, 1983.

Augustine, Aurelius, *Augustine: Earlier Writings*, ed. Burleigh, J.H.S., Philadelphia: The Westminster Press, 1953. Includes *Concerning the Nature of the Good* (*De Natura Boni),*and *On Free Choice of the Will.*

Auxier, Randall E., "The Death of Darwinism and 'The Limits of Evolution'," Paper presented to a conference on Philosophy and Biology, draft of 9/21/06 (subsequently published).

Ayres, C.E., "The Value Economy," in. Lepley, ed., *Value, A Cooperative Inquiry*, New York: Columbia Univ. Press, 1949, ch. 2.

Baier, Kurt, "The Point of View of Morality," *Australasian Journal of Philosophy*, 32/2, 1954, pp 104-135; repr. in Cahn, S., and Markie, P., *Ethics: History, Theory and Contemporary Issues*, Oxford University Press, 1998 ed.

Barber, Benjamin R., *Jihad vs. McWorld: Terrorism's Challenge to Democracy*, New York: Ballantine Books, 1995.

Bayles, Michael D., *Ethics and Population*, Cambridge, Mass: Schenkman Publ. Co., 1976

Beal, J.M., and Ownbey, Marion. "Cytological Studies in Relation to the Classification of the Genus *Calochortus*, III," *The Botanical Gazette*, 104/4, June 1943, 553-562.

Beauchamp, Tom L., and Walters, Leroy, eds. *Contemporary Issues in Bioethics*, (5th ed.) Belmont, CA: Wadsworth, 1999, (6th ed.) 2003.

Beardsley, Monroe. "Intrinsic Value," *Philosophy and Phenomenological Research,* 26, 1965, p. 6 ff.

Beckerman, Wilfred, *Two Cheers for the Affluent Society: A Spirited Defense of Economic Growth*, St. Martin's Press, 1974.

Bentham, Jeremy. *Principles of Morals and Legislation.*

Bendix, R., *Max Weber: An Intellectual Portrait*, London: Mathuen & Co., 1959.

Berger, Peter L., Luckman, Thomas, *The Social Construction of Reality: A Treatise in the Sociology of Knowledge*, New York: Knopf Publishing Group, 1967

Bernstein, Richard, *Radical Evil: A Philosophical Interrogation*, Malden, Mass, Polity Publishers/Blackwell, 2002.

Berry, Thomas, "The Viable Human," originally publ. in *Revision*, 9/2, 1987; repr in Zimmerman, et al., *Environmental Philosophy,* 1998, p 183 ff.

Berry, Wendell, "The Body and the Earth," Ch. 7 of *The Unsettling of America*, San Francisco: Sierra Club Books, 1977.

Blanshard, Brand, *Reason and Goodness*, New York: Humanities Press, 1961.

Bookchin, Murray, "Social Ecology Versus Deep Ecology," *Socialist Review*, 88/3 (1988), 11-29; (repr. in Pojman, 2001, p. 168 ff.)

Bourke, Vernon J. *History of Ethics*, Garden City: Doubleday/Image, 1968 (1970).

Bradley, Ben, "The Value of Endangered Species," *Journal of Value Inquiry*, 35/1, 2001, 43-56

Brand, Ralf, *Synchronizing Science and Technology with Human Behavior*, London: Earthscan, 2005.

Brandom, Robert, ed., *Rorty and His Critics*, Oxford and Malden, Mass.: Blackwell Publ., 2000.

Brock, Gillian and Reader, Soran, "Needs Centered Ethical Theory," *Journal of Value Inquiry*, 36/4, 425-34.

Burnham, James, *The Managerial Revolution*, Bloomington: Indiana University Press, 1960.

_______, *The Machiavellians, Defenders of Freedom*, New York: The John Day Co. Inc., 1943, reissued by : Henry Regnery, Chicago, 1963.

_______, "Pax Atomica" *National Review*, 1969.

Cafaro, Phillip and Staples, Winthrop III, "The Environmental Argument for Reducing Immigration into the United States," *Environmental Ethics*, 31 (Spr. 09)

Calhoun, John B., "Crowding Into the Behavioral Sink," from "Population Density and Social Pathology," *Scientific American*, 206:3, (1962) 139-0148; in Hock, R., *Forty Studies that Changed Psychology*, 6th ed., New Jersey: Pearson/Prentice Hall,

Callihan, Daniel, "Ethics and Population Limitation," in Bayles, Michael D., *Ethics and Population*, Cambridge, Mass: Schenkman Publ. Co., 1976

Callicott, J. Baird, *In Defense of the Land Ethic, Essays in Environmental Philosophy*, Albany: State University of New York Press, 1989.

______, "Animal Liberation a Triangular Affair," *Environmental Ethics*, Vol. 2, #4, 1980, p. 316, (reprinted in *In Defense of the Land Ethic,* ch. 1).

______, "The Case Against Moral Pluralism," *Environmental Ethics* 12/2, 1990.

______, "Do Deconstructive Ecology and Sociobiology Undermine Leopold's Land Ethic?" in Zimmerman, M., et al., 1998, p. 145 ff.

_______, "Elements of an Environmental Ethic: Moral Considerability and the Biotic Community," reprinted in *In Defense of the Land Ethic*, ch. 4, p. 63.

_______, "Intrinsic Value, Quantum Theory and Environmental Ethics," ch. 9 of *In Defense of the Land Ethics*, p. 157 ff.

_______, "Non Anthropocentric Value Theory and Environmental Ethics," *Amer. Phil Quart*., 21/4, 1984, p. 299.

_______, "On the Intrinsic Value of Non-human Species," ch. 8 of *In Defense*, p. 131.

_______, "Rolston on Intrinsic Value: a Deconstruction," *Env. Eth*., 14/2, 1992.

_______, "The Conceptual Foundations of the Land Ethic," in Zimmerman et al., 1998, p. 101 ff.

_______, "Whither Conservation Ethics?" *Conservation Biology* 4: 15-20, 1990; (repr. in Armstrong and Botzler, eds. 2004, p. 383 ff.).

Camus, Albert, *The Myth of Sisyphus and Other Essays by Albert Camus*, Justin O'Brien, (Translator), New York: Knopf Publishing Group/Vintage, 1991

Caplan, Arthur, ed., *The Sociobiology Debate*, Harper-Collins Publ., 1978.

Carson, Rachel, *Silent Spring*, and "The Obligation to Endure," from *Silent Spring*,

Chandler, Alfred D. Jr. *The Visible Hand: The Managerial Revolution in American Business* (Cambridge, MA; Belknap, Harvard University Press, 1977)

Cheney, Jim, and Weston, Anthony. "Environmental Ethics as Environmental Etiquette: Towards an Ethics-Based Epistemology," *Environmental Ethics,* 21, 2, 115-34, (repr. in Armstrong and Botzler, 2004, p. 95 ff.)

Cicero, M. T., *Academica*, Loeb Classical Library series.

______, *De Officiis/OnDuties,* trans. Edinger, H.G., Indianapolis: Bobbs-Merill, 1974.

Clement, Roland C., "Watson's Reciprocity of Rights and Duties," *Environmental Ethics,* Vol. 1, #4, 1979.

Cohen, Carl, "The Case for the Use of Animals in Biomedical Research," *The New England Journal of Medicine*, vol. 315, pp. 78-80. (Reprinted in Cahn and Markie, eds., *Ethics, History, Theory and Contemporary Issues* (Oxford University Press, 1998).

Collier, Kenneth W., and Kalish, Richard A., *Exploring Human Values: Psychological and Philosophical Considerations*, Monterrey, CA: Brooks/Cole Publ./Wadsworth, 1981.

Conly, Sarah, "The Right to Precreation: Merits and Limits," *American Philosophical Quarterly*, 42/2, April 2005, 105-115.

Confucius (see Kong fu-tse).

Cooper, John. *Reason and Human Good in Aristotle*, Cambridge: Harvard Univ. Press, 1975.

Copleston, Frederick. *A History of Philosophy*, (in Nine Volumes), Garden City: Doubleday/Image, 1946-74.

Cronin, John, and Kennedy, Robert F. Jr., *The Riverkeepers*, New York: Touchstone/Simon and Schuster, 1999.

Daly, Herman E., "Sustainable Growth? No Thank You," in *The Case Against the Global Economy, and For a Turn Toward the Local*, ed. by Mander and Goldsmith, San Francisco, Sierra Club Books, 1996, ch. 16, p. 192 ff.

Darwin, Charles, *Origin of Species*, Amherst: Prometheus Books, 1991, sixth edition.

DeGrazia, David, "The Ethics of Animal Research: What are the Prospects for Agreement," *Cambridge Quarterly of Healthcare Ethics*, 8 (1999), 23-24; reprinted in Beauchamp and Walters, eds., 2003, pp. 418 ff..

Derringh, Frank, W., "Technology, Nature and the Alleged Duty of Human Survival," in Hershock, P. D., Stepaniants, M and Ames, R., eds., *Technology and Cultural Values: On the Edge of the Third Millenium*, (from the 8th East-West Philosophers Conference), Honolulu: Univ. of Hawaii Press, 2003, pp. 26-45.

Descartes, Rene, *Discourse on Method*, and *Meditations on the First Philosophy*, in one vol., *The Philosophical Works of Descartes*, trans. and ed. by Haldane and Ross, Cambridge: Cambridge Univ. Press, 1968.

Des Jardins, Joseph, R., *Environmental Ethics: An Introduction to Environmental Philosophy*, Belmont, Ca.: Wadsworth Publ., 2001 (3rd ed.).

de-Shalit, Avner, "Is Liberalism Environment-Friendly?" *Social Theory and Practice*, 21/2 (Summer 1995), repr in Zimmerman, et al., 1998, p. 386 ff.

De Silva, Lily, "The Buddhist Attitude Toward Nature," from *The Buddhist Attitude Toward Nature*, Sri Lanka: The Buddhist Publication Society, 1987; (repr. in Pojman, 2001, p. 256 ff.).

Dewey, John, and Tufts, James, *Ethics* (New York: Henry Holt, 1908, 1932 revision) [M5/L7].

Dewey, John, *Reconstruction in Philosophy* (Boston: Beacon, 1920, 1972) [M12: 77-204].

______, *Human Nature and Conduct* (New York: Modern Library, 1922, 1957) [M14].

______, "The Ethics of Animal Experimentation," *Atlantic Monthly,* 138, Sept. 1926, pp. 343-6; [L: 98-103].

______, *Experience and Nature* (New York: Dover, 1925, (1958 ed. which follows the 1929 "second edition") [L1].

_______, "The Meaning of Value," *Journal of Philosophy*, 22, Feb. 1925, pp. 126-33; reprinted in *John Dewey, The Later Works*, ed. by Boydston, J., Carbondale: So. Illinois Univ. Press, 1984, pp. 98-103. [1925b]

_______, "Value, Objective Reference and Criticism," *Philos. Review*, 34, July 1925, pp. 313-32; reprinted in *John Dewey, The Later Works*, ed. by Boydston, J., Carbondale: So. Illinois Univ. Press, 1984, pp. 98-103 [1925c].

_______, "The Ethics of Animal Experimentation," *Atlantic Monthly,* 138, Sept. 1926, pp. 343-6; reprinted in *John Dewey, The Later Works*, ed. by Boydston, J., Carbondale: So. Illinois Univ. Press, 1984, pp. 98-103.

______, *The Quest for Certainty, A Study of the Relation of Knowledge and Action* (New York: Capricorn/ Putnam, 1929, 1960 ed.) [L4].

______, "Three Independent Factors in Morals," orig. publ. (in French), in the *Bulletin de la société Française de philosophie*, 30 (Oct.-Dec. 1930) 118-27. English version trans. by Boydston, Jo Ann, in *Educational Theory*, 16, July 1966, 198-209 [Dewey 1930].

_______, *Liberalism and Social Action,* reprinted in *John Dewey, The Later Works*, ed. by Boydston, J., Carbondale: So. Illinois Univ. Press, 1984.

_______, *The Public and Its Problems,* Chicago: Swallow Press, 1927 (1953).

______, *Freedom and Culture,* New York: Capricorn Books, 1939.

______, "Theory of Valuation," from the *International Encyclopedia of Unified Science* (Chicago: Univ. of Chicago Press, Vol. II, #4, 1939) [L13: 189-254].

_______, "Ambiguity of Intrinsic Good," Journ. Of Philosophy, XXXIX, June 42, pp. 328-30, reprinted in *John Dewey, The Later Works*, ed. by Boydston, J., Carbondale: So. Ill. Univ. Press, 1984.

______, "The Field of Value," in Lepley, R., ed., *Value, a Cooperative Inquiry* (New York: Columbia Univ. Press, 1949) ch. 3, p. 64 ff. [L16: 343-357].

Diamond, Jared, *The Third Chimpanzee: the Evolution and Future Development of the Human Animal*, New York: Harper-Collins, 1992.

Diehm, Christian, "Minding Nature: Val Plumwood's Critique of Moral Extensionism," *Environmental Ethics*, 32/1 (Spring 2010), p. 3 ff.

Dillingham, Catherine K., Newton, Lisa H., *Watersheds 3: Ten Cases in Environmental Ethics*; Belmont: Wadsworth, 2002.

diZerega, Gus, "Empathy, Society, Nature and the Relational Self: Deep Ecology and Liberal Modernity," *Social Theory and Practice*, 21/2, (repr. in Gottlieb, R.S., ed., *The Ecological Community*, 1997, p. 56 ff.)

Dobson, Andrew. "Biocentrism and Genetic Engineering," *Environmental Values*, 4, 227-239, 1995.

Dodds, Eric R. *The Greeks and the Irrational* (Berkeley: Univ. of California Press, (1951) 1973 ed.).

Dunham, Barrows, *Man Against Myth*, 1947 (Kessinger Publ. Reprint 2005).

Dwivedi, O.P., "*Satyagraha* for Conservation: Awakening the Spirit of Hinduism," Ethics of Environment and Development, Engel, J.R. and Engel, J.G. eds., London: Bellhaven Press, 1990; (repr. in Pojman, 2001, p. 250 ff.).

Ehrenfeld, David, *The Arrogance of Humanism*, Oxford University Press, 1978.

Ehrlich, Paul R. and Anne H. *Betrayal of Science of Reason, How Anti-Environmental Rhetoric Threatens our Future*, Washington: Island Press, 1998. (Washington/Covelo: Shearwater, 1996)

______, *Extinction*, NY: Random House, 1981

Eldredge, Niles, *Dominion, Can Nature and Culture Co-exist?* New York: Henry Holt and Co., 1995.

Ellul, Jacques. *The Technological Society*, trans. J. Wilkenson, New York: A.A. Knopf, 1964.

Elstein, Daniel. "The Social Construction of Species and the Moral Indefensibility of Speciesism," paper presented to the Society for the Study of Ethics and Animals, Winter 2003 APA Conference Washington, DC.

______, "Species as a Social Construction: Is Species Morally Relevant?," in *Animal Liberation Philosophy and Policy Journal*, I, 1, 2003.

Emmet, Dorothy, *Function, Purpose and Powers: Some Concepts in the Study of Individuals and Societies*, Philadelphia: Temple University Press, 1958 (1972)

Engels, Frederick. "Letter to Joseph Bloch" of 9/21/80. Reprinted in Marx, Karl and Engels, Frederick, *Basic Writings on Politics and Philosophy*, ed. Feuer, L.S., Garden City: Anchor/Doubleday, 1959. p. 397 ff.

Epictetus, *Enchiridion* or *Manual*. Repr. in a vol. with Marcus Aurelius' Meditations, trans. Geo. Long, (South Bend, Ind.: Gateway edit., 1956).

Epicurus' "Letter to Menoeceus," (from Diogenes Laërtius' *Lives of the Eminent Philosophers*, X, 27)

Feinberg, Joel, ed. *Moral Concepts*, Oxford Univ. Press, 1970.

______, "The Nature and Value of Rights," *Journal of Value Inquiry*, vol. 4, 1970, pp. 245-257; (repr. in Cahn, S., and Markie, P., *Ethics: History, Theory and Contemporary Issues* (Oxford University Press, 1998 ed.).

_______, "The Rights of Animals and Unborn Generations," in Blackstone, W. T., ed., *Philosophy and Environmental Crisis*, 1974, pp. 43-68.

_______, "Psychological Egoism," from *Reason and Responsibility*, ed. by Joel Feinberg, Belmont, CA: Wadsworth Publ. 1978. (Repr. in Cahn, S., and Markie, P., *Ethics: History, Theory and Contemporary Issues* (Oxford University Press, 1998 ed.) p. 557 ff.

Findlay, John N., *Axiological Ethics*, London: Macmillan, 1970.

Foot, Phillipa, "Moral Arguments," *Mind*, LXVII, 1958, 502-513; repr. in Margolis, J., *Contemporary Ethical Theory: A Book of Readings*, New York: Random House, 1966.

Foreman, Dave, *Confessions of an Ecowarrior*, New York: Crown Trade Publications, 1991.

Foucault, Michel, *Power/Knowledge, Selected Interviews and Other Writings 1972-1977*, ed. and trans. Gordon, C., New York: Random House/Pantheon, 1972/1980.

______, *Ethics, Subjectivity and Truth*, P. Rabinow, ed., R. Hurley, and others, trans., New York: The New Press, 1994 (1997).

Fox, Warwick, "The Deep Ecology–Ecofeminism Debate and Its Parallels," in Zimmerman, et al., 1998, 227 ff.

Frank, Thomas, *What's the Matter with Kansas? How Conservatives Won the Heart of America*, New York, Henry Holt and Co. 2004

Frankena, William. *Ethics*, Englewood Cliffs: Prentice-Hall, 1963.

_______, "The Naturalistic Fallacy," *Mind*, XLVIII, 1939, pp. 103-114; repr. in Margolis, J., *Contemporary Ethical Theory: A Book of Readings*, New York: Random House, 1966.

Fesmire, Steve, "Ecological Imagination," *Environmental Ethics*, 32/2 (Summer 2010), p 183 ff.

Franklin, Ben, *Autobiography*

Frantzen, Anton, "Dissolving the Explanatory Gap: Do Fish Feel Pain?"

Frey, Raymond G., "Why Animals Lack Beliefs and Desires," in *Animal Rights and Human Obligations*, ed. T. Regan and P. Singer (Englewood Cliffs: Prentice Hall, 1976) P II, sect. 11.

Friedman, Thomas. *Hot, Flat, and Crowded: Why We Need a Green Revolution—and How It Can Renew America*, New York: Farrar, Strauss and Giroux, 2008, p. 252-3.

Friskics, Scott, "The Twofold Myth of Pristine Wilderness: Misreading the Wilderness Act in Terms of Purity," *Environmental Ethics*, 30/4 (Winter 2008), 381-399.

Frondisi, Risieri. *What is Value-An Introduction to Axiology*, trans. Lipp, La Salle: Open Court Publ., 1971.

Fuller, Robert C., "American Pragmatism Reconsidered: William James Ecological Ethic," *Environmental Ethics,* Vol. 14, #2, 1992.

Galbraith, John Kenneth, *The New Industrial State*, Boston: Houghton-Mifflin, 1967/1971. See also *The Affluent Society*

Gans, Gerald F., "What is Deontology?" *Journal of Value Inquiry*, 35/1, March 2001, 27-42.

Gardiner, Patrick, ed., *Theories of History*, New York: The Free Press/Macmillan, 1959.

Garnett, A. Campbell, "Intrinsic Good : its Definition and Referent," in Lepley, R., ed., *Value, a Cooperative Inquiry*, New York: Columbia Univ. Press, 1949.

Gerber, Lisa, "What is So Bad About Misanthropy," *Environmental Ethics*, 24/1, Spring 2002, 41-56.

Glasser, Harold, "Demystifying the Critiques of Deep Ecology," from *Rethinking Deep Ecology*, Witoszek, R., ed., University of Oslo, 1996; repr. in Zimmerman, et al., 1998, p. 212 ff.

Gleitman, Henry, Fridlund, Alan J. and Reisburg, Daniel, *Basic Psychology*, New York: W. W. Norton, 2000 (5th edition)

Godfrey-Smith, William, "The Value of Wilderness," *Environmental Ethics,* Vol. I, #4, 1979, p. 309.

Goldman, Michael, "A Transcendental Defense of Speciesism," *Journal of Value Inquiry*, 35/1, March 2001, 59-69.

Goodland, Robert. "Growth Has Reached its Limit," in *The Case Against the Global Economy, and For a Turn Toward the Local*, ed. by Mander and Goldsmith, San Francisco, Sierra Club Books, 1996, ch. 18, p. 197 ff.

Goodman, Nelson, *Ways of Worldmaking*. Indianapolis, Hackett Publ. 1978

Goodman Paul, "Can Technology be Humane?" from *The New Reformation: Notes of a Neolithic Conservation*, repr. in Teich, A., ed. *Technology and the Future*, Boston: Bedford/St. Martin's, 2000 (8th ed.), pp. 90-103.

Goodpaster, Kenneth . "On Being Morally Considerable," *Journ. Phil.* 75, 1978, (repr. in Zimmerman et al., 1998, p. 56 ff.; and also in Pojman and Pojman, 2008, p. 154 ff.).

Goodpaster, Kenneth E. and Sayre, K.M., eds. *Ethics and the Problems of the 21st Century*, Notre Dame: University of Notre Dame Press, 1979.

Gore, Al, *Earth in the Balance: Ecology and the Human Spirit*, New York: Rodale, 1992/2006.

Gorke, Martin, *The Death of Our Planet's Species: A Challenge to Ecology and Ethics*, trans. Nevers, P, Washington D.C.: Island Press, 2003.

Gottlieb, Roger S., ed., *The Ecological Community*, New York: Routledge, 1997.

Gouinlock, James, *John Dewey's Philosophy of Value*, New York: Humanities Press, 1972.

______, ed., *The Moral Writings of John Dewey*, New York: Macmillan/Hafner, 1976.

Gould, Steven J., "The Golden Rule—A Proper Scale for Our Environmental Crisis," *Natural History*, Sept. 1990, (repr. in Pojman, 2001, p. 206 ff.).

Green, Thomas H., *Prolegomena to Ethics*, New York: Thomas Y. Crowell Co., 1969 ed.

Greenstein, Harold, "Intrinsic Values and the Explanation of Behavior," *Journ. of Value Inquiry*, Vol. 6, #4, 1973, p. 304.

Guevara, Sergio, and LaBorde, Javier, "The Landscape Approach: Designing New Reserves for the Protection of Biological and Cultural Diversity in Latin America," *Environmental Ethics*, 30/3 (Fall 2008), 251-262.

Gunter, Pete, A.Y., "Six Scientific Alternatives to Darwinism," *The Pluralist*, 1:11 (2006), 13-30.

Guha, Ramachandra, "Radical Environmentalism and Wilderness Preservation: A Third World Critique," *Environmental Ethics* 11, Spring 1989; (repr. in Pojman, 2001, p. 271 ff.)

Gunn, Alistair, "Why Should We Care About Rare Species," *Environmental Ethics*, Vol. II, #1, 1980.

Guyer, Paul "Ends of Reason and Ends of Nature: The Place of Teleology in Kant's Ethics," *Journ of Value Inq.*, 36/2-3, 2002, 161-186

Hacking, Ian, *The Social Construction of What?* Cambridge: Harvard University Press, 2000.

Hall, Matthew, "Plant Autonomy and Human-Plant Ethics, *Environmental Ethics*, 31/2 (Summer 2009), 169-181.

Hampshire, Stuart. "Fallacies in Moral Philosophy," *Mind*, LVIII, 1949, pp. 466-482; repr. in Margolis, J., *Contemporary Ethical Theory: A Book of Readings*, New York: Random House, 1966.

Hardin, Garrett, *Living within Limits: Ecology, Economics and Population Taboos*, Oxford University Press, 1993

_______, "Lifeboat Ethics," from "Living on a Lifeboat," *Bioscience* 24 (1974); (repr. in Pojman, 2001, p. 356 ff.).

_______, "The Tragedy of the Commons," in Bayles, Michael D., *Ethics and Population*, Cambridge, Mass: Schenkman Publ. Co., 1976. Orig. printed in Hardin and Baden, J., *Managing the Commons*, W.H. Freeman and Co., 1977.

Hare, Richard M. *The Language of Morals*, Oxford: Oxford Univ. Press, 1952 (1967).

______, *Freedom and Reason*, Oxford: Oxford Univ. Press, 1963 (1972).

Harris, Sam, *The End of Faith: Religion, Terror and the Future of Reason*, New York: W. W. Norton, 2004

Harris, Sheldon, "Factories of Death," from *Factories of Death: Japanese Biological Warfare, 1932-45 and the American Cover-Up*, 1994, Routledge; repr. in Beauchamp and Walters, eds., 1999.

Hartshorne, Charles. "The Rights of the Subhuman World," *Environmental Ethics*, I/1, 1979, p. 49.

Hawken, Paul, "A Declaration of Sustainability," originally printed in *Utne Reader*, #59 (September/October 1993). 54-61. (repr. in Zimmermann et al. eds., 1998 p. 375 ff.).

Hayakawa, Sam. *Language in Action*, New York: Harcourt, Brace and Co., 1941.

Heidegger, Martin. *Being and Time*. trans. by Macquarrie, J., and Robinson, E., New York: Harper and Row,1962.

______, *The Question of Being*, trans. by Kluback, W. and Wilde J, New Haven: College and University Press, 1955.

_______, *The Question Concerning Technology and Other Essays*, ed. Lovitt, New York: Harper and Row, 1977.

Held, Virginia, "Feminist Transformations of Moral Theory," orig. publ. in *Philosophy and Phenomenological Research*, Fall 1990, (Supplement), pp. 321-344. (Repr in Cahn and Markie, *Ethics*, p. 684)

Hettinger, Ned, "Comments on Holmes Rolston's 'Naturalizing Values'," (repr. from a conference presentation in Pojman, 2001, p. 86 ff.).

Hettinger, Ned and Throop, Bill, "Refocusing Ecocentrism: De-emphasizing Stability and Defending Wildness," in Pojman, L., ed., *Environmental Ethics*, 2001, 5th ed.

Hickman, Larry. *Pragmatism as Post-Postmodernism: Lessons from John Dewey*, NewYork: Fordham UP, 2007.

Hilde, Thomas C., and Thompson, Paul B., eds, *The Agrarian Roots of Pragmatism*, Nashville: Vanderbilt University Press, 2000

Hinde, Robert A., *Animal Behavior: A Sunthesis of Ethology and Comparative Psychology*, New York: McGraw-Hill, 1970 (2nd ed).

Hobbes, Thomas, *Leviathan*. Indianapolis: Hackett Publishing Co., 1994.

Hoff, Christina, "Kant's Invidious Humanism," *Environmental Ethics,* Vol. 5, #1, pp. 63-70.

Horn, Miriam and Krupp, Fred, *Earth: the Sequel, The Race to Reinvent Energy and Stop Global Warming*, New York: W. W. Norton, 2008.

Howard, Ross and Perly, Michael, *Acid Rain: the Devastating Impact on North America*, New York: McGraw-Hill, 1980.

Hughes, J. Donald, "The Ancient Roots of our Ecological Crisis," from *Ecology in Ancient Civilizations*, Albuquerque: University of New Mexico Press, 1975, pp. 147-56; (repr. in Armstrong and Botzler, 2004, p. 160 ff.).

Hughes, Thomas, "Technological Momentum" in Smith, Merritt Roe and Leo Marx, eds., *Does Technology Have History?* (Cambridge: MIT Press, 1994) pp. 101-113. Repr in Teich, A., ed. *Technology and the Future*, Boston: Bedford/St. Martin's, 2000 (8th ed.)

Hume, David. *Treatise of Human Nature*, reprinted in *Hume's Moral and Political Philosophy*, ed. Aiken, Darien: Hafner, 1970).

_______, *An Enquiry Concerning the Principles of Morals,* reprin in *Hume's Moral and Political Philosophy*, ed. Aiken, Darien: Hafner, 1970.

Huxley, Aldous, *Brave New World*, NY: Modern Library, 1946/1956.

Izzi Deen, Mawil Y., "Islamic Environmental Ethics, Law and Society," *Ethics of Environment and Development*, Engel, J.R. and Engel, J.G. eds., London: Bellhaven Press, 1990; (repr. in Pojman, 2001, p. 260 ff.).

James, William. *Radical Empiricism*. (Reprinted in one volume with *A Pluralistic Universe*,) New York: E. P. Dutton, 1971.

_______, *Pragmatism and Other Essays*, 1910, New York: Washington Square Press, (1975)

_______, *The Principles of Psychology*, New York: Dover Publications, 1890/1950 in 2 vols.

_______, *Psychology, the Briefer Course*, New York: (Henry Holt, 1892) Harper and Row 1961.

Jenkins, C., and Pimm, Stuart L., "Sustaining the Variety of Life," *Scientific American*, 293/3 (Sept 2005), 66-73 and the map, "Special Places," p. 70.

Jessup, Bertram "On Value," in Lepley, R., ed., *Value, a Cooperative Inquiry*, New York: Columbia Univ. Press, 1949.

Jonas, Hans, *The Imperative of Responsibility, In Search of an Ethics for the Technological Age,* Chicago: Univ. of Chicago Press, 1984.

_______, "Technology and Responsibility: Reflections on the New Task of Ethics." This essay originally appeared in *Social Research* in 1973. From *Philosophical Essays* by Hans Jonas.

Jones, Van, with Ariane Conrad, *The Green Collar Economy: How One Solution Can Fix Our Two Biggest Problems*, New York: Harper One/Harper Collins; 2008.

Jünger Ernst, "Die totale Mobilmachung" (in German; "Total Mobilization"). In *Krieg und Krieger*, ed./ by E. Jünger, Berlin: Junker und Dünnhaupe, 1930. Quoted in Heidegger, M., *The Question of Being*, New Haven: College and University Press, 1958.

Kahn, Herman. *Thinking the Unthinkable*, NY: Harper/Avon, 1962.

Kant, Immanuel. *Groundwork of the Metaphysics of Morals*, Abbott trans., London: Longman, Green and Co., 1785 (1873).

______,*Critique of Practical Reason*, trans. Beck, L., Indianapolis: Bobbs-Merrill, 1788, (1956, 1976 ed.).

______, *Lectures on Ethics*, trans. Infield, L., Indianapolis: Hackett, 1775-80 (1930, 1963 ed.)

______, *The Metaphysics of Morals* Part I, trans. Ladd, J., as *The Metaphysical Elements of Justice* (Indianapolis: Bobbs-Merrill, 1965). Part II trans. Ellington, J., as *The Doctrine of Virtue*, Indianapolis, Hackett, 1797b (1994).

______, "What is Enlightenment" ("Was ist Aufklarung?") trans. by Louis White Beck, New York: Macmillan, 1963.

______, "Idea for a Universal History with a Cosmopolitan Purpose," in *Kant: Political Writings*, trans by H.B. Nisbet, Cambridge Univ. Press.

Kates, R.W., "Sustaining Life on Earth," *Scientific American*, October 1994. Repr in *Society, Ethics and Technology*, ed. Winston, M. and Edelbach, R., Belmont, CA: Wadsworth 2000

Katz, Eric, and Light, Andrew, eds. *Environmental Pragmatism*, New York: Routledge, 1996.

Katz, Eric. "Imperialism and Environmentalism," *Social Theory and Practice*, 21/2, (repr. in Gottlieb, 1997, pp. 163 ff.).

________, "Searching for Intrinsic Value: Pragmatism and Despair in Environmental Ethics," reprinted in *Environmental Pragmatism*, 1996, ch. 15, p. 307 ff. Reprinted from *Environmental Ethics,* Vol. 9, #3, 1987.

Kauffman, Stuart, and Clayton, Philip, "On emergence, agency, and organization," *Biology and Philosophy*, (2006) 21: 501-521

Kelbessa, Workineh, "The Rehabilitation of Indigenous Environmental Ethics in Africa," D*iogenes.* 2005; 52: 17-34.

_______, *The Utility of Ethical Dialogue for Marginalized Voices in Africa*. Available on-line.

Khatchadourian, Haig, "'Intrinsic' and 'Instrumental' Good: an Untenable Dichotomy," *Journ. of Value Inquiry,* Vol. 4, #3, 1969, p. 172.

Kheel, Marti "Nature and Feminist Sensitivity," in *Animal Rights and Human Obligations,* Englewood Cliffs: Prentice Hall, 1976, p. 256 ff.

Kim, Sung-Hae, "The Immortal World: The *Telos* of Daoist Environmental Ethics," *Environmental Ethics*, 30/2 (Summer 2008) 135-157.

Kimbrell, Andrew "Biocolonialization, The Patenting of Life and the Global Market in Body Parts," ch. 11 in *The Case Against the Global Economy*, ed. by Mander, J. and Goldsmith, E., San Francisco: Sierra Club Books, 1996, ch. 11.

King, Rev. Dr. Martin L. "Letter from a Birmingham Jail," reprinted in Estey, G. and Hunter, D., eds., *Nonviolence: A Reader in the Ethics of Action*, Waltham: Xerox College Publications, 1971.

Kirkman, Robert, "Through the Looking Glass: Environmentalism and the Problem of Freedom," *Journal of Value Inquiry*, 36/1, 2002, 27-41.

Kluckholn, Clyde, "Ethical Relativity, Sic et Non," in John Ladd, ed., *Ethical Relativism*, Belmont, CA: Wadsworth Publ., 1973.

Kockelmans, Joseph, *Contemporary European Ethics*, Garden City, NY: Anchor/Doubleday, 1972.

Kong fu-tse (Confucius), *Analects*. Oxford University Press edition, Dawson translation,

Krebs, Al, and Lehman, Karen. "Control of the World's Food Supply," in *The Case Against the Global Economy*, ed. by Mander, J. and Goldsmith, E., San Francisco: Sierra Club Books, 1996, P. 1, ch. 10.

Krugman, Paul, "How Did Economists Get It So Wrong?" *New York Times*, 9/6/09

Kuklick, Bruce, *A History of Philosophy in America*, Oxford: Clarendon Press, 2001.

Ladd, John ed., *Ethical Relativism*, Belmont, Ca., Wadsworth Publ., 1973.

Laertius, Diogenes, *Lives and Opinions of Eminent Philosophers*.

Laird, John, *The Idea of Value*, Cambridge University Press, 1929.

Lakoff, George and Johnson, Mark, *Philosophy in the Flesh, the Embodied Mind and Its Challenge to Western Thought*, New York: Basic Books, 1999.

La Porte, Joseph, "Species as relations: examining a new proposal," *Biology and Philosophy* (2006), 21: 381-393

Laurence, Stephen, and Macdonald, Cynthia, *Contemporary Readings in the Foundations of Metaphysics*, (Oxford/Malden, Mass.: Blackwell, 1998).

Lee, Dorothy. "Culture and the Experience of Value," in Maslow, A., ed., *New Knowledge of Human Values*, Chicago: Henry Regnery Company, 1959/1970, p. 165 ff.

Lemos, Noah M. *Intrinsic Value, Concept and Warrant*, Cambridge: Cambridge Univ. Press, 1994.

Leopold, Aldo, *A Sand County Almanac and Sketches Here and There*, Oxford University Press, 1949/1968.

_______, "The Land Ethic," from *A Sand County Almanac: And Sketches Here and There*.

Lepley, Ray, ed. *Value, a Cooperative Inquiry*, New York: Columbia Univ. Press, 1949.

Levinas, Emmanuel, *Ethics and Infinity; Conversations with Philippe Nemo*. Trans. by Richard A. Cohen. Pittsburg: Duquesne Univ. Press, (1982)/1985.

Lewis, Clarence I. *An Analysis of Knowledge and Valuation*, (The Paul Carus Lectures Series 7), La Salle, Ill.: Open Court, 1946, (1962 ed.)

_____, *Values and Imperatives*, ed. Lange, J., Stanford: Stanford University Press, 1969.

Lewis, Hunter, *A Question of Values: Six Ways We Make the Personal Choices that Shape Our Lives*, San Francisco: Harper, 1990

Lifton, Robert J. "Sterilization and the Nazi Biomedical Vision," from *The Nazi Doctors*, Basic Books, 1986 (repr in Beauchamp and Walters, eds., 1999, p. 533 ff.).

Light, Andrew and McKenna, Erin, eds. *Animal Pragmatism, Rethinking Human-Non-Human Relationships*, Bloomington: Indiana University Press, 2004.

Light, Andrew. "Materialists, Ontologists and Environmental Pragmatists," *Social Theory and Practice*, 21/2, (repr. in Gottlieb, 1997, pp. 255 ff.).

Lipset, Seymour M. *Political Man: The Social Bases of Politics*, Garden City: Anchor/Doubleday, 1963.

Liszka, James J., *Moral Competence: An Integrated Approach to the Study of Ethics*, Upper Saddle River, NJ, Prentice Hall, 2nd Edition, 2002.

Locurto, Charles, *Sense and Nonsense About IQ: The Case for Uniqueness*, Greenwood Publ. Group, 1991

Lucretius, *De Rerum Natura*,
Machan, Tibor, "Why Human Beings May Use Animals," *Journal of Value Inquiry*, 36/1, 2002, 9-14.
MacIntyre, Alysdair. *After Virtue*, University of Notre Dame Press, 1984.
Mackay, Donald S., "Pragmatism," ch. 31 in Ferm, V., ed., *A History of Philosophical Systems*, New York: Philosophical Library, 1950, pp. 381-404.
Mander, Jerry and Goldsmith, Edward, eds., *The Case Against the Global Economy*, San Francisco: Sierra Club Books, 1996.
Manes, Christopher, *Green Rage: Radical Environmentalism and the Unmaking of Civilization,* Boston and New York: Back Bay Books, 1990.
Mannheim, Karl, *Ideology and Utopia*, trans. Wirth, L. and Shils, E., New York: Harvest/ Harcourt, Brace and World, 1936/1964.
Marcuse, Herbert. *One Dimensional Man* (Boston: Beacon, 1964 & 1967.
Margolis, Joseph. *Contemporary Ethical Theory: A Book of Readings*, New York: Random House, 1966.
Marx, Karl. *The German Ideology*, NY: International Publ., 1947, trans. anon.
Marx, Leo. "Does Improved Technology Mean Progress?" first published in *Technology Review,* January 1987, pp. 33-41, 71.
Maslow, Abraham, ed., *New Knowledge of Human Values*, Chicago: Henry Regnery Company, 1959/1970.
_______, *Toward a Psychology of Being*, Insight Van Nostrand, 1962.
Mathews Freya, "Planetary Collapse Disorder: The Honeybee as Portent of the Limits of the Ethical," *Envir. Ethics*, 32/3 (Fall 2010), p. 353 ff.
McCumber, John. *Time in the Ditch: American Philosophy and the McCarthy Era*, Northwestern University Press, 2000.
McDermott, John. "Technology, the Opiate of the Intellectuals," from *New York Review of Books*, July 31, 1969. Repr. in Teich, *Technology and the Future*, 2000, 8th ed.
McDonald, Hugh, "Creative Actualization as a Theory of Value" *Contemporary Pragmatism*, December 2006.
______, *Creative Actualization, a Meliorist Theory of Value*, Editions Rodopi, 2012.
______, "Does Nature Exist?" *Contemporary Philosophy,* Vol. XXI, #5 & 6, Sept./Oct and Nov./Dec 2000.
_______, *John Dewey and Environmental Philosophy*, Albany: SUNY Press, 2003.
______, *Political Philosophy and Ideology*, Lanham: University Press of America, 1997.
______, *Radical Axiology, a First Philosophy of Values*, Editions Rodopi, 2004.
______, *Speculative Evaluations, Essays on a Pluralistic Universe*, Editions Rodopi, 2012.
______, "The Ethics of Definitions," ined.

______, "Towards a Deontological Environmental Ethic," *Environmental Ethics* Winter (Nov.) 2001.

McKibben, Bill, "A Special Moment in History," *Atlantic*, 1998, repr. in Pojman and Pojman, 2008, p. 378 ff.

______, "The City at Hyperspeed: A Challenge to Think Pretty Damn Fast" presented at the 28th Annual Conference of the Society for Advancement of American Philosophy.

_______, *The End of Nature*, New York: Anchor, 1989

______, "Year One: Climate Chaos Has Arrived," *Sierra*, Jan/Feb 2006

McVaugh, Rogers. *Flora Novo-Galliciana*, Vol. 15, *Bromeliaceae to Dioscoreaceae,* Ann Arbor: The Univ. of Michigan Herbarium, 1989.

Meadows, Donella H. and the Club of Rome Staff, *The Limits to Growth: A Report for the Club of Rome's Project on the Predicament of Mankind*, Universe Books, 1972.

Meadows, Donella H., "Biodiversity: The Key to Saving Life on Earth," from *The Land Steward Letter*, Summer 1990; (repr. in Pojman, 2001, p. 206 ff.).

Merchant, Carolyn. "Environmental Ethics and Political Conflict: A View from California," *Environmental Ethics*, Vol. 12, #1, 1990.

Merton, Robert K. *Social Theory and Social Structure*, NY: Free Press, 1968.

Michels, Roberto, *Political Parties*, trans. Paul, E. & C., NY: Free Press/Macmillan, 1968.

Mill, John Stuart. "Nature," from *Three Essays on Religion*, New York: Henry Holt, 1874; repr. in Pojman, 2001, p. 89 ff.).

_______, *Principles of Political Economy*, 1848.

______, *Utilitarianism*, in *The Utilitarians*, Garden City: Anchor/Doubleday, 1863, (1973), (publ. with *On Liberty*, and Bentham's *Principles of Morals and Legislation*).

Miller, Peter. "Value as Richness: Toward a Value Theory for an Expanded Naturalism in Environmental Ethics," *Environmental Ethics*, Vol. 4, #2, 1982.

Mitchell, E.T., "Values, Valuing and Evaluation," in Lepley, ed.. *Value, a Cooperative Inquiry*, New York: Columbia Univ. Press 1949, p. 202 ff.

Molseed, Elwood, *The genus Tigridia (Iridaceae) of Mexico and Central America.* Berkeley: Univ of California Press, 1970. (doct. dissertation)

Moncrief, Lewis W., "The Cultural Basis of Our Environmental Crisis," Science, Vol. 170 (Oct. 1970), pp. 508-512; (repr. in Pojman, 2008, p. 19 ff.).

Moore, George E. *Principia Ethica*, Cambridge: Cambridge Univ. Press, 1903 (1959 ed.).

_______, *Ethics*, Oxford: Oxford Univ. Press, 1912 (1965).

_______, "Is Goodness a Quality?" from *Philosophical Papers*, New York: Macmillan, (1959) 1963.

_______, "Nature of Moral Philosophy," ch. 10 of *Philosophical Studies*, London: Routledge, 1922 (1958).

_______, "The Conception of Intrinsic Value" in *Philosophical Studies*, London: Routledge, 1922 (1958).

Münsterberg, Hugo, *The Eternal Values*, Boston: Houghton-Mifflin, 1909.

Murdoch, Iris, *The Sovereignty of Good*, New York: Ark/Routledge, 1970/1985.

Murdy, William H. "Anthropocentrism: A Modern Version," *Science*, 187, pp. 1168-72, 1975; repr. in Armstrong and Botzler, 2004, p. 280 ff.

Murphey, Murray J., *C. I. Lewis, The Last Great Pragmatist*, Albany: State University of New York Press, 2005.

Murphy, Arthur E., *The Theory of Practical Reason*, (Paul Carus Lectures 1955), LaSalle, Illinois: Open Court, 1964.

Naess, Arne, "Ecosophy T: Deep Versus Shallow Ecology," repr. in Pojman and Pojman, 2008 (from Tobias, M., ed., *Deep Ecology*, Santa Monica, CA: IMT Productions, 1985).

_______, "The Deep Ecological Movement: Some Philosophical Aspects" *Philosophical Inquiry* 8/1-2 (1986) (repr. in Zimmerman, et al., 1998, p. 193 ff.).

Narveson, Jan. "Moral Problems of Population," in Bayles, 1976.

Nash, Roderick F. *The Rights of Nature*, Madison, University of Wisconsin Press, 1989.

Neville, Robert C., *Reconstruction of Thinking*, Albany: SUNY Press, 1981.

Nielsen, Kai, "On Ascertaining What is Intrinsically Good," *Journ. of Value Inquiry*, Vol. 10, #2, 1976.

Nietzsche, Friedrich. *The Birth of Tragedy,* trans F. Golffing, Garden City, N.Y.: Doubleday/Anchor, 1956.

_________, *Beyond Good and Evil: Prelude to a Philosophy of the Future,* trans. Kaufmann, New York: Vintage Press/Random House, 1966.

_______, *On the Genealogy of Morals*, trans. Kaufmann, in an edition with *Ecce Homo*, New York: Vintage Press/Random House, 1969.

______, *The Will to Power*, trans. Kaufman, (N.Y.: Vintage, 1968)

Nobis, Nathan, "Carl Cohen's "Kind" Arguments *FOR* Animal Rights and *AGAINST* Human Rights," paper presented at the Eastern APA.

Norton, Bryan. *Why Preserve Natural Variety?*, Princeton: Princeton Univ. Press, 1987.

_______, "The Constancy of Leopold's Land Ethic," in Katz and Light, *Environmental Pragmatism* 1996, ch. 5.

_______, "Integration and Reduction," ch. 6 of Katz and Light, *Environmental Pragmatism*, 1996.

Ogungbemi, Segun, "An African Perspective on the Environmental Crisis," in Pojman, L., ed., *Environmental Ethics*, 2001, p. 265 ff.

O'Neill, John, "Time, Narrative and Environmental Politics," in Wenz, P. S., "Environmentalism and Human Oppression," (repr. in Gottlieb, 1997, p. 22 ff.)

Orwell, George. *Animal Farm*, NY: Signet, 1964 ed.

_______,*1984*, NY: Signet/New American Library, 1948/1961.

Osborne, Harold. *Foundations of the Philosophy of Value*, Cambridge: Cambridge at the Univ. Press, 1933.

Ott, P., "Value as Practice and the Practice of Value: Dewey's Value Theory for Environmental Ethics," *Envir Ethics,* 32, 3, Fall 2010, 285 ff.

Ownbey, Marion. "A Monograph of the Genus *Calochortus*," *Annals of the Missouri Botanical Garden*, 27/4, #4, 11/40, 1940.

Packard, Vance, *The Hidden Persuaders*, New York: Simon and Schuster, 1976.

______, *The Status Seekers*, New York: Simon and Schuster, 1969.

______, *The Waste Makers*, New York: Simon and Schuster, 1978.

Pappas, George F. "To Be or to Do, John Dewey and the Great Divide in Ethics," *Hist. Phil. Quart.*, 14:4, 1997, 447-472.

Pareto, Vilfredo, *The Rise and Fall of the Elites: an Application of Theoretical Sociology*, Totowa NJ: Bedminster Press, 1968.

Parker, Kelley. "Pragmatism and Environmental Thought," in Katz and Light, *Environmental Pragmatism,* 1996, ch. 1.

_______, "The Values of a Habitat," *Environmental Ethics,* Vol. 12, #4, 1990.

Parsons, Talcott, *The Structure of Social Action*, NY: The Free Press, 1937/1968.

Paton, Herbert J., *The Categorical Imperative, a Study in Kant's Moral Philosophy*, Philadelphia: Univ. of Pennsylvania Press, (1947), 1971.

Pearson, Veronica, "Population Policy and Eugenics in China," *British Journal of Psychiatry*, 167, July 1995, pp. 1-4 (repr. in Beauchamp and Walters, eds., 1999, p. 541).

Pellegrino, Edmund D. "The Virtuous Physician and the Ethics of Medicine," in Shelp, E. ed., *Virtue and Medicine: Exploration in the Character of Medicine*, D. Reidel Publ., 1985, pp. 243-55.

Pepper, Stephen. "Observations on Value from an Analysis of Simple Appetition," in Lepley, R., ed., *Value, a Cooperative Inquiry*, 1949. ch. 13.

______, "A Brief History of General Theory of Value," in Ferm, V., ed., *A History of Philosophical Systems*, New York: Philosophical Library, 1950.

Perry, Ralph B., *Realms of Value, A Critique of Human Civilization*, Cambridge: Harvard University Press, 1954.

Peterson, Anna L., "Environmental Ethics and the Social Construction of Nature," *Environmental Ethics,* 21. 4, pp. 339-57; (repr. in Armstrong and Botzler, 2004, p. 87 ff.).

Phihar, Evelyn B., "The Justification of an Environmental Ethic," *Environmental Ethics,* Vol. 5, #1, pp. 47-61.

Pico della Mirandola, Giovanni, "Essay on the Dignity of Man" [*De Hominis Dignitate*] trans by Sears, Jane, New York: Peter Lang, 1984.

Pihlström, Sami, *Pragmatic Moral Realism, A Transcendental Defense* (Amsterdam/New York: Rodopi, Value Inquiry Book Series, 2005

Pini, Prof. Giorgio, "The Logics of Duns Scotus and the Pragmatists," Sept. 2006, a paper read before the New York Pragmatist Forum.

Pirsig, Robert M., *Zen and the Art of Motorcycle Maintenance, An Inquiry into Values*, New York: William Morrow and Co. Publ., 1974 (Bantam ed. 1975)

Plato, *Crito Phaedo. Gorgias*, *Meno.Republic*, *Gorgias*, *Sophist* or *Statesman* and *Laws* in *Collected Dialogues*, Cairns, H. and Hamilton, E., eds., Princeton: Princeton University Press, 1973 ed.

Pojman, Louis P., ed. *Environmental Ethics: Readings in Theory and Application*, Belmont, CA: Wadsworth, 2001, 3rd ed.

_______, and Pojman, Paul, ed. *Environmental Ethics: Readings in Theory and Application*, Belmont, CA: Wadsworth, 2008, 5th ed.

Polin, Raymond. "Values and Truths," ch. 7 of *Contemporary European Ethics*, ed. Kockelmans, Garden City, Doubleday/Anchor, 1972.

Popper, Karl, *The Open Society and its Enemies*, New York: Harper and Row, 1963 ed. See also his *The Poverty of Historicism*.

Potter, Vernon J., *Charles S. Pierce on Norms and Ideals*, Worcester: University of Massachusetts Press, 1967.

Povilitis, Anthony J., "On Assigning Rights to Animals and Nature," *Environmental Ethics,* Vol. 2, #1, 1980, p. 67-71.

Prall, David, W., *A Study in the Theory of Value*, Berkeley: Univ. of Calif. Publications in Philosophy, Vol. 3, #2, pp. 179-290, 1921.

Pritchard, Harold A., "Does Moral Philosophy Rest on a Mistake?" Mind, XXI (1912) 487-499. Reprinted in Margolis, J., ed., *Contemporary Ethical Theory*, 1966.

Putnam, Hillary. *Ethics without Ontology*, Cambridge, MA: Harvard U P, 2004.

Rachels, James, "The Challenge of Cultural Relativism," from *Elements of Moral Philosophy*, McGraw-Hill, 1978, pp. 15-29; repr. in Cahn, S., and Markie, P., *Ethics: History, Theory and Contemporary Issues*, Oxford University Press, 1998 ed.

_______, "Darwin, Species and Morality," in *Animal Rights and Human Obligations*, Regan, T., Singer, P., eds., part 3, sect. 3.

Radest, Howard, "Modern Humanism," paper delivered to the Society for Advancement of American Philosophy convention, Michigan State Univ. 3/14/08.

Rand, Ayn, *Atlas Shrugged*. New York : Plume, 1992.

Rashdall, Hastings, *The Theory of Good and Evil*, Oxford, 1907.

Rauber, Paul, “Why Not Nukes?” *Sierra* Magazine, 92/1, Jan./Feb. 2007, p. 37

Rawls, John, “Two Concepts of Rules,” *The Philosophical Review*, LXIV, 195, 3-32; repr. in Margolis, J., *Contemporary Ethical Theory: A Book of Readings*, New York: Random House, 1966.

Regan, Tom, and Singer, Peter, eds., *Animal Rights and Human Obligations*, Englewood Cliffs: Prentice Hall, 1976.

Regan, Tom. “Animal Rights and Human Wrongs,” *Environmental Ethics* II/2, (1980) p. 99 ff. (repr. in Zimmerman et al., 1998, p. 41 ff.).

_______, “The Nature and Possibility of an Environmental Ethic” *Environmental Ethics*, III/2, 1981, pp. 19-34.

_______, “The Case for Animal Rights,” in *Animal Rights and Human Obligations*, ed. T. Regan and P. Singer.

______, “The Radical Egalitarian Case for Animal Rights,” excerpted from *In Defense of Animals*, ed. Singer, repr. In Pojman, 2001, p. 40 ff.

_______, “Why Death Does Harm to Animals,” in *Animal Rights and Human Obligations*, Regan, T., Singer, P., eds., Englewood Cliffs: Prentice Hall, 1976 (c), p. 153 ff.

Reisner, Marc, *Cadillac Desert, The American West and its Disappearing Water*, New York: Viking/Penguin, 1986/1993.

Rescher, Nicholas. *Introduction to Value Theory*, Englewood Cliffs: Prentice-Hall, 1969.

______, *Unpopular Essays on Technological Progress*, 1980.

Rice, Phillip. “Science, Humanism and the Good,” in Lepley, ed., *Value, A Cooperative Inquiry*, New York: Columbia Univ. Press, 1949, ch. 14.

Richmond, Samuel A., “On Replacing the Notion of Intrinsic Value,” *Journ. of Value Inquiry,* Vol. 10, #3, 1976 p. 205.

Rieppel, Olivier, and Dearney, Maureen, “The poverty of taxonomic characters,” *Biology and Philosophy* (2007) 22: 95-113

Riesman, David. *The Lonely Crowd: A Study of the Changing American Character*, New Haven: Yale Univ. Press, 1950/1961.

Roberts, Paul, *The End of Oil*, Boston: Houghton-Mifflin, 2005

Rolston, Holmes III, “The Value of Species,” in *Animal Rights and Human Obligations*, T. Regan and P. Singer eds., 1976, p. 252 ff.

_______, “Are Values in Nature Subjective or Objective?” *Env. Eth.* 4/2, 1982, 125-151 (repr. in *Philosophy Gone Wild*, 1989).

______, “Challenges in Environmental Ethics,” from *Ecology, Economics, Ethics: the Broken Circle*, New Haven: Yale University Press, 1991 (repr. in Zimmerman, et al., 1998, p. 124 ff.).

______, *Environmental Ethics, Duties to and Values in the Natural World* (Philadelphia: Temple Univ. Press, 1988)

_______, “Feeding People versus Saving Nature,” (repr. in Gottlieb, 1997, p. 208 ff.

_______, "Naturalizing Values: Organisms and Species," in Pojman, L., ed., *Environmental Ethics*, 2001, p. 76 ff.

______, *Philosophy Gone Wild: Environmental Ethics*, Amherst: Prometheus Books, 1989.

Rorty, Richard. *Philosophy and Social Hope*, New York, Penguin, 1999.

_____, *Truth and Progress, Philosophical Papers, Vol. 3*, Cambridge University Press, 1998.

_____, *Philosophy as Cultural Politics, Philosophical Papers, Vol. 4*, Cambridge University Press, 2007.

Rosenthal, Sandra B. and Buchholz, Rogene, A., "How Pragmatism is an Environmental Ethic," in Light, A. and Katz, E., *Environmental Pragmatism,* New York: Routledge, 1996.

Ross, Sir W. David, *The Right and the Good*, London: Oxford at the Clarendon Press, 1930 (1961)

Rousseau, Jean Jaques, "Discourse on the Arts and Sciences," in *The Essential Rousseau*, trans. Bair, L., NY: New American Libr., 1974.

_______, "On the Sciences and Arts," taken from the "Final Reply" in *Collected Writings of Jean-Jacques Rousseau*, vol. 1, ed. by Masters, R. D. and Kelly, C., Hanover NH: University Press of New England, 1992, pp. 110-99.

Routley, Richard (Sylvan), "Is There a Need for a New, an Environmental Ethic?" in Zimmerman, et al., 1998, p. 17 ff.).

_______, "Against the Inevitability of Human Chauvanism" *Ethics and Problems of the 21st Century,* ed. Goodpaster and Sayre (University of Notre Dame Press, 1979), 36-59.

Rowe, Jonathan, "The Common Good," *Sierra*, 90/4 (July/Aug. 2005), 54 ff.

Ruhlen, Merritt. *A Guide to the World's Languages.* Stanford University Press, 1987.

Russow, Lilly-Marlene, "Why Do Species Matter?" *Environmental Ethics*, Vol.3, #2, pp. 101-112.

Sagoff, Mark, "The Ecological Critique of Agricultural Biodiversity," excerpted from "Biotechnology and Agriculture: The Common Wisdom and Its Critics" *Indiana Journal of Global and Legal Studies*, 9.3 2001 (Indiana Univ. Press); (repr. in Armstrong and Botzler, 2004, p 480 ff.).

______, "Genetic Engineering and the Concept of the Natural" *Philosophy and Public Policy Quarterly*, Spring/ Summer 2001, Vol. 21 (2/3), pp 2-10. Repr. in Winston and Edelbach, 2000.

Sahakian, William S. *Ethics, An Introduction to Theories and Problems*, New York: Harper and Row/Barnes and Noble, 1974.

Sahlins, Marshall, *Stone Age Economics*, Aldine Transaction, 1972

Samuelsson, Lars, "Environmental Pragmatism and Environmental Philosophy: A Bad Marriage," *Environmental Pragmatism*, 32/4, Winter 2010, 405-415.

Samuelson, Lars, "On the Demarcation Problem and the Possibility of Environmental Ethics: a Refutation of 'A Refutation of Environmental Ethics," *Envir. Ethics*, 32/3 (Fall 2010), p. 247 ff.

Saydah, J. Roger. *The Ethical Theory of Clarence Irving Lewis*, Athens: Ohio Univ. Press, 1969.

Scheler, Max. *Formalism in Ethics and a Non-Formal Ethics of Values*, trans. Frings and Funk, Evanston: Northwestern Univ. Press, 1916/1973.

______, *Man's Place in Nature*, New York: the Noonday Press, 1928/1961, trans. by Hans Meyerhoff.

Schiller, Ferdinand Canning S., *Humanism, Philosophical Essays*, New York: Macmillan, 1903.

Schönfeld, Martin, "The Green Kant: Environmental Dynamics and Sustainable Policies," repr. in Pojman and Pojman, eds., 2008, p. 49 ff.

Schroyer, Trent, *The Critique of Domination: the Origins and Development of Critical Theory*, Boston: Beacon, 1973.

Schumpeter, Joseph, *Capitalism, Socialism and Democracy*, New York: Harper and Row, 1962 (1942).

Schürmann, Reiner. *Heidegger on Being and Acting: From Principles to Anarchy* Bloomington: Indiana Univ. Press, 1990.

Schweitzer, Albert. *Civilization and Ethics*, trans. J. Naish, repr. in *Animal Rights and Human Obligations*, ed. T. Regan and P. Singer, Englewood Cliffs: Prentice Hall, 1976, sect. 10. (Part repr as "Reverence for Life," in Pojman, 2001, p. 95 ff.).

Scully, Matthew, *Dominion, the Power of Man, the Suffering of Animals and the Call to Mercy*, New York: St. Martin's Press, 2002.

Searle, John R., *Rationality in Action*, (the Jean Nicod Lectures) Cambridge: MIT Press, 2001.

Sellars, Wilfrid, *Empiricism and the Philosophy of Mind*, Cambridge: Harvard University Press, 1956/1997.

Sessions, George, "Ecocentrism, Wilderness and Global Ecosystem Protection," in Zimmerman, et al., *Environmental Philosophy* (Upper Saddle River, NJ: Prentice Hall, 1998, p. 245 ff.).

Shaner, David E. and Duval, R. Shannon, "Conservation Ethics and the Japanese Intellectual Tradition," *Environmental Ethics*, Vol. 11, #3, 1989, pp. 197-214.

Shook, John, "American Humanism, Past and Future," paper presented to SAAP, Michigan State University, March, 2008.

Sidgewick, Henry. *The Methods of Ethics*, N.Y.: Dover, 1966.

Simberloff, Daniel, "Nature, Natives, Nativism and Management: Worldviews Underlying Controversies in Invasion Biology," *Envir Ethics,* 34, 1, Spr 2012, p. 5 ff.

Simmons, Aaron, "Two Arguments Against Biological Interests," *Envir. Ethics*, 32/3 (Fall 2010), p. 229 ff.

Singer, Peter, "All Animals are Equal," reprinted in Zimmerman, et al., *Environmental Philosophy,* 1998, p. 26 ff..

_______, *Animal Liberation*, New York: New York Review Press, 1975.

_______, "A Utilitarian Defense of Animal Liberation," from *Animal Rights and Human Obligations*, repr. In Pojman, *Environmental Ethics* 5th ed., 33 ff.

______, "A Utilitarian Population Principle," in Bayles, Michael D., *Ethics and Population*, Cambridge, Mass: Schenkman Publ. Co., 1976

______, "Famine, Affluence and Morality," reprinted in Cahn and Markie, eds., *Ethics, History, Theory and Contemporary Issues* (Oxford U.P. 1998).

Skinner, Burrhus F., *Walden II*, Bantam, 1972.

Smith, Ian, "The Role of Humility and Intrinsic Goods in Preserving Endangered Species: Why Preserve the Humpback Chub?" *Environmental Ethics*, 32/12, Summer 2010, p. 165 ff.

Smith, John E., *The Spirit of American Philosophy*, London and New York: Oxford University Press, 1963.

Smith, Peter and Jones, O.R. *The Philosophy of Mind, an Introduction*, Cambridge University Press, 1986.

Sorel, Georges. *Reflections on Violence*, trans. Hulme, T.E., NY: Peter Smith Publ., 1941 ed.

Speth, James G., *Red Sky at Morning: America and the Crisis of the Global Environment,* New Haven: Yale University Press, 2004.

Stafford, Sue, "Intellectual Virtue in Environmental Virtue Ethics," *Environmental Ethics*, 32/3 (Fall 2010), p. 339 ff.

Stamos, David N., *The Species Problem: Biological Species, Ontology and the Metaphysics of Biology*, Lanham: Lexington Books, 2003.

Stephens, Piers, "Nature and Human Liberty, The Golden Country in George Orwell's 1984 and an Alternative Conception of Human Freedom," *Organization and Environment*, 17/1, 2004, 76-98 .

Stevenson, Charles L., "The Emotive Meaning of Ethical Terms," *Mind*, XLVI, 1938, pp 10-31, also in *Facts and Values*, New Haven: Yale University Press, 1963; repr. in Margolis, 1966.

Sterba, James, "Abortion, Distant Peoples and Future Generations," publ. in the *Journal of Philosophy*, 77/7 (July 1980), pp. 424-440.

Stone, Christopher D., *Earth and other Ethics: The Case for Moral Pluralism* New York: Harper and Row, 1987.

Stohr, Karen and Wellman, Christopher, "Recent Work in Virtue Ethics," *American Philosophical Quarterly*, Jan. 2002, 49 ff.

Stokkink, Karin, *Cultural Relativism*, unpublished essay submitted to and accepted by the philosophy faculty of Lehman College, CUNY, and winner of the first Weinstock Award.

Strauss, Leo, *Natural Right and History*, Univ. of Chicago Press, 1974.

Stuart, Henry W., "Dewey's Ethical Theory," in Schilpp, P.A., ed. *The Philosophy of John Dewey*, New York: Tudor Publ., 1939 (1951), pp. 293-333 (sect. 10).

Sumner, William G., "Folkways," repr. in Ladd, John ed., *Ethical Relativism*, Belmont, Ca., Wadsworth Publ., 1973, p. 23 ff.

Taylor, Bob P., "John Dewey and Environmental Thought: A Response to Chaloupka," *Environmental Ethics*, Vol. 12, #2, 1990.

Taylor, Paul W., "The Ethics of Respect for Nature," *Environmental Ethics,* Vol. 3, #3, 1981.

_______, "In Defense of Biocentrism," *Env. Eth.*, Vol. 5, #3, 1983, pp. 237-243.

Taylor, R., *Action and Purpose*, Englewood Cliffs: Prentice-Hall, 1966.

Thompson, Jenna, "A Refutation of Environmental Ethics," reprinted in Hinman, L., ed., *Contemporary Moral Issues: Diversity and Consensus* (Upper Saddle River: Prentice Hall, 2000), pp. 576-586.

Throop, William, "The Rationale for Environmental Restoration," in Gottlieb, ed., 1997, p. 39 ff.

Toulmin, Steven. *An Examination of the Place of Reason in Ethics*, Cambridge: Cambridge Univ. Press, 1970.

Toynbee, Arnold. *A Study of History,* Somervell Abridgement, Oxford University Press, 1946.

Travers, Martin. *Critics of Modernity: The Literature of the Conservative Revolution in Germany, 1890-1933*. Bern and New York: Peter Lang Publishing, 2001

Trenberth, Kenneth. "Stronger Evidence of Human Influence on Climate—The 2001 IPCC Assessment," from "Stronger Evidence of Human Influence on Climate," by Kevin E. Trenberth, *Environment Magazine*, Vol. 43 (May 2001) pp. 10-19. Published by Heldref Publications, Washington, D.C., 2001.

UNESCO "Universal Declaration on the Human Genome and Human Rights," in Beauchamp and Walters, 1999, p. 555 ff.

Vallentyne, Peter, "The Teleological-Deontological Distinction," *Journal of Value Inquiry*, Vol. 21, #1, 1987.

Van den Haag, Ernest. "In Defense of the Death Penalty," reprinted in Cahn and Markie, eds., *Ethics, History, Theory and Contemporary Issues* (Oxford U.P. 1998).

Veblen, Thorstein. "Why is Economics Not an Evolutionary Science," and "The Preconceptions of the Classical Economists," (orig. from *The Place of Science in Modern Civilization*) in *The Portable Veblen*, Max Lerner, ed., New York: Viking, 1948/1969, p. 215 ff.

______, *The Engineers and the Price System*, New York: Augustine M. Kelley Reprints, 1921/1965. Repr. in *The Portable Veblen*, Max Lerner, ed., New York: Viking, 1948/1969.

______, "Sabotage," in Veblen, T., *The Engineers and the Price System*, Repr. in *The Portable Veblen*, Max Lerner, ed., New York: Viking, 1948/1969.

Wang, Henry, "Rethinking the Validity and Significance of Final Causation: From the Aristotelian to the Peircean Teleology," *Transactions of the Charles S. Peirce Society,* Vol. 41, #3, Summer 2005, 603-625.

Warren, Mary A., "A Critique of Regan's Animal Rights Theory," *Between the Species*, 2/4 (Fall 1987); (repr. in Pojman, 2001, p. 46 ff.).

Weber, Max, *From Max Weber: Essays in Sociology*, trans. and ed. by C. Wright Mills and H.H. Girth, NY: Oxford Univ. Press, 1946.

______, *The Protestant Ethic and the Spirit of Capitalism*, trans. by Stephen Kalberg, Englewood Cliffs: Prentice Hall, 1977

Wellbank, Joseph H., "Is a New Definition of Ethical Naturalism Needed?" *Journ. of Value Inquiry*, Vol. 8, #1, 1974, p. 46 ff.

Wenz, Peter S., "Environmentalism and Human Oppression," in Gottlieb, 1997, p. 3 ff.

______, "Just Garbage: Environmental Injustice" from *Faces of Environmental Racism: Confronting Issues of Global Justice*, Wenz and Westra, L., eds., Lanham, Maryland: Rowman and Littlefield, 1995; (repr. in Armstrong and Botzler, 2004, p. 106 ff.).

Werkmeister, William H. *Historical Spectrum of Value Theories*, Lincoln: Johnsen Publ. Co., 1970.

Werth, Barry, *Banquet at Delmonico's: Great Minds, the Gilded Age and the Triumph of Evolution in America*, New York: Random House, 2009

Weston, Anthony, "Before Environmental Ethics," *Env. Eth.*, 14/4, 1992; repr. in Katz and Light, eds., *Environmental Pragmatism,* New York: Routledge, 1996, ch. 7, p. 139.

______, "Beyond Intrinsic Value, Pragmatism in Environmental Ethics," *Environmental Ethics*, 7/4, 1985; (repr. in Katz and Light, 1997.)

White, Lynn, "The Historical Roots of our Ecologic Crisis," *Science*, 155: 1203-7; (repr. in Armstrong and Botzler, 2004, 219 ff.).

White, Morton, "Value and Obligation in Dewey and Lewis," *Philosophical Review*, 58, (1949), 322.

Whyte, William H., *The Organization Man*, New York: Clarion/Simon and Schuster, 1956

Wilkins, John, "The dimensions, modes and definitions of species and speciation," *Biology and Philosophy* (2007) 22: 247-266.

Willard, L.Duane, "On Preserving Nature's Aesthetic Features," *Environmental Ethics,* Vol. 2, #4, 1980.

Wilshire, Bruce, *Get 'Em All! Kill 'Em!: Genocide, Terrorism, Righteous Communities* Lanham, MD: Rowman & Littlefield Publishers, 2004.

Wilson, Edward, O., "The Little Things that Run the World," *Conservation Biology*, 1: 344-46; (repr. in Armstrong, and Botzler, 2004, p. 46 ff.).

Wilson, Holly L., "The Green Kant: Kant's Treatment of Animals," (repr. in Pojman and Pojman, 2008, p.

Windelband, Wilhelm, *A History of Philosophy*, trans. Tufts, J.H. (NY: Harper and Bros., (1901) 1958).

______, "History and Natural Science," (Rectorial Address of 1894) trans. G. Oakes, in *History and Theory*, xix, 1980, pp. 165-185.

Winner, Langdon, "Technologies as Forms of Life" From *Epistemology, Methodology and the Social Sciences*, Cohen and Wartofsky, eds., Kluwer Academic Publishers, pp. 249-263.

Winston, M. and Edelbach, R., eds., *Society, Ethics and Technology*, Belmont, CA: Wadsworth 2000.

Wittfogel, Karl A. *Oriental Despotism*, New Haven: Yale Univ. Press, 1957/1967.

Wittgenstein, Ludwig. *Philosophical Investigations*, trans. Anscombe, N.Y.: Macmillan, (1953) 1970.

Zimmerman, Michael. "The Critique of Natural Rights and the Search for a Non-Anthropocentric Basis for Moral Behavior," *Jour. Value Inquiry*, Vol. 19, #1, 1985.

________, "Ecofascism: A Threat to American Environmentalism?" *Social Theory and Practice*, 21/2, (repr. in Gottlieb, 1997, p. 229 ff.).

Zimmerrnan, Michael, Callicott, J. Baird, Sessions, George, Warren, Karen and Clark, John, eds. *Environmental Philosophy, From Animal Rights to Radical Ecology*, Upper Saddle River, NJ: Prentice Hall, 1998.

"Zeno," from Diogenes Laërtius' *Lives of the Eminent Philosophers*, VII, 1

Magazine, Newspaper, on-line articles and movies

Anderegg, William, "Diagnosis Earth: the Climate Change Debate," *Thought and Action*, V. 26, Fall 2010.

Angier, Natalie, "Pigs Prove to Be Smart, if Not Vain," *New York Times*, 10/6/2009. [animal intelligence]

_______, "Nut? What Nut? The Squirrel Outwits to Survive," *New York Times*, July 5, 2010

Associated Press, "Top Scientists Warn of Water Shortages and Disease Linked to Global Warming," *New York Times*, 3/12/07

Barringer, Felicity, "White Roofs Catch on as Energy Cost Cutters," *New York Times*, 7/30/09

Barrow, Wayne, "Study Calls for 'Energy Revolution'," Associated Press/*New York Times*, 6/6/08.

Belson, Ken, "Air Storage," *New York Times,* Published: 8/26/2008.

Bhanoo, Sindya N., "Elephants Benefit From Elder Matriarchs' Experience," *New York Times*, March 16, 2011

Blakemore, Bill, "North Pole Meets South Pole: Earth is Melting at Both Ends," ABC News.com, 3/3/06.

Blechman, Moisha, "An Examination of Nuclear Power," City Sierran, Fall 2011. Also contains articles on fracking.

Broder, John M., "Climate Change Seen as Threat to U.S. Security," *The New York Times*, 8/7/2009

Carey, Benedict, "Brainy Parrot Dies, Emotive to the End." *The New York Times*, 9/11/2007:

Dean, Cornelia, "Coral Reefs and What Ruins Them," *New York Times*, Feb. 26,2008.

Dreifus, Claudia, "Listening to the Climate Models, and Trying to Wake Up the World: A Conversation with Jerry Mahlman," *New York Times*, 12/19/03.

Fountain, Henry, "Fish Use UV Patterns to Tell Species Apart," *New York Times*, March 2, 2010

Galbraith, Kate, "Efficiency Drive Could Cut Energy Use 23% by 2020, Study Finds," *New York Times*, 7/30/09.

Gettleman, Jeffrey, "Even the Camels Are Dying," *New York Times*, 10/8/09.

Gillis, Justin, Jonas, G. and Wheaton, S., "Norman Borlaug, Plant Scientist Who Fought Famine, Dies at 95," *New York Times*, 9/14/09

Gorman, James, "Scientists Predict Widespread Extinction by Global Warming," *New York Times*, 1/9/04.

Gutfreund, Prof. Owen, "Pick on the Big Guys," *New York Times*, 4/9/08.

Horowitz, Alexandra, and Shea, Ammon, "Think You're Smarter Than Animals? Maybe Not," *New York Times,* August 20, 2011

Judson, Olivia, "Where Tasty Morsels Fear to Tread," *New York Times*, 9/29/2009

______, "'Leopard Behind You!'" *New York Times*, October 6, 2009

Kanter, James and Revkin, Andrew, "Scientists Detail Climate Changes, Poles to Tropics," *New York Times*, 4/7/07.

Kaufman, Frederick, "Wasteland: A Journey Through the American Cloaca," *Harper's Magazine*, February 2008.

Kelley, Katie, "Emblem of the West is Dying and No One Can Figure Out Why," *New York Times*, 9/27/06.

Krugman, Paul, "How Did Economists Get It So Wrong?" *New York Times*, 9/6/09

______, "It's Easy Being Green," *New York Times*, 9/25/09

LaFraniere, Sharon, "Africa's World of Forced Labor," *New York Times*, 10/29/06.

MacFarquhar, Neil, "Refugees Join List of Climate-Changing Issues, *New York Times*, 5/29/09.

McKibben, Bill, "Energizing America," *Sierra*, Jan-Feb 2007, 92/1, 30-31.

McManus, Reed, "Negawatt Power," *Sierra*, Jan-Feb 2007, 92/1 35-36.

Northrup, James, "Fracking is Literally a Dirty Bomb," *Sierra Atlantic*, V. 37, Winter 2010.

Rauber, Paul, "Why Not Nukes? We Might as Well Burn Money," *Sierra*, Jan-Feb 2007, 92/1 37-38.

Revkin, Andrew C., and Rosenthal, Elizabeth, "Panel Issues Bleak Report on Climate Change," *New York Times*, 2/2/07.

_______, "Science Panel Calls Global Warming 'Unequivocal'," *New York Times*, 2/3/ 07

Revkin, Andrew C., and Seelye, Katherine Q., "Panel Tells Bush Global Warming is Getting Worse: President Sought Report on Issue After Rejecting World Pact," *New York Times*, June 2001.

______, "Report by the E.P.A. Leaves Out Data on Climate Change," *New York Times*, 6/19/03.

Revkin, Andrew C., "Climate Talks Shift Focus to How to Deal with Changes," *New York Times*, 11/4/02.

______, "Commercial Fleets Reduced Big Fish by 90%, Study Says," *New York Times*, 5/15/03.

______, "Global Warming is Expected to Raise Hurricane Intensity," *New York Times*, 10/1/04.

______, "Big Arctic Perils Seen in Warming, Survey Finds," *New York Times*, 11/1/04.

______, "Steps to Limit Global-Warming Gas," *New York Times*, 9/25/05

______, "Yelling 'Fire' on a Hot Planet," *New York Times*, 4/24/2006.

______, "Federal Study Finds Accord on Warming," *New York Times*, 5/9/06.

______, "Emissions Already Affecting Climate, Report Finds," *New York Times*, 4/6/07.

______, "What's the Right Number of Americans?" *New York Times*, 8/4/2009.

Ritchie, John, "Toxic Legacies" Canadian TV Documentary on pesticide damage to human health, including estrogen mimicry.

Robbins, Jim, "Hunting Habits of Wolves Change Ecological Balance in Yellowstone, *New York Times*, 10/18/2005.

Rosenthal, Elizabeth, "As the Climate Changes, Bits of England's Coast Crumble," *New York Times*, 5/4/07.

______, "Likely Spread of Deserts to Fertile Land Requires Quick Response, U.N. Report Says," *New York Times*, 6/28/07.

______, "An Amazon Culture Withers as Food Dries Up," *New York Times*,7/25/09.

Saulny, Susan, "Holding Out, to Last Tiny Isle, as Cajun Land Sinks Into Gulf "*New York Times*, 8/24/08

Schell, Orville, "The Thaw at the Roof of the World," *New York Times*, 9/26/09.

Seltenrich, Nate, "A Matter of Survival," *Sierra*, 96, 6, Nov-Dec 2011, 42-45 (Solar power w/o killing desert animals

Sierra, Jan-Feb 2007, 92/1 special edition on energy:

Slater, Dashka, "This Much Mercury" *Sierra*, 96, 6, Nov-Dec 2011, 46-53. Mercury from coal and its effects.

Snell, Marilyn B., "Can Coal be Clean—when Pigs Fly," *Sierra*, Jan-Feb 2007, 92/1 32-34.

Than, Ker and Thompson, Andrea, "Frightening Forecast," Live Science/Imaginova Corp., 4/22/07.

"Troubled Waters," City Sierran, 26/3, 10-12.

Wade, Nicholas, "Humans Have Spread Globally, and Evolved Locally," *New York Times*, June 26, 2007.

______, "Is 'Do Unto Others' Written Into Our Genes?" *New York Times*, 9/18/2007.

______, "Darwin, Ahead of His Time, Is Still Influential," *New York Times*, 2/9/09.

______, "Boom! Hok! A Monkey Language Is Deciphered" *New York Times*, December 8, 2009

_______, "Sit. Stay. Parse. Good Girl!" *New York Times*, January 17, 2011

Whittelsey, Frances C., "The Birds and the Breeze: Making Wind Power Safe," *Sierra*, Jan-Feb 2007, 92/1 39-40.

Yardley, William, "Engulfed by Climate Change, Town seeks Lifeline," *New York Times*, 5/27/07.

Yoon, Carol R., "Scientists Say Orangutans Can Exhibit 'Culture'," *New York Times*, January 3, 2003

Zimmer, Carl, "Humans May Have Limiting Effect on the Origin of (New) Species," *New York Times*, 5/24/06.

______, "Can Answers to Evolution Be Found in Slime?" *New York Times,* October 3, 2011

Cox, Stan, "Turf Wars," http://www.alternet.org/envirohealth/28361/ , AlterNet. 11/17/05

"Social Epistemology,"*Stanford Encyclopedia of Philosophy* on-line

www.eden-foundation.org/project/desertif.html.

www.ipcc.ch

Movies:

"An Inconvenient Truth"

"Blade Runner"

"Gattica."

"The Sixth Day."

ABOUT THE AUTHOR

Hugh McDonald (Ph.D. New School for Social Research) is Associate Professor of Philosophy at a campus of the City University of New York (CUNY). He has taught at several schools in the New York City area. He is the author of *Political Philosophy and Ideology* (University Press of America, 1996), *Radical Axiology, a First Philosophy of Values* (Rodopi, Value Inquiry Book Series, 2004), *John Dewey and Environmental Ethics* (SUNY Press, 2004), *Creative Actualization: A Meliorist Theory of Values (*Rodopi, Value Inquiry Book Series, 2011), and *Speculative Evaluations: Essays on a Pluralistic Universe* (Rodopi, Value Inquiry Book Series, 2012). He is co-editor, with Dr. John Shook, of *F. C. S. Schiller on Pragmatism and Humanism: Selected Writings, 1891-1937,* (Prometheus Books, 2007). He is also the editor of *Pragmatism and Environmentalism*, (Rodopi/Contemporary Pragmatism Book Series, 2012). His numerous journal articles span ethics, value theory, pragmatism, and the history of philosophy. His interests are in historical pragmatism, where he is one of the few F.C.S. Schiller scholars, value theory, and applied ethics, particularly environmental ethics and biotechnology. He is the co-founder of the New York Pragmatist Forum, a group that meets monthly to discuss issues in American philosophy. In 2009 he was elected Scholar on Campus by his peers at City Tech.

Prof. McDonald is also the founder of the Calochortus Society, a group devoted to a beautiful family of native American wildflowers, and past editor of its newsletter. He has also written articles in botany and horticulture journals. An avid wildflower photographer, he maintains a web site of his published and unpublished photographs.

Index

actions, 4, 6, 23, 32, 34-37, 48, 74, 76, 78, 85, 87-89, 92, 93-95, 98-101, 105, 114, 117, 132, 137, 141, 144, 151-152, 160-167, 169-173, 176-177, 179-184, 186, 188, 191-192, 198, 209, 212, 217, 223, 232, 243, 250, 269, 301, 314

activities, 69, 76, 86-87, 95, 123, 158, 170, 197, 222, 230, 256, 307, 313, , 355, 375, 395

air, soil and water, 2 -4, 9, 15, 26, 40, 42, 44, 49, 62, 73, 79, 80, 111, 127-128, 131, 135, 144, 147, 149, 151, 154, 189, 209, 223, 232-233, 238-239, 243, 298, 312, 325, 333, 335, 337, 342-345, 349-350, 358, 365-366, 368, 373, 378, 381, 383, 391

altruism, altruistic, 78, 110, 157, 186, 194, 203, 208

analysis, 83, 101, 171, 194, 227, 311
    cost-benefit a., 138

animals, 3, 13, 20-30, 38-39, 42, 44-48, 50, 53-62, 64-65, 71, 73, 75-76, 79, 86-88, 95, 104, 110, 112, 115, 121, 123-124, 127-128, 130-139, 144, 148-150, 162, 170, 172-173, 199, 202-203, 205, 208-209, 212, 216, 218-219, 233-234, 236, 241, 254, 262, 273, 277-280, 284, 286-287, 290, 296-300, 317-318, 330-331, 334-335, 337, 341-342, 349-352, 366-367, 373, 381-382

anthropocentric, 2-4, 8-9, 13-14, 19, 24, 38, 40, 47, 51, 53-58, 62, 65, 67-71, 81, 83, 85, 108, 110-111, 113, 120, 123, 128, 137, 143, 147, 154, 198, 209, 229, 236, 238, 260-261, 269, 275-279, 280-282, 286, 289-293, 295, 299, 303-305, 315, 323, 342, 355, 387, 389

arbitrary, 56, 57, 58, 103, 104, 135-136, 197, 208, 261, 264, 267, 393

arguments, 61
    extension a., 3, 8, 11-13, 18, 47, 53-55, 63, 70, 72-73, 127-128, 130-131, 137, 152-153, 269, 304-305, 386

Aristotle, Aristotelian, 4, 14, 21, 35, 56, 64, 66, 85-88, 93-95, 98, 100-101, 104-106, 111, 114, 117, 123, 142, 157, 168-169, 176, 185, 187-189, 192-193, 216, 228, 254, 256, 262, 281, 285, 289, 295, 307, 311, 317, 320

Augustine, Aurelius, 96, 98, 99

autonomy, 6, 28, 32, 47, 64, 92, 94, 96, 102-103, 112, 118-119, 154, 166, 170, 178, 185, 210, 239, 247-248, 250-251, 255

axiology, axiological, 6, 31, 38, 41, 50, 143, 148, 183, 305, 307, 311

basis, 2, 5, 10, 18-19, 22, 26, 33, 35-39, 43-44, 48, 51, 53, 58-59, 61-62, 73, 79, 98-100, 109, 123, 129-130, 137-138, 141, 175, 203-204, 207, 209-210, 215-216, 219, 223, 229, 239, 242, 244, 253, 262, 268, 270-271, 273, 277, 303, 318, 326, 366, 381, 385-386

bearer, 31, 129-130, 148, 194
beauty, beautiful, 18-19, 36, 46, 49, 59, 79, 85, 87, 96-97, 111, 113, 123, 128, 151, 152, 153, 172, 180, 200, 203, 209, 213, 217, 222-223, 229, 241, 256, 258, 302, 310-311, 315, 333, 338, 351, 355, 356, 359, 378, 388
belief, 24, 51-52, 55-56, 189-190, 287
Bentham, Jeremy, 18, 208, 214, 228
biodiversity, 13, 25-26, 115, 120, 122, 150, 153, 209, 224, 235, 301, 325, 338, 351, 365, 367, 368-369, 378
biology, biological, 2, 56, 63, 88, 90, 145, 227, 248, 261, 273-275, 278, 282-283, 286, 289, 291-294, 296-297, 299, 301, 304, 307, 312-314, 321, 348, 359-360, 389, 392-393, 396
biosphere, 2-3, 7-9, 11, 19, 26-29, 35, 41-44, 47-51, 53-55, 64, 66, 71-73, 76, 78-79, 81, 113, 129, 131, 135, 144, 154-155, 181, 183, 197, 199, 210, 232-233, 243, 254, 260, 262, 264, 267, 298, 301, 334, 337-338, 342, 345-346, 361, 367-368, 377-378, 383-384
body, 28, 36, 41, 89-90, 97, 172, 200, 284, 289
Bookchin, Murray, 81, 270

Callicott, J. Baird, 3, 46-47, 70-71, 77, 128, 130, 132, 135-137, 139, 208, 261, 299
Cartesian, 5-6, 8, 12, 31, 42, 56, 91, 176, 191, 205, 344, 389
category, 4, 6, 19, 21, 25, 31, 36, 49, 66, 68-69, 75, 78, 89, 91, 93, 106, 108, 114, 146, 151, 166, 169, 171, 199, 201, 205, 212, 222, 255, 274-275, 282, 286, 296, 298, 303, 309-310, 313-315, 323, 345, 351
cause(s), 25, 39, 45, 47, 61, 79, 82, 129, 159, 166, 176, 178, 187-189, 210, 213, 238, 266, 269-270, 287, 292, 297, 302, 304, 312, 319, 322, 325, 329, 338-340, 344-345, 365, 374, 391, 395
c. and effect, 188
change(s), 4-6, 20-21, 24, 26, 45, 53, 55, 57, 70, 86, 92, 98-100, 104, 109, 116-117, 142, 152, 171, 176, 180, 188, 199, 218, 223, 225, 227, 233, 238, 247, 250, 262, 280, 283, 290, 301, 306, 310, 323, 329, 332, 344, 346, 349, 352, 361, 384, 387, 395
character, 4, 6, 11-14, 32, 36, 39, 64, 75, 85-125, 129, 132, 142, 144, 146, 158, 170-171, 176-178, 180, 183, 195, 211, 216, 223, 233, 282, 313, 317, 324-325, 327, 336, 352, 367, 379, 389
choice, 5, 7, 10-11, 13, 31, 47, 49, 97, 99, 101-102, 112, 169-170, 179, 202, 204, 208, 223, 231, 237, 244, 246, 249-250, 261, 278, 317, 322, 361, 380
Christianity, 13, 110, 248, 262, 308, 336, 366
circumstances, 20, 22, 24-25, 34, 47, 75, 79, 92, 98, 101-106, 109, 112, 122, 139-142, 157-158, 165, 171, 177-178, 187, 189, 191-192, 257, 305, 313, 341, 349, 388
coercion, problem of, 7, 14, 243, 267, 268
Cohen, Carl, 59, 60, 61
communication, 68, 145, 149, 232, 237, 240, 377
community, communities, 22, 26-28, 33, 41-42, 45, 49, 70, 76, 90, 127, 154, 164, 185, 205,

240, 246, 269, 353, 359, 387-388
biotic c., 128
condition(s), 2, 4, 6-9, 12, 20, 22, 26-30, 32-36, 40-48, 50, 58, 62-64, 71-73, 76-80, 82, 91, 93, 104, 106-109, 112-115, 123-124, 127-128, 131, 139, 144, 147, 149, 153, 160, 163, 166-171, 174-176, 181-184, 187, 189, 201-202, 205, 209-213, 229, 232-233, 243, 252-258, 260, 264, 274, 278-279, 284, 292-293, 298-301, 311, 313-314, 316, 320, 323-327, 342, 344, 346, 351-352, 360, 365, 372, 376, 378, 381-382, 384, 388, 393, 396
Confucius (see Kong fu-tse)
conscience, 62, 93, 99, 176, 215, 217, 326, 329, 336
consciousness, 23, 47, 58, 70, 88, 96, 122, 133, 202, 271, 307, 322
consequence(s), 12, 38, 72-76, 80, 138, 141, 159-161, 163, 173, 175, 178, 181, 183-184, 188, 219, 246, 257, 330, 341-342, 379
environmental c., 51, 75- 76, 144, 158, 183, 269, 340
social c., 6, 32, 92, 93, 223, 389
consequentialism, onsequentialist, 14, 74-75, 93, 129, 139, 157-161, 163-164, 166, 173-174, 208, 215, 238
pragmatic c., 14, 160, 224
conservation, 243, 307, 318-321, 326, 346, 350-351, 354, 363-364, 373
c. for future generations 351-362
consumer society, 14, 59, 63, 81, 113, 116-118, 199, 207, 210, 219, 260, 307, 316, 324, 337, 373
context, 4, 13, 22, 35, 41, 51, 62, 81, 109-110, 113, 121, 136, 142, 161, 178-180, 185, 196, 201, 211, 214, 220, 226, 228, 247, 268, 271, 278, 294, 303, 316-317, 319
contract, 145, 146, 207
corporate, 119, 159, 203, 228, 313, 315, 337, 374
cosmopolitan, cosmopolitanism, 5, 7, 8-15, 28, 31, 34-35, 40, 50-54, 62, 65-67, 70-74, 76-79, 80-83, 85, 94, 109-110, 113-114, 117-118, 121, 123-124, 129-132, 139, 143-145, 148, 153-154, 157-158, 162, 165, 183-184, 210, 229, 238, 245, 249, 251-255, 259, 261-262, 264, 269, 302, 325, 351-352, 358, 363-364, 366, 383, 387-389
c. ethics, 9-15, 40, 51, 52, 54, 65, 71, 73, 76-77, 82, 94, 109-110, 114, 143, 154, 157-158, 246, 251, 262, 264, 358, 383, 387
c. virtue, 11, 85, 110
creation, 8, 10, 19, 21, 27-28, 31, 55, 56, 59, 68, 88, 106, 280, 310, 333, 335, 376
creative actualization, 7, 12, 14, 19, 20-21, 25-30, 32-33, 36-38, 40-46, 49-50, 69, 76-78, 86, 94, 99, 103, 106, 141, 143, 148, 158, 160, 165, 167-170, 172, 175-177, 180-183, 191, 200, 202, 222, 387
c. a. of moral values, 14, 33, 37, 158, 172
Cronin, John, 145, 242
culture(s)cultural, 4-10, 14, 32-33, 36, 49, 51, 52, 53, 60, 64-67, 69-73, 79, 85, 87, 90, 91, 94, 96-98, 100, 104, 107, 110-111, 113, 117, 121, 124, 138, 140-141, 146, 151, 154, 162,

167, 169-172, 175-178, 183, 207, 232, 235, 238-239, 245-257, 259-262, 264, 266, 272-273, 276, 285-287, 290, 295, 307, 311, 313- 314, 326. 331-332, 334-336, 339-340, 347-348, 350-352, 364-365, 370, 381, 390-391
  c. of death, 207, 339-340, 348, 350, 352
  Western c., 6, 62, 347
custom, 24, 53, 67

Darwin, Charles, 22, 23, 27, 56, 60, 64, 82, 110, 149, 279, 280, 283, 285, 292, 297, 299, 305, 391, 394
decision(s), 70, 104 143, 150, 158, 163, 175, 178, 184, 187-188, 208, 216, 231, 236, 238, 246, 262, 269, 376
definition, 21, 32, 39, 48, 50, 54, 57, 60, 64, 68, 81, 160, 164, 172, 194, 276-278, 280, 282-283, 291, 297, 321, 351, 391
deliberation, 11, 22, 37, 39, 49, 83, 88, 101, 142, 144, 178, 180, 182, 317
  moral d., 12, 74-75, 93, 98, 101, 105, 140, 157-158, 163, 165-166, 169, 183
democracy, 251
deontological, 11, 12, 39, 49 , 75, 94, 142-143, 157-160, 162-164, 166, 178, 219, 251, 267
Descartes, René, 2, 56, 92, 190, 391
  see also, Cartesian
Des Jardins, Joseph, R., 67, 132, 376
desirable, 74, 107, 221, 227, 256, 312, 334
desire(s), desiring, 4, 5, 19, 23, 25, 30-31, 35, 49, 55, 62, 76, 92, 97, 99, 101, 105-107, 114-117, 137-138, 179, 185, 187-196, 200-201, 206-207, 216-217, 220-221, 225-226, 229, 271, 308, 311-313, 321, 323, 336, 347, 351, 368, 379, 382
  human d., 121
destiny, 4, 106, 154, 223, 233, 261, 333, 337
devaluation, 32, 62, 65, 71, 200, 203-204, 222-223
  d. of the world, 65
development, 6, 8, 11, 13-14, 29, 31-33, 53-54, 64, 66, 76, 81-82, 85-89, 91-92, 94-96, 99-106, 110-111, 117, 125, 136, 142, 148, 153, 158, 170-171, 203, 211, 213, 216, 232, 235, 249, 317, 345, 349, 352, 355-357, 361, 364, 380, 389
  economic d., 79, 150, 151, 324
Dewey, John, 2, 12, 14, 21, 33, 43, 66, 75, 94, 97, 100-101, 104, 113, 129, 132, 138, 141-144, 159-160, 165-166, 174, 178, 180, 185, 188, 191, 221, 244, 271, 351, 386, 388
Diamond, Jared, , 64, 271, 331, 334, 349, 350
difference, 23, 28, 39, 44-45, 57, 59, 62, 71, 90, 103, 135, 140-141, 157-158, 165, 168, 177, 187, 193, 200, 231, 247, 254-255, 262, 270, 273-274, 276, 284-287, 290, 295, 305, 324, 392, 396
duty, 4-5, 11-13, 15, 21, 31, 33-34, 38-40, 42-46, 48-51, 53-54, 61, 65, 73-76, 78-79, 89, 91-92, 94, 98, 104, 110, 113, 119, 121, 123-124, 129, 132-134, 137, 140, 143-144, 146-147, 149, 151-152, 154, 157-163, 165, 168-175, 177, 179-184, 189, 192, 203, 205-206, 216-217, 219, 223, 232, 239, 249-252,

261, 264, 268, 271, 302, 311, 317, 320, 326, 333, 351-352, 355-356, 361, 363, 365

earth, 4, 8-9, 13, 29, 54, 56, 73, 80-81, 121, 131, 135, 151-152, 211, 219, 231, 255, 260, 265, 284, 302, 307, 324, 333, 338, 342-343, 346, 351, 364, 369, 377, 381, 387, 396
ecocentrism, 3, 10, 51, 54, 62-64, 66-67, 71, 78, 81, 109, 124, 152, 154, 197
ecologists, 394
 deep e., 67
ecology, 51, 67, 70, 111, 152, 167, 241, 332, 376, 382
 e. for society, 362-375
economic(s), 4, 14-15, 29, 33, 35-36, 40, 63, 69, 70, 75, 78-80, 82, 85, 92, 97, 100, 107-108, 112, 114, 116, 118, 120, 123, 128, 138, 146, 150-151, 153, 171, 174, 195, 200, 203-204, 208, 220-223, 225-229, 233, 237-243, 245, 248, 258, 282, 287, 301, 307-322, 324, 327, 330-331, 333, 338, 345-347, 349, 353, 358, 362, 364, 368, 378, 381, 384
ecosystems, 53, 64, 73, 151-153, 269, 301, 355-357, 368, 378
effect, 4, 38, 44, 80-81, 99, 108, 111, 119, 121, 145, 159, 166, 187, 202, 224, 227, 242, 244, 258, 270, 279-280, 318, 322, 327, 340, 342, 344-345, 349, 353, 364, 367, 374, 380, 382, 395-396
ego, egoism, 38, 92, 96, 119, 172, 175, 186, 191, 199, 204, 206, 215, 217, 223, 227, 256, 333
electricity, 15, 111, 224, 242, 302, 314, 321, 340-341, 347, 359, 362-363, 370, 373-375, 377, 382
element(s), 11-12, 14, 28, 33, 41, 43, 74-75, 86, 93, 97, 100-102, 104, 127, 141-143, 151, 168, 170, 177-178, 183-184, 190, 212, 220, 253, 262, 278, 285, 291, 297, 361
emotion(s), 5, 31, 86, 191, 311
empiricism, 3, 92
end(s), 4, 20-21, 29, 33, 35, 48, 58, 69, 78, 93, 95, 101, 114, 117, 119-120, 132, 141, 149, 151, 159, 169, 175, 179, 181, 187, 191-193, 195-196, 200-201, 205, 212-217, 220-222, 227, 230, 235, 241, 245, 248, 252, 258, 269, 274, 276, 290, 302, 307-312, 314-315, 320-321, 323, 332, 342, 348, 362-363, 367, 376, 383-384
 e in itself, 20, 308, 310, 315
 fixed e., 7
energy, 80-81, 111, 115, 123, 127, 226, 237, 241, 282, 290, 302, 319-320, 325, 335-339, 342-343, 345, 354, 359-360, 362, 364, 367-368, 370-375, 377, 380
environment, 2-5, 7-9, 11-15, 19, 27-29, 33, 36-37, 39-45, 48-50, 51, 53-57, 60, 62-66, 68-71, 73-75, 77-83, 94-95, 97, 102, 107-109, 111-115, 117, 119-124, 127-128, 130-132, 135, 142-144, 147, 157-159, 162-164, 171-172, 177, 182-184, 189, 199, 204, 207, 209-210, 213, 216-217, 229-233, 237-239, 241-243, 249, 254-255, 258-261, 264, 266-268, 270-273, 291, 298, 301-304, 307313, 316-323, 325-326, 329, 333-336, 338-339, 342-344, 346,

349-352, 366-371, 373, 375, 379, 380-382, 385-389, 391-393, 395
e. as a whole, 9, 12, 41-43, 80, 94, 112, 119, 130, 144, 229, 238, 268, 318, 385
impact on the e., 65, 80, 82, 89, 93, 108, 114-115, 117-118, 120, 135, 142, 150-151, 162-163, 180, 184, 231-232, 234, 236, 239-241, 248, 265, 268, 304, 307, 319-323, 325, 327, 338, 341-342, 345, 358-360, 365, 369-370, 373, 375-376, 379, 382, 385
social e., 102
physical e., 33
environmentalism, 3, 66
Environmental Justice, see Justice, environmental
Environmental Philosophy, 1-15, 385-389
Environmental P. as a System, 385-389
Environmental Ethics to Environmental P., 385-388
Epicurus, 106, 144, 176
epistemology, epistemological, 2, 36--37, 51, 56, 81, 108, 205, 248, 256, 274, 334, 388-389
epoch(s), 8, 26, 42
equality, 34-35, 37-38, 59, 71, 78, 82, 105, 118-119, 131-132, 166, 193, 206, 247, 250, 252, 260, 269, 382
essence, 53, 59, 61, 100, 106, 109, 179, 193, 289, 307
esthetic(s), 36, 106, 202, 256, 308, 353, 373-374
ethic(s), ethical 2-15, 30-41, 43-44, 47,48-50, 51-55, 57-79, 81-83, 85-86, 89-97, 99, 101-102, 104, 106-107, 109-113, 117-118, 120-121, 124, 128-134, 137-144, 146, 148, 151, 153-155, 157-160, 163-171, 173, 176-179, 181, 183-184, 185-188, 191-192, 196-199, 204-208, 210-212, 215-216, 219, 227, 229, 232, 238-240, 243, 245-247, 249-256, 258-267, 269, 271, 301-303, 307, 310-311, 317, 323-326, 342, 349, 357, 365, 366, 368, 375, 377-379, 381-384, 385-389
anthropocentric e., 9, 40, 51, 57-58, 62, 65, 71, 81, 137, 191, 260-261, 387
applied e., 9, 385
consequentialist e., 75, 93, 166, 173, 215
economics and e., 225-230
Environmental e., 2, 9-10, 13-14, 36, 50-51, 53-55, 63-64, 66, 70-72, 75, 77-78, 114, 124, 137-139, 141, 143, 158, 173, 185, 197-199, 204, 210, 229, 255, 260, 262, 338, 375, 385-388
Deontological E. e, 37
E. E. as Cosmopolitan, 51-83
Environmental E. as First Philosophy, 385-389 ....
see also Environmental Philosophy
e. of character, 85-125
e. of duty, 17-50
elements of e, 51-84
e. as a way of life, 105-109
e. relativism, 273-286
management e., 54, 153, 378
principles of e., 155
situation e., 140-148, 177
evaluation(s), 4-5, 11-12, 14, 29-31, 36, 49, 55, 61, 67-68, 73-77, 83, 97, 104, 106, 109-110, 114, 131, 138, 141, 143, 145, 151, 158, 160-165, 168-169, 171-172, 175-179, 181-182, 184, 186-188, 190-191, 196-197, 199-202, 205-206, 221-222, 228-229, 262, 268, 271, 298,

303-304, 307-308, 314, 317, 319, 366, 385-388
critical e., 4, 8-9, 36, 52, 67, 70, 77, 97, 104, 109-110, 131, 137, 165, 176, 186, 188, 190-191, 193, 195-196, 199, 205, 221, 229, 249, 252, 259, 262-263, 266, 310
event, 188, 240, 293, 356
evidence, 22-23, 25, 57, 59, 100, 122, 134-135, 150, 160, 187-189, 193, 227, 241, 250, 265, 275, 277-279, 281-287, 294, 321, 329, 333, 335, 383
evil, 11, 29, 30, 39, 60, 93-94, 102, 117, 119-120, 123, 132, 146, 162, 171, 181, 208, 223, 243-245, 252, 265, 273, 303-307, 340, 342, 354, 386
genocide as e., 301-307
evolution, 8, 31, 46, 56, 60, 250, 255, 274-276, 282, 294, 297, 337, 356-357, 378-379
experience, 22, 25, 60, 152, 160, 166, 167, 192, 219, 235, 237, 255, 294, 302, 329, 349, 367
experiment(s), 22, 24, 142

fact(s), 4, 38, 45, 49, 61, 91, 117, 120, 122, 181, 194, 233, 245, 271, 282, 307, 323, 341, 362, 364, 396
feeling(s), 5, 19, 23, 31, 65, 86, 92, 97, 101, 104-105, 142, 187-192, 196, 211, 213, 215-217, 220-221, 225-226, 383
Feinberg, Joel, 185
field, 34, 56, 63, 66, 73, 81, 89, 93, 101, 138, 270, 282, 315, 333, 385
first principles, 37
flux, 5, 31, 36, 92
food chain, 62, 128, 132, 202, 300, 343, 361-362
force(s), 25, 64, 72, 105, 107, 129, 136, 195, 238, 251, 284, 291, 293, 303, 315, 331, 339
Foreman, Dave, 332, 344, 355-357, 378-379
forest(s), forest(s), 2, 54, 62-63, 69, 76, 103, 112, 137, 152, 197, 219, 232, 233, 235, 242, 279, 304-305, 314, 318, 327, 332-333, 335, 338, 340-343, 345, 347, 353, 356-359, 362-363, 369, 379-384
destruction of f. , 15, 81, 111, 223, 242, 345, 347, 372, 381
f. and acid rain 1, 15, 80, 111, 123, 151, 224, 302, 340, 342, 345, 347, 370, 372-373, 382
for its own sake, for their own sake, 33, 93-94, 106, 115, 152, 161, 174, 183, 223, 258, 308, 315, 352-353, 357-358, 361, 366, 381
form, 4-5, 7-9, 12, 14, 18, 21, 23-26, 29, 33, 36-39, 44-46, 49-51, 52, 55-59, 66, 69, 71, 73, 75, 78, 86, 93, 95, 98, 102, 108, 110, 114, 116-118, 127, 133, 135-136, 140-141, 157-160, 164, 166, 168, 171, 175-176, 182, 187-191, 193-194, 208, 211, 214-215, 217, 220, 225-226, 228-229, 239, 240, 242, 246, 248-250, 255-256, 258-261, 267, 273-274, 276, 280-281, 283-284, 293-294, 297, 306, 308-309, 311-312, 315, 317-318, 323, 331, 343-344, 346, 362-363, 365, 371, 375-376, 379-380, 382, 385, 387, 394
for the sake of, 46, 61, 69, 82, 112, 120, 147, 152, 165, 174, 179-183, 231, 254, 258, 270,

315, 321, 325, 327, 344, 352-353, 366, 383
fossil fuels, 4, 151, 208, 241, 320, 334, 339-340, 343, 348, 360, 367, 371, 373, 380
foundational, foundationalism 14, 36, 82, 165, 173, 175-176, 212, 226, 229, 268, 301
framework, 2, 50, 71, 76-77, 91, 140, 157-158, 162, 249, 268, 386-387
Frankena, William, 178, 203
free will, 11, 13, 85-98, 101, 165, 178
f. w. and moral self: development 85-98
freedom, 7, 28, 39, 47, 64, 90, 98-99, 116, 119, 129, 132, 166-167, 238, 240, 251, 260, 268, 270, 298, 323, 324, 360, 363, 369, 371, 382
Friedman, Thomas., 342., 360, 369, 374
function, 20-21, 27-28, 41, 43, 45, 75, 77, 85-87, 91, 95, 97, 110, 127, 131-133, 150, 169, 206, 214, 228, 252, 256-257, 271, 290, 294, 300, 352, 382, 385
future, 4, 12, 24, 36-37, 73, 75-79, 82-83, 91-92, 97-98, 104, 116, 120-123, 142, 148, 160-161, 163, 176, 181-182, 188, 233-237, 241, 257, 276, 282, 294, , 309, 316, 323, 325, 339-342, 347, 349, 363, 371, 386, 388
f. generations, 7, 10, 13, 48-49, 53, 71, 78-79, 111, 118, 120, 123, 146-147, 151-152, 155, 204, 207, 209, 223, 231, 237, 242-243, 258, 270-271, 301-302, 319-321, 326-327, 344, 351-352, 360, 366-368, 370, 380-381
conservation for f. g, 361-364
Galbraith, John Kenneth, 315, 316, 347
genus, genera, 50-51, 71, 77, 89, 120, 262, 274-275, 281, 285-286, 295-296, 362, 393-394
global warming, 15, 59, 69, 74, 78-82, 115, 149, 207-209, 219, 223, 235, 237, 240, 242, 255, 258, 269, 270, 301-302, 305, 307, 310, 312, 315, 320, 322, 325, 334-335, 338, 340, 342-343, 345, 348, 354, 360-361, 363, 367, 369-371, 374-377, 380, 395
goal, 4, 7, 11, 13, 20, 22, 53, 85, 87-88, 92, 95, 97-98, 101, 103-104, 107-108, 119-120, 163, 168-169, 171, 176, 181, 187-188, 195, 211, 256, 299, 308, 315, 320, 339, 355, 363, 369, 374, 379, 382, 385-386
good(s), 7, 9-10, 12, 18-21, 26-27, 29-34, 36-38, 40-47, 53, 58, 60-61, 72, 74-77, 79, 85, 87-95, 97-107, 113-117, 120, 124, 129, 138-39, 142, 144-147, 150-51, 157, 159-160, 162, 165-167, 169, 171-176, 179-183, 186, 188, 190-192, 196-202, 204, 206, 208-209, 211-212, 214-219, 221-227, 229, 231-233, 238, 243, 245-246, 252, 256, 258, 260, 263, 267-269, 271, 281, 284, 295, 299-301, 305-306, 307, 309, 311-315, 318-319, 324-327, 331, 336-337, 340, 342, 349, 352, 360-361, 363, 366-367, 382, 386-388
highest g., 18, 29, 33, 38, 40, 58, 173, 192, 201, 203, 214, 224, 310
instrumental g., 310
moral g., 88, 169, 173
g. of their own, 197, 198

see also, value
Gore, Albert, 235, 367
Gorke, Martin, 27, 28, 45, 68, 282, 291, 378
Gould, Steven J., 278
greenhouse effect, 322
ground, grounding, 28, 33, 36, 42, 54-55, 79, 93, 95, 107, 109-110, 119, 141, 145, 147, 159, 165-166, 171, 173, 175-177, 186-192, 194-196, 200, 204, 213, 216, 237, 241-242, 247, 249, 279, 289-290, 299, 301, 316, 340-341, 345, 368, 372, 387
g. and consequent, 36, 159, 188-190, 247
ultimate g., 188, 306, 386
growth, 25, 69, 104, 116, 148, 233, 242, 245, 258, 307, 315, 320, 324-325, 327, 379, 396

habit, 22, 92, 98, 100-101, 104, 142, 161
habitats, 5, 11, 25-26, 40, 43, 45, 47, 49-50, 57-58, 63-64, 66, 69-70, 74, 77, 79-81, 112, 114-115, 118-120, 123, 136, 139, 148, 151-152, 183-184, 209-210, 219, 232-236, 238, 241-244, 258, 260-261, 267, 269, 278-279, 281, 298-302, 313, 317-318, 324-326, 329-331, 333-334, 338, 342-344, 350, 351-353, 355-356, 358-362, 365, 369, 372, 374, 380, 384, 387, 395
unique h., 152, 255
Hall, Matthew, 47
happiness, 2, 20, 38, 62, 80, 89-91, 142-144, 146, 173-175, 183, 185, 192-193, 197, 203, 208, 211-220, 224-226, 228, 236, 307, 310, 337, 363
Hardin, Garrett, 163, 207, 232, 234, 243, 268, 271, 325, 339, 363, 365, 377
harmony, 4, 12, 51, 74, 82, 111, 119, 176, 385
Harris, Sam, 203, 246
health, 19, 27, 41, 43, 45, 46, 107, 116, 145, 198, 202, 208, 210, 211, 231, 238, 239, 241, 243, 249, 267, 310, 318, 340, 362, 367, 373, 388
hedonism, 21, 30, 31, 80, 97, 110, 139, 157, 158, 192, 208, 211, 212, 213, 214, 215, 216, 217, 218, 220, 225, 227, 229, 322
Hegel, Georg Wilhelm Friedrich, 72, 160, 189, 295, 313
Held, Virginia, 6, 32, 96, 124, 210
Hettinger, Ned, 27
hierarchy, hierarchical, 5, 8, 13, 18-21, 29-32, 34-35, 40, 46, 55, 58, 73, 75, 100, 114, 118-119, 128, 131, 143, 151, 154, 164, 179, 181, 193, 201, 246-247, 259, 263, 282, 332
historical, 6, 28, 51, 53, 76, 92, 101, 109, 142, 147, 171-172, 210, 247, 250, 255, 266, 277-278, 294, 297, 299, 307-308, 316, 337, 345-346, 352, 388
history, histories, 8, 18, 69, 92, 100-101, 107, 110, 117, 120, 124, 140, 198-199, 205, 170, 172, 176, 235, 241-242, 245, 253, 255-256, 266, 291, 303, 308, 323, 330, 334, 338, 345, 379, 387
h. of philosophy, 92, 176, 177, 198
Hobbes, Thomas, 98, 104, 130, 145, 186, 205, 206, 207, 268, 269, 381
holism, holistic, 10, 12, 14, 25-28, 40-41, 44-45, 68, 75, 77, 80, 89, 93, 111, 115, 125, 128, 131-

132, 135-136, 144, 148, 150-151, 154, 162-163, 173-174, 185, 189, 197, 204-205, 232-233, 239, 267, 269, 273, 345, 387
human, 2-5, 7-15, 18-23, 25, 27, 30, 32-35, 38, 41-45, 47-50, 52-73, 75-77, 79-83, 85-87, 90-91, 94-95, 99-100, 103-104, 107-110, 112-113, 115, 117-118, 120-124, 127-130, 132-136, 139, 141, 143-145, 148-155, 162, 171, 177, 179, 184, 187, 193, 197, 199-200, 202, 204-205, 209, 211, 213, 216-217, 219, 222-223, 227, 229, 232-233, 235-236, 238, 241, 243, 249, 253-254, 258, 260-262, 265, 268-269, 271-279, 284-285, 290, 292, 295, 299-302, 304-305, 307, 309, 313-314, 323-325, 329-331, 333-334, 338-346, 348-350, 352-353, 357, 365, 367-368, 371, 375, 377-381, 383, 384-386, 388-389, 391-396
  h. expansion, 14, 15, 80, 184, 238, 343, 380
humanism, 55-56, 58, 62, 65, 67, 83, 107, 325
Hume, David,, 24, 180, 187, 189, 191-192, 195
hypothesis, 48, 250, 274

idea, 2, 18, 37, 47, 51, 53-55, 60-61, 63, 65, 70, 95, 98, 100, 117, 169, 181, 198, 242, 273-274, 311, 316, 318, 333, 346, 348, 378, 383, 389
ideal(s), 7, 85, 91, 104, 111, 228, 274, 324, 357, 369
idealism, 69, 274
identity, 4, 26, 31, 52, 54, 70, 81, 83, 90, 119, 157, 170, 177, 247-248, 254, 259, 260-263, 266, 272, 281, 290, 295, 298, 323
  normative i., 6, 31, 196, 295- 296, 299
  species i., 26, 179, 277, 296
ideological, 50, 54, 56-57, 62, 69, 83, 130, 139, 249, 258, 260, 342, 347-348
imagination, 12, 24, 62, 77, 96, 160, 199, 309, 312, 334, 369
immanent, 294
individuals, 4-6, 14, 22, 31-33, 35, 39-40, 45-50, 72, 77-78, 86, 90-92, 94-96, 102-103, 116, 119, 124, 131, 136, 139, 148, 149, 153-154, 162, 164, 166, 171, 183, 195-197, 201, 203, 210, 214-217, 227, 232, 237-239, 246-253, 256-257, 259, 262, 264, 268, 273-274, 276-277, 280-284, 286-287, 289-299, 303, 306, 321, 326, 346, 349, 366, 389, 391
instance(s), 26, 130, 140-142, 179, 225, 247, 255, 262, 291, 295, 296
instincts, 160
institution(s), 108, 227, 239, 249, 307
  reform of i., 378-379
instrumental, 2, 18-19, 27-28, 31, 41, 43, 55, 91, 183, 186, 189, 198-199, 201, 260, 300, 310
integrity, 102, 112, 128, 253, 385
intellect, intellectualism, 100, 193, 214
intelligence, 22, 23, 24, 47, 97, 112, 115, 133
intent, 75, 160-161, 165-166, 168-169, 181, 183
interaction, 2, 27-28, 41, 44, 103, 124, 127, 158, 198-199, 210, 273, 292, 343, 347, 388-389

interest(s), 5, 31, 38, 50, 54-55, 78, 102, 120, 137, 147, 162, 174, 185, 191, 194, 197, 200-204, 206-209, 220-228, 238, 242, 258, 263, 271, 310, 315, 360, 381
 satisfaction of i., 185, 197, 202, 211, 220-229
intrinsic, 11, 13-14, 18, 20-21, 25, 28, 30-32, 36, 38-43, 45-46, 49, 55, 58, 61, 68-69, 73, 77, 95, 115, 129, 131-133, 135-136, 144, 148-150, 152-154, 172-174, 182-184, 189, 197-204, 212-214, 216, 229, 233, 248, 269, 299-300, 302, 310, 316, 365, 375, 393
 see also intrinsic value

James, William, 6, 22, 24, 29, 64, 76, 94
judgment, 24, 40, 58, 76, 88, 96, 108, 178, 183, 186, 188, 190, 192, 214, 254, 274, 285, 286, 300, 303, 307, 340
justice, 4, 13, 34-35, 48, 51, 64, 69, 71, 73-74, 77-83, 89-91, 103, 105, 107, 111, 114, 118-120, 137, 142, 146-147, 151, 157, 171, 174, 183, 185, 192, 197, 206, 208, 210, 212, 216, 218, 229, 234, 236, 251, 256, 270, 315, 323-326, 340, 342, 352, 358
 environmental j., 80
 social j., 80
Environmental J. and Social J., 80-84
justification, 30, 32, 64, 67, 78, 107, 143, 166, 173-174, 181, 183, 187-188, 191, 195, 203, 211, 214, 218, 245, 357

Kant, Immanuel, 8, 10, 12-14, 35-36, 38-39, 48, 64, 70, 72, 75, 91, 93-95, 98-99, 103, 105, 110, 113, 132, 137, 139, 141, 143, 151, 157-172, 174, 176-182, 185, 189, 190-192, 194, 206-207, 216, 261, 265, 271
Kennedy, Robert F. Jr., 145, 242
King, Rev. Dr. Martin L., 13, 51, 59, 105, 110, 138, 306
knowledge, 2-5, 18-19, 31, 36-38, 50, 68, 91, 98, 100-101, 106-107, 114, 190, 192-193, 246, 248, 255, 256, 286, 292, 301, 334, 340, 345, 348, 386, 388-389
Kong fu-tse, 94, 313
Krupp, Fred, 371

Ladd, John, 35, 247
landscape(s), 152, 203, 214, 243, 330, 356
 intrinsic value of l., 148-155, 361-364
language, 22-24, 38, 52, 57, 69, 86, 88, 90, 93, 107, 128, 147, 149, 168, 171, 195, 222, 236, 247-248, 254, 285, 295-296, 305
laws, 25, 108, 130, 133, 137, 157, 167, 187, 217, 244, 267-269, 292-294, 296, 332, 363, 365, 380
Leopold, Aldo, 3, 27, 42, 54, 65, 68, 70-71, 127, 128, 378-379
Lewis, Clarence Irving, 12, 14, 28, 75, 159-164, 185, 190
life, 2-4, 6, 8-9, 11, 13- 15, 19, 22-29, 31, 35, 39-40, 42-44, 46-48, 50, 51, 54-55, 58, 60, 63-65, 69-0, 72-81, 83, 86, 89-90, 93-94, 96, 101, 103, 105-112, 114-117, 121-124, 128-133, 137-138, 144-148, 151, 161, 167,

170-173, 175-176, 179, 183-185, 191-192, 197, 200-204, 206-209, 211, 213, 215-216, 219-224, 228-229, 231, 234-238, 240-244, 248, 253, 258, 260-261, 263-264, 266-268, 270-271, 294, 300-302, 305, 307-318, 320-325, 327 life, 331, 335, 337, 338, 340, 342, 343, 344, 345, 347, 349, 350-352, 354, 360, 363-364, 366-368, 370, 376, 380, 382, 384, 385-389, 393
human, l., 2, 7-8, 35, 43, 47 62, 65, 80-81, 109, 113, 123-124, 227, 229, 233, 238, 260, 300, 309, 325, 334, 340, 342, 350, 386
l. as a whole, 264, 302, 321, 368
moral l., 11, 35, 64, 74-75, 89, 93, 106, 167, 170, 253
threats to all l., 340-346
ways of l., 4 52, 70-72, 77, 106, 124, 263-264
Light, Andrew, 185
linguistic, 23, 52, 168, 295, 334, 389
Locke, John, 8, 145
logic, logical, 4-5, 8, 10, 21, 23, 31, 37-38, 53-55, 90, 92, 94, 141, 166-167, 174-175, 187, 189-192, 194-196, 206, 211, 250, 261-262, 274-275, 289, 291, 345, 348-349, 379, 382

MacIntyre, Alysdair, 85, 93, 95
making, 8, 14, 36-37, 77, 94, 97, 108, 147, 154, 177, 182-183, 254, 257, 274, 287, 323, 325, 373, 395
management, managerial, 54, 117, 137, 153, 154, 228, 242, 313, 355-356, 377-382
m. of wildlands, 379-384
Manes, Christopher, 332, 380
materialism, materialists, 104, 191, 229, 247, 290, 309, 322, 333, 384
matter, 4, 26, 37, 81, 90, 102-103, 108, 112, 134, 136, 139-140, 142, 157-158, 162, 164, 167, 178, 182, 190-191, 200, 236, 240, 243, 246, 274, 278, 281, 288, 296, 320, 333, 336, 343, 361, 367, 381
McKibben, Bill, 136, 379
Meadows, Donella H, 26, 258
meaning, 20, 105, 109, 129, 142, 186, 188-190, 194, 196, 205, 236, 274-275, 295, 299, 319, 321, 386
means, 8-9, 20, 22-25, 28-29, 32, 34-35, 38, 40, 43-44, 47, 51, 53-54, 74-75, 77, 79, 85, 87, 94, 96, 99, 103, 106-108, 115-116, 119, 122, 137, 141, 142, 150, 151, 154, 158-159, 161, 164, 167, 174, 176, 178-180, 183, 186-188, 190-193, 195-196, 200, 204, 206, 210, 212, 214, 216-217, 220, 225-226, 229, 243, 253, 258, 260-261, 263, 266, 271, 273-274, 284, 289, 298, 301-302, 304, 307-312, 315-320, 322-323, 326, 330, 333, 335, 337, 344, 346, 348-349, 352, 359-360, 366-368, 381-383
medieval, 55, 56, 129
mental, 6, 18, 32, 50, 95, 99, 101, 389
metaphysics, metaphysical, 5-6, 18, 31, 37, 50, 56, 85, 91-92, 95, 98-99, 100, 110, 165, 175, 191, 289, 292, 296, 334, 346, 385-386
method(s), 99, 142, 165, 197, 346
Mill, John Stuart, 98, 132, 138, 142, 157, 177, 179, 185, 191-

192, 195, 208, 211-212, 214-219, 227-228, 231, 300
mind, 23, 30, 37, 89, 97, 109, 122, 191, 203, 213, 222, 268 296, 307, 310, 321, 373, 388
model, 2-6, 11, 14-15, 18, 24, 30-32, 36, 42, 46-47, 55, 57, 63-64, 71, 81, 93, 95, 101-107, 109, 111, 116, 122-123, 127, 129, 131-132, 138, 145-146, 154, 166-168, 173, 183, 187, 193, 195, 204,-205, 208, 223, 225-228, 239, 246-247, 260, 263, 266, 269, 290, 295, 307, 310-314, 317, 324, 368, 384, 386-387, 389
monism, 32, 137, 142-143
Moore, George, E., 138, 221
moral(s), 4, 8, 10-11, 15, 30, 33, 35-38, 46, 53, 60, 62, 74-75, 78, 86, 105, 107-108, 128, 138, 142-143, 159, 162-163, 166, 168, 170-172, 175, 177, 179-184, 192, 205, 212, 223, 228, 240, 246, 253, 306, 308-309, 311, 338, 386
  holistic m., 128
  m. considerability, m. consideration, 7, 11, 13, 15, 34, 40, 46, 53-54, 56, 59, 61-63, 65, 70-73, 75, 128-131, 133, 136, 142-143, 149, 163, 171-174, 182-184, 189, 210, 238, 259, 264, 269, 304-305, 385-386
  m. limits, 11, 75, 78, 106, 148, 169, 171, 180, 183, 191, 216, 249, 254, 268, 312, 315, 386
  m. pluralism, 51, 72, 137, 143
  m. progress, 3, 6, 11, 13, 15, 40, 51, 53, 55, 63, 65, 70-71, 73, 76, 81, 83, 114, 127, 138, 142, 163, 171-172, 183, 208, 210, 231, 255-256, 262, 266, 269-270, 309, 388
  m. responsibility, 11, 120, 369
  m. sentiments, 105, 191, 216
  m. virtue*, 92, 93, 94, 96, 114, 120*, 170
Murdoch, Iris, 37
Murphy, Arthur E., 191

Naess, Arne, 111
Nash, Roderick F., 229
Native Americans, 61, 137, 167, 176, 199, 265, 303, 305, 330-333, 335, 337, 364
naturalism, 55, 66
naturalistic fallacy, 174
nature, 4, 8, 13, 42, 45, 52-57, 59, 60-61, 65-67, 69, 72-73, 87, 95, 100, 104, 109-110, 123, 127-128, 135, 137, 139, 145, 154, 161, 163-165, 167, 173, 190, 194, 202, 204-205, 262, 285, 289, 309, 329, 333, 345-346, 381-382
  human n., 52, 60, 87, 95, 103-104, 193, 216-217, 223, 254, 314
needs, 2, 15, 60, 62, 110, 113, 116, 145, 197, 224-225, 245, 261, 264, 307, 309, 313-315, 321, 339, 347, 351, 360, 378-379
Newton, Sir Issac, 138, 296
niches, 25, 26, 35, 301
Nietzsche, Friedrich, 5, 8, 31, 40, 73, 157, 256, 259, 263, 290
nominalism, 273, 289-296, 298
norm(s), 5, 11-12, 38, 44-45, 59-60, 67, 71-73, 86, 91, 93, 104, 106, 115, 118, 129, 134, 140, 146, 161, 166-167, 171-172, 175, 177, 190, 195-196, 200, 214-215, 223, 225, 229, 236-237, 247, 249, 251-252, 254, 283, 285, 287, 292, 295-296, 303, 310, 315, 325, 339, 366
  cultural n., 4, 141, 154, 239, 249, 252

social n., 6, 32, 88-90, 98, 104, 106, 161, 257, 263, 389
normative, 6, 11, 19, 27-28, 31, 35-36, 86, 96, 99, 167-168, 177, 181, 187, 189, 190, 192, 195-196, 228, 283, 290, 293-296, 299, 313, 315
Norton, Bryan, 138, 185, 229
novelty, novelties, 19, 20, 142, 160, 167

object, 24, 29, 31, 91, 99, 169, 191, 203, 218, 348, 373
objective, objectivity, 3, 6, 18, 91, 294, 326, 388
obligation, obligatory, 2, 10, 14, 33-35, 39, 43-50, 58, 63-64, 70, 72, 77-78, 80, 91, 124, 128, 130, 141, 146-147, 153, 158, 160, 166, 173-175, 177, 179, 183, 208, 211, 215-216, 229, 233, 238, 301, 318, 320, 365-366, 371, 376
moral o., 46, 53, 161, 163, 173, 175, 180-181, 217
o. to species, 43
o. to the environment, 41
old-growth forest, 63, 76, 197
opinion, 37, 122, 198, 246, 257, 341
organic, 26-27, 41, 44, 64, 89, 127, 294, 343, 361
organism, 27, 41, 69, 197-198, 291-294, 297-298
organization, 54, 305, 309, 332, 347, 381-382
origin, 4-7, 18, 104, 107, 114, 130, 176, 185, 188-191, 193-194, 196, 204, 227, 247, 261, 280, 297, 333-334, 346, 386
ought(s), 2, 5, 31, 33, 36-38, 45, 48-49, 67, 70, 72-73, 75, 78, 81, 105, 128, 130, 140, 157-158, 161, 167-168, 171, 174-175, 180-182, 189-191, 197, 203, 210, 228, 249-251, 264, 385, 387
overgrazing, 45, 317, 336
overpopulation, 74, 80, 82-83, 108, 114-115, 121, 209, 219, 232, 235-238, 241, 245, 255, 260, 266-267, 269, 305, 307, 332, 338, 354, 363-364, 366, 381
ozone, 15, 80, 111, 120, 123, 219, 223, 346, 347, 371

paradigm, 39, 95, 161, 289, 310, 315
part, 2-3, 5, 9-11, 13-14, 20, 23, 25-27, 36-37, 41-45, 50, 52, 56-57, 59, 62, 65-66, 69, 71, 74-77, 81-83, 85, 90, 92, 95, 97, 100-101, 104, 108-111, 113, 127-128, 130-132, 135-136, 143-147, 153-155, 161, 163, 166-167, 179, 182-183, 188-189, 198, 204-205, 211-212, 214, 216, 222-223, 233, 238, 240, 248, 250, 253, 255, 257, 261, 263-264, 266-267, 269, 271, 273, 275-276, 289-290, 293-294, 297, 299-301, 307, 315, 318, 323, 325, 332, 335, 338, 340, 343, 345, 354, 363, 375-377, 381, 385, 387
past, 10, 37, 79, 83, 92, 104, 116, 140, 160, 174, 176-177, 210, 231, 233, 236-237, 250, 255, 263, 282, 294, 309, 323, 329, 333, 349-350, 354, 360, 371, 389, 394, 396
Peirce, Charles S., 36, 205, 287, 294
perspective, 7-10, 12, 42, 53, 56, 61, 66-67, 70, 79, 83, 102, 108, 130, 146, 238, 244, 262-263,

267, 269, 270, 280, 298, 300, 321, 342, 385
perspective, 209
phenomena, phenomenal, 138, 165, 181, 296
philosopher(s), 52, 107, 165, 253, 264
environmental p., 205
philosophic, 6, 11, 50, 107, 198
philosophy, philosophies, 2-4, 6, 8-11, 15, 31, 36, 51, 55-56, 64, 67, 69, 71, 73-74, 76-77, 81-83, 86, 95, 101, 105, 108, 127-128, 132, 138, 147, 165, 168-169, 171, 175, 187, 206, 260, 263-264, 273, 289, 298, 303, 333, 351, 375, 385-389
ancient p., 11, 97, 139
Greek p., 85, 261, 345
Environmental p. as a system, 85, 92, 132, 167, 197-198, 218, 235, 245, 251, 286, 290, 310, 314, 362, 365, 374, 379, 385-388
p. of the subject, 101, 175
political p., 3, 145
physics, 37, 138, 167, 170, 181, 286, 294, 296, 389
plant(s), 25-26, 47, 49, 71, 87, 127, 135, 209, 236, 275, 281-283, 286, 291, 294, 298, 300, 321, 335, 352-354, 359-361, 372, 391-395
Plato, Platonic, 18, 20, 55, 77, 86-87, 94-95, 97, 99, 101, 104, 106, 111, 113-114, 124, 132, 137, 157, 169, 176, 189, 191-193, 202, 263, 281, 285, 295, 346, 368
pleasure, 5, 18, 30-31, 33, 78, 90, 96, 139, 150, 157-158, 162, 176, 192, 200, 203, 208, 211-219, 224-225, 228-229, 237, 252, 315, 349, 367
pluralism, 12-13, 21, 32, 136, 138-139, 143, 183, 283, 302
moral p., 51, 72, 137, 143
plurality, 32, 40, 48, 66, 73
politics, 4, 35, 67, 81, 83, 117, 119, 128-130, 259-260, 315, 323, 376
pollution, 13, 15, 28, 49, 80, 82, 109, 111, 115, 123, 145, 151, 210, 219, 223, 226, 242-243, 302, 317, 321, 328, 340-342, 347, 349, 350-351, 362, 370, 372-373, 375, 379, 381-383
population, 2, 14, 81-83, 134, 151, 163, 231-246, 258, 264, 266-272, 284, 311, 321, 324, 334, 336, 338-342, 351, 354, 356, 362-367, 370, 380-381, 394
p. growth, 82, 231, 235, 244, 267, 271, 340, 365
possible, 4, 7-8, 12, 32, 37, 50, 67, 76, 78, 86, 89, 92-93, 104, 106-107, 116, 131, 139, 149, 153-154, 160-161, 163, 169-170, 180, 182-184, 192, 204, 213, 221, 241, 245-246, 256-257, 268, 272, 307, 317-318, 321, 326-327, 334, 344, 355, 358, 361-362, 368, 374, 379, 388-389
potency, 77, 98-99, 104, 249
potential, potentialities, 23, 36-37, 44, 76, 86-87, 89, 98-99, 102, 144, 150, 154-155, 168-170, 198, 200-201, 212, 232, 234, 244, 269, 304, 316, 319, 325, 342, 353, 365, 386, 394, 396
power, 64, 76, 85, 87, 98-99, 101, 108, 122, 129, 138, 145, 189, 191, 196, 226, 242-243, 251-252, 266, 268, 306, 308-309, 311, 319, 329, 333, 339-340, 345, 348-349, 360, 368, 370-375, 382

practical, 3, 15, 21-23, 35, 37-38, 70, 72, 90-91, 93, 101, 105, 114, 140-142, 149, 154, 159, 164-165, 168, 173, 182, 186-188, 190-193, 195-196, 224-225, 228, 244, 250, 255, 269, 273, 276, 287, 299-300, 339, 351-352, 362, 368, 383, 386, 388-389
practice, 5, 7, 10, 31, 33, 35-37, 49, 52, 58, 63-64, 72-73, 88, 100, 106-107, 121, 136, 141, 160, 165-169, 172, 178, 181-182, 192, 216, 218-219, 250-253, 265, 272, 281, 285, 287, 293, 298, 304, 311-312, 315-316, 322, 326-327, 333, 344, 354, 357, 365-366, 377, 380, 385-387
pragmatic, 12, 14, 19, 21, 44, 69, 75, 100, 105, 129, 134, 138, 160, 163, 178, 224, 232, 246, 249, 284, 287, 291, 303, 310, 312. 315
Prall, David, 174, 220
predators, 22, 24-25, 28, 37, 41, 46, 62, 71, 78, 80, 110, 133-136, 150, 214, 277, 298, 300, 304, 337, 343, 351, 355, 358-359, 361-362, 392, 394-396
prey, 24-25, 27, 37, 43, 45, 59, 80, 116, 132, 233, 235, 285, 298, 300, 337, 343, 351, 361, 395
principle, 4, 7-8, 11, 18-19, 34, 53, 62, 65, 67, 71-76, 79, 82, 90, 92, 99-100, 102, 108, 110, 113-114, 119, 128, 133, 138-143, 145, 153, 157, 159, 162, 164, 167, 170, 175, 177-180, 183-184, 186, 188, 192, 196-199, 206-208, 212, 214-219, 225-227, 247, 252, 254, 268, 270, 284-285, 287, 293, 295, 298, 316, 318, 321, 345, 361, 365, 383e, 386, 387
p. of utility, 226
p. of participation, 73-74
problematic, 3, 5, 12, 50-52, 66, 67, 91, 93, 110, 141, 164, 177, 188, 190, 202, 214-216, 221, 236-238, 269, 284, 290, 331, 349, 388-389
process(es), 4, 19-20, 23, 28-29, 46, 69, 79, 85-86, 96, 99-100, 104, 117, 199, 201, 280, 313, 318, 338, 341, 343, 353, 356, 364, 368, 380, 388
project(s), 53, 96-97, 160, 164, 175, 299, 331, 355
property, 22, 47, 49, 127, 133, 139, 145-146, 199, 242, 243, 259, 267, 296, 298, 305, 333, 335, 366
psychology, 6, 23, 31, 95, 195, 388
philosophical p., 190
purpose, 64, 103, 114, 133, 150, 176, 208, 223, 231, 245, 311, 322, 351, 363, 381

quality, 15, 38, 45, 69, 74, 81, 83, 100, 110, 114, 123, 137, 147, 151, 172, 215, 221, 224, 231, 236-240, 243-245, 264, 266-268, 270-271, 303, 309-310, 313, 340, 363-364, 366, 370, 381-382, 384, 386, 392, 394

Rachels, James, 6, 253
rare, rarity, 13, 15, 29, 44-45, 49, 79, 90, 111, 139, 149-152, 210, 212, 216-217, 231, 233, 236, 238, 241, 244, 269, 273, 281, 287, 297, 299, 302, 314, 324, 331, 335, 345, 352-353, 355, 358, 365, 391-396
rationalist(s), 97-98, 193

rationality, 22, 23, 52, 87, 99, 141, 159, 164, 186-187, 189-190, 192, 194-196, 209, 215, 225
realist(s), 129, 137
reality, 18, 81, 191, 273-275, 293-294, 296, 334, 346, 383
realm, 4, 36, 129, 136, 167-168, 178, 180, 194, 206, 262, 308
reason, 5, 11, 14, 22-23, 31, 33-34, 37-38, 52, 57, 59, 74, 90, 92, 97, 101, 107, 109, 131, 134, 137, 140-141, 144, 146, 149-150, 159, 161, 164, 168, 173, 176, 178, 185-197, 202, 207, 215, 225, 229, 231, 233, 244-245, 254, 281-282, 287, 305-306, 311, 345, 357, 368, 382, 393
reciprocity, 73-74
Regan, Tom, 68, 128, 219, 318, 331
Reisner, Marc, 235, 383
relation(s), relational, 2, 4-5, 7, 12-15, 18-19, 21, 27-28, 30, 32-35, 37-38, 40-44, 46, 51, 58, 64, 66, 73-74, 77, 85, 89, 93-94, 99, 109, 111, 113-114, 117, 124, 128-132, 136, 138, 141-142, 144, 146, 149, 153-154, 157-158, 166, 168, 173, 175-176, 178-181, 183-185, 189, 201, 204-205, 211, 214, 220-222, 228, 232, 239, 241, 242, 247, 251, 253, 256, 259, 269, 281, 283, 289, 292-293, 296-298, 300, 305, 314, 344, 350, 386-389, 391, 393, 396
  external r., 65, 176, 293
  instrumental r., 19, 41
  internal r., 4, 92, 113, 175, 205, 293, 388
  means-end r., 188
  social r., 34, 40, 71, 73, 91, 133, 157, 201, 205, 240, 257-258, 273, 317
relativism, 66, 87, 92, 140-141, 195, 203, 214, 245, 246, 247, 249, 250, 251, 252, 253, 254, 255, 256, 257, 262, 263, 285
  cultural r., 14, 66, 247, 249, 250, 251, 252, 253, 255, 259, 273, 285, 286
  historical r, 250, 255
  individual r., 208, 245, 246, 208, 257, 258
religion, 2, 10, 35, 48, 70, 76, 96, 232, 248, 265-266, 271, 305, 333, 336, 345, 364-365, 388
result, 15, 20, 26, 29, 34, 40, 44-45, 48, 65, 69, 80, 93, 96, 99, 124, 144, 150, 152, 158-159, 165, 168, 170, 175, 178, 188, 203, 205, 222-224, 227, 234, 236-237, 240-242, 245, 251, 254, 257-258, 260, 266, 270, 274, 299, 301, 305, 312, 318, 321-322, 324, 329, 337-342, 345, 353, 359, 364, 366-367, 369, 378-380, 392, 395
revaluation, 8-15, 50, 65, 73-74, 78, 83, 102, 111, 114, 117, 123, 158, 198, 206, 226, 264, 296, 302, 308, 350, 352, 385, 387
right(s), 4, 7-8, 10, 14, 33-34, 44, 48-49, 52-53, 61-63, 65, 67-68, 70, 75, 78, 83, 88, 91, 93-94, 96, 99, 103-105, 109, 120-122, 124, 128-129, 132-133, 137-138, 140, 142, 144-148, 154, 157, 159-161, 165, 167-168, 170-171, 173-174, 179, 181-182, 185, 191-192, 196, 206, 212, 214, 217-218, 224, 231-232, 236, 239-240, 242-246, 250, 256, 259, 265, 267-269, 295, 304, 306, 313, 322, 325,

333, 357, 360, 363, 365-367, 370-371
bearer of r., 130
human r., 12, 55, 128-133, 153, 155, 236, 268, 305, 325
natural r., 145, 146, 149, 236
r. to unlimited procreation, 231-272
wild r., 127-136
Roberts, Paul, 4, 237
Rolston, Holmes, III, 27, 46, 70, 71, 208, 365, 385
Rorty, Richard, 22, 390-391
Ross, Sir W. David, 157, 159, 161-162, 179
Routley, Richard (Sylvan), 197
rule(s), 7, 13, 33-34, 48-49, 57, 108, 134, 140-142, 161, 164, 177-179, 187, 191, 196, 212, 216, 245, 265, 321, 332, 346-347, 359, 382-383
see also, principles

satisfaction(s), 185, 191, 197, 202, 211, 215, 220-229, 315, 329
Schiller, Ferdinand Canning S., 36
Schweitzer, Albert, 60
science(s), 6, 24, 36-37, 56, 68, 110, 123, 140, 161, 167, 178, 187-188, 192-193, 195, 228, 233, 237, 241, 246, 250, 266, 277, 278, 279, 286, 289, 291, 294, 297, 298, 299, 303, 341, 345, 346, 348, 396
Scully, Matthew, 24, 383
Searle, John R., 187, 196
self-interest, 4, 2, 9, 14, 38, 50, 63, 74, 78, 83, 96, 119-120, 137-138, 141, 143, 150-151, 166, 185-186, 191-211, 215, 220-222, 224-229, 238, 246-247, 256, 258, 263, 269, 300, 312, 323, 342, 352
rational self-interest as oxymoron, 185-230
self-realization, 21, 38, 158
senses, 24, 67, 91, 97, 129, 153, 190, 266, 298
Simberloff, Daniel, 27, 135, 382
Singer, Peter, 64, 65, 68, 185, 271
situation, 5, 14, 24, 76, 88, 93-94, 96, 99, 101, 109, 112, 120, 124, 135, 140-142, 162, 164, 167, 174, 177-178, 195, 205-206, 224, 233-234, 236, 238, 244, 261, 263, 320, 343, 362, 376
problematic s., 177, 238
situation e., see ethics, situation
skeptics, skepticism, 241, 246, 249, 388
social, 4, 6-8, 13-15, 22, 32-34, 36, 40, 51, 60, 64, 71, 73, 75, 78, 80-83, 88-94, 96, 98, 100-103, 106-108, 110, 112, 119-120, 130, 133, 141, 146-148, 157-158, 161, 165-166, 169-171, 174-176, 180. 201, 203, 205-207, 210, 215-217, 223, 231-232, 236, 238-240, 242-243, 248, 252-253, 257-259, 262-263, 266, 268, 270, 273-280, 285-287, 290, 292, 299, 301, 307-308, 310-317, 320, 323, 326-327, 330, 339, 342, 347-348, 363-364, 381, 388-389, 314
social constructivism, 273
s. construction of species, 273, 287
society, 4, 5, 7, 13, 33-34, 36, 41-42, 47, 59, 68, 71, 74, 76, 78, 81, 82, 85, 89-94, 96-97, 100-103, 106, 108, 113-117, 119-121, 123, 131, 133-134, 138, 141-142, 145-146, 148, 153-154, 161, 165-166, 170, 175, 178, 182-183, 185, 202-203, 207, 209, 215, 217-219, 223, 225-228, 231-233, 239-240, 242-244, 248-253, 255-258, 260, 267, 269, 271, 273-275, 277-279, 285-286, 292, 305-306, 308, 310-313, 315-316, 323-324, 326, 329, 332-333,

336, 338-342, 344, 346-348, 351, 353, 359, 362, 366, 371, 374, 376, 381-382, 384
solipsism, 3, 253, 388
species, 2-4, 10-15, 19, 21-31, 34-35, 37, 39-51, 53-66, 68-71, 73-74, 77-78, 80-82, 86-90, 94-96, 104, 110-112, 114-115, 118-120, 122-123, 127-136, 139, 144-145, 147-155, 160, 162-163, 173, 179, 182-183, 185, 193, 197, 200, 202, 204-205, 208-213, 216-218, 220, 222-224, 226, 232-233, 235-236, 238-239, 241-244, 246, 251, 254-256, 258-262, 264-265, 267-269, 273-287, 289, 291-305, 307-308, 311-312, 316, 323-326, 329-338, 341-344, 346, 349, 351-356, 358-362, 365, 367-371, 375, 378-383, 385-387, 389, 391-396
climax s., 65, 352, 379, 393
endangered s., 49, 82, 134, 149, 150, 212, 217, 273, 299, 352, 355, 377, 396
rights of s., 127-136
s. difference, 42, 56-57, 59, 262, 275, 278, 281, 283, 285-287, 292, 299
s. preservation, 9, 71, 79, 135, 148-149, 151, 220, 229, 299- 300, 352
threatened s., 394
See also inherent value of s.
speciesist, 11, 38, 47, 51, 57, 172, 260, 302
Speth, James G., 242, 339
spirit, 8, 52, 65, 99, 106, 111, 118, 158, 206, 214, 247, 266, 302, 310, 383
Stamos, David N., 282, 283, 284, 292, 293, 297
standard(s), 4, 8, 10-12, 27, 36, 51-54, 60, 66, 67, 69, 73-74), 89, 97, 99, 103-104, 119, 130, 136, 148, 151, 161-162, 165, 169, 172, 177, 186-192, 194-196, 201, 203, 214, 216, 221, 224-225, 229, 236-237, 249, 253, 280, 282, 290, 293, 303, 307, 309, 311, 313-315, 317, 320-322, 326, 335, 337, 346-347, 349, 354, 369, 372, 382, 391
state, 7, 27, 68, 99, 104, 108, 113, 114, 129-130, 132, 145, 152-153, 199, 224, 231, 235-236, 239, 242, 249, 267-268, 276, 280, 307, 313, 316, 346, 352, 355, 357, 361, 391
Stone, Christopher D., 137-139, 140, 143, 343
structure(s), 18, 25, 30, 35, 55, 78, 159-161, 164, 192, 247, 253
subject, 3-6, 10, 18-21, 25, 29-32, 38, 42, 56, 65, 73-74, 92, 95-96, 99, 113, 117, 130, 165, 175, 176, 187, 191-192, 200, 204-205, 211, 213, 221, 225, 240, 256, 269, 293, 333, 335, 347, 361, 364, 383, 388-389
subjective, subjectivity, 3, 5-6, 20-21, 30-32, 38, 42-43, 56, 58, 62, 86, 91-92, 96, 105, 176, 199, 201, 203-205, 211, 213, 221, 238, 263, 388
subordination, 19, 29-30, 32, 35, 38-39, 42-43, 46, 48, 72-73, 75, 80, 122, 151, 159, 331-333
summum bonum, 12, 18-21, 29-33, 41, 68, 122, 143, 172-173, 183, 201, 211-212, 214, 222, 225, 310
survival, 2, 5-6, 19, 25-26, 30, 40-46, 48, 50, 62-63, 69, 74, 76, 127, 131, 148, 150, 154, 162, 179, 232, 253, 257, 269, 297, 300-301, 310, 313-314, 359,

369, 381, 388-389, 391, 394-395
symbiosis, 3, 8, 25, 34, 42, 80, 115, 232-233, 239, 264, 277, 290, 301, 344-345
system(s), 85, 28, 132, 167, 197-198, 218, 235, 245, 251, 286, 290, 303, 310, 314, 362, 365, 374, 379, 388

Taylor, Paul W., 70, 123
technology, technological, 7, 15, 49, 65, 71, 80, 88, 93, 109, 140, 178, 232, 234, 239, 241, 244, 266, 309, 312, 315, 320, 329, 334, 337-339, 345-349, 368, 375, 380, 382
t. as cause of prosperity 346-352
t. as threat to environment 384-386
teleological, 23, 93, 95, 139, 157, 159, 163, 182, 257, 294
temporal, 99, 255, 293-294, 297, 388
theory, theories, 3, 5, 11-12, 14, 18-21, 27, 30-31, 33, 35-36, 38, 40, 42, 44, 47, 56, 59, 62, 66-69, 77, 80, 88, 94-95, 97, 101-102, 105-107, 109, 114, 128-129, 138-139, 141-142, 144-146, 151, 157-158, 165-168, 173-174, 177-178, 185-188, 192-193, 195-196, 198-206, 208-229, 232, 241, 247, 253, 267, 374, 273, 275, 278-279, 289, 292-293, 296, 309, 312-313, 333, 341, 346, 386-388
biocentric t, 71
ethical t., 4, 5, 35, 38, 67, 137, 139, 143, 151, 157, 166, 232
t. of value(s), 5, 12, 19-20, 27, 30-31, 33, 42, , 56, 62, 77, 80, 92, 94, 106, 139, 165, 171, 174, 182, 185, 192-193, 195, 198-200, 202-203, 206, 208, 211-216, 220, 222-223, 225-226, 228, 256
anthropocentric t.o.v., 55, 58
conativism, 174, 191, 202, 220, 228, 312
t..and practice, 35, 107, 109
separation of t. and p., 36, 168
therapy, 196
thinking, thought, 4, 5, 23, 25, 31, 56, 67, 77, 85, 104, 107, 110, 114, 117, 127-128, 146, 167, 169, 195, 207, 187-189, 196, 199, 243, 254, 266, 276, 279-280, 294, 307, 311, 314, 317, 333, 353, 377-378
Thompson, Jenna, 197, 198, 199, 205, 208, 222
thrift, 14, 115, 124, 307, 315-321, 324-327, 352-353, 359, 366, 368, 370, 373, 377
Toulmin, Steven, 187, 188
tradition, 7, 47, 48, 70, 83, 108, 129-130, 132, 157, 165, 169, 185, 261, 266, 318, 334-335, 386
tribalism, 2, 9, 52, 57, 65-66, 250, 254, 259, 262
truth, 4-5, 36-37, 78, 82, 89, 96, 99, 106-108, 110, 114, 121-124, 140, 161, 179, 181, 234, 242, 244, 249-251, 253, 263, 275, 303, 310, 313, 332, 336, 339, 346, 357, 378
Tufts, James, 64, 94, 180, 244

universal(s), universality, 2, 6-7, 9-14, 32, 34-35, 38-40, 48, 50-51, 51-54, 57-60, 62, 65, 67, 70-73, 78-79, 83, 86, 90-91, 102-104, 107, 110, 114, 120, 124, 129-133, 138-141, 143, 145-147, 153, 155, 157-158, 162-167, 170-171, 177-178, 183, 187, 189-190, 193-198, 206-207, 215, 219, 221, 229, 247, 249, 250-254, 260-262,

264, 268, 270, 276, 282, 285, 289-290, 294-296, 320, 334, 348, 365, 382, 385-386, 389
universe, 8, 39, 48, 51, 56, 65, 68, 70, 167, 205, 266. 292-294, 332, 342, 379, 389
pluralistic u., 48, 298
Utilitarian, Utilitarianism, 9, 14, 33, 38-39, 142, 160, 163, 173-174, 178, 185, 192, 206, 208, 211, 214-220, 225, 227-229, 243, 261, 300
Economic U., 185, 225
immorality of U., 210
utility, 25, 78, 90, 107, 162, 197, 203, 214, 225-226, 229, 238, 312, 320-321, 329

valuable for its own sake, 315
valuable in itself, 151
valuation, 11, 30, 50, 182
value(s), 2-5, 7, 9, 11-14, 18-21, 24-33, 35-46, 48-50, 54-58, 61-63, 65-80, 87, 89, 91-93, 96-98, 102, 106-107, 109, 112, 122-124, 128-131, 136-137, 139, 142-145, 147-152, 155, 157-158, 162, 165-176, 179-183185, 189-193, 197, 198, 199-206, 208-209, 211-212, 215, 220-225, 227-229, 233, 235, 238, 241, 243, 246-256, 258-259, 262, 269, 271 value(s), 286, 300, 302, 303, 304, 307-313, 315-316, 320, 324, 331, 333, 338, 340-341, 344, 349, 352, 357-358, 363, 367, 371, 374, 381-382, 386-389
anthropocentric v., 4, 8, 19, 57
creative actualization as a theory of v, *see* creative actualization
cultural v., 5, 9, 79, 98, 171, 247-249, 252-254, 256, 264
extrinsic v., 21, 32, 129, 183, 189, 247, 308, 310, 311, 347
hierarchy of v., 8, 18, 31
inherent v., 13-14, 21, 25, 28, 31, 36, 41, 45-46, 68, 115, 131, 148, 150 189, 200, 204, 209, 212-213, 167, 182-183, 259, 301, 324, 352, 357
instrumental v., 18, 27-28, 31, 43, 55, 91, 300
(*see also* means)
intrinsic v., 11, 13, 18, 20-21, 25-26, 28, 31-32, 36, 38-42, 45-46, 49, 55, 58, 61, 68, 73, 95, 129, 131-133, 135-136, 144, 148-150, 152-154, 172-174, 182-184, 233, 269, 299-300, 302, 310, 316, 365
i.v. of other species, 17-28
relation of v. to obligation, 12, 14, 157, 173, 180, 211, 214
v. as bringing about, 181
value-free, 228 , 257, 312
v. theory, see theory of value
wild v., 124
see also for the sake of, for its own sake
value judgments, 31, 199, 259
valuer, 31, 221
vegetarian, 79, 115. 135, 345
virtue, 11-14, 18, 29-30, 33, 37, 39, 52, 75, 77, 85-89, 91-105, 107, 109-114, 116-124, 132-133, 140, 142-144, 148, 151, 157-159, 161, 163, 169-171, 175-176, 180-181, 183, 189, 192-193, 212, 216, 228, 233, 243, 306-307, 311-313, 315-318, 320-327, 352-353, 359, 366, 368, 373, 377, 384, 387, 396

web of relations, 90
Weber, Max, 188, 316, 336
wetlands, 279, 281, 352, 358-359, 383, 392, 394
whole, 2-5, 9-10, 12-15, 22, 26-29, 35-36, 38, 41-46, 48-54, 56, 60, 62, 64-67, 69, 71-73, 75-79,

81, 83, 86, 89-90, 95, 97-98, 100, 102-103, 106, 109, 111-113, 124, 127-128, 130-133, 135-136, 139, 143-147, 150, 153-155, 162-163, 173, 181, 183, 198, 200-202, 204, 206, 214, 223, 226, 229e, 232-233, 238-239, 243, 251, 254-255, 258, 260-261, 264, 269, 272-273, 277, 280, 284, 290-291, 298-301, 303, 306,309, 323, 338-340, 347, 350, 352, 359, 369, 374, 377, 380, 385-387, 389
wild, 3, 12-13, 41-42, 44, 46-48, 71-72, 111, 114, 117-118, 120-121, 123-124, 128-137, 139, 143-144, 148-149, 151, 153-155, 162-163, 173, 183, 231, 235-238, 240, 243, 261, 268, 271, 276, 282, 301-302, 307, 317, 329-330, 332, 338, 341-343, 349, 352, 354-358, 360-362, 365-366, 369, 374, 378, 382-383, 389, 394, 396
wildlife, 49, 77-78, 82, 153, 233, 236-237, 318, 348, 355, 360, 372, 377
wisdom, 2, 49, 88, 91, 101, 106, 109, 112, 114, 117, 124, 176, 192, 311, 313, 341, 350-351, 375, 386, 389
world, 2-4, 6, 8-12, 15, 19-22, 25-26, 29-30, 32, 36-38, 40, 42, 45, 48, 50, 51-52, 54-56, 58-59, 63, 65-71-77, 79, 81-83, 86, 90-91, 95, 97, 99, 106, 108, 113-115, 118, 123, 128-129, 132, 135, 139, 144, 146-147, 153-154, 158, 160-163, 165, 167-169, 172, 174-176, 179, 181-184, 198-199, 201, 203-204, 208-210, 212-213, 217, 223, 227, 229 world, 234, 237-238, 244-245, 248, 250, 254, 258-259, 263-265, 269-270, 272, 274, 282, 289-290, 292-295, 297-299, 301-302, 307, 312, 315, 319-320, 325-326, 329-330, 333-334, 337, 340-342, 347-348, 349, 354, 357-358, 360, 364, 366, 368-371, 373, 375, 377-378, 382, 384, 385-389, 393
world view, 51, 55, 68, 102, 266
worth, 32, 42, 58, 75-78, 85, 87, 89, 97-98, 103, 116, 164-166, 169, 171-172, 175-176, 179, 204, 206-207, 213, 217-218, 221, 229, 247, 251, 253, 301, 309, 311, 313-314, 353, 368, 384
human w., 68, 83
instrumental w., 189, 198, 199, 201
intrinsic w., 189, 197-202, 204, 212-214, 216, 229

VIBS

The **Value Inquiry Book Series** is co-sponsored by:

Adler School of Professional Psychology
American Indian Philosophy Association
American Maritain Association
American Society for Value Inquiry
Association for Process Philosophy of Education
Canadian Society for Philosophical Practice
Center for Bioethics, University of Turku
Center for Professional and Applied Ethics, University of North Carolina at Charlotte
Central European Pragmatist Forum
Centre for Applied Ethics, Hong Kong Baptist University
Centre for Cultural Research, Aarhus University
Centre for Professional Ethics, University of Central Lancashire
Centre for the Study of Philosophy and Religion, University College of Cape Breton
Centro de Estudos em Filosofia Americana, Brazil
College of Education and Allied Professions, Bowling Green State University
College of Liberal Arts, Rochester Institute of Technology
Concerned Philosophers for Peace
Conference of Philosophical Societies
Department of Moral and Social Philosophy, University of Helsinki
Gannon University
Gilson Society
Haitian Studies Association
Ikeda University
Institute of Philosophy of the High Council of Scientific Research, Spain
International Academy of Philosophy of the Principality of Liechtenstein
International Association of Bioethics
International Center for the Arts, Humanities, and Value Inquiry
International Society for Universal Dialogue
Natural Law Society
Philosophical Society of Finland
Philosophy Born of Struggle Association
Philosophy Seminar, University of Mainz
Pragmatism Archive at The Oklahoma State University
R.S. Hartman Institute for Formal and Applied Axiology
Research Institute, Lakeridge Health Corporation
Russian Philosophical Society
Society for Existential Analysis
Society for Iberian and Latin-American Thought
Society for the Philosophic Study of Genocide and the Holocaust
Unit for Research in Cognitive Neuroscience, Autonomous University of Barcelona
Whitehead Research Project
Yves R. Simon Institute

Titles Published

Volumes 1 - 237 see www.rodopi.nl

238. Stefano Franchi and Francesco Bianchini, Editors, *The Search for a Theory of Cognition: Early Mechanisms and New Ideas.* A volume in **Cognitive Science**

239. Michael H. Mitias, *Friendship: A Central Moral Value.* A volume in **Ethical Theory and Practice**

240. John Ryder and Radim Šíp, Editors, *Identity and Social Transformation, Central European Pragmatist Forum, Volume Five.* A volume in **Central European Value Studies**

241. William Sweet and Hendrik Hart, *Responses to the Enlightenment: An Exchange on Foundations, Faith, and Community.* A volume in **Philosophy and Religion**

242. Leonidas Donskis and J.D. Mininger, Editors, *Politics Otherwise: Shakespeare as Social and Political Critique.* A volume in **Philosophy, Literature, and Politics**

243. Hugh P. McDonald, *Speculative Evaluations: Essays on a Pluralistic Universe.* A volume in **Studies in Pragmatism and Values.**

244. Dorota Koczanowicz and Wojciech Małecki, Editors, *Shusterman's Pragmatism: Between Literature and Somaesthetics.* A volume in **Central European Value Studies**

245. Harry Lesser, Editor, *Justice for Older People,* A volume in **Values in Bioethics**

246. John G. McGraw, *Personality Disorders and States of Aloneness (Intimacy and Aloneness: A Multi-Volume Study in Philosophical Psychology, Volume Two),* A volume in **Philosophy and Psychology**

247. André Mineau, *SS Thinking and the Holocaust.* A volume in **Holocaust and Genocide Studies**

248. Yuval Lurie, *Wittgenstein on the Human Spirit.* A volume in **Philosophy, Literature, and Politics**

249. Andrew Fitz-Gibbon, *Love as a Guide to Morals*. A volume in **Ethical Theory and Practice**

250. Ronny Miron, *Karl Jaspers: From Selfhood to Being*. A volume in **Studies in Existentialism**

251. Necip Fikri Alican, *Rethinking Plato: A Cartesian Quest for the Real Plato*. A volume in **Philosophy, Literature, and Politics**

252. Leonidas Donskis, Editor, *Yet Another Europe after 1984: Rethinking Milan Kundera and the Idea of Central Europe*. A volume in **Philosophy, Literature, and Politics**

253. Michael Candelaria, *The Revolt of Unreason: Miguel de Unamuno and Antonio Caso on the Crisis of Modernity*. A volume in **Philosophy in Spain**

254. Paul Richard Blum, *Giordano Bruno: An Introduction*. A volume in **Values in Italian Philosophy**

255. Raja Halwani, Carol V. A. Quinn, and Andy Wible, Editors, *Queer Philosophy: Presentations of the Society for Lesbian and Gay Philosophy, 1998-2008*. A volume in **Histories and Addresses of Philosophical Societies**

256. Raymond Angelo Belliotti, *Shakespeare and Philosophy: Lust, Love, and Law*. A volume in **Philosophy, Literature, and Politics**

257. Jim Kanaris, Editor, *Polyphonic Thinking and the Divine*. A volume in **Philosophy and Religion**

258. Michael Krausz, *Oneness and the Displacement of Self: Dialogues on Self-Realization*. A volume in **Interpretation and Translation**

259. Raymond Angelo Belliotti, *Jesus or Nietzsche: How Should We Live Our Lives?* A volume in **Ethical Theory and Practice**

260. Giorgio A. Pinton, *The Conspiracy of the Prince of Macchia & G. B. Vico*. A volume in **Philosophy, Literature, and Politics**

261. Mechthild E. Nagel and Anthony J. Nocella II, Editors, *The End of Prisons: Reflections from the Decarceration Movement*. A volume in **Social Philosophy**

262. Dorota Koczanowicz, Leszek Koczanowicz, and David Schauffler, Editors, *Discussing Modernity: A Dialogue with Martin Jay*. A volume in **Central European Value Studies**

263. Pekka Mäkelä and Cynthia Townley, Editors, *Trust: Analytic and Applied Perspectives*. A volume in **Nordic Value Studies**

264. Krzysztof Piotr Skowroński, *Beyond Aesthetics and Politics: Philosophical and Axiological Studies on the Avant-Garde, Pragmatism, and Postmodernism*. A volume in **Central European Value Studies**

265. David C. Bellusci, *Amor Dei in the Sixteenth and Seventeenth Centuries*. A volume in **Philosophy and Religion**

266. Vasil Gluchman, Editor, *Morality: Reasoning on Different Approaches*. A volume in **Ethical Theory and Practice**

267. Jakob Lothe and Jeremy Hawthorn, Editors, *Narrative Ethics*. A volume in **Philosophy, Literature, and Politics**

268. Greg Moses and Gail Presbey, Editors, *Peace Philosophy and Public Life: Commitments, Crises, and Concepts for Engaged Thinking*. A volume in **Philosophy of Peace**

269. Bartholomew Ryan, *Kierkegaard's Indirect Politics: Interludes with Lukács, Schmitt, Benjamin and Adorno*. A volume in **Philosophy, Literature, and Politics**

270. Patricia Hanna, Editor, *Reality and Culture: Essays on the Philosophy of Bernard Harrison*. A volume in **Interpretation and Translation**

271. Piotr Nowak, *The Ancients and Shakespeare on Time: Some Remarks on the War of Generations*. A volume in **Philosophy, Literature, and Politics**

272. Brian G. Henning and David Kovacs, Editors, *Being in America: Sixty Years of the Metaphysical Society*. A volume in **Histories and Addresses of Philosophical Societies**

273. Hugh P. McDonald, *Environmental Philosophy: A Revaluation of Cosmopolitan Ethics from an Ecocentric Standpoint*. A volume in **Studies in Applied Ethics**

www.ingramcontent.com/pod-product-compliance
Lightning Source LLC
La Vergne TN
LVHW020650300926
849161LV00004B/877